ADVANCED CALCULUS

An Introduction to
Classical Analysis

ADVANCED CALCULUS
An Introduction to Classical Analysis

LOUIS BRAND

DOVER PUBLICATIONS, INC.
Mineola, New York

Bibliographical Note

This Dover edition, first published in 2006, is an unabridged republication of the 1967 revised printing of the work first published by John Wiley & Sons, Inc., New York, 1955.

Library of Congress Cataloging-in-Publication Data

Brand, Louis, 1885–
    Advanced calculus : an introduction to classical analysis / Louis Brand.
        p. cm.
    Originally published: New York : Wiley, 1955.
    Includes index.
    ISBN 0-486-44548-8 (pbk.)
    1. Calculus—Textbooks. I. Title.

QA303.2.B73 2006
515—dc22

                                                                        2005053704

Manufactured in the United States of America
Dover Publications, Inc., 31 East 2nd Street, Mineola, N.Y. 11501

Preface

This book presents a course in analysis dealing essentially with functions of a real variable. Since certain portions of real variable theory can be fully explained and comprehended only by a consideration of functions of a complex variable, one chapter has been devoted to the latter. This deals primarily with topics that are closely related to problems in the real domain, such as the removal of value barriers in real functions, disclosing relations not otherwise apparent, explaining the interval of convergence in power series, and finally the use of complex integration in computing real integrals.

The opening chapter sketches the structure of the system of real and complex numbers. This may be provisionally omitted and referred to as occasion arises. The next chapter deals at length with the convergence of sequences and series. Thus some of the basic concepts of analysis are introduced in their simplest setting, through functions of an integral variable. In the following chapters on functions of a real variable, the theory is constantly motivated and illustrated with examples. Theorems are stated in a way that is most likely to prove useful in applications, and the proofs, where lengthy, have been broken down into parts so that the logical thread is more readily followed. A course in advanced calculus should offer no hiding place for difficulties; they are frankly encountered where they arise and are dealt with as simply as possible without sacrifice of rigor. Once overcome, a difficulty often suggests a method for later use, and such parallel arguments are always carefully exploited.

The fifth chapter gives a brief and self-contained introduction to vectors and the important differential invariants, gradient, divergence, and rotation. They are used in the sequel whenever the subject matter is clarified or simplified. Since the use of vectors is now almost universal in many parts of mathematical physics, the applied mathematician should be thoroughly conversant with their properties. This applies with full force to those basic theorems that convert volume into surface integrals, surface into line integrals. These constitute a natural generalization of the fundamental theorem of the calculus, which expresses an integral over an interval in terms of function values at the end points.

Since many of the subtlest and refined parts of analysis are concerned with the reversal of order in limiting processes, an entire chapter has been devoted to such considerations. This rounds out the early chapter on sequences and series and postpones many difficult parts of analysis to a point when the student has acquired some mathematical maturity. The final chapter deals with Fourier series; and this ends with Gibbs' phenomenon, which affords a happy illustration of the role of precise definitions and sharp concepts in mathematics.

Because very full cross references are given in this book, an article as well as page number is given at the top of each page. Equations are numbered serially (1), (2), $\cdots$ in each article. A reference to an equation in another article is made by giving article and number to the left and right of a point; thus (88.3) means article 88, equation (3). This style of reference is also applied to theorems, examples, and problems. Figures are given the number of the article in which they appear followed by a serial letter; thus Fig. 160c is the third figure in article 160.

The final article in each chapter gives a brief summary of its essential content. The student should carefully observe the distinction between *necessary* and *sufficient* conditions. If the assumption of a certain property P leads deductively to a condition C, the condition is *necessary*. But, if the assumption of the condition C leads deductively to the property P, the condition is *sufficient*. Thus we have symbolically

$$P \to C \quad \text{(necessary)}, \qquad C \quad \text{(sufficient)} \to P.$$

When $P \rightleftarrows C$, the condition C has the admirable quality of being both *necessary and sufficient*.

In order to hold the book to a reasonable size, some important matters were reluctantly omitted. Thus differential equations and the summability of divergent series are but casually mentioned. Moreover only Riemann integrals are considered, since the Lebesgue integral and others are now customarily introduced in the theory of functions.

Finally I wish to thank my colleagues, Professors Jaeger, Jurkat, Justice, Lipsich, Lubin, and Merriman, for their suggestions in matters of exposition and for various problems. I am also indebted to Professor C. R. Wylie, Jr. of the University of Utah for permission to reprint his poem *Paradox,* which expresses so felicitously the essential nature of mathematics.

LOUIS BRAND

University of Cincinnati
March 14, 1955

Contents

Chapter 1. THE NUMBER SYSTEM

Chapter 2. SEQUENCES AND SERIES

Contents

Chapter 3. FUNCTIONS OF A REAL VARIABLE

Chapter 4. FUNCTIONS OF SEVERAL VARIABLES

Contents   ix

Contents

Chapter 11. FUNCTIONS OF A COMPLEX VARIABLE

Chapter 12. FOURIER SERIES

Contents

PARADOX

Not truth, nor certainty. These I foreswore
In my novitiate, as young men called
To holy orders must abjure the world.
'If . . . , then . . . ,' this only I assert;
And my successes are but pretty chains
Linking twin doubts, for it is vain to ask
If what I postulate be justified,
Or what I prove possess the stamp of fact.

Yet bridges stand, and men no longer crawl
In two dimensions. And such triumphs stem
In no small measure from the power this game,
Played with the thrice-attenuated shades
Of things, has over their originals.
How frail the wand, but how profound the spell!

Clarence R. Wylie, Jr.

CHAPTER 1

The Number System

1. Groups. An *equivalence relation* ($=$) between mathematical elements $a, b, c, \cdots$ (such as numbers, matrices, permutations) must be

E_1 (*reflexive*): $a = a$;

E_2 (*symmetric*): $a = b$ implies $b = a$;

E_3 (*transitive*): $a = b$, $b = c$ imply $a = c$.

An *operation* ($\circ$) relates every ordered pair of elements a, b to a third element $a \circ b = c$. If $a' = a$, $b' = b$, and $a' \circ b' = c'$, the operation is said to be *well defined* if $c' = c$.

A *group* is composed of a finite or infinite set S of elements, an equivalence relation, and a well-defined operation ("multiplication") which obeys the four postulates M_1 to M_4:

M_1 (Closure). If a and b are in S,

$$(1) \qquad a \circ b \text{ is in } S.$$

M_2. The operation is *associative*:

$$(2) \qquad (a \circ b) \circ c = a \circ (b \circ c).$$

M_3. The set S includes an *identity element* e such that

$$(3)' \qquad e \circ a = a.$$

M_4. Every element a has an *inverse* a^{-1} in S such that

$$(4)' \qquad a^{-1} \circ a = e.$$

The group *may* satisfy the additional postulate

M_5. The operation is *commutative*:

$$(5) \qquad a \circ b = b \circ a.$$

In this event the group is called *commutative* or *abelian*.

From the postulates M_1 to M_4 we now make a series of important deductions. We shall call the operation "multiplication," and write ab for $a \circ b$ in the proofs.

1

I. *The cancelation law for left factors:*

(6) $c \circ a = c \circ b$ implies $a = b.$

Proof. Multiply (6) by c^{-1} on the left. From (2), (4)′, and (3)′:

$$(c^{-1}c)a = (c^{-1}c)b, \qquad ea = eb, \qquad a = b.$$

II. *The identity element may be a right factor:*

(3)″ $a \circ e = a.$

Proof. $a^{-1}(ae) = (a^{-1}a)e = ee = e = a^{-1}a$; hence $ae = a$ from (6).

III. *Element a is an inverse of* a^{-1}:

(4)″ $a \circ a^{-1} = e.$

Proof. $a^{-1}(aa^{-1}) = (a^{-1}a)a^{-1} = ea^{-1} = a^{-1}e$; hence $aa^{-1} = e$ from (6).

IV. *The cancelation law for right factors:*

(7) $a \circ c = b \circ c$ implies $a = b.$

Proof. Multiply (7) by c^{-1} on the right. From (2), (4)″, and (3)″:

$$a(cc^{-1}) = b(cc^{-1}), \qquad ae = be, \qquad a = b.$$

V. *A group has only one identity element.*
Proof. If $xa = a = ea$, $x = e$ from (7).

VI. *An element of a group has only one inverse.*
Proof. If $xa = e = a^{-1}a$, $x = a^{-1}$ from (7).

VII. *The inverse of* $a \circ b$ *is* $b^{-1} \circ a^{-1}$.
Proof. $(b^{-1}a^{-1})(ab) = b^{-1}(a^{-1}a)b = b^{-1}eb = b^{-1}b = e.$

VIII. *The equations* $a \circ x = b$ *and* $y \circ a = b$ *have the unique solutions*
$x = a^{-1} \circ b, y = b \circ a^{-1}.$
Proof. The correctness of the given solutions follows by direct substitution; their unicity follows from the cancelation laws.

A group is the simplest mathematical system of any consequence. Its essential properties may be summarized as follows:

A group is closed under an associative, well-defined operation $(\circ)$ *and has a unique identity element e such that*

(3) $a \circ e = e \circ a = a;$

moreover, every element a has a unique inverse a^{-1} *such that*

(4) $a^{-1} \circ a = a \circ a^{-1} = e.$

Note that the operation is commutative in the special cases (3) and (4). *But in general the operation need not be commutative* (cf. Ex. 3).

Example 1. The matrices

$$e = \begin{pmatrix} 1 & 0 \\ 0 & 1 \end{pmatrix}, \quad a = \begin{pmatrix} -1 & 0 \\ 0 & -1 \end{pmatrix}, \quad b = \begin{pmatrix} 1 & 0 \\ 0 & -1 \end{pmatrix}, \quad c = \begin{pmatrix} -1 & 0 \\ 0 & 1 \end{pmatrix}$$

form a finite group under matrix (row-by-column) multiplication. The matrix e is the identity, and each matrix is its own inverse. Note that $ab = c$, $bc = a$, $ca = b$, and that the group is commutative.

Example 2. The integers (positive, negative, and zero) form a group under addition $(+)$; for the integers are closed under addition, addition is associative, the identity element is zero, and the integers $\pm n$ are inverse to each other. Since $a + b = b + a$, the group is commutative.

The integers do not form a group under multiplication; although we have closure, the associative law, and the identity element 1, an element other than ± 1 has no inverse (reciprocal).

Example 3. Permutation Groups. Consider the six permutations of 1, 2, 3, namely:

$$p_1 = 123, \quad p_2 = 231, \quad p_3 = 312, \quad p_4 = 132, \quad p_5 = 321, \quad p_6 = 213.$$

We regard each p_i as replacing 123 by the digits in its symbol. Thus we may write more explicitly

$$p_2 = \frac{123}{231} = \frac{231}{312} = \frac{213}{321} \quad \text{etc.,}$$

where the upper digits are carried into the corresponding digits below. Such "fractions" are not altered when the upper and lower digits are subjected to the same permutation.

The *product* of two permutations $p_i p_j$ is defined as the permutation of 123 obtained by first applying p_i, then p_j. For example,

$$p_2 p_4 = \frac{123}{231} \frac{123}{132} = \frac{123}{231} \frac{231}{321} = \frac{123}{321} = p_5.$$

This "fraction" notation shows that such products are associative; that is, $(p_i p_j)p_k = p_i(p_j p_k)$. But the commutative law does not hold in general; thus,

$$p_4 p_2 = \frac{123}{132} \frac{123}{231} = \frac{123}{132} \frac{132}{213} = \frac{123}{213} = p_6.$$

The six elements $p_1, \cdots, p_6$ form a group; for under the product operation just defined, postulates M_1 and M_2 are fulfilled, p_1 is the identity element, and each p_i has a unique inverse defined by the "reciprocal" of its fraction symbol; thus,

$$p_2^{-1} = \frac{231}{123} = \frac{123}{312} = p_3.$$

The *multiplication table* for this group, where p_i is replaced by i, is as follows

	1	2	3	4	5	6
1	1	2	3	4	5	6
2	2	3	1	5	6	4
3	3	1	2	6	4	5
4	4	6	5	1	3	2
5	5	4	6	2	1	3
6	6	5	4	3	2	1

The table shows that p_1, p_2, p_3 alone form a group—a commutative *subgroup* of the given group.

Similar considerations show that the $n!$ permutations of 1, 2, 3, $\cdots$, n form a non-commutative group under the product operation just defined. Moreover the n cyclic permutations $1\,2\cdots n$, $2\,3\cdots 1$, $\cdots$, form a commutative subgroup.

2. Fields. A *field* is composed of a set S of elements, an equivalence relation, and *two* well-defined operations, addition ($+$) and multiplication ($\cdot$), which satisfy the following postulates.

A. *The set S forms a commutative group under addition, whose identity is called zero.*

M. *The set S with zero omitted forms a commutative group under multiplication.*

D. *Multiplication is distributive with respect to addition:*

$$(1) \qquad a \cdot (b + c) = a \cdot b + a \cdot c.$$

The identity under addition is written 0 ("zero"), and the additive inverse of a is written $-a$ and called the *negative* of a:

$$(2)\,(3) \qquad 0 + a = a, \qquad -a + a = 0.$$

The identity under multiplication is written 1 ("unity"), and the multiplicative inverse of a is written a^{-1} and called the *reciprocal* of a:

$$(4)\,(5) \qquad 1 \cdot a = a, \qquad a^{-1} \cdot a = 1.$$

The field postulates may now be displayed as follows:

$A_1.$ $a + b$ belongs to S;	$M_1.$ $a \cdot b$ belongs to S;
$A_2.$ $(a + b) + c = a + (b + c)$;	$M_2.$ $(a \cdot b) \cdot c = a \cdot (b \cdot c)$;
$A_3.$ $0 + a = a$;	$M_3.$ $1 \cdot a = a$;
$A_4.$ $-a + a = 0$;	$M_4.$ $a^{-1} \cdot a = 1$, $(a \neq 0)$;
$A_5.$ $a + b = b + a$;	$M_5.$ $a \cdot b = b \cdot a$;

$$\text{D. } a \cdot (b + c) = a \cdot b + a \cdot c.$$

In a field both cancelation laws,

$$(6) \qquad c + a = c + b \quad \text{implies} \quad a = b,$$

$$(7) \qquad c \cdot a = c \cdot b \quad \text{implies} \quad a = b \text{ if } c \neq 0,$$

are valid (§ 1). Note that zero is not an element of the multiplicative group.

Zero has the important property

$$(8) \qquad a \cdot 0 = 0$$

for all elements of the field. For, from $0 + b = b$ and D, we have

$$a \cdot 0 + a \cdot b = a \cdot b = 0 + a \cdot b,$$

and on canceling $a \cdot b$ we get (8).

Equation (8) shows that zero has no reciprocal. This explains the exclusion of zero from the multiplicative group. We also have the important

THEOREM. *If a, b are elements of a field, $a \cdot b = 0$ when and only when at least one factor is zero.*

Proof: If $b = 0$, the equation is satisfied for any a.
If $b \neq 0$, multiply the equation by its reciprocal b^{-1}; then

$$(a \cdot b) \cdot b^{-1} = 0, \qquad a \cdot (b \cdot b^{-1}) = 0, \qquad a = 0.$$

Subtraction $(-)$ is the operation inverse to addition: To subtract b from a is to find an element $x = a - b$ such that

$$x + b = a.$$

If we add $-b$ to both members, we have $x = a + (-b)$; hence,

$$(9) \qquad\qquad a - b = a + (-b).$$

Subtracting an element is the same as adding its negative.

Division $(/)$ is the operation inverse to multiplication; to divide a by b is to find an element $x = a/b$ such that

$$x \cdot b = a.$$

If we multiply both members by b^{-1}, we have $x = a \cdot b^{-1}$; hence,

$$(10) \qquad\qquad a/b = a \cdot b^{-1}.$$

Dividing by an element is the same as multiplying by its reciprocal. In particular, when $a = 1$, $1/b = b^{-1}$.

Division by zero is impossible. For *no* element x can satisfy $x \cdot 0 = a$ if $a \neq 0$; whereas if $a = 0$ *any* element x satisfies $x \cdot 0 = 0$.

3. Integers. The unending set of counting numbers $1, 2, 3, \cdots$ is called the *positive integers*. Every positive integer n has a unique successor $n + 1$; and every set of positive integers that contains 1 and the successors of every member of the set contains *all* the positive integers. This property of the positive integers enables us to prove theorems by *finite induction*:

If a theorem is true in the case $n = 1$, and if the truth of the theorem for any n implies its truth for $n + 1$, then the theorem is true for all positive integers.

The positive integers may be characterized by a set of postulates, and, after addition and multiplication are defined, the following properties may be deduced.† Addition and multiplication are associative, commutative,

---

† Landau, E., *Grundlagen der Analysis*, Leipzig, 1930. This book gives a detailed account of the logical structure of the real and complex number system. The positive integers are dealt with in Chapter 1.

MacDuffee, C. C., *Introduction to Abstract Algebra*, Wiley, New York, 1940, § 1.

and connected by the distributive law; and both cancelation laws are valid.

If a and b are positive integers, $a - b$ and a/b have no positive integral values in infinitely many cases. We shall now remedy this lack of closure in subtraction and division by adding new "numbers" to the system of positive integers.

In order to obtain closure under subtraction, we form ordered pairs (a, b) of positive integers. We call (a, b) simply an *integer* and proceed to define equality, addition, and multiplication of integers. If we tentatively identify (a, b) with $a - b$, the known properties of $a - b$ when $b < a$ suggest the definitions that follow.

Equality of integers is defined by

(1) $$(a, b) = (c, d) \quad \text{only when} \quad a + d = b + c.$$

This definition satisfies the three requirements of an equivalence relation (§ 1). E_1 and E_2 are obvious; as to E_3, assume (1) and

$$(c, d) = (e, f); \quad \text{that is,} \quad c + f = d + e;$$

then $a + d + c + f = b + c + d + e$, and, on canceling $c + d$,

$$a + f = b + e; \quad \text{that is,} \quad (a, b) = (e, f).$$

If x is any positive integer, (1) shows that

(2) $$(a, b) = (a + x, b + x).$$

Addition and *multiplication* are now well defined by the equations,

(3) $$(a, b) + (c, d) = (a + c, b + d),$$

(4) $$(a, b) \cdot (c, d) = (ac + bd, ad + bc),$$

where, as usual, ab means $a \cdot b$. When (a, b), (c, d) are replaced by their equals $(a + x, b + x)$, $(c + y, d + y)$, these definitions give a sum and product equal to those above.

It is now easy to prove that addition and multiplication of *integers* also obey the commutative, associative and distributive laws. The proofs depend upon the corresponding laws for positive integers.[†]

The integers (a, b) are of three types:

$$\text{I} \quad (a + x, x); \qquad \text{II} \quad (x, x); \qquad \text{III} \quad (x, a + x).$$

---

† MacDuffee, C. C., *op. cit.*, p. 6. Weiss, M. J., *Higher Algebra for the Undergraduate*, Wiley, New York, 1949, p. 8.

The integers $(a + x, x)$ of type I are isomorphic with the positive integers. This means that, if we set up the many-to-one correspondence,

(5) $(a + x, x) \leftrightarrow a,$

then, from (3) and (4),

$$(a + x, x) + (b + y, y) = (a + b + u, u) \leftrightarrow a + b,$$

$$(a + x, x) \cdot (b + y, y) = (a \cdot b + v, v) \leftrightarrow a \cdot b.$$

Thus $(a + x, x)$ may be regarded as another notation for the positive integer a.

The integers (x, x) of type II all represent the same number, namely, the identity for addition; for, from (2),

$$(a, b) + (x, x) = (a + x, b + x) = (a, b).$$

We call (x, x) *zero* and write

(6) $(x, x) = 0.$

Zero also has the property (2.8) for

$$(x, x) \cdot (a, b) = (y, y) = 0.$$

The integers $(x, a + x)$ of type III are the *negatives* of $(a + x, x)$ of type I; for

$$(x, a + x) + (a + x, x) = (y, y) = 0$$

is the defining equation (2.3) for negatives. Hence, in accordance with (5), we write

(7) $(x, a + x) = -a,$

and call $-a$ a *negative integer*. The *rules of sign*,

(8) $(-a)b = a(-b) = -ab,$ $(-a)(-b) = ab,$

are simple consequences of (5) and (7); for example,

$$(-a)(-b) = (x, a + x) \cdot (y, b + y) = (ab + z, z) = ab.$$

Since every integer (a, b) has a negative (b, a), subtraction is always possible in the system of integers; and, from (2.9),

$$(a, b) - (c, d) = (a, b) + (d, c) = (a + d, b + c).$$

In particular the difference of two positive integers is given by

(9) $a - b = (a + x, x) + (y, b + y) = (a, b).$

The multiplication of integers is reduced to that of *positive* integers by the rules of sign (8).

The division a/b is only possible when a is an integral multiple of b. Moreover no integer except 1 and -1 has an integral reciprocal. *The system of integers, though closed under subtraction, is not closed under division.*

Since the integers form a group under addition, the cancelation law (2.6) for addition is valid. Although the integers do not form a group under multiplication, the cancelation law (2.7) still holds; for this law, known to be valid for positive integers, may be extended to negative integers by means of equations (8).

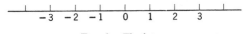

FIG. 3.   The integers

The integers may be represented by an endless set of equidistant points on a directed line. The positive sense, and the points 0 and 1 are chosen at pleasure; the points representing all other integers are then determined (Fig. 3). To add a to b, count b points away from a in the sense given by the sign of b; the end point is $a + b$. Integers that are negatives of each other are symmetric about 0.

The integers do not form a field since postulate M_4 is not satisfied; for no integers except ± 1 have a reciprocal. If we replace M_4 by a weaker postulate, namely, the cancelation law,

$m_4.$ $c \cdot a = c \cdot b$ implies $a = b$ if $c \neq 0,$

the integers satisfy the postulates

$$A_1, A_2, A_3, A_4, A_5; \qquad M_1, M_2, M_3, m_4, M_5; \quad D.$$

Any mathematical system of this type is called an *integral domain*.

4. Rational Numbers.   To remedy the lack of closure of integers under division we attempt to fill the gaps between the integers by constructing new numbers. Just as the integers were formed from pairs of positive integers, we now construct the *rational numbers* or *fractions* (a, b) from ordered pairs of integers a, b with $b \neq 0$.† We proceed to define equality, addition, and multiplication of fractions. If we tentatively identify (a, b) with a/b, the known properties of a/b when a is an integral multiple of b suggest the definitions that follow.

---

† Since the pairs (a, b) of § 3 are now denoted by $a - b$, we abandon this notation. The pairs (a, b) we now introduce have a different meaning.

Equality of fractions is defined by

(1) $(a, b) = (c, d)$ only when $ad = bc$.

This definition satisfies the requirements E_1, E_2, E_3. If x is any non-zero integer, (1) shows that

(2) $(a, b) = (ax, bx)$.

Addition and multiplication are well defined by the equations

(3) $(a, b) + (c, d) = (ad + bc, bd)$,

(4) $(a, b) \cdot (c, d) = (ac, bd)$.

When (a, b), (c, d) are replaced by their equals (ax, bx), (cy, dy), these definitions give a sum and product equal to those above. It is now easy to prove that addition and multiplication of fractions obey the commutative, associative, and distributive laws; the proofs depend on the corresponding laws for integers.

The fractions of the form (ax, x) are isomorphic with the integers a, Thus, if we set up the many-to-one correspondence

(5) $(ax, x) \leftrightarrow a$,

then, from (3) and (4),

$$(ax, x) + (by, y) = ((a + b)z, z) \leftrightarrow a + b,$$
$$(ax, x) \cdot (by, y) = (abz, z) \leftrightarrow a \cdot b.$$

Thus (ax, x) may be regarded as another notation for the integer a; and we shall write

(6) $(ax, x) = a$, $(-ax, x) = -a$,

and, in particular,

(7) $(0, x) = 0$, $(x, x) = 1$.

The fractions (a, b) and $(-a, b)$ are *negatives*; for

$$(a, b) + (-a, b) = (0, b^2) = 0.$$

If a and $b \neq 0$, the fractions (a, b) and (b, a) are *reciprocals*; for

$$(a, b) \cdot (b, a) = (ab, ba) = 1.$$

Thus all fractions have negatives, and all but zero have reciprocals. Consequently, addition and division of fractions is always possible and unique, except division by zero. Thus we have

$$(a, b) - (c, d) = (a, b) + (-c, d),$$
$$(a, b)/(c, d) = (a, b) \cdot (d, c).$$

In particular, the quotient of two integers is given by

(8) $a/b = (ax, x) \cdot (y, by) = (a, b)$.

It is now clear that the rational numbers form a *field*. Hence both cancelation laws (2.6) and (2.7) are valid; and a product can vanish only when a factor is zero.

5. Properties of Rational Numbers. We shall now denote the rational numbers by single letters $a, b, c, \cdots$. Since $p/q = -p/(-q)$, we can write any rational number r as a fraction with a positive denominator; then r is called *positive* or *negative* according to the sign of its numerator.

A set of elements is said to be *ordered* if there is a relation $<$ such that:

O_1. When $a \neq b$, either $a < b$ or $b < a$;

O_2. $a < b, b < c$ imply $a < c$.

Thus the integers on a directed line (Fig. 3) are ordered when $<$ means "precedes."† The relation $a < b$ may also be written $b > a$.

If a and b are two unequal rational numbers, then either $a - b$ or $b - a$ is positive. We now order the rational numbers by defining $<$ so that

(1) $a < b$ when $b - a$ is positive.

If $b - a$ and $c - b$ are positive,

$$c - a = (c - b) + (b - a) \text{ is positive.}$$

Thus $<$ satisfies both requirements O_1, O_2 of an order relation. Consequently:

The rational numbers form an ordered set.

The following properties of the relation $<$ are simple consequences of (1).

(2)   If a is positive, then $a > 0$.

(3)   If $a < b$, then $a + x < b + x$.

(4)   If $a < b, x > 0$, then $ax < bx$.

(5)   If $a < b, x < 0$, then $ax > bx$.

We can now add the rational fractions to the line that carries the integers (Fig. 3). Thus $-5/3$ is represented by the point $1/3$ of the way from 0 to -5. On this diagram the relation $<$ between two rational points means "precedes." However, *no rational point on the line has an immediate predecessor;* for, between any two rational points a, b, no matter how close, there is always another, and therefore an infinite number. In fact, the point $c = \frac{1}{2}(a + b)$ is rational and just halfway between a and b; for, if $a < b$,

$$c - a = b - c = \tfrac{1}{2}(b - a);$$

---

† A point A precedes B on a directed line when A is on the "negative" side of B.

and this process of bisection can be continued indefinitely. This property is expressed by saying:

The rational numbers form a dense set.

All positive fractions p/q can be arranged in successive groups for which $p + q = 2, 3, 4, 5, \cdots$; thus,

$$\tfrac{1}{1}; \ \tfrac{1}{2}, \tfrac{2}{1}; \ \tfrac{1}{3}, \tfrac{2}{2}, \tfrac{3}{1}; \ \tfrac{1}{4}, \tfrac{2}{3}, \tfrac{3}{2}, \tfrac{4}{1}; \ \tfrac{1}{5}, \tfrac{2}{4}, \tfrac{3}{3}, \tfrac{4}{2}, \tfrac{5}{1}; \cdots,$$

$$1 \quad 2 \ 3 \quad 4 \ * \ 5 \quad 6 \ 7 \ 8 \ 9 \quad 10 \ * \ * \ * \ 11 \cdots.$$

Each fraction is repeated indefinitely often; but if we omit repetitions (marked *) we obtain a sequence in which each positive fraction occurs just once. This sequence of fractions may be put in one-to-one correspondence with the positive integers as shown; if $r_n \leftrightarrow n$, we have

$$r_1 = 1, \quad r_2 = \tfrac{1}{2}, \quad r_3 = 2, \quad \cdots, \quad r_{10} = \tfrac{1}{5}, \quad r_{11} = 5, \quad \cdots.$$

When sufficiently extended, this sequence determines r_n uniquely for any n.†

All rational numbers (positive, negative, and zero) may also be ranged in a linear sequence and put in one-to-one correspondence with the positive integers; thus

$$0 \quad r_1 \quad -r_1 \quad r_2 \quad -r_2 \quad r_3 \quad -r_3 \quad \cdots,$$

$$1 \quad 2 \quad\ 3 \quad\ 4 \quad\ 5 \quad\ 6 \quad\ 7 \quad \cdots.$$

When an infinite set can be put in one-to-one correspondence with the counting numbers, we say it is *countable* (or *denumerable*). Therefore:

The rational numbers form a countable set.

The rational fractions p/q can be expressed as decimals by "long division." If the denominator q contains no prime factors other than 2 or 5, the decimal for p/q will *terminate*; otherwise, the decimal will be *periodic*; that is, eventually a group of digits will repeat without end. This is clear from the process of long division of q into p; for, after the digits in p have been exhausted and zeros are carried down, only the $q - 1$ remainders of q can appear. After at most $q - 1$ divisions, a remainder r will appear for a second time, and thereafter all remainders will repeat indefinitely in the same order. If there are $n(<q)$ different remainders $r_1, r_2, \cdots, r_n$, and

$$10r_i = qb_i + r_{i+1} \quad (i = 1, 2, \cdots, n), \qquad r_{n+1} = r_1,$$

then the period of p/q will consist of the digits $b_1 b_2 \cdots b_n$. For example,

$$\tfrac{1}{3} = 0.\dot{3}, \qquad \tfrac{1}{5} = 0.2\dot{0}, \qquad \tfrac{1}{7} = 0.\dot{1}4285\dot{7},$$

---

† Cf. Harrington, W. J., A Note on the Denumerability of the Rational Numbers, *Am. Math. Monthly*, vol. 58, no. 10, 1951, for a constructive correspondence between the rational numbers and the positive integers.

the dots marking the period (the digit 0 is the period of a terminating decimal). This periodicity is obviously independent of the scale of notation; thus, in the binary scale,

$$^1/_3 = 0.\dot{0}\dot{1}, \quad ^1/_5 = 0.\dot{0}01\dot{1}, \quad ^1/_7 = 0.\dot{0}0\dot{1}.$$

Conversely we shall see (§ 22) that every periodic decimal can be represented as a rational fraction. We thus have the theorem:

A digital fraction is rational when and only when it is periodic.

Note that, if the digital fraction terminates, it may be put in periodic form in *two* ways. Thus a terminating *decimal* may be given the period 0, or the period 9; for example,

$$^1/_5 = 0.2\dot{0} \quad \text{or} \quad 0.1\dot{9}.$$

In spite of the density of rational numbers, they do not exhaust the *real* numbers, the numbers that can be represented by decimals; for any nonperiodic decimal such as $0.101001000 \cdots$ cannot be rational. Such numbers are called *irrational*.

A number that is the root of an algebraic equation

(6) $$a_0 x^n + a_1 x^{n-1} + \cdots + a_{n-1} x + a_n = 0$$

whose coefficients a_i are integers is called an *algebraic number*. Every rational number p/q is algebraic for it satisfies $qx - p = 0$. However, an algebraic number is not in general rational. The proof that certain algebraic numbers are not rational often follows from the

THEOREM (Gauss). *Any rational root of the equation* (6) *with integral coefficients and* $a_0 = 1$ *must be an integer that divides* a_n *exactly.*

Proof. Let $x = p/q$ be a root where p and q are integers without a common factor. Putting $x = p/q$ in the equation (with $a_0 = 1$) and multiplying it by q^{n-1}, we obtain

$$-\frac{p^n}{q} = a_1 p^{n-1} + a_2 p^{n-2} q + \cdots + a_{n-1} p q^{n-2} + a_n q^{n-1}.$$

Here $-p^n/q$ is a fraction in its lowest terms, while the right member is a sum of integers and therefore an integer. This is impossible unless $q = 1$ and $x = p$; and, since

$$-p(p^{n-1} + a_1 p^{n-2} + \cdots + a_{n-1}) = a_n,$$

p is a divisor of a_n.

Example. 1. If n is an integer which is not a perfect square, the equation $x^2 - n = 0$ can have no integral roots and hence no rational roots. The root $\sqrt{n}$ must therefore be irrational. For example $\sqrt{2}$ and $\sqrt{3}$ are irrational.

Example 2. The equation

$$f(x) = x^3 + x - 5 = 0$$

has a root between 1 and 2; for $f(1) = -3, f(2) = 5$. Since this root is not integral, it must be irrational.

The algebraic numbers form a countable set.

Proof. Equation (6) can always be written so that $a_0 > 0$ and the integral coefficients have no common factor. Then assign to each equation a positive integer

$$N = a_0 + |a_1| + \cdots + |a_n| + n, \quad \text{its } index.$$

There are only a finite number of equations having a given index N; and each of those equations has at most n different roots. Therefore there are only a finite number $f(N)$ of algebraic numbers that satisfy any equations of index N.[†] We can now arrange *all* algebraic numbers in a sequence by starting with $f(2)$ of index 2, then $f(3)$ of index 3, and so on. Discarding duplicates, we can then put the algebraic numbers in one-to-one correspondence with the counting numbers.

We shall see in the next article that the totality of real numbers is *not countable*. Hence there must be real numbers that are not algebraic; such numbers are called *transcendental*. Both e and π are transcendental; the rather intricate proofs were first given by Hermite (1881) and Lindemann (1882), respectively. It is much easier to prove that e and π are irrational. The irrationality of e is proved in § 66; as to π, see Niven's proof.[‡]

6. Real Numbers. We shall now define irrational numbers without reference to their decimal representation.

Let us suppose that by some rule we can separate the *rational* numbers into two classes (a) and (A) such that

(i)   every number belongs to either (a) or (A);

(ii)   there are numbers in each class;

(iii) any number a of (a) is less than any number A of (A).

In brief, the classification is exhaustive, neither class is empty, and $a < A$. Such a separation is called a *cut* (or a *Dedekind cut*§) in the rational numbers and written $a \,|\, A$.

There are now just three possibilities:

1. *Class* (a) *contains a greatest number* a_M. The classes then consist of the rationals

$$a \leqq a_M, \qquad A > a_M.$$

[†] The reader may verify that $f(2) = 1, f(3) = 3, f(4) = 9$.

[‡] Niven, I., A Simple Proof that π Is Irrational, *Bull. Am. Math. Soc.*, vol. 53, 1947, p. 509.

§ Dedekind, *Stetigheit und irrationale Zahlen*, Braunschweig, 1872.

2. *Class* (A) *contains a smallest number* A_m. Now the classes consist of the rationals

$$a < A_m, \qquad A \geqq A_m.$$

Note that *both* a_M and A_m cannot exist at the same time; for $a_M < A_m$ by (iii), and the rational number $\frac{1}{2}(a_M + A_m)$ lies between a_M and A_m and hence belongs to neither class.

Cases 1 and 2 are easily realized; we need only choose a rational number r and define the classes by

$$a \leqq r, \quad A > r; \qquad \text{or} \qquad a < r, \quad A \geqq r.$$

In either case we say that the number r determines the cut and that *the cut* $a \mid A$ *is rational*. If we always put r in the upper class, we may dispense with case 1; then there is a one-to-one correspondence between the rational numbers and the cuts they produce.

3. *Class* (a) *contains no greatest number, class* (A) *no smallest number.* In this case we say that the *cut* $a \mid A$ *is irrational* and defines an *irrational number* α. This number fills the gap between the rational classes (a) and (A); and we write $\alpha = a \mid A$.

For example, let class (a) consist of all negative rationals and those positive rationals whose square < 2; and class (A) of those positive rationals whose square > 2. Then $a \mid A = \sqrt{2}$, an irrational number (Ex. 5.2).

We now define a *real number* α to be a cut $a \mid A$ in the rational numbers that is rational in case 2, irrational in case 3.

The real numbers $\alpha = a \mid A$, $\beta = b \mid B$ are said to be *equal* when class (a) contains the same numbers as class (b); we then write $\alpha = \beta$. When α and β are rational, both equal the smallest number in the upper class; the new definition of equality thus agrees with the old.

When class (a) is a part of class (b), we write $\alpha < \beta$. Evidently, if $\alpha \neq \beta$, either (a) is part of (b) or vice versa; thus, if $\alpha \neq \beta$, either $\alpha < \beta$ or $\beta < \alpha$.

Next let $\alpha = a \mid A$, $\beta = b \mid B$, $\gamma = c \mid C$. Then, if $\alpha < \beta$ and $\beta < \gamma$, we have $\alpha < \gamma$; for (a) is a part of (b), (b) a part of (c), and hence (a) is a part of (c). Consequently real numbers fulfil the requirements O_1 and O_2 of § 5:

The real numbers form an ordered set.

When α and β are rational, the new definition of $<$ coincides with the old. Moreover if $\alpha = a \mid A$, any $a < \alpha$, any $A \geqq \alpha$ ($A > \alpha$ when α is irrational). Consequently, when $\alpha < \beta$, we can always find a rational between α and β; for the definition of $<$ shows that we can find a rational $r \neq \alpha$ that is in (A) as well as in (b), and hence $\alpha < r < \beta$.

If we separate all *real* numbers into two classes (a), (A) which satisfy the conditions (i), (ii), (iii), we have a *cut $a \mid A$ in the real numbers.* Such a cut, however, does not lead to a further extension of the number concept. For now we shall see that case 3 is impossible: There are only *two* possibilities instead of *three.*

DEDEKIND'S THEOREM. *A cut $a \mid A$ in the real numbers corresponds to the real number $a' \mid A'$, where (a') and (A') consist of the rationals in (a) and (A), respectively.*

Proof. Let the cut $a' \mid A' = \alpha$, a rational or irrational real number. Every rational number $< \alpha$ is in class (a') and hence in (a). Moreover, for every irrational number $\gamma < \alpha$, there exist rationals r such that $\gamma < r < \alpha$; but we have just seen that r is in (a), and hence γ is also in (a). Thus, every real number $< \alpha$ is in class (a). Similarly every real number $> \alpha$ is in class (A). But by (i) *all* real numbers must be in either (a) or (A); hence α must either be the greatest number in (a) or the least number in (A). Thus the cut $a \mid A$ in the reals always "corresponds" to a real number, whereas a cut in the rationals sometimes "corresponds" to a rational number, but not always.

The real numbers include the rational numbers and like the latter form a dense set (§ 5). But, unlike the rational numbers, *the real numbers are not countable.* Cantor's indirect proof is based on the following ingenious "diagonal process." If the real numbers between 0 and 1 are countable, they can be written in a sequence $\alpha_1, \alpha_2, \alpha_3, \cdots$. Let us write their decimals:

$$\alpha_1 = 0 . a_{11} a_{12} a_{13} \cdots ,$$
$$\alpha_2 = 0 . a_{21} a_{22} a_{23} \cdots ,$$
$$\alpha_3 = 0 . a_{31} a_{32} a_{33} \cdots ,$$

filling in zeros if a decimal terminates. Now form a decimal

$$\beta = 0 . b_1 b_2 b_3 \cdots ,$$

where b_n is any digit from 1 to 8 that differs from a_{nn}. Then β is a nonterminating decimal between 0 and 1 which differs from all the α's: for β differs from α_1 in the first decimal place, from α_2 in the second, and so on. Therefore, no sequence $\alpha_1, \alpha_2, \alpha_3, \cdots$ can include *all* the real numbers between 0 and 1.

Thus there are at least two kinds of infinite sets; the countable sets, which includes the integers, the rational numbers, and algebraic numbers, and the noncountable set of real numbers: the *real continuum.*

If two finite or infinite sets can be put in one-to-one correspondence, they are said to have the same *cardinal number.* The cardinal number of a finite set is simply the number of its elements. The cardinal number of all infinite sets that are countable is denoted by $\aleph_0$ (*aleph-null*). They have the same manyness in the sense that the elements of two such sets may be put in one-to-one correspondence. The cardinal number of the real continuum is denoted by C; and we write $\aleph_0 < C$ since the real numbers contain the integers as a subset but cannot be placed in one-to-one correspondence with them. However, all real numbers and the real numbers in the interval $(0, 1)$ have the same cardinal number C; for the transformation $x' = 1/(1 + e^{-x})$ puts the entire axis of reals $-\infty < x < \infty$ in one-to-one correspondence with the segment $0 < x' < 1$.

We have seen that the rational numbers may be represented as points on the line that carries the integers (Fig. 3) and that these points form a dense set; between any two, however close, there are infinitely many others. Nevertheless these rational points do not exhaust the points in the line; in fact, the points filling the gaps between them are far more numerous since they have a greater cardinal number. The points in these gaps correspond to the irrational numbers. This assumption is known as the

CANTOR-DEDEKIND AXIOM. *The real numbers can be put in one-to-one correspondence with the points of an infinite straight line: the axis of reals.*

In view of this axiom, real numbers are often called "points."

Cantor has shown how to form a succession of infinite sets whose cardinal numbers form an increasing sequence: $\aleph_0, \aleph_1, \aleph_2, \cdots$. In fact, if A is set and B the set of all its subsets (including the null set and A itself), it can be shown that B has a greater cardinal number than A.[†] The question now arises: Is C less than, equal to, or greater than $\aleph_1$? This is the famous *continuum problem* which is unsolved to this day. Cantor conjectured that $C = \aleph_1$; more recently Lusin believed that $C > \aleph_1$ was more likely. But definitive proofs are lacking.

7. Operations with Real Numbers. If $\alpha = a \,|\, A$, $\beta = b \,|\, B$ are real numbers, we proceed to define $\alpha + \beta$, $\alpha - \beta$, $\alpha\beta$, α/β as definite cuts in the rational numbers.

Addition. Consider classes (c), (C) formed by the numbers $c = a + b$, $C = A + B$. Since $a < A$, $b < B$, we have $c < C$. Moreover, the density of the rationals shows that $A - a$ and $B - b$ can be made arbitrarily small.[‡] Hence the same is true of

$$C - c = (A - a) + (B - b).$$

It follows that there is at most *one* rational number r that does not belong

---

† Cf. Courant and Robbins, *What Is Mathematics?*, Oxford University Press, 1941 p. 84.

‡ To give a formal proof, let ε be a small positive rational and a_0 a number of class (a). Then the sequence $a_0 + \varepsilon$, $a_0 + 2\varepsilon$, $a_0 + 3\varepsilon$, $\cdots$ is unbounded and its numbers will eventually lie in class (A). If $a_0 + n\varepsilon$ is the first of these, its predecessor lies in class (a) and their difference is ε.

either to (c) or (C); for, if r' were a second (say $r' > r$), $C - c$ could not be less than $r' - r$. In case r exists, we add it to class (C). Then in every case the classes (c) and (C) include all the rational numbers and form a cut $c \mid C$ which we define as $\alpha + \beta$. In brief,

(1) $$a \mid A + b \mid B = (a + b) \mid (A + B).$$

This definition includes as a particular case the addition of rational numbers. Moreover, the validity of the commutative and associative laws of addition for real numbers follows at once from these laws for rational numbers:

(2) (3) $$\alpha + \beta = \beta + \alpha, \qquad (\alpha + \beta) + \gamma = \alpha + (\beta + \gamma).$$

Negative. When $\alpha = a \mid A$, we define its negative as

(4) $$-\alpha = -A \mid -a,$$

with the proviso that, when α is rational, $-\alpha$ belongs to the upper class $(-a)$. Since a rational α is the smallest number in (A), this proviso prevents the lower class $(-A)$ of the cut (4) from having $-\alpha$ as largest number. Evidently we have $-(-\alpha) = \alpha$. Moreover,

(5) $$\alpha + (-\alpha) = 0;$$

for the cut

$$(a - A) \mid (A - a) = -(A - a) \mid (A - a) = 0,$$

since it separates the positive from the negative rational numbers.

Subtraction. The difference $\xi = \alpha - \beta$ is the solution of the equation $\beta + \xi = \alpha$. On adding $-\beta$ to both members we find that

(6) $$\alpha - \beta = \alpha + (-\beta).$$

Thus subtraction is reduced to addition.

Multiplication. We first consider two positive numbers $\alpha = a \mid A$, $\beta = b \mid B$ given by cuts in the *positive* rational numbers. If $c = ab$, $C = AB$, $c \mid C$ is also a cut in the positive rationals since $c < C$ and

$$C - c = (A - a)B + a(B - b)$$

can be made arbitrarily small. We now define

(7) $$(a \mid A) \cdot (b \mid B) = ab \mid AB.$$

When one or both factors is negative, their product is given by the definitions

(8) (9) $$(-\alpha)\beta = \alpha(-\beta) = -\alpha\beta, \qquad (-\alpha)(-\beta) = \alpha\beta;$$

thus the *rules of sign* (3.8) are extended to all real numbers.

Finally we define

(10) $$\alpha \cdot 0 = 0 \cdot \alpha = 0 \quad \text{for any } \alpha.$$

From these definitions we may now prove that the multiplication of real numbers is commutative, associative, and distributive with respect to addition:

(11) (12) $\alpha\beta = \beta\alpha,$ $(\alpha\beta)\gamma = \alpha(\beta\gamma),$

(13) $\alpha(\beta + \gamma) = \alpha\beta + \alpha\gamma.$

Reciprocal. If the positive number $\alpha = a \mid A$ is a cut in the positive rational numbers, its *reciprocal* is

(14) $$\frac{1}{\alpha} = \frac{1}{A} \bigg| \frac{1}{a},$$

with the proviso that, when α is rational, $1/\alpha$ belongs to the upper class $(1/a)$. For

(15) $$\alpha \cdot \frac{1}{\alpha} = \frac{a}{A} \bigg| \frac{A}{a} = 1,$$

since the cut $a/A \mid A/a$ separates the positive numbers <1 from those >1. The reciprocal of $-\alpha$ is now $-1/\alpha$ as we see from (9).

Division. If $\beta \neq 0$, the quotient $\xi = \alpha/\beta$ is the solution of the equation $\beta \cdot \xi = \alpha$. On multiplying both members by $1/\beta$, we find that

(16) $$\xi = \frac{\alpha}{\beta} = \alpha \cdot \frac{1}{\beta}.$$

Thus division is reduced to multiplication.

When $\beta = 0$, α/β is not defined.

It can now be shown that all the operations in the algebra of rational numbers may be carried over to the algebra of real numbers. The reader is referred to Landau, *Grundlagen der Analysis*, for details.

We note finally that the real numbers form a *field* and partake in all the properties given in § 2.

8. Complex Numbers. For the real number $\alpha \neq 0$,

$$\alpha \cdot \alpha = (-\alpha)(-\alpha) > 0;$$

hence a negative real number has no real square root. Consequently, we again enlarge the number system so that square roots always exist. Indeed we shall see that in the system of *complex numbers* every number other than zero has exactly n nth roots.

We define a *complex number* as an ordered pair of real numbers (α, α').

Equality of complex numbers is defined by

(1) $(\alpha, \alpha') = (\beta, \beta')$ only when $\alpha = \beta,$ $\alpha' = \beta'.$

This equivalence relation is an identity and is obviously reflexive, symmetric, and transitive (§ 1).

Addition and *multiplication* of complex numbers are defined by the equations

(2) $$(\alpha, \alpha') + (\beta, \beta') = (\alpha + \beta, \alpha' + \beta'),$$

(3) $$(\alpha, \alpha') \cdot (\beta, \beta') = (\alpha\beta - \alpha'\beta', \alpha\beta' + \alpha'\beta).$$

Addition is obviously commutative and associative; and it is easy to verify that multiplication is commutative, associative, and distributive with respect to addition.

Complex numbers of the form $(\alpha, 0)$ are isomorphic with the real numbers α; for, if we set up the one-to-one correspondence $(\alpha, 0) \leftrightarrow \alpha$, we have, from (2) and (3),

$$(\alpha, 0) + (\beta, 0) = (\alpha + \beta, 0) \leftrightarrow \alpha + \beta,$$

$$(\alpha, 0) \cdot (\beta, 0) = (\alpha\beta, 0) \leftrightarrow \alpha\beta.$$

In practice, we write $(\alpha, 0) = \alpha$ and regard the real numbers as a subset of the complex numbers. In particular we write $(0, 0) = 0$, $(1, 0) = 1$; moreover, for any real number λ,

(4) $$\lambda(\alpha, \alpha') = (\lambda, 0) \cdot (\alpha, \alpha') = (\lambda\alpha, \lambda\alpha')$$

from (3). Thus we have

(5) (6) $$0 + (\alpha, \alpha') = (\alpha, \alpha'), \qquad 1 \cdot (\alpha, \alpha') = (\alpha, \alpha').$$

We write the complex number $(0, 1) = i$; then, from (3), we have

(7) $$i^2 = (0, 1) \cdot (0, 1) = (-1, 0) = -1.$$

Hence i is a square root of -1. Moreover, from (2) and (4), we have, for any complex number,

$$(\alpha, \alpha') = (\alpha, 0) + (0, \alpha') = \alpha + \alpha'(0, 1),$$

(8) $$(\alpha, \alpha') = \alpha + i\alpha'.$$

Complex numbers in this form may be added and multiplied formally as in real algebra provided i^2, when it occurs, is replaced by -1; for

(2)' $$(\alpha + i\alpha') + (\beta + i\beta') = \alpha + \beta + i(\alpha' + \beta'),$$

(3)' $$(\alpha + i\alpha') \cdot (\beta + i\beta') = \alpha\beta - \alpha'\beta' + i(\alpha\beta' + \alpha'\beta)$$

give the results required by the definitions (2) and (3). It is now clear *why* these definitions imply the commutative, associative, and distributive laws; for these laws are valid in real algebra and will remain valid after we put $i^2 = -1$.

The complex numbers satisfy all the requirements of a *field*. They form a group under addition with the identity *zero*; the additive inverse of $\alpha + i\alpha'$ is its negative $-\alpha - i\alpha'$. With zero omitted, they also form a group under multiplication with the identity *one*; the multiplicative inverse of $\beta + i\beta' \neq 0$ is its reciprocal

$$(9) \qquad \frac{1}{\beta + i\beta'} = \frac{\beta - i\beta'}{\beta^2 + \beta'^2};$$

for the product of this number and $\beta + i\beta'$ is 1. Moreover division by $\beta + i\beta'$ is reduced to multiplication by its reciprocal; thus the quotient

$$(10) \qquad \frac{\alpha + i\alpha'}{\beta + i\beta'} = \frac{(\alpha + i\alpha')(\beta - i\beta')}{\beta^2 + \beta'^2} = \frac{\alpha\beta + a'\beta'}{\beta^2 + \beta'^2} + i\,\frac{\alpha'\beta - \alpha\beta'}{\beta^2 + \beta'^2}.$$

The complex numbers therefore have all the properties of a field. Thus the cancelation laws (2.6) and (2.7) hold in complex algebra. Moreover *the product of two complex numbers is zero when, and only when, one of the factors is zero.* A direct proof of this important property is readily given. For, from (3)′, we see that $(\alpha + i\alpha')(\beta + i\beta') = 0$ implies that

$$\alpha\beta - \alpha'\beta' = 0, \qquad \alpha\beta' + \alpha'\beta = 0;$$

and, on squaring and adding these equations, we get

$$(\alpha^2 + \alpha'^2)(\beta^2 + \beta'^2) = 0.$$

Hence one factor must vanish; thus, if

$$\alpha^2 + \alpha'^2 = 0, \qquad \alpha = \alpha' = 0 \quad \text{and} \quad \alpha + i\alpha' = 0.$$

The numbers $\alpha + i\alpha'$ and $\alpha - i\alpha'$ are said to be *conjugates*. Note that their sum 2α and product $\alpha^2 + \alpha'^2$ are both real. Moreover, we can readily verify the

THEOREM. *The conjugate of the sum, difference, product, and quotient of two complex numbers is equal to the respective sum, difference, product, and quotient of their conjugates.*

As a field, the complex numbers are closed under addition, subtraction, multiplication, and division—excluding division by zero. Thus a polynomial in $z = \xi + i\xi'$ with complex coefficients,

$$(11) \qquad w = P(z) = a_0 z^n + a_1 z^{n-1} + \cdots + a_{n-1} z + a_n$$

is again a complex number. Moreover, if $\bar{z} = \xi - i\xi'$ denotes the conjugate of z, the above theorem shows that

$$\bar{w} = \bar{a}_0 \bar{z}^n + \bar{a}_1 \bar{z}^{n-1} + \cdots + \bar{a}_{n-1} \bar{z} + \bar{a}_n.$$

When the coefficients a_i are real, $\bar{a}_i = a_i$ and $\bar{w} = P(\bar{z})$; hence, if z_1 is a root of $P(z) = 0$, $w = 0$, $\bar{w} = 0$, and $\bar{z}_1$ is also a root. *The complex roots of a polynomial with real coefficients occur in conjugate pairs.*

The complex numbers are dense in the complex plane but, unlike the real numbers, are not ordered. Naturally they are not countable since the real numbers are not; but surprisingly they have the same cardinal number C as the real continuum.

9. Complex Numbers as Vectors. According to the Cantor-Dedekind axiom the real numbers may be put in one-to-one correspondence with the points of an infinite line after the points O and U representing 0 and 1 are chosen. We draw this *axis of reals* horizontally and choose the point U to the right of O; the points representing positive and negative

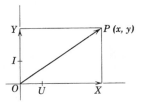

FIG. 9a. The complex plane

real numbers then lie to the right and left of O, respectively (Fig. 9a). The real numbers x are now in one-to-one correspondence with the points X on the axis of reals. We prefer, however, to regard this correspondence as one between numbers x and vectors $\overrightarrow{OX}$; thus

$$1 \leftrightarrow \overrightarrow{OU}, \qquad x \leftrightarrow \overrightarrow{OX}.$$

We now draw a second line through O perpendicular to the axis of reals and choose the point I above O so that $OI = OU$. If we assign the point I to represent i, all points Y on this *axis of imaginaries* represent complex numbers of the form yi, the "pure imaginary" numbers. We again use a vector representation:

$$i \leftrightarrow \overrightarrow{OI}, \qquad yi \leftrightarrow \overrightarrow{OY}.$$

The complex number $x + yi$ is now represented by $\overrightarrow{OX} + \overrightarrow{OY} = \overrightarrow{OP}$ where the vectors are added according to the "parallelogram law." Thus, if P has the rectangular coordinates (x, y) referred to the axes, we have the one-to-one correspondence

(1) $$x + iy \leftrightarrow \overrightarrow{OP}$$

between the complex numbers $x + iy$ and the position vectors of the points $P(x, y)$ in the xy-plane.

If $r =$ length OP and $\theta =$ angle (OU, OP) are the polar coordinates of P,

$$x = r \cos \theta, \qquad y = r \sin \theta \qquad (r \geq 0),$$

and we obtain the *polar form*

(2) $$x + iy = r (\cos \theta + i \sin \theta)$$

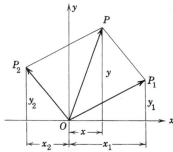

Fig. 9b. Addition of complex numbers

of the complex number. The real, nonnegative number r is called the *absolute value* or *modulus* of $x + iy$ and written $|x + iy|$; and θ, taken positive in the counterclockwise sense, is called the *angle* of $x + iy$. Thus we have

(3) $$r = |x + iy| = \sqrt{x^2 + y^2},$$
$$\tan \theta = y/x.$$

If we denote complex numbers by single letters, $z = x + iy$, we have (Fig. 9b)

(4) $$z_1 + z_2 = (x_1 + x_2) + i(y_1 + y_2) \leftrightarrow \overrightarrow{OP} = \overrightarrow{OP_1} + \overrightarrow{OP_2};$$

hence, $z_1 + z_2$ corresponds to the vector sum of $\overrightarrow{OP_1}$ and $\overrightarrow{OP_2}$. Since $OP_1 + OP_2 \geq OP$, we have the *triangle inequality*:

(5) $$|z_1| + |z_2| \geq |z_1 + z_2|.$$

To find the product $z_1 z_2$ we use the polar form (2); thus

$$z_1 z_2 = r_1 r_2 (\cos \theta_1 + i \sin \theta_1)(\cos \theta_2 + i \sin \theta_2);$$

or, since

$$\cos (\theta_1 + \theta_2) = \cos \theta_1 \cos \theta_2 - \sin \theta_1 \sin \theta_2,$$
$$\sin (\theta_1 + \theta_2) = \sin \theta_1 \cos \theta_2 + \cos \theta_1 \sin \theta_2,$$

(6) $$z_1 z_2 = r_1 r_2 [\cos (\theta_1 + \theta_2) + i \sin (\theta_1 + \theta_2)].$$

The angle of $z_1 z_2$ is $\theta_1 + \theta_2$; and the modulus

(7) $$|z_1 z_2| = |z_1| |z_2|.$$

Complex addition and multiplication thus correspond to simple geometric operations: *To add complex numbers, add their vectors; to multiply complex numbers, multiply the lengths of the vectors and add their angles.* The commutative, associative, and distributive laws now have a simple geometric interpretation.

If $\bar{z} = x - iy$ denotes the conjugate of z, the reciprocal

(8) $$\frac{1}{z} = \frac{\bar{z}}{z\bar{z}} = \frac{\bar{z}}{|z|^2}.$$

Thus, if $|z| = 1$, $z^{-1} = \bar{z}$.

From (6) we have *De Moivre's theorem* for the product of n equal factors:

(9) $$z^n = r^n (\cos n\theta + i \sin n\theta).$$

We may now show that a nonzero complex number

$$Z = R (\cos \Theta + i \sin \Theta)$$

has exactly n nth roots: namely, the n complex numbers having $r = \sqrt[n]{R}$ as modulus and

(10) $$\theta_1 = \frac{\Theta}{n}, \quad \theta_2 = \frac{\Theta + 2\pi}{n}, \quad \cdots, \quad \theta_n = \frac{\Theta + (n-1)2\pi}{n}$$

as angles. Of these the root with the smallest angle is called the *principal root* and is denoted by $\sqrt[n]{Z}$ or $Z^{1/n}$. Thus, if x is real and positive, $\sqrt{x}$ denotes its *positive* square root (angle 0); the negative root is written $-\sqrt{x}$ (angle π).

Finally we state an existence theorem often called

THE FUNDAMENTAL THEOREM OF ALGEBRA. *A polynomial equation with complex coefficients always has a complex root.*

Since complex numbers include the real numbers, this root may be real. A simple proof is given in § 208.

10. Development of the Number System. Starting with the positive integers and two operations, addition and multiplication, we have seen that both operations were commutative and associative and were connected by the distributive law. The positive integers are closed under addition and multiplication, but the inverse operations are not always possible. When the positive integers are augmented by the negative integers and zero, subtraction is always possible; and, when these integers are further augmented by the fractions, both subtraction and division are possible, excluding division by zero. The resulting system of rational numbers—the periodic decimals—is thus closed under addition, subtraction, multiplication, and division (except by zero); moreover this system is *ordered*, *dense*, and *countable*. At each stage in the expansion of the number system,

positive integers → integers → rational numbers,

addition and multiplication are redefined so that the five basic laws (two commutative, two associative, and one distributive) maintain their validity. Moreover the new numbers include the old, and the new operations contain the old as special cases. The central principle underlying this expansion is that of *permanence of form*; the laws of reckoning are kept intact when new numbers are introduced.

Although the rational numbers are dense, they still do not include the nonperiodic decimals or *irrational numbers*. When the latter are adjoined to the rational numbers, we obtain the system of real numbers. The real numbers are likewise closed to addition, subtraction, multiplication, and division (except by zero); they also form an ordered and a dense system, but they are no longer countable. We assume, however, that they may be put in one-to-one correspondence with the points of an infinite line, an *axis of reals* (the Cantor-Dedekind axiom).

Real numbers may be raised to any positive integral power; but the inverse operation of root extraction is not always possible. Thus a negative number has no real square root; more generally, there are polynomial equations with real coefficients that have no real roots. To remedy these defects we once more expand the number system to include the *complex numbers*: ordered pairs of real numbers. Once again addition and multiplication are redefined so as to maintain the five basic laws. The complex numbers may be put in one-to-one correspondence with the points of a plane, or with the totality of vectors $\overrightarrow{OP}$ issuing from a fixed origin or zero point O. Complex numbers are added *vectorially*; and, to find their product, we multiply their lengths and add their angles. These new definitions apply to *all* numbers, the integers, the rationals, the reals, and finally the complex. Moreover, they cast a strong light on such a mysterious rule as minus times minus is plus.† They enable us to invariably extract roots, and indeed each complex number other than zero has exactly n nth roots. More generally, a polynomial equation with complex coefficients always has a complex root and indeed exactly n roots when a repeated root α is counted as many times as the factor $x - \alpha$ occurs in the equation.

The rational, the real, and the complex numbers are all examples of mathematical systems called *fields*. In a field the additive identity 0 (zero) has the property $x \cdot 0 = 0$ for any element x. Moreover a product is zero only when one of the factors is zero.

Consider now the method of making these successive extensions of the number system. The integers were introduced as ordered pairs of positive

---

† The negative reals $-x$, $-y$ have the moduli x, y, the angle π; hence $(-x)(-y)$ has the modulus xy, the angle 2π.

integers, the rational numbers as ordered pairs of integers. Both extensions left the new system of numbers *countable*: Their cardinal number still remained $\aleph_0$, the same as that of the positive integers. The next extension, by means of Dedekind cuts in the rational numbers, gave the real continuum whose cardinal number $C > \aleph_0$. The final extension from real to complex numbers, again achieved by ordered pairs, did not alter the cardinal number C but resulted in the loss of order. The end result, the field of complex numbers, is not only a logical and imaginative creation of the first magnitude but a structure of inestimable value in geometry, physics, and indeed in all of science where quantitative considerations are paramount. We conclude with an outline of this development.

$\aleph_0$ Positive integers (ordered)
with ordered pairs $(a, b) = a - b$,

$\downarrow$

$\aleph_0$ Integers (integral domain; ordered)
with ordered pairs $(a, b) = a/b$,

$\downarrow$

$\aleph_0$ Rational numbers (field; ordered, dense)
with Dedekind cuts $a \mid A$,

$\downarrow$

C Real numbers (field; ordered, dense)
with ordered pairs $(a, b) = a + ib$,

$\downarrow$

C Complex numbers (field; dense).

PROBLEMS

1. If a, b, c are elements of a group and ab means $a \circ b$, show that $axba = c$ has a unique solution x.

2. Which of the following sets are groups? Why?

    (*a*) All even integers under addition?

    (*b*) All integers under subtraction?

    (*c*) All rational numbers under addition; under multiplication?

    (*d*) All complex numbers z for which $|z| = 1$ under multiplication?

    (*e*) All irrational numbers and zero under addition?

3. Show by exhibiting a multiplication table that the following sets are groups:

    (*a*) The integers 0, 1, 2, 3 under addition mod 4.

    (*b*) The integers 1, 2, 3, 4 under multiplication mod 5.

    (*c*) The permutations $p_1 = 1234$, $p_2 = 2143$, $p_3 = 4321$, $p_4 = 3412$.

4. If a group has an even number of elements, show that at least one element besides the identity is its own inverse.

5. Assuming that addition and multiplication of *positive* integers obey the commutative, associative, and distributive laws, prove these laws for integers in general. [Use (3.3) and (3.4).]

6. Assuming that addition and multiplication of integers obey the commutative, associative, and distributive laws, prove these laws for rational numbers. [Use (4.3) and (4.4).]

7. Give a direct proof that $\sqrt{2}$ is irrational.

[Assume $\sqrt{2} = p/q$, a fraction in its lowest terms; then $2q^2 = p^2$. Hence show that p and q are both even.]

8. From (8.2) and (8.3) show that multiplication of complex numbers is commutative, associative, and distributive,

9. Find the six sixth roots of 1, and show that they form a group under multiplication.

10. The equation $z^3 + 2z - i = 0$ has the root i; find the other roots.

11. Find the cube roots of $2 + 2i$.

12. If the triangle $z_1 z_2 z_3$ in the complex plane is equilateral, show that

$$\frac{z_3 - z_1}{z_2 - z_1} = \frac{z_1 - z_2}{z_3 - z_2} \quad \text{and, hence,} \quad z_1^2 + z_2^2 + z_3^2 = z_2 z_3 + z_3 z_1 + z_1 z_2.$$

If this condition is fulfilled show that the triangle $z_1 z_2 z_3$ is equilateral.

13. Prove that the complex vectors $z_1 = x_1 + iy_1$, $z_2 = x_2 + iy_2$ are perpendicular when and only when $z_1 \bar{z}_2 + \bar{z}_1 z_2 = 0$.

14. Show that a straight line perpendicular to the complex vector c has an equation of the form

$$c\bar{z} + \bar{c}z + k = 0 \quad (k \text{ real}).$$

15. Show that a circle whose center is given by the complex vector c has an equation of the form

$$z\bar{z} - (c\bar{z} + \bar{c}z) + k = 0 \quad (k \text{ real}).$$

$[(z - c)(\bar{z} - \bar{c}) = r^2.]$

16. If a, b, c, d are four points on the unit circle $|z| = 1$, show that the lines ab and cd meet at the point

$$z = \frac{\bar{a} + \bar{b} - \bar{c} - \bar{d}}{\bar{a}\bar{b} - \bar{c}\bar{d}}.$$

[The equation of ab is $z + ab\bar{z} = a + b$; and $\bar{a} = a^{-1}$.]

17. Express $(1 - i\sqrt{3})^5$ as $x + iy$ by using De Moivre's theorem. Check by using the binomial theorem.

18. Show that $|z_1| + |z_2| \geq |z_1| - |z_2|$.

19. Show that $\sqrt{2}|x + iy| \geq |x| + |y|$.

20. If the polynomial $P(z)$ in (8.11) has coefficients a_i in the complex field, its complex zeros occur in conjugate pairs when, and only when, the points a_i lie on a line through $z = 0$.

21. The zeros of the polynomial $P(z)$ in (8.11) and the zeros of its derivative $P'(z)$ have the same centroid, namely $-a_1/na_0$, if $n > 1$.

CHAPTER 2

Sequences and Series

11. Linear Point Sets. A collection of real numbers defined in some specific way corresponds, by the Cantor-Dedekind axiom (§ 6), to a set of points on the axis of reals. Such a *point set* may consist of a finite or an infinite number of points. In what follows we use the word *point* in the sense of *real number*.

As examples of infinite point sets consider

(1) The positive integers n: $1, 2, 3, \cdots$;

(2) all integers $0, \pm 1, \pm 2, \pm 3, \cdots$;

(3) the reciprocals of the positive integers $1/n$;

(4) the fractions $1 - 1/n$;

(5) the rational numbers r in the closed interval $0 \leq r \leq 1$;

(6) the rational numbers in the open interval $0 < r < 1$;

(7) the real numbers x in the closed interval $a \leq x \leq b$.

All of these sets except the last contain a countable infinity of points.

Let S denote a set of points x. If there exists a number G such that $x \leq G$ for every point in S, S is said to have an *upper bound G*.

If there exists a number g such that $x \geq g$ for every point in S, S is said to have a *lower bound g*.

A set having both upper and lower bounds is said to be *bounded*. Evidently every finite set of points is bounded.

In the preceding examples:

Set (1) has the lower bound 1 but no upper bound;

Set (2) has neither upper nor lower bounds;

Sets (3), (4), (5), and (6) have the bounds 0 and 1;

Set (7) has the bounds a and b.

12. Cluster Points. Let r denote a definite real number and δ an arbitrary positive number; then the set of points on the open interval of length 2δ,

$$r - \delta < x < r + \delta,$$

is called a *δ-neighborhood of r.* Thus for every choice of δ we have a corresponding neighborhood of r.

Now let S be any set containing an infinite number of points x. These points may cluster around one or more points, the *cluster points* of the set. Such points are precisely defined as follows:

DEFINITION. *A point ξ, which may or may not belong to a set S, is said to be a cluster point† of S if every neighborhood of ξ, however small, contains a set point other than ξ.*

Evidently any neighborhood of a cluster point, however small, contains an *infinite* number of points of the set. For, if $x_1 \neq \xi$ is a point of S in a δ-neighborhood of ξ, there will be a second point $x_2 \neq \xi$ of S in the smaller δ_1-neighborhood where $\delta_1 = |\xi - x_1|$. Thus postulating a single point of S other than ξ in any neighborhood of ξ is equivalent to postulating infinitely many.

Consider now the point sets given in § 11.

Sets (1) and (2) have no cluster points.

Set (3) has the cluster point 0, not a member of the set.

Set (4) has the cluster point 1, not a member of the set.

Set (5) has all real numbers in the interval $0 \leq x \leq 1$ as cluster points; the irrational cluster points do not belong to the set.

Set (6) has all real numbers in the interval $0 \leq x \leq 1$ as cluster points; 0, 1, and the irrational cluster points do not belong to the set.

Set (7) has all real numbers in the interval $a \leq x \leq b$ as cluster points, and all belong to the set.

The cluster points of a set S constitute its *derived set S′*, which may be empty. If S includes $S′$, the set S is said to be *closed.* Then S either has no cluster points or contains all of its cluster points. In the examples just considered, sets (1), (2), and (7) are closed; for (1) and (2), $S′$ is empty.

If $S′$ includes S, the set S is said to be *dense in itself.* This is the case in the sets (5), (6), and (7).

Finally, if $S′$ and S are identical, the set S is said to be *perfect.* In the examples cited only set (7) is perfect.

13. Bounded Sets. For any bounded set containing infinitely many points we have the fundamental

BOLZANO-WEIERSTRASS THEOREM. *Every bounded, infinite set of points has at least one cluster point,*

Proof. If g and G are bounds of the set, all of its points lie within the

---

† We prefer this briefer term to *point of condensation* or *point of accumulation.* Although *limit point* is also widely used we shall use this term only when the set has a *single* cluster point.

interval $g \leq x \leq G$. Now separate all points x of the real axis into two classes $\{a\}$ and $\{A\}$ by putting x in class $\{a\}$ or $\{A\}$ according as there are a finite or an infinite number of points of S to the *left* of x. These classes evidently satisfy the requirements (i) and (iii) of § 6; and, since g is in $\{a\}$, G in $\{A\}$, (ii) is also satisfied. Hence by Dedekind's theorem, the cut $a \mid A$ defines a unique real number α. The point $\alpha - \varepsilon$ is in class $\{a\}$, $a + \varepsilon$ in class $\{A\}$; hence any ε-neighborhood of α contains an infinite number of points of S and α is a cluster point.

In case S has more than one cluster point, α will be the *least* one in the interval $g \leq x \leq G$. For, as the point x moves to the right from g to G, α is clearly the first cluster point encountered. The *least cluster point* α can be characterized as follows: For every $\varepsilon > 0$,

(1) $\qquad\qquad S$ has $\begin{cases} \text{infinitely many points} < \alpha + \varepsilon, \\ \text{only a finite number} \quad < \alpha - \varepsilon. \end{cases}$

We might also separate all real points x into two classes $\{b\}$ and $\{B\}$ by putting x in class $\{B\}$ or $\{b\}$ according as there are a finite or an infinite number of points of S to the *right* of x. The cut $b \mid B$ again defines a real number β which is a cluster point of S. If S has more than one cluster point, β will be the greatest one in the interval $g \leq x \leq G$. For, as the point x moves to the left from G to g, β is clearly the first cluster point encountered. The *greatest cluster point* β can be characterized as follows: For every $\varepsilon > 0$,

(2) $\qquad\qquad S$ has $\begin{cases} \text{infinitely many points} > \beta - \varepsilon, \\ \text{only a finite number} \quad > \beta + \varepsilon. \end{cases}$

We have now proved

THEOREM 2. *Every bounded, infinite set of points has a least and a greatest cluster point.*†

When $\alpha \neq \beta$, α is sometimes called the *least limit*, β the *greatest limit* of the set.

When the least and greatest cluster points coincide ($\alpha = \beta$), the set has but a single cluster point, which is then called the *limit point* of the set. When λ is a limit point of S, $\lambda = \alpha = \beta$, and, from (1) and (2),

(3) $\qquad\qquad S$ has only a finite number of points $\begin{cases} < \lambda - \varepsilon, \\ > \lambda + \varepsilon. \end{cases}$

Hence at most a finite number of set points lie *outside* of the interval $(\lambda - \varepsilon, \lambda + \varepsilon)$; "almost all" points of the set lie within the interval no matter how small ε is chosen.

---

† See Appendix 1 for an alternative proof which employs binary fractions instead of Dedekind cuts.

The bounded set S always has a *greatest lower bound* (g.l.b.) m characterized by the properties; for any $\varepsilon > 0$,

(4) S has $\begin{cases} \text{all points} & \geq m, \\ \text{at least one point} & < m + \varepsilon. \end{cases}$

If no points of S lie to the left of α, $m = \alpha$ satisfies conditions (4). Then m, like α, may or may not be a point of S. But S may have points to the left of α; if x' is such a point, only a finite number of points of S are $< x'$ and we put m equal to the least of these. Then m again satisfies conditions (4); for m is a point of the set always $< m + \varepsilon$.

Similarly S always has a *least upper bound* (l.u.b.) M characterized by the properties:

(5) S has $\begin{cases} \text{all points} & \leq M, \\ \text{at least one point} & > M - \varepsilon. \end{cases}$

If no points of S lie to the right of β, $M = \beta$ satisfies conditions (5). But S may have points to the right of β; if x'' is such a point only a finite number of points of S are $> x''$ and we put M equal to the greatest of these. Again M satisfies (5).

When m or M is not a cluster point, m is the least point, M the greatest point of S. But in case m or M are cluster points, S may have neither a least or a greatest point. In any case the interval from m to M is the smallest that contains the set; and

(6) $m \leq \alpha \leq \beta \leq M.$

Example 1. The set $1, {}^1/_2, {}^1/_3, \cdots$ has $m = 0$, $M = 1$; there is no least point, but 1 is the greatest point of the set. The only cluster point is 0; hence $\alpha = \beta = 0$ is the limit point of the set.

Example 2. The set ${}^1/_2, {}^2/_3, {}^3/_4, \cdots$ has $m = {}^1/_2$, $M = 1$; there is no greatest point, but ${}^1/_2$ is the least point of the set. The only cluster point is 1; hence $\alpha = \beta = 1$ is the limit point of the set.

Example 3. The set $1/p + 1/q$, where p and q are positive integers, has $m = 0$, $M = 2$; there is no least point, but 2 is the greatest point. The set has 0 and all the fractions $1/p$ as cluster points; $\alpha = 0$ does not belong to the set while $\beta = 1$ does.

PROBLEMS

1. Find m, M, α, β for the following sets, stating in each case which belong to the set:
   (a) the points $1 - (-1)^n/n$ where n is a positive integer;
   (b) the points $x_1 = 3$, $x_{n+1} = 2 - 2/x_n$.
   (c) the points $x_n = 1 + (-1)^n(n^2 + 2)(2n^2 + 1)^{-1}$.
   (d) the points $\pm 1/p \pm 1/q$, where p and q are positive integers;
   (e) the points $1, 2, 3/2, 7/4, \ldots$, where each number after the second is the mean of the two preceding. [Cf. Ex. 15.6.]

2. Show that m, the g.l.b. of a set S, is given by the cut c/C in the real numbers, where class $\{c\}$ includes all numbers having no points of S to their left, class $\{C\}$ includes all numbers having at least one point of S to their left.

3. Define M, the l.u.b. of a set S, by a cut in the real numbers.

4. If $b > a + 1$, find m, M, α, β for the point sets:

$$x_n = \begin{cases} a + n^{-1} \\ b - n^{-1} \end{cases}, \qquad y_n = \begin{cases} b + n^{-1} & (n \text{ odd}) \\ a - n^{-1} & (n \text{ even}) \end{cases},$$

and for the set $x_n + y_n$.

14. Sequences. An ordered countable set of numbers

$$x_1, x_2, x_3, \cdots, x_n, \cdots,$$

not necessarily all different, is called a *sequence* and denoted by $\{x_n\}$. Giving x_n as a function of n defines the sequence, as in the following examples:·

(1) $1, -1, 1, -1, \cdots,$ $\qquad\qquad$ $x_n = (-1)^{n+1};$

(2) $1, 2, 3, 4, \cdots,$ $\qquad\qquad$ $x_n = n;$

(3) $1, 1/2, 1/3, 1/4, \cdots,$ $\qquad\qquad$ $x_n = 1/n;$

(4) $1, 2, 6, 24, \cdots,$ $\qquad\qquad$ $x_n = n!;$

(5) $0.3, 0.33, 0.333, 0.3333, \cdots,$ $\qquad$ $x_n = \frac{1}{3}(1 - 1/10^n);$

(6) $1, -1/2, 1/4, -1/8, \cdots,$ $\qquad$ $x_n = (-1/2)^{n-1};$

(7) $2, 2^{1/2}, 2^{1/3}, 2^{1/4}, \cdots,$ $\qquad$ $x_n = 2^{1/n};$

(8) $2/1, (3/2)^2, (4/3)^3, (5/4)^4, \cdots,$ $\qquad$ $x_n = (1 + 1/n)^n;$

(9) $1/2, 1/3, 1/5, 1/7, 1/11, 1/13, \cdots,$ $\qquad$ $x_n = 1/p_n;$

where p_n is the nth prime number. The sequence $\{p_n\}$ contains infinitely many numbers (Euclid), but a formula for p_n in terms of n is not known nor is likely to become known.†

A sequence may also be defined by a *recurrence relation* when the values of certain initial terms are given. Thus the famous *Fibonacci sequence*,

(10) $\qquad\qquad 1, 1, 2, 3, 5, 8, 13, 21, \cdots,$

in which each term after the second is the sum of the two preceding, is defined by the recurrence relation,

(11) $\qquad\qquad x_n = x_{n-1} + x_{n-2} \qquad (n > 2),$

† Cf. Hardy and Wright, *An Introduction to the Theory of Numbers*, Oxford, 1938, pp. 5–8.

and the initial values $x_1 = 1$, $x_2 = 1$. This sequence plays a role in *phyllotaxy*, the arrangement of leaves along a plant stem.† Surprisingly, the general term of this sequence of integers involves the irrational $\sqrt{5}$:

$$x_n = \frac{1}{\sqrt{5}}\left[\left(\frac{1+\sqrt{5}}{2}\right)^n - \left(\frac{1-\sqrt{5}}{2}\right)^n\right].$$

This follows from the well-known method of solving recurrence relations with constant coefficients (Appendix 2). Thus, if (11) admits solutions of the form $x_n = k^n$, we must have $k^2 = k + 1$; hence the values

$$k_1 = \frac{1+\sqrt{5}}{2}, \qquad k_2 = \frac{1-\sqrt{5}}{2}$$

yield the general solution of (1)

$$x_n = Ak_1{}^n + Bk_2{}^n,$$

where A and B are arbitrary constants. The initial values $x_1 = x_2 = 1$ now give $A = -B = 1/\sqrt{5}$.

15. Limit of a Sequence. The cluster points of a sequence $\{x_n\}$ comprise

(*a*) the cluster points of the point set given by the numbers x_n, and

(*b*) any sequence point that recurs infinitely often.

In both cases any interval about the cluster point contains an infinite number of sequence values.

For example the sequence $\{(-1)^n\}$ corresponds to the point set 1, -1; this *set* has no cluster points, but both 1 and -1 are cluster points of the *sequence*.

When a *bounded* sequence $\{x_n\}$ has a single cluster point ξ it is said to *converge* to the *limit* ξ, and the sequence is termed *convergent*. We then write

$$\lim x_n = \xi \quad \text{or} \quad x_n \to \xi.$$

An unbounded sequence, or a bounded sequence with more than one cluster point, is termed *divergent*. Thus, of the sequences given in § 14, (1), (2), (4), (10) are divergent, the others convergent.

When $x_n \to \xi$, there are only a finite number of sequence points outside of any interval $\xi - \varepsilon < x < \xi + \varepsilon$ where $\varepsilon > 0$ is arbitrary; and, if N is the largest index of these outside points, x_n with $n > N$ will lie inside of this interval. Thus, if $x_n \to \xi$, we can always find an integer N such that

(1) $|x_n - \xi| < \varepsilon$ when $n > N$.

Conversely, when (1) is fulfilled,

† Cf. Thompson, D. W., *On Growth and Form*, Cambridge, 1943, p. 923.

(i) an infinite number of points x_n lie inside of the interval $\xi - \varepsilon < x < \xi + \varepsilon$; and

(ii) only a finite number lie outside.

From (i) we see that ξ is a cluster point of $\{x_n\}$. Moreover (ii) shows that ξ is the *only* cluster point; for, if $\xi' \neq \xi$, ξ' is not a cluster point since the interval $\xi' - \delta < x < \xi' + \delta$, where $\delta < |\xi' - \xi|$, contains but a finite number of sequence points. Thus (1) is a necessary and sufficient condition that $x_n \to \xi$. The limit point ξ need not belong to the sequence $\{x_n\}$.

For example, the sequence $\{1/n\}$ converges to 0; for

$$\frac{1}{n} - 0 < \varepsilon \quad \text{when} \quad n > N,$$

where N is the first integer after $1/\varepsilon$. But the sequence $\{(-1)^n\}$ diverges for it has two cluster points, -1 and 1.

If $x_n \to 0$, $\{x_n\}$ is called a *null sequence*; then

(2) $$|x_n| < \varepsilon \quad \text{when} \quad n > N.$$

The condition (1) may now be stated in the form $x_n \to \xi$ when and only when $\{x_n - \xi\}$ is a null sequence.

If two sequences $\{x_n\}$ and $\{x_n'\}$ differ only in the assignment of the indices 1, 2, 3, $\cdots$ to the same countable set of numbers, they must have precisely the same cluster points. Consequently, if $x_n \to \xi$, then $x_n' \to \xi$; *the limit of a sequence is not changed by a rearrangement of its elements.*

Every convergent sequence is bounded. For, if $\{x_n\}$ is convergent, (1) shows that, when $n > N$,

$$|x_n| = |x_n - \xi + \xi| \leq |x_n - \xi| + |\xi| < \varepsilon + |\xi|.$$

Hence, if G is the greatest of the $N + 1$ numbers $x_1, x_2, \cdots, x_N, \varepsilon + |\xi|$, $|x_n| < G$ for *all* values of n.

Not every bounded sequence is convergent. But every bounded sequence $\{x_n\}$ will have a least and greatest cluster point (§ 13) which we denote by

(3) $$\underline{\lim} \, x_n = \xi, \qquad \overline{\lim} \, x_n = \bar{\xi}.$$

These numbers are called the *least limit* and *greatest limit* of the sequence; they are also denoted by lim inf x_n and lim sup x_n. The g.l.b. and l.u.b. of $\{x_n\}$ are written inf x_n, sup x_n.

This notation is sometimes applied to unbounded sequences. Thus we may write $\xi = -\infty$ if $\{x_n\}$ has no lower bound, $\bar{\xi} = \infty$ if $\{x_n\}$ has no upper bound. This usage, however, is purely conventional for ∞ is not a number.

From § 13 we see that $\underline{\xi}$ and $\bar{\xi}$ may be characterized as follows. Given any $\varepsilon > 0$, the sequence $\{x_n\}$ has a least limit

(4) $\underline{\xi}$ when $\begin{cases} \text{an infinite number of } x_n < \underline{\xi} + \varepsilon, \\ \text{a finite number of } \quad x_n < \underline{\xi} - \varepsilon. \end{cases}$

The sequence has a greatest limit

(5) $\bar{\xi}$ when $\begin{cases} \text{an infinite number of } x_n > \bar{\xi} - \varepsilon, \\ \text{a finite number of } \quad x_n > \bar{\xi} + \varepsilon. \end{cases}$

When $\underline{\xi} = \bar{\xi} = \xi$, (4) and (5) show that all but a finite number of x_n lie in any interval $\xi - \varepsilon < x < \xi + \varepsilon$; then, and only then, $x_n \to \xi$.

Example 1. The sequence $\{r^n\}$, n a positive integer. We first establish *Bernoulli's inequality*.

(6) $(1 + h)^n > 1 + nh$ when $h\,(\neq 0) > -1, \quad n > 1,$

by finite induction (§ 3). Since

$$(1 + h)^2 = 1 + 2h + h^2 > 1 + 2h,$$

(6) holds for $n = 2$. Assume now that (6) holds for some n; then, on multiplying (6) by $1 + h > 0$, we see that it holds also for $n + 1$:

$$(1 + h)^{n+1} > 1 + (n + 1)h + nh^2 > 1 + (n + 1)h.$$

Therefore (6) is true for any integer $n > 1$.
    We can now readily prove that

(7) $r^n \to 0$ when $|r| < 1.$

For, if we write $|r| = 1/(1 + h)$ where $h > 0$,

$$|r|^n = \frac{1}{(1 + h)^n} < \frac{1}{1 + nh} < \frac{1}{nh} \to 0.$$

Moreover,

$$|r|^n \to \infty \quad \text{when} \quad r > 1.$$

For, if we write $r = 1 + h$ where $h > 0$,

$$|r|^n = (1 + h)^n > 1 + nh \to \infty.$$

The symbol ∞ is not a number; $|r|^n \to \infty$ simply means that $|r|^n$ will exceed any number however great when n is sufficiently large.

Example 2. The sequence $\{nr^n\}$, $|r| < 1$.

Put $|r| = 1/(1 + h)$ where $h > 0$; then

$$n\,|r|^n = \frac{n}{(1 + h)^n} = \frac{n}{1 + nh + \tfrac{1}{2}n(n - 1)h^2 + \cdots + h^n}$$

from the binominal theorem. Hence, if $n > 1$ and we retain only the third term in the denominator,

$$n \mid r \mid^n < \frac{2}{(n-1)h^2} \to 0, \quad \text{and}$$

(8) $$nr^n \to 0, \qquad |r| < 1.$$

Example 3. The sequence $\{\log n/n^h\}, h > 0$. From (7) and (8) we can show that

(9) $$\frac{\log n}{n^h} \to 0 \qquad (h > 0).$$

for any logarithm whose base $b > 1$. For, if m is the characteristic of $\log n$,

$$m \leq \log n < m + 1, \qquad b^m \leq n < b^{m+1},$$

$$\frac{\log n}{n^h} < \frac{m+1}{b^{mh}} = (m+1)r^m,$$

where $r = 1/b^h < 1$. Now, as $n \to \infty$, also $m \to \infty$ and $r^m \to 0$, $mr^m \to 0$.

Example 4. The sequence $\{r^{1/n}\}, r > 0$.

   1. When $r = 1$, $r^{1/n} = 1$.

   2. When $r > 1$, $r^{1/n} > 1$ and we can write $r^{1/n} = 1 + h_n$, $h_n > 0$. Then

$$r = (1 + h_n)^n > 1 + nh_n, \qquad h_n < \frac{r-1}{n};$$

hence, $h_n \to 0$ and $r^{1/n} \to 1$.

   3. When $0 < r < 1$, $r^{1/n} < 1$, and we can write $r^{1/n} = 1/(1 + h_n)$, $h_n > 0$. Then

$$r = \frac{1}{(1+h_n)^n} < \frac{1}{1 + nh_n}, \qquad h_n < \frac{\frac{1}{r} - 1}{n};$$

hence, $h_n \to 0$ and $r^{1/n} \to 1$.

Thus, in all cases

(10) $$r^{1/n} \to 1 \quad \text{when} \quad r > 0.$$

Example 5. The sequence $\{\sqrt[n]{n}\}$.

When $n > 1$, $\sqrt[n]{n} > 1$ and $h_n = \sqrt[n]{n} - 1 > 0$. Now

$$n = (1 + h_n)^n = 1 + nh_n + \tfrac{1}{2}n(n-1)h_n^2 + \ldots + h_n^n$$

by the binomial theorem. Hence, if we retain only the third term on the right,

$$n > \tfrac{1}{2}n(n-1)h_n^2, \qquad h_n^2 < \frac{2}{n-1} \to 0;$$

thus $h_n \to 0$, and, consequently,

(11) $$\sqrt[n]{n} \to 1.$$

Example 6. Consider the sequence 1, $^1/_2$, $^3/_4$, $^5/_8$, $^{11}/_{16}$, ..., in which each number after the second is the mean of the two preceding. The sequence is thus defined by the recurrence relation.

(i) $$2x_n = x_{n-1} + x_{n-2}$$

and the initial values $x_1 = 1$, $x_2 = ^1/_2$. If (i) admits solutions of the form $x_n = k^n$, we must have $2k^2 = k + 1$. Admissible values of k are given by the roots $k_1 = 1$, $k_2 = -^1/_2$ of this quadratic. Hence the general solution of (i) is

$$x_n = Ak_1{}^n + Bk_2{}^n = A + B(-\tfrac{1}{2})^n$$

where A and B are arbitrary constants. Putting $x_1 = 1$, $x_2 = ^1/_2$ gives

$$1 = A - \tfrac{1}{2}B, \qquad \tfrac{1}{2} = A + \tfrac{1}{4}B;$$

whence $A = ^2/_3$, $B = -^2/_3$. Thus the sequence is given by

$$x_n = \tfrac{2}{3}[1 - (-\tfrac{1}{2})^n] \quad \text{and} \quad x_n \to \tfrac{2}{3},$$

since $(^1/_2)^n$ is a null sequence (Ex. 1).

Example 7. From a given sequence $\{a_n\}$ we can always form others that tend to the same limit. We give two methods: by arithmetic and by geometric means.

1. *Arithmetic means.* If $a_n \to \alpha$, then

(12) $$b_n = \frac{a_1 + a_2 + \ldots + a_n}{n} \to \alpha.$$

Proof. Let $\alpha = 0$; then we can choose m so that

$$|a_n| < \frac{\varepsilon}{2} \quad \text{when} \quad n > m.$$

Now

$$|b_n| \leqq \frac{|a_1 + \ldots + a_m|}{n} + \frac{|a_{m+1}| + \ldots + |a_n|}{n} < \frac{k}{n} + \frac{\varepsilon}{2},$$

where $k = |a_1 + \ldots + a_m|$. Since k is fixed, we can choose $N > m$ so that

$$\frac{k}{n} < \frac{\varepsilon}{2}, \quad \text{and hence} \quad |b_n| < \varepsilon, \quad \text{when} \quad n > N$$

Therefore $b_n \to 0$.

Let $\alpha \neq 0$; then $a_n - \alpha \to 0$, and, by the case just treated,

$$\frac{(a_1 - \alpha) + \ldots + (a_n - \alpha)}{n} = b_n - \alpha \to 0,$$

and $b_n \to \alpha$. Thus the proof is complete.

2. *Geometric means.* If $a_n > 0$ and $a_n \to \alpha \neq 0$, then

(13) $$c_n = \sqrt[n]{a_1 a_2 \ldots a_n} \to \alpha.$$

Now

$$\log c_n = \frac{\log a_1 + \log a_2 + \ldots + \log a_n}{n}$$

is an arithmetic mean for the sequence $\{\log a_n\}$ which approaches $\log \alpha$.† Hence, by part 1, $\log c_n \to \log \alpha$ and $c_n \to \alpha$.

---

† We assume that $\log x$ is a continuous function when $x > 0$ (Ex. 49.5).

PROBLEMS

1. Apply (2) to show that the sequence $0.3, 0.33, 0.333, \cdots \to \frac{1}{3}$. $[1 - 3x_n = 10^{-n}.]$

2. Find $\lim (\sqrt{n+1} - \sqrt{n})$.

3. If $x_1 = 1$, $x_2 = 2$, and $x_n = \sqrt{x_{n-1}x_{n-2}}(n > 2)$, find $\lim x_n$.   [Find $\log_2 x_n$.]

4. Apply (12) to prove that

$$\left(1 + \frac{1}{2} + \cdots + \frac{1}{n}\right) \Big/ n \to 0.$$

5. If $\{a_n\}$ is a positive increasing (decreasing) sequence, $\{b_n\}$ in (12) is also.

6. With $a_1 = 1$, $a_n = n/(n-1)$ when $n > 1$, apply (13) to prove that $\sqrt[n]{n} \to 1$.

7. Given that $(1 + 1/n)^n \to e$ (19.1), apply (13) to prove that

$$\frac{(n+1)}{\sqrt[n]{n!}} \to e \quad \text{and hence} \quad \frac{\sqrt[n]{n!}}{n} \to \frac{1}{e}.$$

8. If $x_1 = 1$, $x_2 = 1$, and $x_n = x_{n-1} + x_{n-2}$ (the *Fibonacci sequence of* § 14), find the limit of $y_n = x_n/x_{n+1}$.

9. Find the cluster points of each sequence:

$$(a) \quad x_n = \frac{(n^2 + 1)\sin(\pi/n)}{n}; \qquad (b) \quad y_n = \frac{\sin \frac{1}{2}n^2\pi}{1 + \cos \frac{1}{2}n^2\pi}.$$

10. If $x_1 > 0$, $x_{n+1} = 3(1 + x_n)/(3 + x_n)$, show that $x_n \to \sqrt{3}$ monotonely. [If $x_n \to \xi$, ξ must be the positive root of $x = 3(1 + x)/(3 + x)$: namely $\sqrt{3}$. (Why?) Now show that

(i) $x_{n+1} - x_n$ and $x_n - x_{n-1}$ have the same sign;

(ii) $|x_{n+1} - \sqrt{3}| < k |x_n - \sqrt{3}|$, $0 < k < 1$.]

11. Find all cluster points for the sequences: (a) $x_1 > 0$, $x_{n+1} = \sqrt{2 + x_n}$; (b) $y_1 = 6$, $y_{n+1} = 3 - 3/y_n$.

12. Prove that $\limsup (x_n + y_n) \leq \limsup x_n + \limsup y_n$. What is the corresponding relation for $\liminf$?   Test on Prob. 13.4.

16. Operations with Limits. We have seen that $x_n \to \xi$ when $\{x_n - \xi\}$ is a null sequence. Null sequences have the following important properties.

THEOREM 1.   *If $\{x_n\}$ and $\{y_n\}$ are null sequences, $\{x_n + y_n\}$ is a null sequence*

Proof.   Given $\varepsilon > 0$, choose N so that

$$|x_n| < \frac{\varepsilon}{2}, \quad |y_n| < \frac{\varepsilon}{2} \quad \text{when} \quad n > N;$$

then

$$|x_n + y_n| \leq |x_n| + |y_n| < \varepsilon.$$

THEOREM 2.   *If $\{x_n\}$ is null and $\{c_n\}$ a bounded sequence, then $\{c_n x_n\}$ is a null sequence.*

Proof. Since $\{c_n\}$ is bounded, $|c_n| < C$. Given $\varepsilon > 0$, choose N so that

$$|x_n| < \frac{\varepsilon}{C} \quad \text{when} \quad n > N;$$

then

$$|c_n x_n| = |c_n| |x_n| < C \cdot \frac{\varepsilon}{C} = \varepsilon$$

Making use of these theorems we can easily prove

THEOREM 3. *If* $a_n \to \alpha$, $b_n \to \beta$, *then*

(1) $a_n + b_n \to \alpha + \beta$,

(2) $a_n b_n \to \alpha \beta$,

(3) $a_n/b_n \to \alpha/\beta$,   *if* $b_n \neq 0$, $\beta \neq 0$.

Proof. Write $a_n = \alpha + x_n$, $b_n = \beta + y_n$ where $\{x_n\}$, $\{y_n\}$ are null sequences. Then

(1)′ $a_n + b_n - (\alpha + \beta) = x_n + y_n$,

(2)′ $a_n b_n - \alpha \beta = \alpha y_n + \beta x_n + x_n y_n$,

(3)′ $\dfrac{a_n}{b_n} - \dfrac{\alpha}{\beta} = \dfrac{\beta x_n - \alpha y_n}{b_n \beta} = \dfrac{1}{b_n} x_n - \dfrac{\alpha}{\beta b_n} y_n.$

In each case Theorems 1 and 2 show that the right-hand members are elements of a null sequence. This proves the theorem.

In the case $\beta = 0$ excluded in (3) there are several possibilities.

If $\alpha \neq 0$, $\beta = 0$, we find that $|a_n/b_n|$ will ultimately exceed any number however large. If a_n/b_n remains ultimately positive, we write $a_n/b_n \to \infty$; if negative, $a_n/b_n \to -\infty$.

If $\alpha = 0$, $\beta = 0$, the limit a_n/b_n may or may not exist. For example, if

$a_n = 1/n, \quad b_n = 1/n^2:$ $a_n/b_n = n \to \infty;$

$a_n = 1/n, \quad b_n = 1/\sqrt{n}:$ $a_n/b_n = 1/\sqrt{n} \to 0;$

$a_n = 1/n, \quad b_n = 1/(2n+1):$ $a_n/b_n = 2 + 1/n \to 2.$

Theorem 3 enables us to evaluate many limits without resorting to the basic definition of a limit. For example:

$$\frac{n^2 + 3n}{2n^2 + n - 1} = \frac{1 + 3/n}{2 + 1/n - 1/n^2} \to \frac{1}{2},$$

$$\frac{n - 3}{n^2 + 1} = \frac{1/n - 3/n^2}{1 + 1/n^2} \to 0;$$

and, in general, if p and q are the highest powers of n in two polynomials in n forming numerator and denominator,

$$(4) \qquad \frac{an^p + \cdots}{bn^q + \cdots} \rightarrow \begin{cases} 0 & p < q, \\ a/b & \text{when} \quad p = q, \\ \infty & p > q. \end{cases}$$

THEOREM 4. *If $a_n \rightarrow \alpha$, $b_n \rightarrow \beta$, and $a_n > b_n$ for all values of n, then $\alpha \geq \beta$.*

Proof. From (1), $a_n - b_n \rightarrow \alpha - \beta$. Now $\alpha - \beta$ is not negative; otherwise $a_n - b_n$ would be negative for sufficiently large values of n. Hence $\alpha - \beta \geq 0$.

Note that we cannot conclude that $\alpha > \beta$ from $a_n > b_n$. For example, when

$$a_n = \frac{1}{n}, \quad b_n = \frac{1}{n+1}, \qquad a_n > b_n \quad \text{but} \quad \alpha = \beta = 0.$$

17. Fundamental Convergence Criterion. The condition (15.1) for the convergence of a sequence $\{x_n\}$ involves its limit ξ. In the following criterion, due to Cauchy, only the sequence values are needed to prove the existence of a limit.

CAUCHY'S CRITERION. *A necessary and sufficient condition that the sequence $\{x_n\}$ converge to a limit is that, for an arbitrary choice of $\varepsilon > 0$, we can always find a number N, dependent upon ε, such that*

$$(1) \qquad |x_m - x_n| < \varepsilon \quad \text{when} \quad m, n > N.$$

Proof. The condition is *necessary.* For, if $x_n \rightarrow \xi$, we have from (15.1)

$$|x_m - \xi| < \tfrac{1}{2}\varepsilon, \quad |x_n - \xi| < \tfrac{1}{2}\varepsilon, \qquad \text{when} \quad n, m > N,$$

and, hence,

$$|x_m - x_n| = |x_m - \xi + \xi - x_n| \leq |x_m - \xi| + |\xi - x_n| < \varepsilon.$$

The condition is *sufficient.* The sequence $\{x_n\}$ is bounded. For choose a *fixed* $m > N$; then $x_1, x_2, \cdots, x_m$ lie in the interval between the greatest and least of these numbers. Moreover (1) shows that all subsequent numbers $x_{m+1}, x_{m+2}, \cdots$ lie in an interval of length 2ε about x_m.

The bounded sequence $\{x_n\}$ has at least one cluster point ξ (§ 13). Hence there are infinitely many points x_m such that $|x_m - \xi| < \varepsilon$, and, for infinitely many of these, $m > N$. If the sequence has a second cluster point ξ', there must be infinitely many points x_n such that $\left| x_n - \xi' \right|$ $< \varepsilon$, and, for infinitely many of these, $n > N$. Let $\xi - \xi' = d > 0$; then when m and $n > N$,

$$x_m - x_n = (x_m - \xi) + (\xi - \xi') + (\xi' - x_n) > d - 2\varepsilon.$$

Hence, if we take $\varepsilon = d/3$,

$$\left| x_m - x_n \right| > \varepsilon, \qquad m, n > N,$$

in contradiction of (1). Hence ξ is the only cluster point.

Example 1. Consider the sequence

$$s_n = 1 - \frac{1}{2} + \frac{1}{3} - \frac{1}{4} + \cdots + (-1)^{n-1} \frac{1}{n}.$$

If $m > n$,

$$\left| s_m - s_n \right| = \frac{1}{n+1} - \left(\frac{1}{n+2} - \frac{1}{n+3} \right) - \cdots < \frac{1}{n+1}.$$

Hence, if we choose N so that $N + 1 > 1/\varepsilon$,

$$\left| s_m - s_n \right| < \frac{1}{N+1} < \varepsilon \quad \text{when} \quad m, n > N.$$

Hence the sequence converges, evidently to a number between $1/2$ and 1, for

$$s_n = (1 - \tfrac{1}{2}) + (\tfrac{1}{3} - \tfrac{1}{4}) + \cdots = 1 - (\tfrac{1}{2} - \tfrac{1}{3}) - (\tfrac{1}{4} - \tfrac{1}{5}) - \cdots.$$

Example 2. For the sequence

$$s_n = 1 + \frac{1}{2} + \frac{1}{3} + \cdots + \frac{1}{n},$$

$$s_{2n} - s_n = \frac{1}{n+1} + \frac{1}{n+2} + \cdots + \frac{1}{2n} > n \cdot \frac{1}{2n} = \frac{1}{2}.$$

Thus, if $\varepsilon = 1/2$, we cannot satisfy (1) when $m = 2n$; hence the sequence diverges.

18. Monotone Sequences. A sequence is said to be *monotone increasing or monotone decreasing* if for every n

$$x_{n+1} \geqq x_n \quad \text{or} \quad x_{n+1} \leqq x_n,$$

respectively. Both kinds are called *monotone* sequences; and, if

$$x_{n+1} > x_n \quad \text{or} \quad x_{n+1} < x_n$$

for every n, they are *strictly monotone*.

THEOREM. *A bounded, monotone sequence always has a limit point.*

Proof. Let $\{x_n\}$ be monotone increasing. Since it is bounded, it has a cluster point ξ (§ 13). Then no number of the sequence can lie above ξ; for, if $x_m = \xi + \varepsilon$ ($\varepsilon > 0$), the $m - 1$ points $x_1, x_2, \cdots, x_{m-1}$ at most lie in the interval $\xi - \varepsilon < x < \xi + \varepsilon$ and ξ could not be a cluster point. Since no sequence points lie above ξ, there can be no cluster point ξ' above ξ. Nor can a cluster point ξ' lie below ξ; for then ξ would lie above ξ'. Thus ξ is the only cluster point.

If x_n is monotone decreasing, the proof is essentially the same; then no term of the sequence can lie below ξ.

The argument above also establishes the

COROLLARY. *If* $\{x_n\}$ *is monotone increasing and* $x_n \to \xi$, $x_n \leqq \xi$ *for every* n; *and, if strictly increasing,* $x_n < \xi$ *for every* n.

Note that the limit of a strictly monotone sequence is not a number of the sequence.

When $x_n \to \xi$ through increasing or decreasing values, we write $x_n \uparrow \xi$ or $x_n \downarrow \xi$, respectively.

PROBLEMS

1. Let $A_n = a_1 + a_2 + \cdots + a_n$, $B_n = b_1 + b_2 + \cdots + b_n$, and all $b_i > 0$. Then, if the sequence $\{a_n/b_n\}$ is increasing or decreasing, prove that $\{A_n/B_n\}$ has the same property.

$$\left[\frac{a_1}{b_1} < \frac{a_2}{b_2} \text{ implies } \frac{a_1}{b_1} < \frac{a_1 + a_2}{b_1 + b_2} < \frac{a_2}{b_2} \text{ or } \frac{A_1}{B_1} < \frac{A_2}{B_2} < \frac{a_2}{b_2} ; \text{ etc.} \right]$$

2. If $x_n = \sqrt{n^2 + n} - n$, prove that $x_n \uparrow \frac{1}{2}$.

3. If $x_1 > 0$, $x_{n+1} = \sqrt{x_n + 2}$, show that $x_n \to 2$ monotonely.

4. If $a > b > 0$, prove that $\sqrt[n]{a^n + b^n} \downarrow a$.

5. If $x_1 = \sqrt{2}$, $x_n = \sqrt{2x_{n-1}}$, prove that $x_n \uparrow 2$.

6. Prove that $\{x_n\}$ is a null sequence if $\left| x_{n+1}/x_n \right| \to r < 1$.

7. If $x_n = n!/n^n$, prove that $x_{n+1}/x_n \to 1/e$ (19.1) and $x_n \to 0$.

8. If $x_n = a^n/n!$, prove that $x_n \to 0$.

19. The Number *e*.   We shall prove that the sequence having

$$x_n = \left(1 + \frac{1}{n} \right)^n$$

is strictly monotone increasing and bounded. To this end expand $(1 + 1/n)^n$ by the binomial theorem:

$$x_n = 1 + n\frac{1}{n} + \frac{n(n-1)}{2!}\frac{1}{n^2} + \frac{n(n-1)(n-2)}{3!}\frac{1}{n^3} + \cdots$$

$$+ \frac{n(n-1)\cdots 1}{n!}\frac{1}{n^n}$$

$$= 1 + 1 + \frac{1}{2!}\left(1 - \frac{1}{n}\right) + \frac{1}{3!}\left(1 - \frac{1}{n}\right)\left(1 - \frac{2}{n}\right) + \cdots$$

$$+ \frac{1}{n!}\left(1 - \frac{1}{n}\right)\left(1 - \frac{2}{n}\right)\cdots\left(1 - \frac{n-1}{n}\right).$$

As we pass from x_n to x_{n+1}, each term after $1 + 1$ increases, and another term is added; hence $x_{n+1} > x_n$. Moreover, after $1 + 1$ each term of x_n is less than the corresponding term in

$$y_n = 1 + 1 + \frac{1}{2!} + \frac{1}{3!} + \cdots + \frac{1}{n!},$$

and

$$y_n \leqq 1 + 1 + \frac{1}{2} + \frac{1}{2^2} + \cdots + \frac{1}{2^{n-1}} = 1 + 2\left(1 - \frac{1}{2^n}\right) < 3,$$

on summing the geometric progression (cf. § 22). Clearly $2 < x_n < 3$, and, since x_n is bounded and monotone, x_n approaches a limit, denoted by e, which lies between 2 and 3: thus

(1)
$$\left(1 + \frac{1}{n}\right)^n \uparrow e.$$

Since $x_n \leqq y_n < y_{n+1} < 3$, the monotone bounded sequence $\{y_n\}$ approaches a limit $\eta \geqq e$. Now let $m < n$, and consider the first $m + 1$ terms of x_n; since they are all positive,

$$1 + 1 + \frac{1}{2!}\left(1 - \frac{1}{n}\right) + \cdots + \frac{1}{m!}\left(1 - \frac{1}{n}\right) \cdots \left(1 - \frac{m-1}{n}\right) < x_n.$$

Holding m fast and letting n become infinite, this inequality gives

$$y_m = 1 + 1 + \frac{1}{2!} + \cdots + \frac{1}{m!} \leqq e$$

for all values of m; hence, as $m \to \infty$, we have $\eta \leqq e$. This and $\eta \geqq e$ are incompatible unless $\eta = e$.

We have thus proved that

(1)
$$\lim_{n \to \infty} \left(1 + \frac{1}{n}\right)^n = e = 1 + 1 + \frac{1}{2!} + \frac{1}{3!} + \cdots.$$

The infinite series on the right is well adapted for computation; to ten decimal places we find $e = 2.71828\ 18285$,

We also have

(2)
$$\lim_{n \to \infty} \left(1 - \frac{1}{n}\right)^n = \frac{1}{e};$$

for this limit equals

$$\lim_{n \to \infty} \left(1 - \frac{1}{n+1}\right)^{n+1} = \lim \left(\frac{n}{n+1}\right)^{n+1} = \lim \frac{\dfrac{n}{n+1}}{\left(1 + \dfrac{1}{n}\right)^n} = \frac{1}{e}.$$

Moreover, the limit in (1) is still valid when n becomes infinite through *negative* integral values; for, if $m = -n$,

(3) $$\lim_{n \to -\infty} \left(1 + \frac{1}{n}\right)^n = \lim_{m \to \infty} \left(1 - \frac{1}{m}\right)^{-m} = e.$$

More generally, if $x \to \infty$ through any set of real values (§ 44),

(4) $$\lim_{x \to \infty} \left(1 + \frac{1}{x}\right)^x = e.$$

For, if $n \leq x < n + 1$, where n is a positive integer,

$$\left(1 + \frac{1}{n + 1}\right)^n < \left(1 + \frac{1}{x}\right)^x < \left(1 + \frac{1}{n}\right)^{n+1},$$

and both extremes $\to e$ as $n \to \infty$. (Why?)

The number e plays a role in all processes that involve continuous growth or decay, and indeed might well be called the *growth constant*. Consider, for example, the growth of \$1 in one year at 100% annual interest. At simple interest it amounts to \$2; compounded twice a year we have \$1.50 at the half-year, and \$2.25 at the year's end, or $(1 + 1/_2)^2$ dollars. If interest is compounded quarterly, the previous amount is multiplied by $(1 + 1/_4)$ at the end of each interest period; hence at the end of the year the amount is $(1 + 1/_4)^4$ dollars or slightly more than \$2.44. In general, if interest is compounded n times a year, the factor of increase for each interest period is $(1 + 1/n)$, and for n periods the amount is $(1 + 1/n)^n$. Consequently, as n increases indefinitely, the amount increases to the limiting value e. *One dollar at 100% interest, continuously compounded, amounts to e dollars at the end of one year.*

20. Nested Intervals. If $\{x_n\}$ is a monotone increasing sequence and $\{y_n\}$ a monotone decreasing sequence such that

$$x_n < y_n \quad \text{and} \quad y_n - x_n \to 0,$$

then the closed intervals (x_n, y_n) are said to form a *nest of intervals*. Thus in a nest of intervals each interval contains all points of its successor, and the lengths of the intervals form a null sequence.

THEOREM. *A nest of intervals determines a unique real number.*

Proof. If the intervals (x_n, y_n) form a nest, $\{x_n\}$ and $\{y_n\}$ are *bounded* monotone sequences; for $x_1 \leq x_n < y_n \leq y_1$. Hence, $x_n \to \xi$, $y_n \to \eta$ (§ 17), and

$$\eta - \xi = (\eta - y_n) + (y_n - x_n) + (x_n - \xi) \to 0.$$

But, since $\eta - \xi$ is a *fixed* number, $\eta - \xi = 0$; thus x_n and y_n converge to the same limit.

If the numbers x_n, y_n are all rational, the nest (x_n, y_n) is said to be rational. A rational nest determines a real number which may be rational or irrational. For example, the nest

$$(0.3, 0.4), (0.33, 0.34), (0.33, 0.334) \to 1/3;$$

and, since $\sqrt{2} = 1.4142 \cdots$, the nest

$$(1, 2), (1.4, 1.5), (1.41, 1.42), (1.414, 1.415), \cdots \to \sqrt{2}.$$

More generally, if the cut $a \,|\, A$ defines the real number α, we can choose an increasing sequence $\{a_n\}$ from the a's and a decreasing sequence $\{A_n\}$ from the A's so that $A_n - a_n \to 0$. Such sequences can be chosen in infinitely many ways, and they can be chosen rational if desired. All such nests define α uniquely.

Example. Arithmetic-Geometric Mean. Given two positive numbers a_1 and b_1, $a_1 < b_1$, we construct two sequences $\{a_n\}$ and $\{b_n\}$ as follows:

$$a_2 = \sqrt{a_1 b_1}, \qquad b_2 = \tfrac{1}{2}(a_1 + b_1);$$

$$a_3 = \sqrt{a_2 b_2}, \qquad b_3 = \tfrac{1}{2}(a_2 + b_2);$$

and, in general,

$$a_n = \sqrt{a_{n-1} b_{n-1}}, \qquad b_n = \tfrac{1}{2}(a_{n-1} + b_{n-1}).$$

Evidently a_2 and b_2 lie between a_1 and b_1. Moreover, the geometric mean a_2 of a_1, b_1 is less than their arithmetic mean b_2, for

$$b_2 - a_2 = \tfrac{1}{2}(a_1 + b_1 - 2\sqrt{a_1 b_1}) = \tfrac{1}{2}(\sqrt{b_1} - \sqrt{a_1})^2 > 0;$$

hence, $a_1 < a_2 < b_2 < b_1$. Similarly, a_3 and b_3 lie between a_2 and b_2, and $a_2 < a_3 < b_3 < b_2$. Thus, in general,

$$a_1 < a_2 < a_3 < \cdots < b_3 < b_2 < b_1,$$

so that the sequence $\{a_n\}$ is monotone increasing, $\{b_n\}$ monotone decreasing. Thus each interval (a_n, b_n) includes all those of higher index.

The lengths of these intervals form a null sequence. For we have

$$b_n - a_n = \frac{1}{2}(\sqrt{b_{n-1}} - \sqrt{a_{n-1}})^2 = \frac{1}{2}\frac{\sqrt{b_{n-1}} - \sqrt{a_{n-1}}}{\sqrt{b_{n-1}} + \sqrt{a_{n-1}}}(b_{n-1} - a_{n-1}),$$

and, hence,

$$b_n - a_n < \tfrac{1}{2}(b_{n-1} - a_{n-1}).$$

On applying this inequality repeatedly, we finally obtain

$$b_n - a_n < \frac{1}{2^{n-1}}(b_1 - a_1);$$

consequently, $b_n - a_n \to 0$ as $n \to \infty$. The intervals (a_n, b_n) thus form a *nest* and determine a unique number, the *arithmetic-geometric mean* of a_1 and b_1.

PROBLEMS

1. Show that e is determined by the nest (x_n, y_n) where

$$x_n = \left(1 + \frac{1}{n}\right)^n, \qquad y_n = \left(1 + \frac{1}{n}\right)^{n+1}.$$

2. If $x_{n+1} = a/(1 + x_n)$ where $x_1 > 0$ and $a > x_1^2 + x_1$, show that the intervals $(x_1, x_2), (x_3, x_4), \cdots$ form a nest that determines the positive root of $x^2 + x = a$.

Discuss the case when $x_1 > 0$, $0 < a < x_1^2 + x_1$.

[When $a > x_1^2 + x_2$, show that (a) $x_2 - x_1 > 0$; (b) $x_{n+1} - x_n$ and $x_n - x_{n-1}$ have opposite signs; (c) $\left| x_{n+1} - x_n \right| < k \left| x_n - x_{n-1} \right|$ where $k = x_2/(1 + x_2) < 1$; (d) as $n \to \infty$, $x_{n+1} - x_n \to 0$.]

21. Infinite Series. From any sequence $\{a_n\}$ we can construct another by adding its first n elements:

$$(1) \qquad s_n = a_1 + a_2 + \cdots + a_n = \sum_{i=1}^{n} a_i.$$

If $\{s_n\}$ converges to a limit s as n becomes infinite, we write

$$(2) \qquad s = a_1 + a_2 + \cdots = \sum_{i=1}^{\infty} a_i,$$

where the dots imply that the a's are added indefinitely to form an *infinite series*. The limit s when it exists is called the *sum* of the series; however, s is not a sum in the usual sense but the *limit of a sum*. When this limit exists, the series is said to *converge*; otherwise the series is said to *diverge*. Thus the series converges when and only when the sequence $\{s_n\}$ is bounded and has a single cluster point. By definition, *a series and its sum-sequence converge or diverge together*.

From (15.1), $s_n \to s$ when

$$(3) \qquad \left| s_n - s \right| < \varepsilon \quad \text{when} \quad n > N;$$

then $\{s_n - s\}$ is a null sequence. From Cauchy's criterion (§ 17) we have the necessary and sufficient condition for the existence of a unique limit s:

$$(4) \qquad \left| s_m - s_n \right| < \varepsilon \quad \text{when} \quad m, n > N;$$

that is, given $\varepsilon > 0$, there is always a number N which depends on ε, such that for any numbers $m, n > N$, $\left| s_m - s_n \right| < \varepsilon$. When $m = n + 1$, this gives

$$(5) \qquad \left| a_{n+1} \right| < \varepsilon, \quad n > N, \quad \text{or} \quad a_{n+1} \to 0.$$

The terms of a convergent series must form a null sequence.

Thus $a_n \to 0$ is a necessary condition for the convergence of $a_1 + a_2 + \cdots$. That this condition is not sufficient is shown by the divergent series (Ex. 17.2)

$$1 + \frac{1}{2} + \frac{1}{3} + \cdots \quad \text{for which} \quad a_n = \frac{1}{n} \to 0.$$

When the series $a_1 + a_2 + \cdots$ converges to a sum s,

(6) $$r_n = s - s_n = a_{n+1} + a_{n+2} + \cdots$$

is called the *remainder* after n terms. Thus the remainder is an infinite series which also converges. If the given series diverges, the remainder $a_{n+1} + a_{n+2} + \cdots$ obviously diverges. Thus, in studying the convergence or divergence of an infinite series, a finite number of terms at the start may always be discarded.

The associative law may be applied to any *convergent* infinite series. Thus we have the

THEOREM 1. *If the terms of a convergent series are grouped in parentheses in any manner to form new terms, the resulting series will converge to the same sum. The* order *of the terms must not be altered.*

Proof. All the partial sums of the new series occur among the old partial sums s_n for certain values of n. Since (3) holds good for the old series, it is also valid for the new series.

It is not true, however, that, if a series converges after parentheses are inserted, it will still converge when they are removed; thus, the series

$$(1 - 1) + (1 - 1) + \cdots \to 0; \qquad 1 - (1 - 1) - (1 - 1) + \cdots = 1,$$

while $1 - 1 + 1 - 1 + \cdots$ diverges since $\{s_n\}$ has two cluster points, 1 and 0. However, *if the series with parentheses diverges, the original series also diverges*; for, if it converged, it would still converge after grouping terms.

THEOREM 2. *If a series whose separate terms are sums in parenthesis converges to s, then the series with parentheses omitted will also converge to s, provided it converges at all.*

Proof. If the series without parentheses converges to s', inserting parentheses will not change its sum (Theorem 1); hence $s' = s$.

22. Geometric Series. The sum of n terms of the geometric series

(1) $$a + ar + ar^2 + \cdots + ar^{n-1} + \cdots$$

is readily computed; thus

$$s_n = a + ar + ar^2 + \cdots + ar^{n-1};$$
$$rs_n = \quad\;\; ar + ar^2 + \cdots + ar^{n-1} + ar^n,$$

and, on subtraction,

$$(1 - r)s_n = a(1 - r^n).$$

Hence, when $r \neq 1$,

$$(2) \qquad\qquad s_n = a\,\frac{1 - r^n}{1 - r}.$$

When $|r| \geq 1$, $\lim ar^n \neq 0$, and the series diverges.

When $|r| < 1$, $r^n \to 0$ from (15.7), and $s_n \to a/(1 - r)$; hence, the geometric series converges to the sum

$$(3) \qquad a + ar + ar^2 + \cdots = \frac{a}{1 - r}, \qquad |r| < 1.$$

A pure periodic decimal is an infinite geometric series. If the period consists of m digits, the ratio $r = 10^{-m}$; for example,

$$0.\overset{..}{1}\overset{.}{8} = \frac{18}{100} + \frac{18}{10{,}000} + \cdots = \frac{0.18}{1 - 0.01} = \frac{0.18}{0.99} = \frac{2}{11}.$$

23. Positive Series. A series $a_1 + a_2 + \cdots$ is said to be *positive* when all of its terms are positive ($a_n > 0$). Since

$$s_n = s_{n-1} + a_n > s_{n-1} \quad \text{if} \quad a_n > 0,$$

the sequence $\{s_n\}$ in a positive series is always monotone increasing; hence, from § 18 we have the

THEOREM. *A positive series will converge if s_n is bounded $(s_n \to s)$; otherwise it will diverge to infinity $(s_n \to \infty)$.*

The fact that $\{s_n\}$ in a positive series can have but *one* cluster point makes them simpler to handle than series with positive and negative terms. Thus in the series $2 - 1 - 1 + 2 - 1 - 1 + \cdots \{s_n\}$ has *three* cluster points 0, 1, 2; although s_n remains finite, the series diverges "by oscillation." In a positive series either $s_n \to s$ or $s_n \to \infty$.

Both associative and commutative laws are valid in obtaining the "sum" of a positive series. In *any* convergent series the terms may be grouped in parentheses (§ 21); and, if $s_n \to \infty$ in a positive series, obviously the same is true for the series with parentheses.

As to the commutative law, we have

DIRICHLET'S THEOREM. *A positive series converges to the same sum in whatever order the terms are taken.*

Proof. Let $a_1 + a_2 + \cdots \to s$, and let $b_1 + b_2 + \cdots$ be a series consisting of the a's taken in another order. Now

$$s_n' = b_1 + b_2 + \cdots + b_n < s$$

for every b is some a, and the sum of n a's, however chosen is less than s.

Hence $\{s_n'\}$ is bounded and converges to a sum $s' \leqq s$. But, since every a is some b, we can show in exactly the same way that $s \leqq s'$. Hence $s = s'$.

24. Telescopic Series. Every series $\displaystyle\sum_{n=1}^{\infty} a_n$ may be written as a *telescopic series*. For, if s_n denotes its nth partial sum, $a_n = s_n - s_{n-1}$, and the series becomes

$$(1) \qquad s_1 + (s_2 - s_1) + (s_3 - s_2) + \cdots.$$

Moreover, if the series converges to the sum s, and $r_n = s - s_n$ denotes the remainder after n terms, the series may also be written

$$(2) \qquad (s - r_1) + (r_1 - r_2) + (r_2 - r_3) + \cdots.$$

The following theorems show how to construct series that converge to any given sum or that diverge to infinity.

THEOREM 1. *If $\{\alpha_n\}$ is any sequence that converges to α, the series*

$$(3) \qquad \sum_{n=1}^{\infty} (\alpha_n - \alpha_{n+1}) = \alpha_1 - \alpha.$$

Proof. The sum of n terms of (3),

$$s_n = (\alpha_1 - \alpha_2) + (\alpha_2 - \alpha_3) + \cdots + (\alpha_n - \alpha_{n+1}) = \alpha_1 - \alpha_{n+1},$$

converges to $\alpha_1 - \alpha$.

THEOREM 2. *If the sequence $\{\beta_n\}$ diverges to infinity, the series*

$$(4) \qquad \sum_{n=1}^{\infty} (\beta_{n+1} - \beta_n) \to \infty;$$

and, if $\beta_n \neq 0$,

$$(5) \qquad \sum_{n=1}^{\infty} \left(\frac{1}{\beta_n} - \frac{1}{\beta_{n+1}} \right) = \frac{1}{\beta_1}.$$

Proof. The sum of n terms of (4),

$$s_n = (\beta_2 - \beta_1) + (\beta_3 - \beta_2) + \cdots + (\beta_{n+1} - \beta_n) = \beta_{n+1} - \beta_1,$$

diverges to ∞. Since $\{1/\beta_n\}$ is a *null* sequence, (5) follows from Theorem 1.

Note that, if $\{\beta_n\}$ is an *increasing* sequence, (4) and (5) are *positive* series.

Example 1. With $\beta_n = n$, we obtain the convergent series

(6)
$$\sum_{n=1}^{\infty} \left(\frac{1}{n} - \frac{1}{n+1} \right) = \sum_{n=1}^{\infty} \frac{1}{n(n+1)} = 1.$$

Example 2. With $\beta_n = \log n$, we obtain the divergent series

(7)
$$\sum_{n=1}^{\infty} [\log(n+1) - \log n] = \sum_{n=1}^{\infty} \log\left(1 + \frac{1}{n}\right) \to \infty.$$

Example 3. With $\alpha_n = n^2/(n^2 + 1)$, we have

$$\alpha_n - \alpha_{n+1} = - \frac{(2n+1)}{(n^2+1)((n+1)^2+1)} ;$$

Hence, from (3),

$$\sum_{n=1}^{\infty} \frac{2n+1}{(n^2+1)((n+1)^2+1)} = - \left(\frac{1}{2} - 1 \right) = \frac{1}{2}.$$

PROBLEMS

1. With $\beta_n = \alpha + n - 1$ $(\alpha > 0)$, show that

$$\sum_{n=1}^{\infty} \frac{1}{(\alpha + n - 1)(\alpha + n)} = \frac{1}{\alpha},$$

2. With $\alpha_n = (-1)^{n+1}/n$, show that

$$\sum_{n=1}^{\infty} (-1)^{n+1} \frac{2n+1}{n(n+1)} = 1.$$

3. With $\beta_n = r^{n-1}$ $(r > 1)$ in (5), show that

$$\sum_{n=1}^{\infty} \frac{1}{r^n} = \frac{1}{r-1}. \quad \text{Check by (22.3)}$$

4. Show that

$$\frac{1}{2!} + \frac{2}{3!} + \frac{3}{4!} + \cdots = 1.$$

5. Deduce

$$\sum_{n=1}^{\infty} \frac{1}{n(n+2)} = \frac{3}{4}, \qquad \sum_{n=1}^{\infty} \frac{(-1)^{n+1}}{n(n+2)} = \frac{1}{4},$$

from

$$\frac{2}{n(n+2)} = \frac{1}{n} - \frac{1}{n+2}.$$

6. Deduce

$$\sum_{n=1}^{\infty} \frac{1}{n(n+1)(n+2)} = \frac{1}{4}$$

from

$$\frac{2}{n(n+1)(n+2)} = \frac{1}{n(n+1)} - \frac{1}{(n+1)(n+2)}.$$

7. Prove that

$$\sum_{n=1}^{\infty} \frac{1}{n(n+1)\cdots(n+p)} = \frac{1}{pp!}.$$

8. Prove that

$$\sum_{n=1}^{\infty} \frac{1}{(2n-1)(2n+5)} = \frac{23}{90}.$$

9. Prove that

$$\sum_{n=1}^{\infty} \frac{1}{(4n^2-1)} = \frac{1}{2}.$$

10. Given that

$$\sum_{n=1}^{\infty} \frac{1}{n^2} = \frac{\pi^2}{6} \quad (223.15), \text{ prove that}$$

$$\sum_{n=1}^{\infty} \frac{n+1}{n(n+2)^2} = \frac{\pi^2-3}{12} . \quad \text{[Use partial fractions.]}$$

11. If $a_n > 0$ and $\sum_{n=1}^{\infty} a_n = s$, show that

$$\sum_{n=1}^{\infty} b_n = (s - \sqrt{r_1}) + (\sqrt{r_1} - \sqrt{r_2}) + (\sqrt{r_2} - \sqrt{r_3}) + \cdots$$

converges *more slowly* to s; that is, $a_n/b_n \to 0$.

12. If $a_n > 0$ and $\sum_{n=1}^{\infty} a_n \to \infty$, show that

$$\sum_{n=1}^{\infty} b_n = \sqrt{s_1} + (\sqrt{s_2} - \sqrt{s_1}) + (\sqrt{s_3} - \sqrt{s_2}) + \cdots$$

diverges *more slowly* to ∞; that is, $b_n/a_n \to 0$.

13. When are the series expansions valid:

$$(a) \ \frac{x}{x-1} = 1 + \frac{1}{x} + \frac{1}{x^2} + \cdots ; \qquad (b) \ \frac{x}{x-1} = -x - x^2 - x^3 - \cdots ?$$

Derive (b) from (a).

25. Comparison Tests. Series may be tested for convergence or divergence by comparing them with series whose behavior is known. The simplest comparison test is given in the

THEOREM.   *Let $\sum a_n$ and $\sum A_n$ be two positive series for which $a_n \leqq A_n$ when $n > N$.   Then*

(i)   *if $\sum A_n$ converges, $\sum a_n$ converges;*

(ii)   *if $\sum a_n$ diverges, $\sum A_n$ diverges.*

Proof. If s_n and S_n denote partial sums of $\sum a_n$ and $\sum A_n$, then $s_n \leqq S_n$ when $n > N$.

(i)   If $S_n \to S$, then s_n is bounded $(s_n \leqq S)$;

(ii)   If $s_n \to \infty$, then $S_n \to \infty$.

Corollary.

If $\sum a_n$ converges and $0 < k_n < K$, $\sum k_n a_n$ converges.

If $\sum a_n$ diverges and $k_n > K > 0$, $\sum k_n a_n$ diverges.

Proof. Compare $\sum k_n a_n$ with $\sum K a_n$.

26. Cauchy's Condensation Test. *If the positive sequence $\{a_n\}$ is monotone decreasing, the series,*

$$(1) \qquad a_1 + a_2 + a_3 + a_4 + \cdots = \sum_{n=1}^{\infty} a_n,$$

$$(2) \qquad a_1 + 2a_2 + 4a_4 + 8a_8 + \cdots = \sum_{m=0}^{\infty} 2^m a_{2^m},$$

both converge or both diverge.

Proof. Let s_n and S_n denote the sum of n terms in the series,

(1) $a_1 + a_2 + a_3 + a_4 + a_5 + a_6 + a_7 + a_8 + \cdots + a_{15} + \cdots,$

(2)′ $a_1 + a_2 + a_2 + a_4 + a_4 + a_4 + a_4 + a_8 + \cdots + a_8 + \cdots.$

Series (2) will converge or diverge according as (2)′ converges or diverges (§ 23).

Since the terms of (2)′ equal or exceed those of (1) written above,

$$(3) \qquad\qquad s_n \leqq S_n.$$

Hence s_n is bounded when S_n is bounded.

Now compare the series $2\sum a_n$ with (2)′:

(4) $a_1 + a_1 + a_2 + a_2 + a_3 + a_3 + a_4 + a_4 + a_5 + a_5$
$$\qquad\qquad\qquad + \cdots + a_8 + a_8 + \cdots,$$

(2)′ $a_1 + a_2 + a_2 + a_4 + a_4 + a_4 + a_4 + a_8 + a_8$
$$\qquad\qquad\qquad + \cdots + a_8 + a_8 + \cdots.$$

Since the terms of (4) equal or exceed those of (2)' written beneath,

(5) $$2s_n > S_{2n-1}.$$

Hence S_n is bounded when s_n is bounded.

The sequences $\{s_n\}$ and $\{S_n\}$ are thus both bounded or both unbounded; therefore series (1) and (2)' both converge or both diverge.

ABEL'S THEOREM. *If the positive sequence $\{a_n\}$ is monotone decreasing and $\sum a_n$ converges, then*

(6) $$na_n \to 0.$$

Proof. If series (1) converges, so also will (2), and hence $2^m a_{2^m} \to 0$ (21.5). Now, if $2^m \leqq n < 2^{m+1}$, then $a_n \leqq a_{2^m}$;

$$na_n < 2^{m+1} a_{2^m} = 2^m a_{2^m} + 2^m a_{2^m} \to 0.$$

Thus, for a series of positive terms that steadily decrease, not only is $a_n \to 0$ necessary for convergence but even $na_n \to 0$.

The utility of this test is due to the fact that the behavior of the condensed series is often more apparent than that of the original. The theorem may be considerably generalized.†

When the series converge, $s_n \to s$ and $S_n \to S$; hence, on letting $n \to \infty$ in (3) and (5), we obtain (since the inequalities *widen*)

(7) $$\tfrac{1}{2}S < s < S.$$

Example 1. The p-series $\displaystyle\sum_{n=1}^{\infty} \frac{1}{n^p}$. The condensed series

$$\sum_{m=0}^{\infty} \frac{2^m}{(2^m)^p} = \sum_{m=0}^{\infty} \left(\frac{1}{2^{p-1}}\right)^m$$

is geometric (§ 22) with the

ratio $\quad r = \dfrac{1}{2^{p-1}} \quad$ and sum $\quad S = \dfrac{1}{1 - 1/2^{p-1}}.$

Since $r < 1$ only when $p > 1$, we have the important result:
The series $\Sigma 1/n^p$ converges when and only when $p > 1$.
In particular, the *harmonic series*

(8) $$1 + \tfrac{1}{2} + \tfrac{1}{3} + \tfrac{1}{4} + \cdots$$

diverges, as previously shown in Ex. 17.2.

In case of convergence the sum is written

(9) $$\zeta(p) = \sum_{n=1}^{\infty} \frac{1}{n^p}, \qquad p > 1,$$

---

† Knopp, K., *Theory and Application of Infinite Series*, London, 1928, p. 120.

and known as *Riemann's zeta function*; and, from (7),

(10)
$$\frac{1}{2} S < \zeta(p) < S = \frac{1}{1 - 1/2^{p-1}}$$

Thus $1 < \zeta(2) < 2$, $^2/_3 < \zeta(3) < ^4/_3$. We shall show in § 223 that $\zeta(2) = \pi^2/6$, $\zeta(4) = \pi^4/90$; and, in general, a closed expression for $\zeta(p)$ is known when p is an even integer. When p is an odd integer, $\zeta(p)$ has not yet been simply expressed; to this day no closed expression for $\zeta(3)$ is known.

The p-series (9) converges when the *constant* exponent $p > 1$. But, when p is a *variable* > 1, the convergence is no longer assured. Thus, when $p = 1 + \dfrac{1}{n}$ the series

$$\sum_{n=1}^{\infty} n^{-(1+1/n)}$$ diverges, as we see on comparison with $\sum_{n=1}^{\infty} n^{-1}$, using the test of § 27.

Example 2. The series $\displaystyle\sum_{n=2}^{\infty} \frac{1}{n(\log n)^p}$.

The condensed series is now

$$\sum_{m=1}^{\infty} \frac{2^m}{2^m(\log 2^m)^p} = \frac{1}{(\log 2)^p} \sum_{m=1}^{\infty} \frac{1}{m^p},$$

a multiple of the p-series. Thus the given series, like the p-series, converges only when $p > 1$. When $p = 1$, we obtain *Abel's series*,

(11)
$$\sum_{n=2}^{\infty} \frac{1}{n \log n} = \frac{1}{2 \log 2} + \frac{1}{3 \log 3} + \cdots,$$

which diverges, but even more slowly than the harmonic series (8), since

$$1/n \log n < 1/n \quad \text{when} \quad n > 2.$$

Example 3. The series $\displaystyle\sum_{n=3}^{\infty} \frac{1}{n \log n \log \log n}$.

The condensed series

$$\sum_{m=2}^{\infty} \frac{2^m}{2^m \log 2^m \log \log 2^m} = \sum_{m=2}^{\infty} \frac{1}{m \log 2 \log (m \log 2)}$$

has larger terms than Abel's series since $\log 2 < 1$ and must therefore diverge. The given series diverges even more slowly than Abel's series since

$$1/n \log n \log \log n < 1/n \log n \quad \text{when} \quad n > 15.$$

27. Limit Form of Comparison Test. *If $\sum a_n$ and $\sum b_n$ are positive series and the ratio*

(1) $$\frac{a_n}{b_n} \to \lambda > 0,$$

the series both converge or both diverge.
If $a_n/b_n \to 0$ and $\sum b_n$ converges, $\sum a_n$ also converges.
If $a_n/b_n \to \infty$ and $\sum b_n$ diverges, $\sum a_n$ also diverges.
Proof. When $\lambda > 0$, choose ε so that $0 < \varepsilon < \lambda$; then

$$\lambda - \varepsilon < \frac{a_n}{b_n} < \lambda + \varepsilon \quad \text{when} \quad n > N;$$

$$(\lambda - \varepsilon)b_n < a_n < (\lambda + \varepsilon)b_n, \qquad n > N.$$

If $\sum b_n$ converges, $(\lambda + \varepsilon)\sum b_n$ and also $\sum a_n$ converge; if $\sum b_n$ diverges, $(\lambda - \varepsilon)\sum b_n$ and also $\sum a_n$ diverge.
When $a_n/b_n \to 0$,

$$\frac{a_n}{b_n} < \varepsilon \quad \text{when} \quad n > N \quad \text{and} \quad a_n < \varepsilon b_n;$$

hence, if $\sum b_n$ converges, $\varepsilon \sum b_n$ and also $\sum a_n$ converge.
When $a_n/b_n \to \infty$,

$$\frac{a_n}{b_n} > G \quad \text{when} \quad n > N \quad \text{and} \quad a_n > G b_n;$$

hence, if $\sum b_n$ diverges, $G\sum b_n$ and also $\sum a_n$ diverge.
From this test we next deduce the useful
POLYNOMIAL TEST. *If $P(n)$ and $Q(n)$ are polynomials of degree p and q, the series,*

(2) $$\sum_{n=1}^{\infty} \frac{P(n)}{Q(n)} = \sum_{n=1}^{\infty} \frac{cn^p + \cdots}{dn^q + \cdots}, \qquad c, d \neq 0,$$

converges when, and only when $q > p + 1$.
Proof. Since the series (2) converges or diverges with

$$\frac{d}{c} \sum_{n=1}^{\infty} \frac{P(n)}{Q(n)} = \sum_{n=1}^{\infty} \frac{n^p + \cdots}{n^q + \cdots},$$

we may test this series instead.   When n is large enough, its terms are all positive and we may apply the limit test (1), taking

$$a_n = \frac{n^p + \cdots}{n^q + \cdots}, \qquad b_n = \frac{n^p}{n^q}.$$

Since

$$\frac{a_n}{b_n} = \frac{n^{p+q} + \cdots}{n^{p+q} + \cdots} \to 1, \tag{16.5}$$

the series $\sum P/Q$ and $\sum 1/n^{q-p}$ converge or diverge together; and from Ex. 26.1 the latter converges when and only when $q - p > 1$.

Example 1. When a, b, c, are positive, the series

$$\sum_{n=1}^{\infty} \frac{1}{(an + b)^p} \quad \text{and} \quad \sum_{n=1}^{\infty} \frac{1}{(an^2 + bn + c)^{p/2}}$$

converge only when $p > 1$.   To show this apply the test (1) with the p-series $\sum 1/n^p$ for comparison.

Example 2. The polynomial test shows that the series

$$\sum_{n=1}^{\infty} \frac{n}{(n + 1)(n + 2)} \text{ diverges;} \quad \sum_{n=1}^{\infty} \frac{n + 1}{n(n + 2)^2} \text{ converges.}$$

28. Cauchy's Root Test.   *If in the positive series* Σa_n,

(1) $\sqrt[n]{a_n} \leq r < 1$ *when* $n > N$,   *the series converges*;

(2) $\sqrt[n]{a_n} \geq 1$ *for infinitely many* n, *the series diverges*.

Proof.   When $n > N$, in case (1), $a_n \leq r^n$, the general term of a convergent geometric series.

In case (2), $a_n \geq 1$ for infinitely many n.   Then $\lim a_n \neq 0$; for $a_n \to 0$ implies that $a_n < \varepsilon$ when $n > N$.

Unless $\sqrt[n]{a_n}$ ultimately remains less than a *fixed number* $r < 1$, we cannot conclude that $\sum a_n$ converges.   Thus the requirement $\sqrt[n]{a_n} < 1$ will not ensure convergence.   For example, in the *divergent* series $\sum 1/n$,

$$\sqrt[n]{a_n} = 1/\sqrt[n]{n} < 1, \quad \sqrt[n]{a_n} \to 1 \quad \text{from below (15.11).}$$

But the convergent series $\sum 1/n^2$ behaves in precisely the same way. Thus, if $\sqrt[n]{a_n} \to 1$ *from below*, no conclusion can be drawn from the root test.   But, if $\sqrt[n]{a_n} \to 1$ *from above*, part (2) of the theorem shows that $\sum a_n$ diverges.

When the sequence $\{\sqrt[n]{a_n}\}$ is bounded, $\sqrt[n]{a_n}$ will always have a *greatest limit* $\bar{\rho}$ (§ 15).   When $\{\sqrt[n]{a_n}\}$ is not bounded, we write conventionally $\bar{\rho} = \infty$.   The root test now leads to the important

THEOREM.   If $\bar{\rho} = \overline{\lim} \sqrt[n]{a_n}$ and

$$\bar{\rho} < 1, \qquad \sum a_n \text{ converges;}$$
$$\bar{\rho} > 1, \qquad \sum a_n \text{ diverges;}$$
$$\bar{\rho} = 1, \qquad \text{no conclusion.}$$

Proof. From (15.5) there will be a finite number of $\sqrt[n]{a_n} > \bar{\rho} + \varepsilon$, an infinite number of $\sqrt[n]{a_n} > \bar{\rho} - \varepsilon$.

When $\bar{\rho} < 1$, choose ε so that $\bar{\rho} + \varepsilon < 1$. Only a finite number of roots can exceed $\bar{\rho} + \varepsilon$; hence, if N is their highest index, $\sqrt[n]{a_n} \leqq \bar{\rho} + \varepsilon$ when $n > N$, and the series converges.

When $\bar{\rho} > 1$, choose ε so that $\bar{\rho} - \varepsilon > 1$. Since an infinite number of roots exceed $\bar{\rho} - \varepsilon$, the series diverges.

When $\bar{\rho} = \infty$, a_n is not bounded and the series diverges.

Note.   When $\sqrt[n]{a_n} \to \rho$, then $\bar{\rho} = \rho$ (§ 15).

29. d'Alembert's Ratio Test.   *Let* $\tau_n = a_{n+1}/a_n$ *be the term-ratio in the positive series* $\sum a_n$; *then, if*

(1) $\tau_n \leqq r < 1$,   *when* $n > N$,   *the series converges;*

(2) $\tau_n \geqq 1$ *when* $n > N$,   *the series diverges.*

Proof. Discard the first N terms of the series and write $b_m = a_{N+m}$. When (1) holds, $b_{m+1} \leqq r b_m$ ($m = 1, 2, 3, \cdots$), and

$$b_{m+1} = b_1 \frac{b_2}{b_1} \frac{b_3}{b_2} \cdots \frac{b_{m+1}}{b_m} \leqq b_1 r^m,$$

the general term of a convergent geometric series of ratio $r < 1$; hence $\sum b_m$ (and also $\sum a_n$) converges.

When (2) holds, $b_{m+1} \geqq b_m$ and $\{b_m\}$ is monotone-increasing; since $\lim b_m \neq 0$, $\sum b_m$ diverges.

In actual practice it is usually more convenient to use the

THEOREM.   *If* $\underline{\tau}$ *and* $\bar{\tau}$ *are the least and greatest limits of the term-ratio* $\tau_n = a_{n+1}/a_n$, *and*

$$\bar{\tau} < 1, \quad \sum a_n \text{ converges;} \qquad \underline{\tau} > 1, \quad \sum a_n \text{ diverges.}$$

Proof. If $\bar{\tau} < 1$, choose ε so that $r = \bar{\tau} + \varepsilon < 1$. Then only a finite number of τ_n exceed r, and condition (1) is fulfilled.

If $\underline{\tau} > 1$, choose ε so that $\underline{\tau} - \varepsilon > 1$. Then all but a finite number of τ_n exceed $\underline{\tau} - \varepsilon$, and condition (2) is fulfilled.

Note that, when τ_n converges to a limit τ, $\underline{\tau} = \bar{\tau} = \tau$.

Example 1. Consider the series

$$1 + b + bc + b^2c + b^2c^2 + \cdots + b^nc^{n-1} + b^nc^n + \cdots,$$

where $0 < b < c$.

The ratio $a_{n+1}/a_n = b$ and c alternately; hence $\underline{\tau} = b$, $\bar{\tau} = c$, and the series converges when $c < 1$, diverges when $b > 1$. But the ratio test fails when $b < 1 < c$.

The root $\sqrt[n]{a_n} \to \sqrt{bc}$; for both

$$(b^n c^{n-1})^{1/2n} \quad \text{and} \quad (b^n c^n)^{1/(2n+1)} \to (bc)^{1/2}.$$

Thus the root test shows that the series converges when $bc < 1$, diverges when $bc > 1$. When $bc = 1$, the series $1 + b + 1 + b + \cdots$ is obviously divergent.

This example shows that the ratio test is less general than the root test. In fact, it can be shown that

(3) $$\qquad\qquad a_{n+1}/a_n \to \tau \quad \text{implies} \quad \sqrt[n]{a_n} \to \tau.\dagger$$

Example 1 shows that the converse is not true. When both limits exist, they are necessarily equal.

In practice, the root and ratio tests are among the most useful and have about the same range of applicability. But the ratio test, although less general, is usually easier to apply, for a_{n+1}/a_n is often a simpler function of n than $\sqrt[n]{a_n}$. To ensure convergence the ratio or the root must eventually remain less than a *fixed number* $r < 1$. When a_{n+1}/a_n or $\sqrt[n]{a_n} \to 1$ from below, we can draw no conclusion; thus the divergent series $\sum 1/n$ and the convergent series $\sum 1/n^2$ have this property.

Example 2. When $0 < b < c < 1$, the series

$$b + c + b^2 + c^2 + b^3 + c^3 + \cdots$$

is certainly convergent; for the odd terms and the even terms separately form convergent geometric series (cf. Dirichlet's theorem, § 23).

The root test confirms the convergence, for

$$(c^n)^{1/2n} \to \sqrt{c} < 1, \qquad (b^{n+1})^{1/(2n+1)} \to \sqrt{b} < 1.$$

But the ratio test does not apply since successive ratios are alternately > 1 and < 1; in fact,

$$\frac{c^n}{b^n} = \left(\frac{c}{b}\right)^n \to \infty, \qquad \frac{b^{n+1}}{c^n} = b\left(\frac{b}{c}\right)^n \to 0 \qquad (15.7).$$

Example 3. In the series $\displaystyle\sum_{n=0}^{\infty} \frac{x^n}{n!}$ (where $0!' = 1$ by definition)

$$\frac{a_{n+1}}{a_n} = \frac{x^{n+1}}{(n+1)!} : \frac{x^n}{n!} = \frac{x}{n+1} \to 0,$$

and the series converges for all $x > 0$.

---

† Goursat-Hedrick, *A Course in Mathematical Analysis*, vol. 1, Boston, 1904, p. 334.

From (3) we must have

$$\sqrt[n]{a_n} = \frac{x}{\sqrt[n]{n!}} \to 0 \quad \text{and hence} \quad \sqrt[n]{n!} \to \infty;$$

but this limit is not obvious.

Example 4. In the series $\sum_{n=1}^{\infty} \frac{x^n}{n}$

$$\frac{a_{n+1}}{a_n} = \frac{x^{n+1}}{n+1} : \frac{x^n}{n} = x\,\frac{n}{n+1} \to x,$$

and the series converges when $0 < x < 1$, diverges when $x > 1$. When $x = 1$, the ratio test is inconclusive; but then the series is harmonic and divergent (Ex. 26.1).

The root test gives the same results:

$$\sqrt[n]{a_n} = \frac{x}{\sqrt[n]{n}} \to x \quad \text{for} \quad \sqrt[n]{n} \to 1; \tag{15.11}$$

it is also inconclusive when $x = 1$.

PROBLEMS

1. Show that $\sum_{n=1}^{\infty} 1/n^2$ converges by comparison with $\sum_{n=1}^{\infty} 1/n(n+1)$ [Ex. 24.1].

2. Show that $\sum_{n=1}^{\infty} 1/n$ diverges by comparison with $\sum_{n=1}^{\infty} \log(1 + 1/n)$ [Ex. 24.2].

3. Test Σa_n for convergence or divergence by (27.1) when

$$a_n = \frac{n}{n^2 + 2}; \quad \left(\frac{n-1}{n^2+1}\right)^{1/2}; \quad \frac{n+3}{n^3 - n}; \quad \sin\frac{c}{n}; \quad n^{-(n+1)/n}; \quad \tan\frac{c}{n^2}.$$

4. Test Σa_n for convergence or divergence by the root test when

$$a_n = \frac{1}{(\log n)^n}; \quad \frac{n}{2^n}; \quad \frac{n^2}{c^n}(c > 0); \quad \frac{n!}{n^n} \quad \text{[Prob. 15.7]}.$$

5. Test the series Σa_n for convergence or divergence by the ratio test when

$$a_n = \frac{c^n}{n!}; \quad n^2 e^{-n}; \quad \frac{n}{2^n}; \quad \frac{n^2}{c^n}(c > 0); \quad \frac{(\log n)^2}{(\log 2)^n}.$$

6. Show that $\sum_{n=2}^{\infty} (\log n)/n^p$ converges when $p > 1$, diverges when $0 < p \leqq 1$.

[When $p > 1$, let $1 < q < p$, and compare with $\Sigma 1/n^q$, using (27.1).]

7. Show that $\sum_{n=2}^{\infty} 1/(\log n)^p$, $p > 0$, diverges.

8. Show that $\sum_{n=1}^{\infty} n^k x^n$ (k arbitrary) converges when $0 < x < 1$, diverges when $x > 1$.

9. Show that $\sum\limits_{n=0}^{\infty} n!x^n/(2n)!$ converges for all positive values of x.

10. Does the series $\sum\limits_{n=2}^{\infty} \dfrac{1}{(\log n)^{\log n}}$ converge or diverge?

$[(\log n)^{\log n} = n^{\log \log n}]$.

30. Kummer-Jensen Tests. Both the root and ratio test depend on a comparison with geometric series and are inconclusive when the root or ratio approaches 1 from below. To obtain sharper tests we turn to the versatile telescopic series for comparison; for these are capable of representing any series whatever (§ 24).

KUMMER'S TEST. *If $\{c_n\}$ is any positive sequence, the positive series $\sum a_n$ will converge if*

$$(1) \qquad K_n = c_n - c_{n+1}\frac{a_{n+1}}{a_n} \geq h > 0, \qquad n \geq N.$$

Proof. From (1) we have

$$0 < ha_n \leq c_n a_n - c_{n+1}a_{n+1}, \qquad n \geq N.$$

Hence the positive sequence $\{c_n a_n\}$ decreases monotonely and $c_n a_n \to \lambda$ (Theorem 18); and the telescopic series

$$\sum_{N}^{\infty} (c_n a_n - c_{n+1}a_{n+1}) = c_N a_N - \lambda \qquad \text{(Theorem 24.1)}.$$

Thus $h\sum a_n$ and also $\sum a_n$ converge.

The corresponding divergence criterion is given in

JENSEN'S TEST. *If $\sum 1/c_n$ is a positive divergent series, the positive series $\sum a_n$ will diverge if*

$$(2) \qquad K_n = c_n - c_{n+1}\frac{a_{n+1}}{a_n} \leq 0, \qquad n \geq N.$$

Proof. When (2) holds, $\{c_n a_n\}$ is a monotone increasing sequence. Thus, when $n \geq N$,

$$c_n a_n \geq c_N a_N \quad \text{and} \quad a_n \geq \text{const}/c_n,$$

the general term of a divergent series; hence $\sum a_n$ diverges.

In the "limit form" these tests may be combined in the

THEOREM. *If $\underline{K}$ and $\overline{K}$ are the least and greatest limits of the sequence $\{K_n\}$, and*

$$(3) \qquad \underline{K} > 0, \qquad \sum a_n \text{ converges};$$

$$(4) \qquad \overline{K} < 0, \qquad \sum a_n \text{ diverges with } \sum 1/c_n.$$

Proof. If $\underline{K} > 0$, choose ε so that $h = \underline{K} - \varepsilon > 0$. Then all but a finite number of K_n will exceed h and (1) is fulfilled.

If $\overline{K} < 0$, choose ε so that $\overline{K} + \varepsilon < 0$. Then all but a finite number of K_n will be negative and (2) is fulfilled.

This theorem now yields various tests by specializing the sequence $\{c_n\}$. Note that $\sum 1/c_n$ must diverge in order to use (4).

1. $c_n = 1$, $K_n = 1 - a_{n+1}/a_n$.

This gives d'Alembert's test (§ 29).

2. $c_n = n - 1$, $K_n = n\left(1 - \dfrac{a_{n+1}}{a_n}\right) - 1$.

Hence, if we put

(5) $R_n = n\left(1 - \dfrac{a_{n+1}}{a_n}\right)$,

we obtain

RAABE'S TEST.   When

$$\underline{\lim} R_n > 1, \qquad \sum a_n \text{ converges};$$

$$\overline{\lim} R_n < 1, \qquad \sum a_n \text{ diverges}.$$

3. $c_n = (n - 1) \log (n - 1)$; then $\displaystyle\sum_{3}^{\infty} \frac{1}{c_n}$ is the divergent Abel's series (Ex. 26.2), and

$$K_n = (n - 1) \log (n - 1) - n \log n\, \frac{a_{n+1}}{a_n}$$

$$= (n - 1) \log \frac{n - 1}{n} + \log n\left[n\left(1 - \frac{a_{n+1}}{a_n}\right) - 1\right]$$

Now the limit of the first term is the same as the limit of

$$n \log \frac{n}{n + 1} = \log \frac{1}{(1 + 1/n)^n} \to \log \frac{1}{e} = -1.$$

Hence, if we put

(6) $B_n = \log n\left[n\left(1 - \dfrac{a_{n+1}}{a_n}\right) - 1\right] = \log n \cdot (R_n - 1)$,

the theorem gives

BERTRAND'S TEST.   When

$$\underline{\lim} B_n > 1, \qquad \sum a_n \text{ converges};$$

$$\overline{\lim} B_n < 1, \qquad \sum a_n \text{ diverges}.$$

Example 1. For the series

$$\sum_{n=0}^{\infty} a_n = 1 + \frac{\alpha}{1} + \frac{\alpha(\alpha+1)}{2!} + \frac{\alpha(\alpha+1)(\alpha+2)}{3!} + \cdots,$$

$$\frac{a_{n+1}}{a_n} = \frac{\alpha+n}{n+1}, \qquad R_n = n\left(1 - \frac{\alpha+n}{n+1}\right) = \frac{n}{n+1}(1-\alpha) \to 1-\alpha.$$

Since $a_{n+1}/a_n > 0$ when $n > -\alpha$, the terms ultimately keep the same sign, and we can apply Raabe's test. Hence the series converges when $1-\alpha > 1$ ($\alpha < 0$), diverges when $1-\alpha < 1$ ($\alpha > 0$).

Example 2. In the series

$$\sum_{n=1}^{\infty} \frac{2 \cdot 4 \cdots 2n}{1 \cdot 3 \cdots (2n-1)} \cdot \frac{1}{2n+2},$$

$$\frac{a_{n+1}}{a_n} = \frac{2n+2}{2n+1} \cdot \frac{2n+2}{2n+4} = \frac{4n^2+8n+4}{4n^2+10n+4},$$

$$R_n = n\left(1 - \frac{a_{n+1}}{a_n}\right) = \frac{2n^2}{4n^2+10n+4} \to \frac{1}{2},$$

and the series diverges.

31. Gauss' Test. *If we can write*

(1) $$R_n = n\left(1 - \frac{a_{n+1}}{a_n}\right) = h + \frac{f_n}{n^p} \qquad (p > 0),$$

where h is a constant and f_n is bounded, the positive series $\sum a_n$ converges when $h > 1$, diverges when $h \leq 1$.

Proof. Since $R_n \to h$, Raabe's test covers the cases $h \neq 1$. When $h = 1$, the series diverges by Bertrand's test: for

$$B_n = \log n \cdot (R_n - 1) = f_n \frac{\log n}{n^p} \to 0. \qquad (15.9).$$

Gauss actually used a special case adapted to series in which a_{n+1}/a_n is the quotient of two polynomials having the same term of highest degree; namely,

(2) $$\frac{a_{n+1}}{a_n} = \frac{n^k + bn^{k-1} + \cdots}{n^k + cn^{k-1} + \cdots} \to 1 \qquad (16.4).$$

The d'Alembert test fails, but

$$R_n = n\left(1 - \frac{a_{n+1}}{a_n}\right) = \frac{(c-b)n^k + \cdots}{n^k + \cdots} \to c - b,$$

the dots standing for terms of lower degree. By Raabe's test, $\sum a_n$

converges when $c - b > 1$, diverges when $c - b < 1$. When $c - b = 1$,

$$B_n = \log n \cdot (R_n - 1) = \frac{\log n}{n} \cdot \frac{rn^{k-1} + \cdots}{n^{k-1} + \cdots} \to 0,$$

and the series diverges by Bertrand's test;. note that the last fraction has a finite limit r and is therefore bounded (§ 15). We have thus proved the useful

THEOREM. *When a_{n+1}/a_n has the form* (2), $\sum a_n$ *converges when $c - b > 1$, diverges when $c - b \leq 1$.*

Example. In the *hypergeometric series*

(3)
$$\sum_{n=0}^{\infty} a_n = 1 + \frac{\alpha \cdot \beta}{1 \cdot \gamma} + \frac{\alpha(\alpha + 1) \cdot \beta(\beta + 1)}{1 \cdot 2 \cdot \gamma(\gamma + 1)} + \cdots,$$

α, β, γ are real and none of them is zero or a negative integer.   This condition on α, β keeps the series from terminating, while that on γ avoids division by zero.   The ratio

$$\frac{a_{n+1}}{a_n} = \frac{(\alpha + n)(\beta + n)}{(1 + n)(\gamma + n)} = \frac{n^2 + (\alpha + \beta)n + \alpha\beta}{n^2 + (\gamma + 1)n + \gamma}$$

ultimately becomes and remains positive and the terms have a constant sign. Hence the series (3) will converge or diverge according as $(\gamma + 1) - (\alpha + \beta) > 1$ or ≤ 1: that is, according as $\gamma > \alpha + \beta$ or $\gamma \leq \alpha + \beta$.

PROBLEMS

Test the following series for convergence by Raabe's test:

1.
$$\sum_{n=1}^{\infty} \frac{n!}{(\alpha + 1)(\alpha + 2) \ldots (\alpha + n)} \qquad (\alpha > 0).$$

2.
$$\sum_{n=1}^{\infty} \frac{1 \cdot 3 \ldots (2n - 1)}{2 \cdot 4 \ldots (2n)} \frac{1}{2n + 1}.$$

3.
$$\sum_{n=1}^{\infty} \left(\frac{1 \cdot 3 \ldots (2n - 1)}{2 \cdot 4 \ldots (2n)} \right)^p \quad \text{when} \quad p = 1, 2, 3.$$

4. Test the series in Problem 3 when $p = 2$ by Gauss' test.

5. Show that the series

$$1 + \frac{\alpha \cdot \beta}{\gamma \cdot \delta} + \frac{\alpha(\alpha + 1) \cdot \beta(\beta + 1)}{\gamma(\gamma + 1) \cdot \delta(\delta + 1)} + \cdots,$$

where $\alpha, \beta, \gamma, \delta$ are not zero or negative integers, converges or diverges according as

$$\gamma + \delta - (\alpha + \beta) > 1 \quad \text{or} \quad \leq 1.$$

32. Absolute Convergence.   We next consider *mixed* series with both positive and negative terms.   Such a series $\sum a_n$ is said to be *absolutely convergent* when the positive series $\sum |a_n|$ converges.

THEOREM. *The absolute convergence of a series implies its convergence.*

Proof. If s_n and S_n are the sum of n terms of $\sum a_n$ and $\sum |a_n|$, and $m = n + r$,

$$|s_m - s_n| = |a_{n+1} + a_{n+2} + \cdots + a_{n+r}|$$
$$\leq |a_{n+1}| + |a_{n+2}| + \cdots + |a_{n+r}| = S_m - S_n.$$

Hence $S_m - S_n < \varepsilon$ implies $|s_m - s_n| < \varepsilon$; the theorem now follows from Cauchy's convergence criterion (21.4).

To determine the absolute convergence of a series we have at our service all the tests developed for positive series. In particular, the series $\sum a_n$ is absolutely convergent if

$$\lim \sqrt[n]{|a_n|} < 1 \quad \text{or} \quad \lim \left| \frac{a_{n+1}}{a_n} \right| < 1.$$

Note that this theorem is not true for sequences; thus the sequence $1, -1, 1, -1, \cdots$ does not converge, but $1, 1, 1, 1, \cdots$ converges to 1.

33. Conditional Convergence. If the series $\sum a_n$ converges but $\sum |a_n|$ diverges, $\sum a_n$ is called *conditionally convergent*.

THEOREM. *If a series is conditionally convergent, the series of its positive terms and the series of absolute values of its negative terms are both divergent.*

Proof. In the sum s_n of n terms of the series let $p_{n'}$ denote the sum of the positive terms, $-q_{n''}$ the sum of the negative terms ($n' + n'' = n$). If S_n is the sum of n terms in the series of absolute values,

$$(1) \qquad\qquad s_n = p_{n'} - q_{n''}, \qquad S_n = p_{n'} + q_{n''}.$$

As n becomes infinite, $s_n \to s$, $S_n \to \infty$ by hypothesis; hence,

$$p_{n'} = \tfrac{1}{2}(S_n + s_n) \to \infty, \qquad q_{n''} = \tfrac{1}{2}(S_n - s_n) \to \infty,$$

as stated in the theorem.

On the other hand, if the series converges absolutely, $S_n \to S$, and

$$p_{n'} \to \tfrac{1}{2}(S + s), \qquad q_{n''} \to \tfrac{1}{2}(S - s);$$

that is, the two subseries both converge. Hence we can characterize the behavior of a mixed series as follows:

$$p_{n'} \to p, \quad q_{n''} \to q: \quad \text{absolute convergence;}$$

$$p_{n'} \to \infty, \quad q_{n''} \to \infty: \quad \text{conditional convergence or divergence;}$$

$$\left. \begin{array}{ll} p_{n'} \to p, & q_{n''} \to \infty \\ p_{n'} \to \infty, & q_{n''} \to q \end{array} \right\} : \text{ divergence.}$$

That a series may diverge when both subseries diverge is shown by the example: $1 - 2 + 3 - 4 + \cdots$.

When the convergence is absolute, we have from (1), on letting $n \to \infty$,

$$(2)\ (3) \qquad\qquad s = p - q, \qquad S = p + q.$$

34. Alternating Series. A mixed series is called *alternating* if its positive and negative terms occur alternately. If $\{a_n\}$ is a positive sequence,

$$(1) \qquad \sum_{n=1}^{\infty} (-1)^{n-1} a_n = a_1 - a_2 + a_3 - a_4 + a_5 - a_6 + \cdots$$

is an alternating series. A sufficient condition for convergence is given in

LEIBNITZ'S TEST. *An alternating series whose terms in absolute value form a monotone null sequence is convergent.*

The remainder after n terms is always numerically less than the first term neglected:

$$(2) \qquad\qquad |r_n| < a_{n+1}.$$

Proof. The sums of even index, $s_2, s_4, s_6, \cdots$, are monotone-increasing;† for

$$s_{2n+2} - s_{2n} = a_{2n+1} - a_{2n+2} > 0.$$

The sums of odd index, $s_1, s_3, s_5, \cdots$, are monotone-decreasing;† for

$$s_{2n+1} - s_{2n-1} = -a_{2n} + a_{2n+1} < 0.$$

The intervals $(s_2, s_1), (s_4, s_3), \cdots, (s_{2n}, s_{2n-1})$ form a *nest* (§ 20); for

$$s_{2n-1} - s_{2n} = a_{2n} > 0 \quad \text{and} \quad a_{2n} \to 0.$$

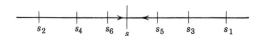

FIG. 34 Convergence of alternating series.

These nested intervals determine a unique number s, the common limit of both even and odd sums, and therefore the sum of the series (Fig. 34); and since s_n and s_{n+1} lie on opposite sides of s,

$$|r_n| = |s - s_n| < |s_{n+1} - s_n| = a_{n+1}.$$

---

† These facts are *visible* on writing the series in the two ways:

$$(a_1 - a_2) + (a_3 - a_4) + (a_5 - a_6) + \cdots,$$
$$a_1 - (a_2 - a_3) - (a_4 - a_5) - \cdots.$$

Example 1. The alternating series

$$(3) \qquad \sum_{n=1}^{\infty} (-1)^{n-1} \frac{1}{n} = 1 - \frac{1}{2} + \frac{1}{3} - \frac{1}{4} + \cdots$$

converges conditionally since $\{1/n\}$ is a monotone null sequence; but $\Sigma 1/n$, the harmonic series, diverges. Note that $p_n \to \infty, q_n \to \infty$.

Example 2. The alternating series

$$\sum_{n=1}^{\infty} (-1)^{n-1} \frac{n+2}{2n+1} = \frac{3}{3} - \frac{4}{5} + \frac{5}{7} - \frac{6}{9} + \cdots$$

diverges; $\{(n+2)/(2n+1)\}$ is monotone-decreasing but its limit is $\frac{1}{2}$.

Example 3.

$$1 - \frac{1}{2} + \frac{1}{2^2} - \frac{1}{3} + \frac{1}{3^2} - \frac{1}{4} + \frac{1}{4^2} - \cdots$$

diverges since p_n converges but $q_n \to \infty$.

Example 4. The series

$$\frac{1}{\sqrt{2}} - \frac{1}{2} + \frac{1}{\sqrt{3}} - \frac{1}{3} + \frac{1}{\sqrt{4}} - \frac{1}{4} + \cdots$$

diverges since $\Sigma(1/\sqrt{n} - 1/n)$ diverges (Theorem 21); for, on comparison with the divergent series $\Sigma 1/\sqrt{n}$, we have

$$\left(\frac{1}{\sqrt{n}} - \frac{1}{n} \right) : \frac{1}{\sqrt{n}} = 1 - \frac{1}{\sqrt{n}} \to 1. \qquad (\S\ 27)$$

Note that $p_n \to \infty, q_n \to \infty$.

Example 5. The alternating series

$$\sum_{n=1}^{\infty} (-1)^{n-1} \frac{n+2}{n+1} = \frac{3}{2} - \frac{4}{3} + \frac{5}{4} - \frac{6}{5} + \cdots$$

diverges since $\lim a_n \neq 0$. By inserting parentheses about every pair of terms we obtain the positive series

$$\frac{1}{2 \cdot 3} + \frac{1}{4 \cdot 5} + \frac{1}{6 \cdot 7} + \cdots = \sum_{n=1}^{\infty} \frac{1}{2n(2n+1)},$$

which converges by the polynomial test (§ 27). Thus by inserting parentheses we have made a divergent series converge.

PROBLEMS

Test the series for convergence:

1. $\frac{1}{3} - \frac{2}{5} + \frac{3}{7} - \frac{4}{9} + \cdots$.

2. $\dfrac{a}{1} - \dfrac{b}{2} + \dfrac{a}{3} - \dfrac{b}{4} + \cdots \qquad (a, b > 0)$.

Discuss the cases $a = b$, $a > b$, $a < b$.

3. $1 - \dfrac{1}{2} + \dfrac{1}{3^2} - \dfrac{1}{4} + \dfrac{1}{5^2} - \dfrac{1}{6} + \cdots$.

4. $1 - \dfrac{1}{1 \cdot 2} + \dfrac{1}{2} - \dfrac{1}{2 \cdot 3} + \dfrac{1}{3} - \dfrac{1}{3 \cdot 4} + \cdots$.

5. $1 - \dfrac{1}{2^p} + \dfrac{1}{3^p} - \dfrac{1}{4^p} + \cdots$ $(p > 0)$.

When does the series converge conditionally?

6. $1 + \tfrac{1}{2} - \tfrac{1}{3} + \tfrac{1}{4} + \tfrac{1}{5} - \tfrac{1}{6} + \cdots$.

$$\left[\text{Show that } \sum_1^\infty \left(\frac{1}{3n-2} + \frac{1}{3n-1} - \frac{1}{3n} \right) \text{ diverges.}\right]$$

7. $\dfrac{1}{a} + \dfrac{1}{a+1} - \dfrac{1}{a+2} + \dfrac{1}{a+3} + \dfrac{1}{a+4} - \dfrac{1}{a+5} + \cdots$ $(a > 0)$.

8. $\dfrac{1}{a} - \dfrac{1}{a+1} - \dfrac{1}{a+2} + \dfrac{1}{a+3} - \dfrac{1}{a+4} - \dfrac{1}{a+5} + \cdots$ $(a > 0)$.

9. $(1 - a) - (1 - a^{1/2}) + (1 - a^{1/3}) - (1 - a^{1/4}) + \cdots$ $(a > 0)$.

10. Show that the alternating series

$$\frac{\alpha}{\beta} - \frac{\alpha(\alpha+1)}{\beta(\beta+1)} + \frac{\alpha(\alpha+1)(\alpha+2)}{\beta(\beta+1)(\beta+2)} - \cdots$$

converges when $0 < \alpha < \beta$, diverges when $\alpha > \beta > 0$.
[Call the series $\Sigma(-1)^{n-1}a_n$. When $\alpha < \beta$ show that $a_{n+1} < a_n$ and $a_n < (\alpha/\beta)^n \to 0$; when $\alpha > \beta$, $a_{n+1} > a_n$ and $\{a_n\}$ is not null.]

11. A series $\displaystyle\sum_{n=1}^\infty a_n$ is said to be *summable by the method of Cesàro* (C1) to a "sum"
A when

$$\lim_{n \to \infty} \frac{s_1 + s_2 + \cdots + s_n}{n} = A$$

When Σa_n *converges* to A, its Cesàro-sum will also be A by (15.12). But Cesàro's method also assigns a "sum" to certain divergent series. Verify this fact by finding the Cesàro-sum of the following divergent series:

$$(a) \quad 1 - 1 + 1 - 1 + 1 - 1 + \cdots, \qquad A = \tfrac{1}{2};$$
$$(b) \quad 1 - 1 + 0 + 1 - 1 + 0 + \cdots, \qquad A = \tfrac{1}{3};$$
$$(c) \quad 1 + 0 - 1 + 1 + 0 - 1 + \cdots, \qquad A = \tfrac{2}{3}.$$

Observe that the "dilution" of a divergent series with zeros in general alters its Cesàro-sum, whereas such dilution has no effect on a convergent series.

35. Addition of Series. *Convergent series may be added term by term.*
More precisely, we have the

THEOREM. *If $\sum a_n \to s$ and $\sum b_n \to s'$,*

$$(1) \qquad \sum(a_n + b_n) \to s + s'.$$

Proof. If S_n is the sum of n terms of the series (1),

$$S_n = s_n + s'_n \to s + s'.$$

Corollary 1. If we remove the parentheses from series (1) we still have

$$a_1 + b_1 + a_2 + b_2 + a_3 + b_3 + \cdots \to s + s'.$$

For the partial sums of this series, namely, $s_n + s'_n$ and $s_n + s'_n + a_{n+1}$, approach $s + s'$ since $a_{n+1} \to 0$.

Corollary 2. If $\sum a_n$ and $\sum b_n$ converge absolutely, the same is true of $\sum(a_n + b_n)$; for

$$|a_n + b_n| \leq |a_n| + |b_n|.$$

Any convergent series may be multiplied by a constant: from $\sum a_n = s$ we have $\sum c a_n = cs$. The proof is obvious.

36. Rearrangement of Series. We may now extend Dirichlet's theorem (§ 23) to mixed series that are absolutely convergent.

THEOREM. *An absolutely convergent series converges to the same sum in whatever order the terms are taken.*

Proof. Since the theorem holds for positive series, it holds for both of the positive subseries formed as in § 33. Thus p and q are unchanged by a rearrangement of terms and the partial sums of the rearranged series again converge to $s = p - q$ (33.2).

In sharp contrast to this behavior, *the sum of a conditionally convergent series depends essentially on the order in which its terms are taken.* Even more, convergence may be changed to divergence by a suitable rearrangement of terms.

RIEMANN'S THEOREM. *A conditionally convergent series can be made to converge to any arbitrary value, or even to diverge, by a suitable rearrangement of its terms.*

Proof. The *theory* of the process (if not the practice) is utterly simple. Let σ be the arbitrary sum to which the rearranged series shall converge. Since both $p_n \to \infty$ and $q_n \to \infty$, we can first add just as many (possibly zero) positive terms as are needed to pass σ, then add just enough negative terms to bring the sum below σ, then add just enough positive terms to pass σ again, then just enough negative terms to repass σ again, and so on indefinitely. Since the given series converges, its terms form a null sequence, and we can approach σ by this process as closely as desired; for, if σ_n is the sum of n terms of the series thus constructed, $\sigma - \sigma_n$ is always numerically less than the last term added.

In order to make the given series *diverge* by a rearrangement of terms, carry out the same process when σ, instead of being constant, is assigned the value σ_n of a divergent sequence at the nth step. Thus we may make the series diverge to infinity by choosing $\sigma_n = n$; or we may force s_n to have two cluster points a, b $(a < b)$ by choosing $\sigma_1 = \sigma_3 = \cdots = b$, $\sigma_2 = \sigma_4 = \cdots = a$.

Example 1. The alternating series,

$$(1) \qquad s = 1 - \frac{1}{2} + \frac{1}{3} - \frac{1}{4} + \cdots = \sum_{1}^{\infty} (-1)^{n-1} \frac{1}{n},$$

converges conditionally to a sum s (Ex. 34.1). We shall see later that $s = \log 2$ (69.3).

The rearrangement of (1) in which a positive term is always followed by two negative terms converges to $s/2$:

$$(2) \qquad \tfrac{1}{2}s = 1 - \tfrac{1}{2} - \tfrac{1}{4} + \tfrac{1}{3} - \tfrac{1}{6} - \tfrac{1}{8} + \cdots,$$

To prove this we write (Theorem 21.1)

$$(3) \qquad s = \sum_{n=1}^{\infty} \left(\frac{1}{2n-1} - \frac{1}{2n} \right),$$

$$(4) \qquad \frac{1}{2}s = \sum_{n=1}^{\infty} \left(\frac{1}{4n-2} - \frac{1}{4n} \right).$$

On subtracting these series, we obtain

$$(5) \qquad \frac{1}{2}s = \sum_{n=1}^{\infty} \left(\frac{1}{2n-1} - \frac{1}{4n-2} - \frac{1}{4n} \right).$$

When the parentheses are removed from (5), we obtain (2). To justify this removal we must show that (2) converges (Theorem 21.2). Now (5) states that the partial sums s_n of (2) converge to $s/2$ when $n = 3m$. Moreover s_{3m+1} and s_{3m+2} differ from s_{3m} by terms that form a null sequence and also converge to $s/2$. Thus we have proved (2).

The rearrangement of (1) in which two positive terms are always followed by one negative term converges to $3s/2$:

$$(6) \qquad \tfrac{3}{2}s = 1 + \tfrac{1}{3} - \tfrac{1}{2} + \tfrac{1}{5} + \tfrac{1}{7} - \tfrac{1}{4} + \cdots.$$

To show this, we group the terms of (1) in fours,

$$s = \sum_{n=1}^{\infty} \left(\frac{1}{4n-3} - \frac{1}{4n-2} + \frac{1}{4n-1} - \frac{1}{4n} \right),$$

and add (4); we thus obtain

$$(7) \qquad \frac{3}{2}s = \sum_{n=1}^{\infty} \left(\frac{1}{4n-3} + \frac{1}{4n-1} - \frac{1}{2n} \right).$$

When the parentheses are removed, (7) becomes (6); their removal is justified as before.

We may rearrange (1) into a divergent series in many ways. For example, it can be shown that

$$1 - \frac{1}{2} - \frac{1}{4} - \frac{1}{6} - \cdots - \frac{1}{2^8} + \frac{1}{3} - \frac{1}{2^8 + 2} - \cdots - \frac{1}{2^{16}} + \frac{1}{5} - \cdots$$

'diverges to $- \infty$ (cf. Knopp, *loc. cit.*, p. 141). A simpler rearrangement of a convergent into a divergent series is given in the next example.

Example 2. The alternating series,

(8)
$$1 - \frac{1}{\sqrt{2}} + \frac{1}{\sqrt{3}} - \frac{1}{\sqrt{4}} + \cdots,$$

converges conditionally; for the monotone sequence $\{1/\sqrt{n}\} \to 0$ and $\Sigma 1/\sqrt{n} \to \infty$ (Ex. 26.1).

But the rearranged series, in which two positive terms are always followed by one negative, diverges:

(9)
$$\left(1 + \frac{1}{\sqrt{3}} - \frac{1}{\sqrt{2}}\right) + \left(\frac{1}{\sqrt{5}} + \frac{1}{\sqrt{7}} - \frac{1}{\sqrt{4}}\right) + \cdots \to \infty.$$

In fact, its general term

$$\frac{1}{\sqrt{4n-3}} + \frac{1}{\sqrt{4n-1}} - \frac{1}{\sqrt{2n}} > \frac{2}{\sqrt{4n}} - \frac{1}{\sqrt{2n}} = \left(1 - \frac{1}{\sqrt{2}}\right)\frac{1}{\sqrt{n}}.$$

PROBLEMS

1. Prove that $1 + \frac{1}{2} - \frac{1}{3} + \frac{1}{4} + \frac{1}{5} - \frac{1}{6} + \cdots$ diverges.

$$\left[\text{Show that } \sum_{n=1}^{\infty}\left(\frac{1}{3n-2} + \frac{1}{3n-1} - \frac{1}{3n}\right) \text{ diverges.}\right]$$

2. If $\displaystyle\sum_{n=1}^{\infty} \frac{1}{n^2} = s \left(s = \frac{\pi^2}{6}\right.$ as we shall see later), prove that

$$\sum_{n=1}^{\infty} \frac{(-1)^{n-1}}{n^2} = \frac{1}{2}s; \qquad 1 + \frac{1}{3^2} + \frac{1}{5^2} + \cdots = \frac{3}{4}s.$$

3. If $\displaystyle\sum_{n=1}^{\infty} \frac{1}{n^3} = s,$ prove that

$$\sum_{n=1}^{\infty} \frac{(-1)^{n-1}}{n^3} = \frac{3}{4}s; \qquad 1 + \frac{1}{3^3} + \frac{1}{5^3} \cdots = \frac{7}{8}s.$$

4. If $a > 0$, show that the alternating series

$$a - a^{1/2} + a^{1/3} - a^{1/4} + \cdots \text{ diverges; but that the series}$$

$$(a - a^{1/2}) + (a^{1/3} - a^{1/4}) + \cdots, \qquad a - (a^{1/2} - a^{1/3}) - (a^{1/4} - a^{1/5}) - \cdots,$$

obtained by inserting parentheses, both converge.

5. Prove that

$$\sum_{n=1}^{\infty} (-1)^{n-1} \frac{2n+1}{n(n+1)} = 1.$$

6. Prove that

$$1 - \tfrac{1}{2} - \tfrac{1}{4} - \tfrac{1}{6} - \tfrac{1}{8} + \tfrac{1}{3} - \tfrac{1}{10} - \tfrac{1}{12} - \tfrac{1}{14} - \tfrac{1}{16} + \cdots = 0.$$

[Write the series $\displaystyle\sum_{n=1}^{\infty} \left(\frac{1}{2n-1} - \frac{1}{8n-6} - \frac{1}{8n-4} - \frac{1}{8n-2} - \frac{1}{8n} \right)$ and compute $3s/2 - 3s/2$, using (4) and (7). Here s is given by (1).]

37. Multiplication of Series. The product of two series $\sum a_i$, $\sum b_j$ is obtained by forming all possible terms $a_i b_j$ and finding the limit of their sum as both i and j become infinite. We show first that the set $a_i b_j$ is *countable* by setting up a one-to-one correspondence between $a_i b_j$ and the positive integers. We first display the products $a_i b_j$ in a doubly infinite array:

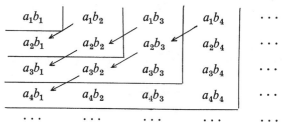

1. If we *count by triangles*, as shown by the arrows, we obtain a simple series,

$$(1) \qquad a_1 b_1 + (a_1 b_2 + a_2 b_1) + (a_1 b_3 + a_2 b_2 + a_3 b_1) + \cdots,$$

in which the index-sum $i + j$ in the successive groups is 2, 3, 4, $\cdots$. The series (1) is called the *Cauchy product* of $\sum a_i$ and $\sum b_j$. It is very important in dealing with *power series* $\sum a_n x^n$ (§ 38).

2. If we *count by squares*, as shown by the broken lines, we again obtain a simple series,

$$(2) \qquad a_1 b_1 + (a_1 b_2 + a_2 b_2 + a_2 b_1) +$$
$$+ (a_1 b_3 + a_2 b_3 + a_3 b_3 + a_3 b_2 + a_3 b_1) + \cdots,$$

in which $i, j \leq 1, 2, 3, \cdots$ in the successive groups. The advantage of this method of counting is that the sum S_n of the first n groups of (2) is

$$(3) \qquad S_n = (a_1 + a_2 + \cdots + a_n)(b_1 + b_2 + \cdots + b_n).$$

When expanded, S_n contains n^2 terms $a_i b_j$.

We can now prove the

THEOREM. *If the series $\sum a_i = A$ and $\sum b_i = B$ converge absolutely, their Cauchy product $\sum c_i = C$ converges absolutely and $AB = C$.*

Proof. If $\sum a_i$ and $\sum b_j$ are convergent *positive* series, series (2) evidently converges to AB; for, if A_n and B_n denote partial sums, we have from (3)

$$S_n = A_n B_n \to AB.$$

Denote the series (1) and (2) *without parentheses* by (1)′ and (2)′. Series (2)′ converges to AB; for, if m is a number between n^2 and $(n+1)^2$, the sum of its first m terms lies between S_n and S_{n+1}, which both converge to AB. Hence series (1)′, a mere rearrangement of (2)′, converges to AB (§ 36). Finally, by grouping the terms of series (1)′ to form (1), we see that series (1) converges to AB (§ 21).

Next let $\sum a_i$ and $\sum b_j$ denote mixed but absolutely convergent series. Then, $\sum |a_i|$ and $\sum |b_j|$ are convergent positive series, and, as we have just shown, the series

$$(1)' \qquad a_1 b_1 + a_1 b_2 + a_2 b_1 + a_1 b_3 + a_2 b_2 + a_3 b_1 + \cdots$$

converges absolutely; for $|a_i b_j| = |a_i| \, |b_j|$. Hence the mixed series (1)′ converges (§ 32); moreover, its sum will not be altered if we rearrange the series into

$$(2)' \qquad a_1 b_1 + a_1 b_2 + a_2 b_2 + a_2 b_1 + a_1 b_3 + \cdots + a_3 b_1 + \cdots.$$

But in *this* form the sum of n^2 terms is $(a_1 + \cdots + a_n)(b_1 + \cdots + b_n)$ and obviously converges to AB. Since series (1)′ converges to AB, the same is true of the Cauchy product (1).

The absolute convergence of both series is a *sufficient* condition for the convergence of their Cauchy product to AB, but is *not necessary*, as we see from

MERTENS' THEOREM. *If one of the two convergent series $\sum a_i$, $\sum b_j$ converges absolutely, their Cauchy product will converge to AB.*

Abel also proved that, if the Cauchy product of two series converges, it *must* converge to AB:

ABEL'S THEOREM. *If the series $\sum a_i$, $\sum b_j$ and their Cauchy product converge to A, B, and C, then $AB = C$.*

We refer the reader to Knopp, *Theory and Application of Infinite Series*, p. 321, for the proof of Mertens' theorem. Abel's theorem will be proved in § 185.

Example 1. The Cauchy product of the geometric series

$$1 + x + x^2 + x^3 + \cdots = \frac{1}{1-x}, \qquad 1 + y + y^2 + y^3 + \cdots = \frac{1}{1-y}$$

in which $|x| < 1$, $|y| < 1$, is

$$1 + (x + y) + (x^2 + xy + y^2) + (x^3 + x^2y + xy^2 + y^3) + \cdots = \frac{1}{(1 - x)(1 - y)}.$$

In particular, if $x = y$,

$$1 + 2x + 3x^2 + 4x^3 + \cdots = \frac{1}{(1 - x)^2}.$$

Example 2. The series for e (§ 19) is

$$e = 1 + \frac{1}{1!} + \frac{1}{2!} + \frac{1}{3!} + \cdots.$$

Hence, on multiplying the series by itself,

$$e^2 = 1 + \frac{(1 + 1)}{1!} + \frac{(1 + 2 + 1)}{2!} + \frac{(1 + 3 + 3 + 1)}{3!} + \cdots$$

$$= 1 + \frac{2}{1!} + \frac{2^2}{2!} + \frac{2^3}{3!} + \cdots.$$

On multiplying the series for e and e^2, we find

$$e^3 = 1 + \frac{3}{1!} + \frac{3^2}{2!} + \frac{3^3}{3!} + \cdots.$$

38. Power Series.   A series of the form

(1)
$$\sum_{n=0}^{\infty} a_n x^n = a_0 + a_1 x + a_2 x^2 + \cdots + a_n x^n + \cdots$$

is called a *power series*. As we advance in the series, the powers of x must *steadily increase*. Every power series converges when $x = 0$. When the sequence $\{a_n\}$ is given, for what *other* real values of x will $\sum a_n x^n$ converge? Except for two doubtful points, this fundamental question is answered by the

CAUCHY-HADAMARD THEOREM. *Let $\bar\rho$ be the greatest limit of the sequence $\{\sqrt[n]{|a_n|}\}$. Then the power series $\sum a_n x^n$*

   converges absolutely when $|x| < 1/\bar\rho$,

   diverges when $|x| > 1/\bar\rho$.

When $\bar\rho = 0$,   *the series converges for all values of x.*

When $\bar\rho = \infty$,   *the series converges only for $x = 0$.*

Proof. From the root test (theorem in § 28) the positive series $\sum |a_n| |x|^n$ will converge or diverge according as

$$\bar\rho |x| < 1 \quad \text{or} \quad > 1.$$

When $\bar\rho = 0$, the series converges for all values of x. Then $\underline\rho = \bar\rho$, and $\sqrt[n]{|a_n|} \to 0$.

When $\bar{\rho} = \infty$, the terms of the series are also unbounded when $x \neq 0$; for

(2) $$\overline{\lim} \, | \, a_n x^n \, |^{1/n} = \bar{\rho} \, | \, x \, |.$$

When $\bar{\rho} \neq 0$ and finite, the series (1) converges absolutely when $| \, x \, | < 1/\bar{\rho}$. When $| \, x \, | > 1/\bar{\rho}$, (2) shows that $\overline{\lim} \, | \, a_n x^n \, | > 1$, and the series will diverge.

Thus the Cauchy-Hadamard theorem determines an *interval of convergence.*

(3) $$-1/\bar{\rho} < x < 1/\bar{\rho}$$

for any power series $\sum a_n x^n$, and this interval is symmetric with respect to $x = 0$. Outside of this interval the series diverges, but, at its end points $x = \pm 1/\bar{\rho}$, the behavior is in doubt. At these points we must apply tests more subtle than the root test in order to reach definite conclusions.

The number $1/\bar{\rho}$ is called the *radius of convergence* of the power series $\sum a_n x^n$. If $\sqrt[n]{| \, a_n \, |}$ approaches a limit ρ, the radius of convergence is $1/\rho$. It is usually simpler, however, to compute the limit of the ratio $| \, a_{n+1}/a_n \, |$ when it exists; for, if

(4) $$\left| \frac{a_{n+1}}{a_n} \right| \to \tau, \quad \text{then} \quad \sqrt[n]{| \, a_n \, |} \to \tau. \tag{29.3}$$

If τ exists, we may apply the d'Alembert test to the series $\sum | \, a_n \, | \, | \, x \, |^n$; then, since

$$\lim \left| \frac{a_{n+1} x^{n+1}}{a_n x^n} \right| = \tau \, | \, x \, |,$$

the series will converge or diverge according as $| \, x \, | < 1/\tau$ or $| \, x \, | > 1/\tau$. Thus the radius of convergence is $1/\tau$.

Example 1. $\displaystyle\sum_{n=0}^{\infty} \frac{x^n}{n+1}$. Since the absolute term ratio $\dfrac{n}{(n+1)} \, | \, x \, | \to | \, x \, |$, the series converges when $| \, x \, | < 1$. When $x = 1$, the series (harmonic) diverges; when $x = -1$, the alternating series converges. The *precise* interval of convergence is thus $-1 \leqq x < 1$.

Example 2. $\displaystyle\sum_{n=0}^{\infty} n! \, x^n$. The absolute term ratio $(n + 1) \, | \, x \, | \to \infty$; hence the series diverges except when $x = 0$. Since $\sqrt[n]{n!} \to \infty$ (Ex. 29.3), we arrive at the same conclusion from the root test.

Example 3. $a + bx + ax^2 + bx^3 + \cdots (0 < a < b)$. The absolute term ratio has the greatest and least limits $b \, | \, x \, | \, /a$ and $a \, | \, x \, | \, /b$; hence the series converges when $| \, x \, | < a/b$, diverges when $| \, x \, | > b/a$ (§ 29). Thus the ratio test gives no information when $a/b < |x| < b/a$.   But, since $\sqrt[n]{a} \to 1$, $\sqrt[n]{b} \to 1$ (15.10), the root test shows that the series converges when $|x| < 1$.   The series diverges when $x = \pm 1$ for the nth term does not approach zero; hence $-1 < x < 1$ is the precise interval of convergence.

Example 4. $\displaystyle\sum_{n=0}^{\infty} \frac{(n+1)^n}{n!} x^n$.   The ratio

$$\frac{a_{n+1}}{a_n} = \frac{(n+1)^n}{n^n} |x| = \left(1 + \frac{1}{n}\right)^n |x| \to e |x| ;$$

hence, the series converges when $|x| < 1/e$.   When $x = \pm 1/e$ the series diverges since $(n+1)^n/n!e^n \to 1$ (Prob. 15.7).

39. Binomial Series.   In this series

$$(1)\quad 1 + kx + \frac{k(k-1)}{2!} x^2 + \frac{k(k-1)(k-2)}{3!} x^3 + \cdots + \binom{k}{n} x^n + \cdots ,$$

the *binomial coefficients* $\binom{k}{n}$ are defined as

$$(2)\quad \binom{k}{0} = 1, \qquad \binom{k}{n} = \frac{k(k-1)\cdots(k-n+1)}{n!} \qquad (n \geqq 1).$$

When k is a positive integer, the series terminates after $k+1$ terms and gives the familiar binomial expansion of $(1 + x)^k$.   When k is not a positive integer or zero, we have an infinite power series, whose convergence we proceed to investigate.

The term ratio,

$$\frac{k-n+1}{n} x = \left(\frac{k+1}{n} - 1\right) x,$$

approaches $|x|$ in absolute value; hence the series converges absolutely when $|x| < 1$, diverges when $|x| > 1$.

When $x = -1$, the term ratio,

$$(3)\qquad\qquad \frac{a_{n+1}}{a_n} = 1 - \frac{k+1}{n},$$

becomes and remains positive when $n > k + 1$.   After this point the terms of the series have the same sign so that the tests for a positive series may be applied.   Using Raabe's test, we have

$$R_n = n\left(1 - 1 + \frac{k+1}{n}\right) = k+1;$$

hence, the series

$$(4)\qquad \sum_{n=0}^{\infty} (-1)^n \binom{k}{n} \quad \begin{array}{ll} \text{converges if} & k > 0; \\ \text{diverges if} & k < 0. \end{array}$$

Remembering that the terms in (4) eventually keep the same sign, we also have proved that the positive series,

(5) $$\sum_{n=0}^{\infty} \left| \binom{k}{n} \right| \quad \begin{array}{l} \text{converges if} \quad k > 0; \\ \text{diverges if} \quad k < 0. \end{array}$$

In view of (3) the terms of this positive series steadily decreases when $k > 0$; hence, from (26.6)

(6) $$n \binom{k}{n} \to 0 \quad \text{when} \quad k > 0.$$

When $x = 1$, the term ratio,

$$\frac{a_{n+1}}{a_n} = \frac{k+1}{n} - 1;$$

hence,

$$\left| \frac{a_{n+1}}{a_n} \right| > 1 \quad \text{if} \quad k+1 < 0; \qquad \left| \frac{a_{n+1}}{a_n} \right| < 1 \quad \text{if} \quad k+1 > 0.$$

If $k + 1 \leq 0$, $a_{n+1} \geq a_n$, and the series $\sum_0^{\infty} \binom{k}{n}$ diverges since $\lim a_n \neq 0$.

If $k + 1 > 0$, a_{n+1}/a_n becomes and remains negative when $n > k + 1$. The terms then alternate in sign and, since $|a_{n+1}| < |a_n|$, steadily decrease in absolute value. Such a series will converge (§ 34) if

(7) $$a_n = \binom{k}{n} \to 0, \qquad k+1 > 0.$$

But from (6) we have

$$(n+1) \binom{k+1}{n+1} = (k+1) \binom{k}{n} \to 0, \qquad k+1 > 0,$$

and, hence, (7) is established. Consequently,

(8) $$\sum_{n=0}^{\infty} \binom{k}{n} \quad \begin{array}{l} \text{diverges if} \\ \text{converges conditionally if} \\ \text{converges absolutely if} \end{array} \quad \begin{array}{l} k \leq -1; \\ -1 < k < 0; \\ k > 0. \end{array}$$

The *character* of the convergence follows from (5).

THEOREM. *The binomial series* $\sum_{n=0}^{\infty} \binom{k}{n} x^n$ *converges absolutely when* $|x| < 1$, *diverges when* $|x| > 1$.

When $x = 1$, *the series*

> *converges absolutely if* $\quad\quad k > 0$;
>
> *converges conditionally if* $\quad -1 < k < 0$;
>
> *diverges when* $\quad\quad\quad\quad k \leq -1$.

When $x = -1$, *the series*

> *converges absolutely if* $\quad k > 0$;
>
> *diverges when* $\quad\quad\quad k < 0$.

We shall see later that, when the binomial series converges, it has $(1 + x)^k$ as its sum.

40. Complex Sequences. A sequence of complex numbers $z_n = x_n + iy_n$ is said to converge to a limit $\zeta = \xi + i\eta$ when

$$(1) \quad\quad\quad |z_n - \zeta| < \varepsilon \quad \text{when} \quad n > N.$$

Since

$$|z_n - \zeta| = |x_n - \xi + i(y_n - \eta)| \leq |x_n - \xi| + |y_n - \eta|$$

from the triangle inequality (9.5), the complex sequence $\{z_n\}$ will converge to ζ when and only when the real sequences $\{x_n\}$ and $\{y_n\}$ converge to ξ and η (§ 15). Thus,

$$(2) \quad\quad x_n \to \xi, \quad y_n \to \eta \quad \text{and} \quad z_n \to \xi + i\eta$$

are completely equivalent statements.

The complex sequence $\{z_n\}$ is null when and only when the real sequence $\{|z_n|\}$ is null; that is, when *both* real sequences $\{x_n\}$, $\{y_n\}$ are null. For example, from (15.7)

$$(3) \quad\quad\quad z^n \to 0 \quad \text{when} \quad |z| < 1.$$

Cauchy's convergence criterion (§ 17) also applies to complex sequences; for, if we choose N so that

$$|x_m - x_n| < \varepsilon, \quad\quad |y_m - y_n| < \varepsilon, \quad\quad m, n > N,$$

then

$$|z_m - z_n| < |x_m - x_n| + |y_m - y_n| < 2\varepsilon, \quad\quad m, n > N.$$

41. Complex Series. If a_n and b_n are real, the series of complex terms $\sum(a_n + ib_n)$ is said to converge to a sum $\alpha + i\beta$ when the complex sequence,

$$(1) \quad\quad\quad S_n = \sum_{1}^{n} (a_n + ib_n) \to \alpha + i\beta.$$

If A_n and B_n denote partial sums of the real series $\sum a_n$ and $\sum b_n$, $S_n = A_n + iB_n$; hence,

$$(2) \quad\quad\quad S_n \to \alpha + i\beta \quad \text{and} \quad A_n \to \alpha, \quad B_n \to \beta$$

are equivalent statements. A complex series will converge when and only when its real series both converge.

A complex series $\sum z_n$ is said to *converge absolutely* when the positive series $\sum |z_n|$ converges. Thus the absolute convergence of complex series is reduced to the convergence of positive series, and this may be tested by our previous criteria. Moreover, if $z_n = a_n + ib_n$,

$$|z_n| = |a_n + ib_n| \leq |a_n| + |b_n|;$$
$$|z_n| \geq |a_n|, \qquad |z_n| \geq |b_n|.$$

Hence, if $\sum z_n$ converges absolutely, the real series $\sum a_n$, $\sum b_n$ also converge absolutely, and conversely. Thus Theorem 32 may be extended:

The absolute convergence of a complex series implies its convergence.

When the series $\sum z_n$ converges absolutely, its terms may be rearranged without altering the sum; for by Theorem 36.1 the corresponding real series have this property.

Convergent complex series may be added term by term. As to multiplication, Theorem 37 applies as well to complex series; in the proof we need only change the word "mixed" to complex: thus,

If the complex series $\sum a_n = A$ and $\sum b_n = B$ converge absolutely, their Cauchy product $\sum c_n = C$ converges absolutely and $AB = C$.

The complex power series,

$$(3) \qquad a_0 + a_1 z + a_2 z^2 + \cdots,$$

converges absolutely when the real series,

$$(4) \qquad |a_0| + |a_1| r + |a_2| r^2 + \cdots, \qquad r = |z|,$$

converges. Hence, if

$$(5) \qquad \bar{\rho} = \overline{\lim} \, |a_n|^{1/n},$$

the Cauchy-Hadamard Theorem 38 shows that the series (3) converges absolutely when $r < 1/\bar{\rho}$, that is, within the circle $r = 1/\bar{\rho}$. Outside of this circle the series (3) diverges; for

$$\overline{\lim} \, |a_n z^n| = |\bar{\rho} r|^n > 1.$$

When $\bar{\rho} = \infty$, the *circle of convergence* $r = 1/\bar{\rho}$ shrinks to the point $z = 0$. When $\bar{\rho} = 0$, the circle expands into the whole complex plane, and the series converges absolutely for all values of z. In brief, the Cauchy-Hadamard theorem holds for complex as well as real series; in the statement in § 38 we need only replace x by z.

We call $1/\bar{\rho}$ the *radius of convergence* of series (3) and also of the corresponding real series (4); the former converges within the *circle* $r = 1/\bar{\rho}$, the latter within the interval that the circle cuts from the axis of reals.

Example 1. The *complex geometric series*,

$$(6) \qquad a + az + az^2 + \cdots + az^{n-1} + \cdots .$$

can be summed as in § 22. Thus,

$$(7) \qquad S_n = a \frac{1 - z^n}{1 - z}, \qquad z \neq 1;$$

and from (40.3)

$$(8) \qquad S_n \to S = \frac{a}{1 - z} \quad \text{when} \quad |z| < 1.$$

The circle of convergence is $r = 1$; since, $\sqrt[n]{|a|} \to 1$ (15.10), this agrees with the Cauchy-Hadamard theorem.

For example,

$$(9) \qquad 1 - z^2 + z^4 - z^6 + \cdots = \frac{1}{1 + z^2}, \qquad |z| < 1.$$

Here the sequence $\sqrt[n]{|a_n|}$ is $1, 0, 1, 0, \cdots$; $\bar{\rho} = 1$, and $r = 1$ is the circle of convergence. The series evidently diverges at all points *on* the circle since $|z^n| = 1$. Note that $1/(1 + z^2)$ becomes infinite when $z = \pm i$, giving a reason why $r = 1$. For the real series $1 - x^2 + x^4 - x^6 + \cdots$, the sum $1/(1 + x^2)$ becomes $1/2$ at ± 1, and its behavior does not account for the divergence at ± 1. In fact, Euler put $1 - 1 + 1 - 1 + \cdots = 1/2$!

Example 2. The series

$$(10) \qquad \sum_{n=1}^{\infty} \frac{z^n}{n} = z + \frac{z^2}{2} + \frac{z^3}{3} + \cdots$$

has the circle of convergence $r = 1$; for $\sqrt[n]{n} \to 1$ (15.11). At $z = 1$ the series diverges; but it converges at all other points of the circle. Thus, when $z = i$, the first four powers are $i, -1, -i, 1$, and the higher powers repeat in this cycle; hence,

$$\sum_{n=1}^{\infty} \frac{i^n}{n} = -\left(\frac{1}{2} - \frac{1}{4} + \frac{1}{6} - \cdots \right) + i \left(1 - \frac{1}{3} + \frac{1}{5} - \cdots \right).$$

Since both real series converge (§ 34,) the series (10) converges when $z = i$ but *not absolutely*. At other points of the circle we have

$$z = \cos \theta + i \sin \theta, \qquad z^n = \cos n\theta + i \sin n\theta. \qquad (9.8)$$

and, hence,

$$\sum_{n=1}^{\infty} \frac{z^n}{n} = \sum_{1}^{\infty} \frac{\cos n\theta}{n} + i \sum_{n=1}^{\infty} \frac{\sin n\theta}{n}.$$

The real series are *Fourier series* (which form the subject matter of Chapter 12) and are shown to converge when $0 < \theta < 2\pi$ in Ex. 154.2. When $\theta = 0$, the sine series is identically zero while the cosine series becomes the divergent harmonic series.

PROBLEMS

1. Show that the series

$$1 + \frac{x}{1!} + \frac{x^2}{2!} + \frac{x^3}{3!} + \cdots, \qquad 1 - \frac{x}{1!} + \frac{x^2}{2!} - \frac{x!}{3!} + \cdots$$

converge for all values of x. Find their Cauchy product.

2. Find the Cauchy product of $1 + x + x^2 + \cdots$ and $1 - 3x + 3x^2 - x^3$.

3. Find the Cauchy product of $1 + x + x^2 + x^3 \cdots$ and $1 - x + x^2 - x^3 + \cdots$. Check the result when $|x| < 1$.   [Cf. (22.3).]

4. Prove that, when $|x| < 1$,

$$(1 + x^2)^{-2} = 1 - 2x^2 + 3x^4 - 4x^6 + \cdots . \quad [\text{Cf. (22.3).}]$$

5. Find the *precise* interval of convergence for the series:

(a) $\displaystyle\sum_{n=1}^{\infty} (-1)^{n-1} \frac{(x-1)^n}{n}$;   (b) $\displaystyle\sum_{n=1}^{\infty} \frac{1}{n}\left(\frac{x-1}{x}\right)^n$;

(c) $\displaystyle\sum_{n=0}^{\infty} \frac{1}{(2n+1)(2x+1)^{2n+1}}$;   (d) $\displaystyle\sum_{n=0}^{\infty} \frac{1}{2n+1}\left(\frac{x-1}{x+1}\right)^{2n+1}$;

(e) $\displaystyle\sum_{n=0}^{\infty} \frac{(n!)^2}{(2n)!} x^n$, $(0! = 1)$.   (f) $\displaystyle\sum_{n=0}^{\infty} \binom{2n}{n} x^n$.

6. Prove that $(1 - x)^{-p-1} = \displaystyle\sum_{n=0}^{\infty} \binom{n+p}{p} x^n$, $\ |x| < 1$.

7. Establish the following results for the *hypergeometric series*

(11) $F(\alpha, \beta, \gamma; x) = 1 + \dfrac{\alpha\beta}{1 \cdot \gamma} x + \dfrac{\alpha(\alpha+1)\beta(\beta+1)}{1 \cdot 2\gamma(\gamma+1)} x^2 + \cdots$

when $\alpha, \beta, \gamma \neq 0, -1, -2. \ldots$.

(i) $|x| < 1$, Conv. abs.

(ii) $|x| > 1$, Div.

(iii) $x = 1$: $r = \gamma - \alpha - \beta$.

       $r > 0$, Conv. abs.; $r \leq 0$, Div.

(iv) $x = -1$: $r = \gamma - \alpha - \beta$.

       $r > 0$, Conv. abs.; $-1 < r \leq 0$, Conv. cond.;

       $r \leq -1$, Div.

8. Prove that the *binomial series* (39.1) is given by

(12) $F(-k, \beta, \beta; -x) = 1 + kx + \dbinom{k}{2}x^2 + \dbinom{k}{3}x^3 + \cdots, \beta \neq 0$.

Hence, show that the convergence tests for the binomial series follow from those in Prob. 7 if we put $r = k$ and interchange the results for $x = 1$ and $x = -1$. Note that $\alpha \neq 0, -1, -2, \cdots$ (Prob. 7) entails $k \neq 0, 1, 2, \cdots$.

9. From the series $(1 - z)^{-1} = \displaystyle\sum_{n=0}^{\infty} z^n$, $|z| < 1$, show that

$$(1 - z)^{-2} = \sum_{n=0}^{\infty} (n+1)z^n; \qquad (1 - z)^{-3} = \sum_{n=0}^{\infty} \frac{1}{2}(n+1)(n+2)z^n.$$

10. If $f(z) = \sum_{n=0}^{\infty} \dfrac{z^n}{n!}$, prove that $f(z)f(z') = f(z+z')$ for arbitrary z and z'.

11. If $f(z) = \sum_{n=0}^{\infty} a_n z^n$, $|z| < 1$, prove that

$$\frac{f(z)}{(1-z)} = \sum_{n=0}^{\infty} b_n z^n, \qquad |z| < 1,$$

where $b_n = a_0 + a_1 + \cdots + a_n$.

12. Show that $\sum_{n=1}^{\infty} \dfrac{z^n}{n^2}$ converges absolutely at all points of its circle of convergence.

13. Show that the series $\sum_{n=0}^{\infty} a^n z^n \; (a > 0)$ does not converge anywhere on its circle of convergence.

14. Find a point on the circle of convergence of the series $\sum_{n=1}^{\infty} (-1)^{n-1} \dfrac{z^n}{n}$ at which it diverges. Write out the corresponding real series at any point $z = \cos\theta + i\sin\theta$ of the circle $r = 1$.

42. Sequences and Series. Every infinite and bounded set of points has a least and a greatest cluster point; if they coincide at λ, λ is a *limit point* of the set; then all but a finite number of points lie in any neighborhood of λ, $\lambda - \varepsilon < x < \lambda + \varepsilon$, however small ε may be.

A *sequence* $\{x_n\}$ is a countable point set. If $\{x_n\}$ is bounded, it will have a *least limit* $\underline{\xi}$ and *greatest limit* $\bar{\xi}$. When $\underline{\xi} = \bar{\xi} = \xi$, $\{x_n\}$ *is said to converge to a limit* ξ; then

$$(1) \qquad\qquad |x_n - \xi| < \varepsilon \quad \text{when} \quad n > N.$$

If $\xi = 0$, the sequence is *null*.

The condition (independent of ξ)

$$(2) \qquad\qquad |x_m - x_n| < \varepsilon \quad \text{when} \quad m, n > N$$

is necessary and sufficient for the convergence of $\{x_n\}$.

A *bounded* monotone sequence always converges. If $\{x_n\}$ is monotone-increasing, $\{y_n\}$ monotone-decreasing, and $\{y_n - x_n\}$ positive and null, the intervals form a *nest* and determine a unique real number, the common limit of $\{x_n\}$ and $\{y_n\}$.

The *infinite series* $\sum a_n = a_1 + a_2 + a_3 + \cdots$ is said to *converge to the limit* s (its "sum") if the partial sums $s_n = a_1 + \cdots + a_n \to s$. The conditions (1) and (2) now become

$$(1)' \qquad\qquad |s_n - s| < \varepsilon, \qquad\qquad n > N;$$
$$(2)' \qquad\qquad |s_m - s_n| < \varepsilon, \qquad\qquad m, n > N.$$

A series that does not converge ($\mathscr{C}$) is said to *diverge* ($\mathscr{D}$); then $\{s_n\}$ is unbounded or has no limit point. If $\sum a_n$ converges, $a_n \to 0$; hence, if $\lim a_n \neq 0$, the series diverges.

For a positive series, $\{s_n\}$ is monotone-increasing; it will converge if s_n is bounded, diverge to infinity if s_n is unbounded.

Positive series ($a_n > 0$) may be tested by comparison with standard series such as

(3) $\quad (a_1 - a_2) + (a_2 - a_3) + (a_3 - a_4) + \cdots = a_1 - \lim a_n;$

(4) $\quad a + ar + ar^2 + \cdots = \dfrac{a}{1-r} \quad \text{if} \ \ |r| < 1, \qquad \mathscr{D} \ \text{if} \ r \geqq 1;$

(5) $\quad 1 + \dfrac{1}{2^p} + \dfrac{1}{3^p} + \cdots = \zeta(p) \quad \text{if} \ \ p > 1, \qquad \mathscr{D} \ \text{if} \ p \leqq 1;$

(6) $\quad \dfrac{1}{2(\log 2)^p} + \dfrac{1}{3(\log 3)^p} + \cdots = \lambda(p) \quad \text{if} \ \ p > 1, \qquad \mathscr{D} \ \text{if} \ p \leqq 1.$

When $p = 1$, (5) and (6) give the harmonic and Abel's series, both divergent.

Basic comparison test: If $0 < a_n \leqq A_n$,
$$\sum a_n \ \ \mathscr{C} \ \ \text{if} \ \ \sum A_n \ \ \mathscr{C}; \qquad \sum A_n \ \ \mathscr{D} \ \ \text{if} \ \ \sum a_n \ \ \mathscr{D}.$$

Limit Test: If $a_n/A_n \to \lambda > 0$, both series $\mathscr{C}$ or both $\mathscr{D}$.

Root Test: $\rho_n = \sqrt[n]{a_n};$
$$\bar\rho < 1 \ \ \mathscr{C}; \qquad \underline\rho > 1 \ \ \mathscr{D}.$$

Ratio Test: $\tau_n = \dfrac{a_{n+1}}{a_n};$
$$\bar\tau < 1 \ \ \mathscr{C}; \qquad \underline\tau > 1 \ \ \mathscr{D}.$$

Raabe's Test: $R_n = n\left(1 - \dfrac{a_{n+1}}{a_n}\right);$
$$\underline R > 1 \ \ \mathscr{C}; \qquad \bar R < 1 \ \ \mathscr{D}.$$

Bertrand's Test: $B_n = \log n \cdot (R_n - 1);$
$$\underline B > 1 \ \ \mathscr{C}, \qquad \bar B < 1 \ \ \mathscr{D}.$$

A *mixed series* $\sum a_n$ of positive and negative terms is said to *converge absolutely* if $\sum |a_n|$ converges. Absolute convergence implies convergence. But $\sum a_n$ may converge when $\sum |a_n|$ diverges; then $\sum a_n$ is said to *converge conditionally*.

An *alternating series* whose terms in absolute value form a monotone null sequence is convergent.

The terms of a convergent series may be grouped in parentheses without altering its sum; and, if absolutely convergent, the terms may be rearranged in any order. The sum of a conditionally convergent series depends essentially on the order of its terms; by a suitable rearrangement of its terms it may be made to converge to an arbitrary value or even to diverge.

As for the algebra of convergent series, if $\sum a_n = A$, $\sum b_n = B$, then

$$\sum c a_n = cA, \qquad \sum(a_n + b_n) = A + B.$$

Moreover, if *one* of the two series converges absolutely, their Cauchy product

$$\sum_{n=1}^{\infty} (a_1 b_n + a_2 b_{n-1} + \cdots + a_n b_1) = AB.$$

If *both* series converge absolutely, the product series does likewise.

The complex series $\sum z_n$ converges absolutely if the positive series $\sum |z_n|$ converges. If $z_n = a_n + ib_n$, the series $\sum z_n$ converges absolutely when and only when the real series $\sum |a_n|$ and $\sum |b_n|$ converge.

A real or complex *power series* $\sum a_n z^n$ has the *radius of convergence* $1/\bar{\rho}$ where

$$\bar{\rho} = \overline{\lim} \sqrt[n]{|a_n|}.$$

When $\bar{\rho} = 0$, the series converges for all values of z; when $\bar{\rho} = \infty$, the series converges only for $z = 0$. Inside of the *circle of convergence* $(r = 1/\bar{\rho})$, the convergence is absolute; outside, the series diverges; on the circumference, the behavior must be determined by special tests.

CHAPTER 3

Functions of a Real Variable

43. Functions. Let the real variable x range over a certain point set R. The set R may consist of discrete points, such as the positive integers $1, 2, 3, \cdots$. However, we shall usually consider a continuous range of x, say the *interval* between two numbers a and b. The interval is said to be *closed* if the end points are included, *open* if they are excluded; thus $a \leq x \leq b$ is a closed interval, while $a < x < b$ is open. The interval may also be *open at one end* as in $a < x \leq b$ or $a \leq x < b$. Finally the interval may be unbounded at either end or both. Such intervals are written

$$a < x < \infty, \qquad -\infty < x \leq b, \qquad -\infty < x < \infty,$$

where the notation shows whether a finite end point is included or excluded. In the last case x ranges over all real numbers: the *real continuum*. We shall also denote intervals such as

$$a \leq x \leq b, \ a < x \leq b, \ a \leq x < b, \ a < x < \infty, \ -\infty < x < \infty$$

by $(a, b), \qquad (a+, b), \qquad (a, b-), \qquad (a+, \infty), \qquad (-\infty, \infty),$

respectively.

Consider now two real variables x and y related by some rule; then, if a *single* definite value of y corresponds to every value of x in the range R, we say that y is a *function* of x over this range and write $y = f(x)$, where f stands for the prescribed rule. The letter used to specify the functional relation is optional; f and F (the initial of "function") are often used, but $g(x)$, $h(x)$, $\varphi(x)$, etc. are all permissible. But in any given problem $f(x)$ always denotes the same relation; a different relation between x and y must be denoted by a different letter.

A sequence $\{y_n\}$ is a function of the integral variable $n = 1, 2, 3, \cdots$; but the index notation is commonly used instead of $y = f(n)$.

The definition of a function does not presuppose an analytic expression connecting the *independent variable* x and the *dependent variable* y. Thus, over the range of positive integers n, y is a function of n if

$$y = \begin{cases} 0 & \text{if} \quad n \text{ is prime,} \\ 1 & \text{if} \quad n \text{ is composite.} \end{cases}$$

This function is well defined, but no analytic expression for it is known. On the other hand, the function

$$y = \begin{cases} 0 & \text{when} \quad n \text{ is even} \\ 1 & \text{when} \quad n \text{ is odd} \end{cases} = \frac{1}{2}\left[1 - (-1)^n \right].$$

When y has just one value c for all values of x, we have the *constant function* $y = c$. In *Dirichlet's function* y has just two values:

(1) $$y = \begin{cases} a & \text{when} \quad x \text{ is rational,} \\ b & \text{when} \quad x \text{ is irrational.} \end{cases}$$

It would seem unlikely that this function, which jumps infinitely often from a to b and back in any interval however small, could have an analytic form; yet it has, and one that is fairly simple (Prob. 44.5).

The essential thing about a function is that, when x is given a value within its range, y is uniquely determined. Some authors admit several values of y to correspond to a single value of x and distinguish between *single-valued* and *multivalued* functions. From our point of view a "multivalued function" is a collection of distinct functions, each of which determines y uniquely when x is given. Thus, if $y^2 = x$, y has two real values when $x > 0$; these are represented by the functions $y = \sqrt{x}$ and $y = -\sqrt{x}$. In this connection note again that $\sqrt{x}$ or $x^{1/2}$ denotes the positive square root; and, in general, when z is complex, $\sqrt[n]{z}$ or $z^{1/n}$ denotes the *principal* nth root of x (§ 9).

If we plot the points x, $y = f(x)$, using rectangular coordinates, we obtain the graph of the function. If, as in a sequence $\{y_n\}$, x ranges over the positive integers n, the graph of $y = f(n)$ is a set of discrete points in the xy-plane. But, if x varies over an interval (a, b), the graph of $y = f(x)$ is a *curve* over this interval having the parametric equations

$$x = t, \qquad y = f(t).$$

Such a curve is cut in only one point by a vertical line, for only one value of y corresponds to a given x.

When $f(x) = f(-x)$, the function is said to be *even*; its graph is then symmetric with respect to the y-axis.

When $f(x) = -f(-x)$, the function is said to be *odd*; its graph is then symmetric with respect to the origin.

The function $y = x^n$ is even or odd according as n is an even or an odd integer.

Example 1. The relation $n! = 1 \cdot 2 \cdot 3 \cdots n$ defines the *factorials* when n ranges over the positive integers. We shall see in § 191 that the *gamma function* $\Gamma(x)$, which is defined when $x > 0$, has the value $n!$ when $x = n + 1$ (191.4). Since $\Gamma(1) = 1$ (§ 191), this suggests that the customary definition $0! = 1$.

Example 2. If $f(x) = x/(x^2 - 1)$,

$$f(2) = \frac{2}{3} \qquad f(0) = 0, \qquad f(-3) = -\frac{3}{8};$$

but $f(1)$ and $f(-1)$ are not defined since division by zero is meaningless. This function is defined in any interval, such as $-1 < x < 1$ or $1 < x < \infty$, which does not include -1 and 1.

Example 3. *Polynomials.* The function

$$y = a_0 x^n + a_1 x^{n-1} + \cdots + a_{n-1} x + a_n$$

is defined over the entire real continuum $-\infty < x < \infty$. If the polynomial contains only even powers of x (the constant term $a_n = a_n x^0$ contains an even power) the function is even; if it contains only odd powers of x, the function is odd.

The *linear* and *quadratic* functions,

$$y = ax + b, \qquad y = ax^2 + bx + c,$$

are important special cases. The graph of the former is a straight line; of the latter a parabola with axis vertical.

Example 4. The equation $x^2 + y^2 = 1$ represents a unit circle about the origin. If we solve it for y, we get two functions,

$$y = \sqrt{1 - x^2}, \qquad y = -\sqrt{1 - x^2},$$

defined in the interval $-1 \leq x \leq 1$. Their graphs are the upper and lower halves of the circle.

Example 5. *Rational Functions* are the quotients of two polynomials:

$$y = \frac{P}{Q} = \frac{a_0 x^n + a_1 x^{n-1} + \cdots + a_n}{b_0 x^m + b_1 x^{m-1} + \cdots + b_m},$$

They are defined at all points where the denominator $Q \neq 0$. At the m roots of $Q = 0$, the function is not defined. When Q reduces to a constant, the rational function is a polynomial.

A rational function is even if P and Q are both even or both odd; it is odd when one polynomial is even, the other odd.

44. Limit of a Continuous Variable. Consider a function $f(x)$ of the real variable x defined over an interval R that includes $x = \xi$. Let $\{x_n\}$ be any sequence of numbers all different from ξ but having ξ as a limit. Then, if the sequence $\{f(x_n)\} \to A$ as $\{x_n\} \to \xi$, no matter how $\{x_n\}$ is chosen, we say that $f(x)$ tends to the limit A as x approaches ξ; and we write

(1) $$\lim_{x \to \xi} f(x) = A.$$

The limit approached by $f(x)$ does not depend at all on the value of $f(\xi)$, and, in fact, $f(\xi)$ need not even be defined. For example,

(2)
$$\lim_{x \to 0} \frac{\sin x}{x} = 1,$$

although $\sin x/x$ is not defined when $x = 0$. This well-known limit, used in finding the derivative of $\sin x$, follows quite simply from Fig. 44 in which $OP = 1$, and arc $AP = x$ (radians). Since the area of the sector OAP lies between the triangular areas OAP and OAB,

$$\frac{1}{2} \sin x < \frac{1}{2} x < \frac{1}{2} \tan x \quad \text{or} \quad 1 < \frac{x}{\sin x} < \frac{1}{\cos x}.$$

Thus $(\sin x)/x$ lies between $\cos x$ and 1, and, as $x \to 0$, $\cos x \to 1$ and $x/\sin x \to 1$.

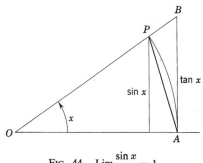

Fig. 44. $\displaystyle\lim_{x \to 0} \frac{\sin x}{x} = 1$

We may avoid the use of sequences in establishing continuity by using the following

THEOREM 1. *In order that $f(x)$ tend to the limit A as x approaches ξ, it is necessary and sufficient that for every $\varepsilon > 0$ we can find a number $\delta > 0$ such that*

(3)
$$|f(x) - A| < \varepsilon \quad \text{when} \quad 0 < |x - \xi| < \delta.$$

Proof. The condition (3) is *sufficient* to ensure (1). For, if $\{x_n\}$ is any sequence not including ξ but having ξ as a limit,

$$0 < |x_n - \xi| < \delta \quad \text{when} \quad n > N,$$

and from (3)

$$|f(x_n) - A| < \varepsilon \quad \text{when} \quad n > N;$$

that is, $f(x_n) \to A$.

Condition (3) is also a necessary consequence of (1). For, if (1) holds and there is an ε for which no corresponding δ can be found, for every null sequence δ_1, δ_2, $\cdots$ there is at least one x in the interval $|x - \xi| < \delta_n$, say x_n, for which $|f(x_n) - A| \geqq \varepsilon$. Now the sequence $\{x_n\} \to \xi$, but the numbers $|f(x_n) - A|$ have ε as a lower bound and cannot form a null sequence. But this contradicts the hypothesis that $f(x_n) \to A$ whenever $x_n \to \xi$.

A simple but important consequence of (3) is given in

THEOREM 2. *If* $\lim\limits_{x \to \xi} f(x) = A \neq 0$, *then* $f(x)$ *will have the same sign as A in some interval about ξ with ξ excluded.*

Proof. Take $\varepsilon = |A|$, and choose δ so that

$$|f(x) - A| < |A| \quad \text{when} \quad 0 < |x - \xi| < \delta.$$

Then $f(x)$, $x \neq \xi$, has the sign of A in the interval $\xi - \delta < x < \xi + \delta$; for the difference between a positive and a negative number is numerically greater than either. We do not assume $f(\xi) = A$.

Since the limit of a continuous variable is defined in terms of sequences, Theorem 16.3 may be extended to give the limits of sums, products, and quotients of functions.

THEOREM 3. *If* $f(x) \to A$, $g(x) \to B$ *as* $x \to \xi$, *then*

$$f(x) + g(x) \to A + B,$$

$$f(x) \cdot g(x) \to A \cdot B,$$

$$f(x)/g(x) \to A/B \quad \text{if} \quad B \neq 0.$$

If $f(x) \to A$ as $x \to \xi$ *from above* $(x > \xi)$ or *from below* $(x < \xi)$, we write

(4) $$\lim_{x \to \xi+} f(x) = A \quad \text{or} \quad \lim_{x \to \xi-} f(x) = A.\dagger$$

When $f(x)$ is only defined in the interval $a < x < b$, the limits of $f(x)$ at the end points a, b are necessarily from above or below.

If $f(x) \to A$ for all *unbounded* increasing sequences $\{x_n\}$, we write

(5) $$\lim_{x \to \infty} f(x) = A.$$

The corresponding condition (3) is now

(6) $$|f(x) - A| < \varepsilon \quad \text{when} \quad x > g;$$

that is, given $\varepsilon > 0$, we can always choose g sufficiently large so that (6) holds. If x becomes infinite through positive or through negative values, we write

(7) $$\lim_{x \to +\infty} f(x) = A \quad \text{or} \quad \lim_{x \to -\infty} f(x) = A.$$

$\dagger$ In the notation of §18, $x \downarrow \xi$ or $x \uparrow \xi$ respectively.

Finally, if $f(x)$ increases without limit as $x \to \infty$, we write

(8) $$\lim_{x \to \infty} f(x) = \infty.$$

We can also show specifically *how* $f(x)$ becomes infinite; thus

$$\lim_{x \to 1-} \frac{1}{x-1} = -\infty, \qquad \lim_{x \to 1+} \frac{1}{x-1} = +\infty.$$

The graph of $y = 1/(x-1)$ is a hyperbola with the vertical asymptote $x = 1$; as x increases through 1, the curve "jumps from $-\infty$ to $+\infty$."

Example 1. $\lim_{x \to 1} \sqrt{x} = 1$. Since

$$\sqrt{|x} - 1| = \frac{x-1}{\sqrt{|x} + 1|} < |x - 1|, \qquad \sqrt{|x} - 1| < \varepsilon \quad \text{when} \quad |x - 1| < \varepsilon.$$

Example 2. $\lim_{x \to \infty} \frac{1}{x} = 0$; for

$$\left| \frac{1}{x} \right| < \varepsilon \quad \text{when} \quad |x| > g = \frac{1}{\varepsilon}.$$

Example 3. $\lim_{x \to 0} \frac{1}{x} = \infty$; for

$$\left| \frac{1}{x} \right| > g \quad \text{when} \quad |x| < \varepsilon = \frac{1}{g}.$$

Example 4. If $f(x) = \dfrac{1}{1 + e^{1/x}} \ (x \neq 0)$,

$$\lim_{x \to 0+} f(x) = 0, \qquad \lim_{x \to 0-} f(x) = 1.$$

For, if

$$x \to 0+, \qquad 1/x \to +\infty, \qquad e^{1/x} \to \infty;$$
$$x \to 0-, \qquad 1/x \to -\infty, \qquad e^{1/x} \to 0.$$

Note that $\lim_{x \to 0} f(x)$ does not exist; for the limiting value depends on the method of approach.

Example 5. When $f(x) = \dfrac{x^n - 1}{x - 1}$ and n is a positive integer,

$$\lim_{x \to 1} \frac{x^n - 1}{x - 1} = n.$$

Here $f(1)$ is not defined. From (22.1), $f(x)$ is the sum of n terms of a geometric series:

$$f(x) = 1 + x + x^2 + \cdots + x^{n-1}, \qquad (x \neq 1).$$

As $x \to 1$, each of the n terms on the right $\to 1$.

Example 6. $f(x) = \dfrac{1 - \cos x}{x^2}$; $f(0)$ is not defined, but, since

$$f(x) = \frac{1 - \cos^2 x}{x^2(1 + \cos x)} = \left(\frac{\sin x}{x} \right)^2 \frac{1}{1 + \cos x},$$

$f(x) \to {}^1\!/_2$ as $x \to 0$.

PROBLEMS

1. Prove that the function

$$y = \lim_{n \to \infty} \frac{1}{1 + n \sin^2 \pi x} = \begin{cases} 1 & \text{when } x \text{ is an integer,} \\ 0 & \text{when } x \text{ is not an integer.} \end{cases}$$

2. The *signum function* is defined by

$$\operatorname{sgn} x = \begin{cases} 1, & x > 0; \\ 0, & x = 0; \\ -1, & x < 0. \end{cases}$$

Show that

$$|x| = x \operatorname{sgn} x, \quad \operatorname{sgn} x = \lim_{n \to \infty} \frac{2}{\pi} \tan^{-1} nx.$$

3. None of the following functions is defined for $x = 0$:

$$(a)\ x \sin (1/x); \qquad (b)\ (\sqrt{1 + x^2} - 1)/x;$$
$$(c)\ x/(1 + e^{-1/x}); \qquad (d)\ (\tan ax)/x.$$

Find their limits as $x \to 0$.

4. The symbol $[x]$ denotes the greatest integer that does not exceed x. If n is an integer, prove that

$$x - [x] \to \begin{cases} 0 & \text{as } x \to n+, \\ 1 & \text{as } x \to n-. \end{cases}$$

5. Prove that *Dirichlet's function* (43.1) is given by

$$y = a + (b - a) \lim_{n \to \infty} \operatorname{sgn} (\sin^2 n! \pi x).$$

45. Continuity at a Point. A function $f(x)$ is said to be continuous at the point $x = \xi$ if

(i) $f(x)$ is defined in some neighborhood of $x = \xi$;

(ii) $f(x) \to A$ as $x \to \xi$ (the limit exists);

(iii) $A = f(\xi)$.

These three requirements are tacitly implied by the single equation

(1) $\lim_{x \to \xi} f(x) = f(\xi).$

Writing this

(1)′ $\lim_{x \to \xi} f(x) = f(\lim_{x \to \xi} x),$

we see that, when $f(x)$ is continuous at ξ, the operations $\lim_{x \to \xi}$ and f are commutative:

At a point of continuity, the limit of the function is the function of the limit.

A function that is not continuous at $x = \xi$ is said to be *discontinuous* at ξ. In particular, if

$$\lim_{x \to \xi+} f(x) = f(\xi+), \qquad \lim_{x \to \xi-} f(x) = f(\xi-)$$

both exist but $f(\xi+) \neq f(\xi-)$, $f(x)$ is said to have a *finite jump* at ξ.

From Theorem 44.1 we may state

THEOREM 1. *In order that $f(x)$ be continuous at $x = \xi$, it is necessary and sufficient that, for every $\varepsilon > 0$, we can find a number $\delta > 0$ such that*

(2) $$\left| f(x) - f(\xi) \right| < \varepsilon \qquad \text{when} \qquad \left| x - \xi \right| < \delta.$$

The requirement $x \neq \xi$ now has no significance, for $f(\xi)$ must exist and (2) is obviously fulfilled when $x = \xi$.

From Theorems 44.2 and 44.3 we have

THEOREM 2. *If $f(x)$ is continuous at ξ and $f(\xi) \neq 0$, $f(x)$ will keep the same sign as $f(\xi)$ in a suitably restricted interval about ξ.*

THEOREM 3. *If $f(x)$ and $g(x)$ are continuous at $x = \xi$, the same is true of*

$$f(x) + g(x), \quad f(x) \, g(x), \quad f(x)/g(x) \quad \text{if} \quad g(\xi) \neq 0.$$

Finally the continuity of composite functions may be inferred from

THEOREM 4. *If $y = f(x)$ is continuous at $x = \xi$ and $F(y)$ is continuous when $y = f(\xi)$, then $F\{f(x)\}$ is continuous at $x = \xi$.*

Proof. Put $g(x) = F\{f(x)\}$; then from (1)′

$$\lim_{x \to \xi} g(x) = F\left\{\lim_{x \to \xi} f(x)\right\} = F\{f(\xi)\} = g(\xi).$$

Example 1. *Rational Functions.* The functions $f(x) = \text{const}$ and $f(x) = x$ are continuous at all points (why?). Hence all integral powers x^n and all polynomials,

$$P(x) = a_0 x^n + a_1 x^{n-1} + \cdots + a_n,$$

are continuous without exception (Theorem 3).

The rational functions $P(x)/Q(x)$ are continuous at all points except the finite number where $Q(x) = 0$; at such points $f(x)$ is not defined.

Example 2. $f(x) = (\sin x)/x$ is not continuous at $x = 0$ for $f(0)$ is not defined. However, if we define $f(0) = 1$, $f(x)$ is continuous at $x = 0$; for $f(x) \to 1$ as $x \to 0$ (44.2).

Example 3. $f(x) = \sin 1/x$ is discontinuous at $x = 0$, no matter how $f(0)$ is defined. For in any neighborhood of $x = 0$, however small, $\sin 1/x$ assumes all values between -1 and 1.

Example 4. The function

$$f(x) = x \sin \frac{1}{x} \quad (x \neq 0), \qquad f(0) = 0$$

is continuous for $x = 0$; for

$$\left| f(x) - f(0) \right| = \left| x \sin \frac{1}{x} \right| \leq \left| x \right| < \varepsilon \quad \text{when} \quad \left| x - 0 \right| < \varepsilon.$$

Example 5. The function

$$f(x) = x \quad (x \text{ rational}), \qquad f(x) = 0 \quad (x \text{ irrational})$$

is continuous at $x = 0$; for

$$|f(x) - f(0)| = |f(x)| \leq |x| < \varepsilon \quad \text{when} \quad |x - 0| < \varepsilon.$$

But $f(x)$ is discontinuous when $x = \xi \neq 0$.

If ξ is rational, then, for all irrational values of x,

$$|f(x) - f(\xi)| = |0 - \xi| = |\xi|.$$

If ξ is irrational, then, for all rational values of $|x| > |\xi|$,

$$|f(x) - f(\xi)| = |x - 0| > |\xi|.$$

If we remember that both rational and irrational points are *dense* in any neighborhood of ξ, it is clear that (2) cannot be satisfied for all points of a neighborhood when $\varepsilon = |\xi|$.

Example 6. Consider the function defined for *positive* values of x:

$$f(x) = 0 \quad (x \text{ irrational}), \qquad f\left(\frac{p}{q}\right) = \frac{1}{q},$$

when p/q is a rational fraction in it lowest terms.

Evidently $f(x)$ is discontinuous when x has a rational value p/q; for, in every neighborhood of p/q, $f(x)$ assumes zero values.

We shall show that $f(x)$ is continuous when x has an irrational value ξ; that is,

$$|f(x) - f(\xi)| = f(x) < \varepsilon \quad \text{when} \quad |x - \xi| < \delta.$$

This is certainly true when x is irrational, for $0 < \varepsilon$. When $x = p/q$, we must show that

$$\frac{1}{q} < \varepsilon \quad \text{when} \quad \left|\frac{p}{q} - \xi\right| < \delta.$$

Given ε, choose an integer $n > 1/\varepsilon$, and plot all positive, irreducible fractions whose denominators do not exceed n. There are only a finite number of such fractions, and one of them, say x', is nearest to ξ. Then, if we choose $\delta = |x' - \xi|$, for all fractions $x = p/q$ still nearer to ξ, we must have $q > n$ and hence $1/q < 1/n < \varepsilon$; but this is precisely the relation above.

This example shows that a function can be continuous on one dense set of points and still be discontinuous on another dense set. It also shows another thing, namely, the potency of the epsilon-delta method of proving continuity. Without this clear-cut criterion we might ponder vaguely about this situation without coming to a sharp conclusion.

46. Continuity in an Interval. The function $f(x)$ is said to be *continuous in an interval* (open or closed) if it is continuous at all points of that interval. For a closed interval $a \leq x \leq b$, continuity at a and b imply that

$$\lim_{x \to a+} f(x) = f(a), \qquad \lim_{x \to b-} f(x) = f(b).$$

At any point ξ of the interval

(1) $$|f(x) - f(\xi)| < \varepsilon \quad \text{when} \quad |x - \xi| < \delta;$$

after $\varepsilon > 0$ is given, the corresponding δ depends on ξ as well as on ε.

This suggests the question: When $f(x)$ is continuous in an interval and ε is given, is it possible to choose a corresponding δ *independent of* ξ, so that (1) *is valid throughout the interval?* If such a $\delta(\varepsilon)$ can be found, $f(x)$ is said to be *uniformly continuous* in the interval in question.

Example 1. Consider $f(x) = x^2$ in the interval $-a \leq x \leq a$. Since

$$|x^2 - \xi^2| = |x - \xi| \, |x + \xi| < 2a \, |x - \xi|,$$

$$|x^2 - \xi^2| < \varepsilon \quad \text{when} \quad |x - \xi| < \frac{\varepsilon}{2a}.$$

Thus, if we choose $\delta = \varepsilon/2a$ (independent of ξ), (1) is valid throughout the interval. The function x^2 is thus uniformly continuous in the given interval.

Example 2. Consider $f(x) = 1/x$ in the interval $0 < h < x < a$. Since

$$\left| \frac{1}{x} - \frac{1}{\xi} \right| = \frac{|\xi - x|}{x\xi} < \frac{|x - \xi|}{h^2},$$

$$\left| \frac{1}{x} - \frac{1}{\xi} \right| < \varepsilon \quad \text{when} \quad |x - \xi| < h^2\varepsilon.$$

If we take $\delta = h^2\varepsilon$ (independent of ξ), we see that the function $1/x$ is uniformly continuous in the open interval $h < x < a$.

Example 3. Consider $f(x) = 1/x$ in the interval $0 < x < a$. Then, if $|x - \xi| = \delta$,

$$\left| \frac{1}{x} - \frac{1}{\xi} \right| = \frac{|\xi - x|}{x\xi} > \frac{\delta}{a\xi}.$$

Hence, no matter how small δ is chosen, the left member can be made arbitrarily large by taking ξ close enough to zero (thus, when $\xi = \delta/a$, the left member > 1). Clearly we cannot satisfy (1) throughout the open interval $0 < x < a$ with any constant value of δ; the function $1/x$, although continuous, is not *uniformly* continuous in this interval.

We note also that $1/x$ is not bounded in the interval; for, given any positive number g, $1/x > g$ when $x < 1/g$.

THEOREM 1. *If a function* $f(x)$ *is continuous in a closed interval, it is uniformly continuous over this interval; that is, if* x, x' *are points of the interval,*

(2) $$|f(x) - f(x')| < \varepsilon \quad \text{when} \quad |x - x'| < \delta,$$

where ε *is arbitrary and* δ *depends on* ε *alone.*

Proof (indirect). Assume that for some ε it is impossible to find a δ so that (2) holds. Then we can choose a monotone null sequence $\delta_1, \delta_2, \cdots$ of positive numbers so that for each δ_n we have numbers x_n, x_n' for which

$$|f(x_n) - f(x_n')| > \varepsilon \quad \text{when} \quad |x_n - x_n'| < \delta_n.$$

The bounded sequences $\{x_n\}$ and $\{x_n'\}$ have a cluster point ξ in common since $x_n - x_n' \to 0$, and ξ will belong to the interval, possibly as an end point, since it is *closed*. Now, in every neighborhood of ξ, however small, there are points x_n, x_n' such that

$$\left| f(x_n) - f(x_n') \right| > \varepsilon.$$

But, since the sequence

$$x_1, x_1', x_2, x_2', \cdots \to \xi,$$

the continuity of $f(x)$ at ξ requires that

$$f(x_1), f(x_1'), f(x_2), f(x_2'), \cdots \to f(\xi);$$

hence, by Cauchy's criterion (§ 17)

$$\left| f(x_n) - f(x_n') \right| < \varepsilon \quad \text{when} \quad n > N.$$

This contradiction proves the theorem.

The hypothesis that the interval be *closed* is essential to the proof; otherwise the cluster point ξ might be at an open end point and thus escape from the interval (Ex. 3).

THEOREM 2. *A function that is continuous over a closed interval is bounded there.*

Proof. Let the interval $a \leq x \leq b$ be divided into n equal subintervals of length less than the δ to ensure (2). Since the variation of $f(x)$ in each subinterval is less than ε, the total variation of $f(x)$ over n subintervals lies between $-n\varepsilon$ and $n\varepsilon$; hence,

$$f(a) - n\varepsilon < f(x) < f(a) + n\varepsilon.$$

This theorem also requires that the interval be closed (Ex. 3).

Example 4. Let a and b be rational numbers, and consider the dense set of rational numbers r in the interval $a \leq r \leq b$. If $f(r)$ is a function defined for all rational numbers in the interval (but not necessarily for the irrationals), we shall say that $f(r)$ is *r-continuous* at a fixed point r' when

(2) $$\left| f(r) - f(r') \right| < \varepsilon \quad \text{when} \quad \left| r - r' \right| < \delta.$$

If $f(r)$ is r-continuous at all rational points of an interval, we say it is r-continuous in the interval. Moreover, if $f(r)$ and $g(r)$ are r-continuous in an interval, we can infer that $f + g, fg, f/g \ (g \neq 0)$ are also r-continuous there.

Now, if ξ is an irrational number between a and b, $r - \xi$ and also $1/(r - \xi)$ are r-continuous in (a, b). Nevertheless $f(r) = 1/(r - \xi)$ can be made to exceed any number however large. We thus see that a function can be r-continuous at all rational points of a closed interval and yet fail to be bounded or uniformly continuous there.

However, if $f(r)$ is *uniformly* r-continuous in the interval $a \leq r \leq b$, then (2) holds *throughout the interval* when r, r' are any two rational points less than a distance δ apart. In this case there is one and only one continuous function $f(x)$ of the real variable x in $a \leq x \leq b$ that coincides with $f(r)$ when $x = r$. This function is defined for irrational x by

(3) $$f(x) = \lim_{r_n \to x} f(r_n),$$

where $\{r_n\}$ is any rational sequence in (a, b) having x as a limit. The proof involves showing that

(i) the limit (3) exists for any sequence $\{r_n\} \to x$;

(ii) the limit (3) is the same for all such sequences;

(iii) $f(x)$ is continuous in the real interval (a, b).

Cauchy's convergence criterion (§ 17) assures (i). As to (ii), let the rational sequence $\{t_n\} \to x$ and

$$g(x) = \lim_{t_n \to x} f(t_n);$$

then $f(x) = g(x)$ follows from

$$|f(x) - g(x)| \le |f(x) - f(r_n)| + |f(r_n) - f(t_n)| + |f(t_n) - g(x)|.$$

Finally, if $\{r'_n\} \to x'$, the uniform continuity of $f(x)$ in (a, b) follows from

$$|f(x) - f(x')| \le |f(x) - f(r_n)| + |f(r_n) - f(r'_n)| + |f(r'_n) - f(x')|.$$

When n is large enough, we may assume that r_n and r'_n lie between x and x'.

For example, the function c^r $(c > 0)$ is uniformly r-continuous in the interval $a \le r \le b$; hence the general power function,

$$c^x = \lim_{r_n \to x} c^{r_n},$$

is continuous in the interval $a \le x \le b$.

47. Bounds of a Continuous Function. In § 13 we saw that a bounded set of points always has greatest lower bound (g.l.b.) m and a least upper bound (l.u.b.) M. These bounds may or may not be points of the set. Thus, the values of the function (Prob. 44.4)

$$\varphi(x) = x - [x], \qquad 1 \le x \le 2,$$

have $m = 0$, $M = 1$; but, though $\varphi(1) = \varphi(2) = 0$, $\varphi(x)$ never equals 1. Note that $\varphi(x)$ is discontinuous at $x = 2$.

The values of a *continuous* function $f(x)$ in a closed interval (a, b) form a bounded set (Theorem 46.2) and have a g.l.b. and l.u.b. These bounds m, M are actually assumed by $f(x)$ at points of the interval and therefore give the least and greatest values of $f(x)$.

THEOREM 1. *A function, continuous in a closed interval, takes on a least and a greatest value at least once in the interval.*

Proof. We shall prove that, for some point x_1 of the interval $a \le x \le b$, $f(x_1) = m$. From (13.4) there is at least one x in the interval for which $f(x) < m + \varepsilon$. Hence, the function

$$F(x) = \frac{1}{f(x) - m}$$

exceeds any number $1/\varepsilon$, however great, in the interval; that is, $F(x)$ is not bounded and consequently not continuous in the interval (Theorem

46.2). But, since $f(x)$ *is* continuous in the interval, $F(x)$ will also be continuous unless $f(x) - m = 0$ for at least one $x = x_1$ (Theorem 45.3); that is, $f(x_1) = m$.

A similar proof, based on the function $1/(M - f(x))$, shows that, for at least one $x = x_2$, $f(x_2) = M$.

Again the closure of the interval is essential. Thus $f(x) = x$ in the open interval $0 < x < 1$ has $m = 0$, $M = 1$, but attains neither of these values; the function has no maximum or minimum value in the interval.

PROBLEMS

1. Show that
$$|\sin x| \leq |x|, \qquad |\sin x - \sin \xi| \leq |x - \xi|.$$
Hence prove that $\sin x$ is continuous at any real value $x = \xi$.

2. Show that Dirichlet's function (43.1) is everywhere discontinuous.

3. If $f(x)$ is continuous in $0 \leq x \leq 2a$ and $f(0) = f(2a)$, prove that $f(x) = f(x + a)$ for some x in the interval $0 \leq x \leq a$.
[Apply Theorem 48.1 to $g(x) = f(x) - f(x + a)$.]
Show that at any instant there are two antipodal points on the equator at which the temperature is the same.

4. For what values of x are the following functions discontinuous?

(a) $\dfrac{1}{1 + 1/x}$;   (b) $\dfrac{1}{\sin x - \sqrt{3}}$;   (c) $\dfrac{1}{\sin x - \cos x}$;

(d) $\dfrac{1}{\tan x + \sqrt{3}}$;   (e) $\dfrac{1}{e^x + 1}$;   (f) $\dfrac{1}{x - [x] - \frac{1}{2}}$.

5. Draw the graph of the function
$$f(x) = \begin{cases} 2 - x, & 1 < x \leq 2, \\ \dfrac{3}{2^n} - x, & \dfrac{1}{2^n} < x \leq \dfrac{1}{2^{n-1}}, \quad n = 1, 2, \cdots, \\ 0, & x = 0. \end{cases}$$

Show that $f(x)$ is defined everywhere on the closed interval $0 \leq x \leq 2$ and takes on each of its values *exactly twice* in this interval.

6. If the function of Prob. 5 is extended to the interval $-2 \leq x \leq 2$ by the further definition,
$$f(x) = \begin{cases} x^2 + 2x, & -2 < x < 0; \\ -1, & x = -2; \end{cases}$$
show that it also has the property stated in Prob. 5.

7. If $g(x)$ is defined in $a \leq x \leq b$ and takes on each of its values exactly twice in this interval, prove that $g(x)$ is discontinuous.
[If $g(x)$ were continuous, show that its greatest (and also least) values could only be assumed at $x = a$ and $x = b$; hence $g(x) = $ const.]

48. Intermediate Values. We begin with a theorem that seems self-evident.

THEOREM 1. *If a function, continuous over a closed interval, differs in sign at its end points, it must vanish at some interior point.*

Proof. Let $f(x)$ be continuous over $a \leqq x \leqq b$ and suppose that $f(a), f(b)$ differ in sign. Divide the interval into subintervals of length $< \delta$ so small that the variation of $f(x)$ in any one is $< \varepsilon$ (Theorem 46.1). If $f(x)$ does not vanish at any point of division, $f(x)$ must change sign in at least one subinterval; and, if x, x' are points of the subinterval for which $f(x), f(x')$ differ in sign,

$$\left| f(x) \right| < \left| f(x) - f(x') \right| < \varepsilon.$$

Thus $1/f(x)$ is not bounded in the interval and consequently not continuous; but, since $f(x)$ *is* continuous, $f(x)$ must vanish for at least one interior point.

Geometrically this theorem seems trivial; it states that the graph of a continuous function $y = f(x)$ between $x = a$ and $x = b$ must cross the x-axes at least once if the ordinates $f(a), f(b)$ differ in sign. However the graph of a continuous function may be impossible to draw or even to visualize. Thus continuous functions exist that have the property *everywhere* which the function

$$f(x) = x \sin \frac{1}{x}, \qquad f(0) = 0$$

has at the point zero; their graphs "wobble" infinitely often in any interval however small so that actually they have no tangents anywhere.†
And after all, the theorem just proved is purely analytic and therefore deserves analytic treatment. Geometric intuition is valuable in suggesting the facts but must not be relied upon for an incontestable proof.

An immediate consequence of Theorem 1 is that a continuous function cannot pass from one value to another without taking on *all* intermediate values.

THEOREM 2. *Let $f(x)$ be continuous over the interval $a \leqq x \leqq b$ and $f(a) \neq f(b)$. Then if μ is any number between $f(a)$ and $f(b)$, $f(x)$ assumes the value μ at least once in the interval.*

Proof. Apply Theorem 1 to the function $F(x) = f(x) - \mu$.

A function $f(x)$ is said to be *monotone-increasing* or *monotone-decreasing* if, when $x_2 > x_1$,

$$f(x_2) \geqq f(x_1) \quad \text{or} \quad f(x_2) \leqq f(x_1),$$

In both cases the function is said to be *monotone*, and, if

$$f(x_2) > f(x_1) \quad \text{or} \quad f(x_2) < f(x_1),$$

strictly monotone.

---

† See the example given by John McCarthy, *Am. Math. Monthly*, vol. 60, 1953, p. 709.

THEOREM 3.   *If $f(x)$ is continuous in the interval $a \leq x \leq b$ and assumes each value between $f(a)$ and $f(b)$ just once, it is strictly monotone in this interval.*

Proof.   Suppose $f(a) < f(b)$; then, if $a < x_1 < b, f(a) < f(x_1) < f(b)$.   To prove this we show that the hypothesis rules out all other possibilities:

(i) $f(x_1) = f(a)$ or $f(x_1) = f(b)$;

(ii) $f(x_1) < f(a)$;

(iii) $f(x_1) > f(b)$.

Case (i) is obviously impossible.   In case (ii), $f(x) = f(a)$ for some x between x_1 and b (Theorem 2).   In case (iii), $f(x) = f(b)$ for some x between a and x_1.

Now, if $a < x_1 < x_2 < b$, we have just seen that $f(x_1) < f(b)$, and consequently $f(x_1) < f(x_2)$.   Thus $f(x)$ is strictly monotone-increasing.

If $f(a) > f(b)$, we can show in the same way that $f(x)$ is strictly monotone-decreasing.

49. Inverse Functions.   Let the function $y = f(x)$ be continuous and strictly increasing in the interval $a \leq x \leq b$; then, if x_1, x_2 lie in this interval,

(1) $x_2 > x_1$ implies $f(x_2) > f(x_1)$,

and conversely.   Let $f(a) = \alpha, f(b) = \beta$; then, if y_1 is any number between α and β, the intermediate-value theorem (Theorem 48.2) shows that, for some value x_1 between a and b, $f(x_1) = y_1$.   Moreover, this value is *unique*; for

(2) $f(x_1) = f(x_2)$ implies $x_1 = x_2$,

since (1) excludes $x_1 \lessgtr x_2$.   Thus the equation $y = f(x)$ sets up the one-to-one correspondence $x \leftrightarrow y$ over the intervals $a \leq x \leq b$ and $\alpha \leq y \leq \beta$.   Hence x is a function of y over $\alpha \leq y \leq \beta$; and, if we denote this *inverse function* by $x = \varphi(y)$, we have the identities

(2) $y = f\{\varphi(y)\},$ $x = \varphi\{f(x)\}.$

If $x_1 = \varphi(y_1), x_2 = \varphi(y_2)$, the converse of (1) states that

(3) $y_2 > y_1$ implies $\varphi(y_2) > \varphi(y_1)$;

that is, the function $\varphi(y)$ is also strictly increasing in the interval $\alpha \leq y \leq \beta$.

To prove the continuity of $\varphi(y)$ in this interval, consider the positive function,

$$F(x) = f(x + \varepsilon) - f(x),$$

where ε is an arbitrary small positive constant. Since $F(x)$ is continuous in the closed interval $a \leq x \leq b - \varepsilon$, it assumes its minimum value δ at some point of this interval (§47, Theorem 1). Hence, if x_1, x_2 are two points for which $|x_1 - x_2| \geq \varepsilon$, then, since $f(x)$ is strictly increasing,

$$|f(x_1) - f(x_2)| = |y_1 - y_2| \geq \delta.$$

Consequently, when $|y_1 - y_2| < \delta$, we must have

$$|x_1 - x_2| = |\varphi(y_1) - \varphi(y_2)| < \varepsilon;$$

that is, $x = \varphi(y)$ is continuous throughout its interval.† Hence the

THEOREM. *If the continuous function $f(x)$ is strictly increasing in the interval $a \leq x \leq b$, the equation $y = f(x)$ defines an inverse function $x = \varphi(y)$ which is also continuous and strictly increasing in the interval $f(a) \leq y \leq f(b)$.*

Remark. If $f(x)$ is strictly decreasing, we may apply the theorem to the increasing function $-f(x)$; than $-\varphi(y)$ is increasing in the interval $-f(a) \leq y \leq -f(b)$ and $\varphi(y)$ is decreasing in the interval $f(b) \leq y \leq f(a)$.

The theorem is obvious from a geometrical point of view, *for both equations,*

$$y = f(x), \qquad x = \varphi(y),$$

are represented by the same curve. If we take x as the independent variable in both functions, the graphs of

$$y = f(x) \quad \text{and} \quad y = \varphi(x)$$

are symmetric to the line $y = x$; for the point (x, y) becomes (y, x) after reflection in this line.

Example 1. The increasing function

$$y = x^2 \qquad (0 \leq x < \infty)$$

has the inverse

$$x = \sqrt{y} \qquad (0 \leq y < \infty),$$

where $\sqrt{y}$ denotes the positive root (§9). Note that the semiparabolas $y = x^2$, $y = \sqrt{x}$ $(x > 0)$ are symmetric to the line $y = x$.

Example 2. The increasing function

$$y = \sin x \qquad \left(-\frac{\pi}{2} \leq x \leq \frac{\pi}{2}\right)$$

has the inverse

$$x = \sin^{-1} y \qquad (-1 \leq y \leq 1),$$

where $\sin^{-1} x$ denotes the *principal branch* of the inverse sine; its values are restricted to the interval $(-\pi/2, \pi/2)$.

---

† Note that we have proved the *uniform* continuity of $\varphi(y)$.

Example 3. The increasing function

$$y = \tan x \qquad \left(-\frac{\pi}{2} < x < \frac{\pi}{2}\right)$$

has the inverse

$$x = \tan^{-1} y \qquad (-\infty < y < \infty),$$

where $\tan^{-1} y$ denotes the principal branch of the inverse tangent; its values are restricted to the interval $(-\pi/2, \pi/2)$.

Example 4. The decreasing function

$$y = \cos x \qquad (0 \leqq x \leqq \pi)$$

has the inverse

$$x = \cos^{-1} y \qquad (-1 \leqq y \leqq 1),$$

where $\cos^{-1} y$ denotes the principal branch of the inverse cosine; its values are restricted to the interval $(0, \pi)$. Note that $\cos^{-1} y$ is likewise decreasing in the interval $(-1, 1)$.

Example 5. *Exponential and Logarithm.* When x is a positive integer n, we define a^n as the product of n equal factors a. For positive integral exponents we can easily prove the *laws of exponents*:

$$a^m \cdot a^n = a^{m+n}, \qquad (a^m)^n = a^{mn}.$$

In order to maintain the validity of these laws for *all* rational exponents, we define

$$a^0 = 1, \qquad a^{p/q} = \sqrt[q]{a^p}, \qquad a^{-r} = 1/a^r.$$

This again is an application of the principle of permanence of form (§ 10). In the following we assume that $a > 0$.

When ρ is an irrational number defined by the cut $r \mid R$ in the rational numbers, we can always find a monotone sequence of rationals $\{r_n\}$ having ρ as a limit. Then $\{a^{r_n}\}$ is also a bounded monotone sequence which has a limit (§ 18); this limit, which is the same for all such sequences, we define as a^ρ. Thus a^x $(a > 0)$ is defined for all real exponents x. The function a^x is positive and monotone in the interval $-\infty < x < \infty$; if

$$a > 1, \qquad a^x \text{ increases from } 0 \text{ to } \infty;$$

$$0 < a < 1, \qquad a^x \text{ decreases from } \infty \text{ to } 0.$$

Finally a^x $(a > 0)$ is everywhere continuous. Continuity at $x = 0$ means that

$$\lim_{h \to 0} a^h = 1 = a^0.$$

When $h \to 0+$, this follows from (15.10); and, since $a^{-h} = 1/a^h$, we have the same limit when $h \to 0-$. Therefore,

$$\lim_{h \to 0} a^{x+h} = \lim_{h \to 0} a^x a^h = a^x,$$

so that a^x is continuous at all real points.

When $a > 1$, $y = a^x$ is continuous and strictly increasing in the interval $-\infty < x < \infty$; hence, the inverse function

$$x = \log_a y \qquad (0 < y < \infty),$$

the *logarithm of y to the base a*, is also continuous and strictly increasing in the interval $0 < y < \infty$. In particular, when $a = e = 2.71828 \cdots$ we have the *exponential function* e^x (or exp x) and its inverse, the *natural logarithm* log x† for which

$$e^{\log y} = y, \qquad \log e^x = x.$$

From the graph of $y = e^x$ we may obtain the graph of $y = \log x$ by a reflection in the line $y = x$ (Fig. 49).

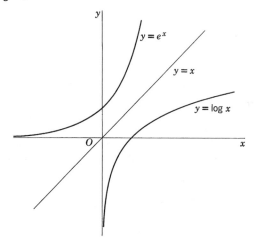

FIG. 49. The exponential and logarithmic curves

50. Derivative. Let $y = f(x)$ be a function defined over an interval $a \leq x \leq b$. At any interior point x we can give x a positive or negative *increment* $\Delta x = h$; the corresponding increment of y is then

$$\Delta y = f(x + h) - f(x).$$

When x is fixed, the *difference quotient*,

(1) $$\frac{\Delta y}{\Delta x} = \frac{f(x + h) - f(x)}{h},$$

is a function of h. If this ratio tends to the same limit as $h \to 0$ in any manner that avoids $h = 0$, the limit is called the *derivative* of $f(x)$ and denoted by dy/dx or $f'(x)$:

(2) $$\frac{dy}{dx} = f'(x) = \lim_{h \to 0} \frac{f(x + h) - f(x)}{h}.$$

---

† The base e is usually omitted. Some authors write ln x to avoid confusion with *common* logarithms to the base 10.

Now consider a fixed point ξ where $f'(\xi)$ exists. Then

$$\frac{f(\xi + h) - f(\xi)}{h} - f'(\xi) = \eta(h)$$

is a function of h and $\eta(h) \to 0$ as $h \to 0$. If we put $x = \xi + h$,

(3) $$f(x) = f(\xi) + h f'(\xi) + h\, \eta(h),$$

$$\lim_{x \to \xi} f(x) = \lim_{h \to 0} \{f(\xi) + h f'(\xi) + h\, \eta(h)\} = f(\xi);$$

hence, $f(x)$ is continuous at the point ξ (45.1). *If a function has a derivative at a given point, it is continuous at this point.* But $f(x)$ may be continuous at x and not have a derivative there, as shown in Exs. 3 and 4.

If the limit of (1) exists when $h \to 0$ through positive or negative values only, we denote the limiting value by $f'_+(x)$ or $f'_-(x)$. These limits are called the *right-hand* and *left-hand derivatives* at x. If either one fails to exist, or if both exist and have different values, then $f'(x)$ does not exist. But, if both have the same value, this value is precisely the derivative $f'(x)$.

Note that $f'_+(\xi)$ and $f'_-(\xi)$ do *not* mean the limits of $f'(x)$ when x approaches ξ from above and below, respectively, namely $f'(\xi+)$, $f'(\xi-)$. Even when the last limits exist and are equal, $f'(\xi)$ may not exist. Consider, for example, the function

$$f(x) = x - [x],$$

where $[x]$ denotes the greatest integer that does not exceed x. Its graph consist of a series of straight segments inclined 45° to the x-axis. When $x = n$, an integer, $f(n) = 0$; now

$$f'(n+) = 1, \qquad f'(n-) = 1;$$

but $f'_+(n) = 1$, and $f'_-(n)$ does not exist, and hence $f(x)$ has no derivative at $x = n$. The discontinuity of $f(x)$ at $x = n$ also precludes the existence of $f'(n)$.

If $f(x)$ has a derivative at every interior point of $a \le x \le b$ and $f'_+(a)$ and $f'_-(b)$ both exist, we say that $f(x)$ is *differentiable in the interval* (a, b). From $f(x)$ we thus obtain the derived function $f'(x)$ over (a, b).

The symbol dy/dx was defined in (2) as the limit of $\Delta y/\Delta x$, but the differentials dx, dy were not given independent meanings, and hence dy/dx is not a quotient. We close this gap by the

DEFINITION. *If the function $y = f(x)$ admits a derivative $f'(x)$ at the point x and Δx is an arbitrary nonzero increment of x, then the differential*

(4) $$dy = f'(x)\, \Delta x.$$

In particular, when $y = x$, $f'(x) = 1$, and

(5) $$dx = \Delta x.$$

From (4) and (5) we now have by an actual division

(6) $$\frac{dy}{dx} = f'(x),$$

in conformity with the notation (2). Note that dy not only depends on the value of x but also on the choice of the increment Δx; that is, dy is a function of the *two* variables x and Δx.

A variable that approaches zero is called an *infinitesimal*. Thus $h = \Delta x$, Δy, dy, and η are all infinitesimals, but Δx is the *principal infinitesimal* since the others all depend on Δx. If η_1 and η_2 are infinitesimals and η_1/η_2 approaches a finite limit K, we say that η_1 and η_2 have the *same order* if $K \neq 0$, but that η_1 is of *higher order* than η_2 if $K = 0$. Now from (3) we have

$$\Delta y = dy + h\eta, \qquad \lim \frac{\Delta y - dy}{\Delta x} = \lim \eta = 0;$$

hence, *the difference $\Delta y - dy$ is an infinitesimal of higher order than Δx.*

Example 1. The Sine and Cosine:

$$f(x) = \sin x, \qquad g(x) = \cos x.$$

If the angle is expressed in radians,

$$f'(0) = \lim_{h \to 0} \frac{\sin h - 0}{h} = 1, \qquad g'(0) = \lim_{h \to 0} \frac{\cos h - 1}{h} = 0;$$

the first limit follows from (44.2), the second from

$$\frac{\cos h - 1}{h} = \frac{\cos^2 h - 1}{h(\cos h + 1)} = -\frac{\sin h}{h} \frac{\sin h}{\cos h + 1}.$$

With the aid of the addition formulas for the sine and cosine we now find

$$f'(x) = \lim_{h \to 0} \frac{\sin(x + h) - \sin x}{h}$$

$$= \lim_{h \to 0} \sin x \frac{\cos h - 1}{h} + \cos x \frac{\sin h}{h} = \cos x;$$

$$g'(x) = \lim_{h \to 0} \frac{\cos(x + h) - \cos x}{h}$$

$$= \lim_{h \to 0} \cos x \frac{\cos h - 1}{h} - \sin x \frac{\sin h}{h} = -\sin x.$$

Hence, if D denotes the differential operator d/dx,

$$D \sin x = \sin\left(x + \frac{\pi}{2}\right), \qquad D \cos x = \cos\left(x + \frac{\pi}{2}\right);$$

the derivatives of $\sin x$ *and* $\cos x$ *are found by increasing* x *by* $\pi/2$. The simplicity of these formulas is due to the choice of the *radian* as the unit of angle; for $(\sin x)/x \to 1$ only when x is expressed in radians.

Example 2. The Natural Logarithm. When $f(x) = \log x$, we have

$$f'(1) = \lim_{h \to 0} \frac{\log(1 + h) - 0}{h} = \lim_{h \to 0} \log(1 + h)^{1/h} = \log [\lim_{h \to 0} (1 + h)^{1/h}]$$

since $\log x$ is a continuous function (§ 49). If we put $t = 1/h$, the last limit equals e by (19.4); hence,

$$f'(1) = \lim_{h \to 0} \frac{\log(1 + h)}{h} = 1.$$

On using this result, we find that

$$f'(x) = \lim_{h \to 0} \frac{\log(x + h) - \log x}{h} = \frac{1}{x} \lim_{h \to 0} \frac{\log(1 + h/x)}{h/x} = \frac{1}{x}.$$

The simplicity of the formula

$$D \log x = \frac{1}{x}$$

is due to the choice of e for the base of logarithms. Since $x = e^{\log x}$,

$$\log_a x = \log x \cdot \log_a e, \qquad D \log_a x = \frac{\log_a e}{x}.$$

Example 3. When $f(x) = |x|$,

$$f'_+(0) = \lim_{h \to 0+} \frac{h - 0}{h} = 1, \qquad f'_-(0) = \lim_{h \to 0-} \frac{|h| - 0}{h} = -1.$$

Since the right-hand and left-hand derivatives at zero differ, $f'(0)$ does not exist. Nevertheless $f(x)$ is continuous at 0. Thus a function may be continuous at a point without having a derivative there.

Example 4. The function

$$f(x) = x \sin \frac{1}{x} \quad (x \neq 0), \qquad f(0) = 0$$

is continuous at $x = 0$ (Ex. 45.4) but has no derivative there. In fact, neither the left-hand or right-hand derivative exists, for

$$\frac{f(h) - f(0)}{h} = \sin \frac{1}{h}$$

does not approach a limit as $h \to 0+$ or $h \to 0-$.

Example 5. For the function

$$f(x) = x^2 \sin \frac{1}{x} \quad (x \neq 0), \qquad f(0) = 0,$$

$$f'(0) = \lim_{h \to 0} \frac{h^2 \sin \frac{1}{h}}{h} = 0.$$

For any $x \neq 0$, the standard differentiation formulas give

$$f'(x) = 2x \sin \frac{1}{x} - \cos \frac{1}{x},$$

but $f'(x)$ does not approach a limit as $x \to 0$. Thus $f'(x)$ exists at all points but is discontinuous at $x = 0$.

51. Increasing and Decreasing Functions.   If the derivative of $f(x)$ is positive when $x = c$,

$$\lim_{h \to 0} \frac{f(c + h) - f(c)}{h} = f'(c) > 0;$$

hence, in suitably restricted neighborhood of c, $f(c + h) - f(c)$, and $h = \Delta x$ have the same sign.   Thus,

$$f(c + h) - f(c) {\scriptstyle > 0 \atop \scriptstyle < 0} \quad \text{when} \quad h {\scriptstyle > 0; \atop \scriptstyle < 0.}$$

Since $f(x)$ increases as x increases through c, the function is said to be *increasing at c*.

Similarly, if $f'(c) < 0$, $f(c + h) - f(c)$ and $h = \Delta x$ have opposite signs; then

$$f(c + h) - f(c) {\scriptstyle < 0 \atop \scriptstyle > 0} \quad \text{when} \quad h {\scriptstyle > 0, \atop \scriptstyle < 0.}$$

and $f(x)$ is said to be *decreasing at c*.

THEOREM.   *According as $f'(c)$ is positive or negative, $f(x)$ is increasing or decreasing at c.*

The condition $f'(c) > 0$ is sufficient that $f(x)$ increase at c, but not necessary.   Thus $f(x) = x^3$ is increasing at 0 but $f'(0) = 0$.   Indeed $f(x)$ may increase at c when $f'(c)$ does not exist as in the case $f(x) = x^{1/3}$ at $x = 0$.

52. The Chain Rule.   Let $u(x)$ be a differentiable function in the interval $a \leq x \leq b$ in which it assumes values in the interval $\alpha \leq u \leq \beta$.   Then, if $y = f(u)$ is a differentiable function of u in the interval $\alpha \leq u \leq \beta$, the compound function,

$$y = f\{u(x)\},$$

has a derivative given by the *chain rule*:

$$(1) \qquad\qquad \frac{dy}{dx} = f'(u)\, u'(x) = \frac{dy}{du}\frac{du}{dx}.$$

Proof.   Let an arbitrary increment Δx produce an increment Δu in the function $u(x)$ while Δu in turn produces an increment Δy in the function $f(u)$.   Then, since both functions have derivatives,

$$\Delta y = \{f'(u) + \eta_2\}\, \Delta u, \qquad \Delta u = \{u'(x) + \eta_1\}\, \Delta x,$$

where η_1 and $\eta_2 \to 0$ as $\Delta x \to 0$, (if $\Delta u = 0$, then $\Delta y = 0$ also, and we put $\eta_2 = 0$). Hence, on eliminating Δu, we have

$$\frac{\Delta y}{\Delta x} = \{f'(u) + \eta_2\}\{u'(x) + \eta_1\},$$

and, on passing to the limit $\Delta x \to 0$, we obtain (1).

If y is compounded of three functions

$$y = f(u), \qquad u = g(v), \qquad v = h(x),$$

y is a function of x through the intermediary functions u and v. Then, if we regard $u = g\{h(x)\}$ as a function of x, we have from (1)

$$\frac{dy}{dx} = \frac{dy}{du}\frac{du}{dx}, \qquad \frac{du}{dx} = \frac{du}{dv}\frac{dv}{dx}$$

and, hence,

$$(2) \qquad \frac{dy}{dx} = \frac{dy}{du}\frac{du}{dv}\frac{dv}{dx},$$

in which the chain has three links. Further generalization is obvious.

We can now show that the equation

$$(3) \qquad dy = f'(x)\, dx,$$

which holds when x is an independent variable, is still true when x depends on an independent variable t. Thus, if

$$y = f\{x(t)\} = F(t),$$

$$dy = F'(t)\, dt = f'(x)\, x'(t)\, dt = f'(x)\, dx.$$

53. Derivative of an Inverse Function. If the function $y = f(x)$ has a derivative $f'(x)$ in an interval $a < x < b$ which is always positive or always negative, then the inverse function $x = \varphi(y)$ exists when y lies between $f(a)$ and $f(b)$ and has the derivative

$$(1) \qquad \frac{dx}{dy} = 1 \bigg/ \frac{dy}{dx}.$$

Proof. Since $f'(x)$ exists in $a < x < b$, $f(x)$ is continuous in this interval; moreover, $f(x)$ increases or decreases throughout the interval according as $f'(x)$ is positive or negative. Hence (§ 49), the inverse function $x = \varphi(y)$ exists and is continuous when y lies between $f(a)$ and $f(b)$. Moreover,

$$\frac{dx}{dy} = \lim_{\Delta y \to 0} \frac{\Delta x}{\Delta y} = \lim_{\Delta x \to 0} 1 \bigg/ \frac{\Delta y}{\Delta x} = 1 \bigg/ \frac{dy}{dx},$$

for the continuity of $x = \varphi(y)$ shows that $\Delta x \to 0$ as $\Delta y \to 0$. Thus at corresponding values of x and y the derivatives of $y = f(x)$ and $x = \varphi(y)$ are reciprocals.

If we write $y = \varphi(x)$, then $x = f(y)$ is the inverse function and

$$(2) \qquad \frac{dy}{dx} = 1 \bigg/ \frac{dx}{dy} = \frac{1}{f'(y)} = \frac{1}{f'\{\varphi(x)\}}.$$

We consider in turn the examples of § 49.

Example 1. The *power function* $y = x^n$ (n a positive integer) increases over $0 \leqq x < \infty$ and has the unique inverse $x = y^{1/n}$ given by the positive nth root. Hence $x = y^n$ ($0 \leqq y < \infty$) has the inverse

$$y = x^{1/n}, \quad \text{and} \quad \frac{dy}{dx} = 1 \bigg/ \frac{dx}{dy} = \frac{1}{ny^{n-1}} = \frac{1}{n} x^{\frac{1}{n}-1}.$$

When n is odd, $y = x^n$ increases over $-\infty < x < \infty$ and has the unique inverse $x = y^{1/n}$ given by the real nth root. If $y = x^{1/n}$, dy/dx has the value above.

Example 2. The *sine* $y = \sin x$ increases over $-\pi/2 \leqq x \leqq \pi/2$ and within this range has a unique inverse $x = \sin^{-1} y$. Hence $x = \sin y$ ($-\pi/2 \leqq y \leqq \pi/2$) has the inverse

$$y = \sin^{-1} x, \quad \text{and} \quad \frac{dy}{dx} = 1 \bigg/ \frac{dx}{dy} = \frac{1}{\cos y} = \frac{1}{\sqrt{1-x^2}},$$

where $\cos y = \sqrt{1-x^2} \geqq 0$ when $-\pi/2 \leqq y \leqq \pi/2$.

Example 3. The *tangent* $y = \tan x$ increases over $-\pi/2 < x < \pi/2$ and has the unique inverse $x = \tan^{-1} y$. Hence $x = \tan y$ ($-\pi/2 < y < \pi/2$) has the inverse

$$y = \tan^{-1} x, \quad \text{and} \quad \frac{dy}{dx} = 1 \bigg/ \frac{dx}{dy} = \frac{1}{\sec^2 y} = \frac{1}{1+x^2}.$$

Example 4. The *cosine* $y = \cos x$ decreases over ($0 \leqq x \leqq \pi$) and has the unique inverse $x = \cos^{-1} y$. Hence $x = \cos y$ ($0 \leqq y \leqq \pi$) has the inverse

$$y = \cos^{-1} x, \quad \text{and} \quad \frac{dy}{dx} = 1 \bigg/ \frac{dx}{dy} = \frac{1}{-\sin y} = -\frac{1}{\sqrt{1-x^2}},$$

where $\sin y = \sqrt{1-x^2} \geqq 0$ when $0 \leqq y \leqq \pi$.

Example 5. The *logarithm* $y = \log x$ increases over $0 < x < \infty$ and has the unique inverse $x = e^y$. Hence $x = \log y$ ($0 < y < \infty$) has the inverse

$$y = e^x, \quad \text{and} \quad \frac{dy}{dx} = 1 \bigg/ \frac{dx}{dy} = \frac{1}{1/y} = y = e^x.$$

Example 6. The *hyperbolic sine*,

$$y = \sinh x = \frac{e^x - e^{-x}}{2},$$

has the derivative

$$\frac{dy}{dx} = \cosh x = \frac{e^x + e^{-x}}{2}.$$

which is always positive; thus $y = \sinh x$ $(-\infty < x < \infty)$ has a unique inverse $x = \sinh^{-1} y$. Hence $x = \sinh y$ $(-\infty < y < \infty)$ has the inverse

$$(3) \qquad y = \sinh^{-1} x, \quad \text{and} \quad \frac{dy}{dx} = 1 \left/ \frac{dx}{dy} \right. = \frac{1}{\cosh y} = \frac{1}{\sqrt{x^2 + 1}},$$

in view of the identity

$$\cosh^2 y - \sinh^2 y = 1.$$

The root $\sqrt{x^2 + 1} = \cosh y > 0$.

The function $y = \sinh^{-1} x$ is found by solving

$$x = \sinh y = \frac{e^y - e^{-y}}{2}$$

for y. From the quadratic in e^y,

$$e^{2y} - 2xe^y - 1 = 0,$$

we obtain the *positive* root,

$$e^y = x + \sqrt{x^2 + 1},$$

and

$$(4) \qquad y = \sinh^{-1} x = \log(x + \sqrt{x^2 + 1}).$$

Example 7. The *hyperbolic cosine*,

$$y = \cosh x = \frac{e^x + e^{-x}}{2}$$

has the derivative

$$\frac{dy}{dx} = \sinh x = \frac{e^x - e^{-x}}{2}$$

which is positive when $x > 0$, negative when $x < 0$. In the interval $0 \le x < \infty$, $y = \cosh x$ is increasing and has the unique inverse $x = \cosh^{-1} y$. Hence $x = \cosh y$ $(y \ge 0)$ has the inverse

$$(5) \qquad y = \cosh^{-1} x, \quad \text{and} \quad \frac{dy}{dx} = 1 \left/ \frac{dx}{dy} \right. = \frac{1}{\sinh y} = \frac{1}{\sqrt{x^2 - 1}}.$$

The root $\sqrt{x^2 - 1} = \sinh y \ge 0$ since $y \ge 0$.

The function $y = \cosh^{-1} x$ $(y \ge 0)$ is found by solving

$$x = \cosh y = \frac{e^y + e^{-y}}{2}$$

for the *positive* value of y. From the quadratic in e^y,

$$e^{2y} - 2xe^y + 1 = 0,$$

we obtain

$$e^y = x \pm \sqrt{x^2 - 1} = \begin{cases} x + \sqrt{x^2 - 1}, \\ (x + \sqrt{x^2 - 1})^{-1}; \end{cases}$$

hence, the positive value of y is

$$(6) \qquad y = \cosh^{-1} x = \log(x + \sqrt{x^2 - 1}).$$

The inverse of $x = \cosh y$ when $y \le 0$ is the negative of this expression.

PROBLEMS

1. From the definition (50.2), compute $f'(0)$ when

$$f(x) = xe^{-x}, \quad \tan x, \quad \frac{x-1}{x+2}, \quad x^3 + 1, \quad \sin x.$$

2. If $f(x) = 0 \ (x \leq 0)$, $f(x) = 1 \ (x > 0)$, find the values of $f'(0)$, $f'_+(0)$, $f'_-(0)$, $f'(0+)$, $f'(0-)$, in case they exist.

3. Find $f'(x)$ when

$$f(x) = e^{\log x}; \quad \log e^{-x}; \quad \log ax; \quad \log |x|;$$

$$\sin^{-1}(1 - x^2)^{1/2}; \quad \tan^{-1}\frac{\cos x}{1 + \sin x}; \quad \tan^{-1}\frac{a+x}{1 - ax}.$$

4. If $f(x) = x + 1 \ (x < 0)$, $f(x) = e^x \ (x \geq 0)$, does $f'(0)$ exist? Is $f'(x)$ continuous at $x = 0$?

5. Compute $f'(x)$ from (50.2) when

$$y = f(x) = \frac{ax + b}{cx + d}.$$

Show that $f(x)$ is an increasing or a decreasing function in any interval not containing $-d/c$ according as $ad - bc > 0$ or < 0. Discuss the case when $ad - bc = 0$.

6. Find the inverse function $x = \varphi(y)$ in Prob. 5 and its derivative $\varphi'(y)$. When is $\varphi(y)$ an increasing function?

7. Show that $x - \sin x$ is an increasing function in any interval.

8. Prove that $\tan x - x$ is an increasing function within the intervals $(-\pi/2, \pi/2)$, $(\pi/2, 3\pi/2)$. Hence show that $\tan x = x$ has exactly one root between $\pi/2$ and $3\pi/2$.

9. Show that $(\sin x)/x$ decreases steadily from 1 to 0 as x increases from 0 to π.

10. If $y = \log(x + \sqrt{x^2 + 1})$ find dy/dx. Find the inverse function $x = \varphi(y)$ and its derivative $\varphi'(y)$.

11. $P(x)$ is a polynomial. If $P(x) = 0$ has a root r of multiplicity n, show that $P'(x) = 0$ has a root r of multiplicity $n - 1$.

12. When the elements of a determinant D are function of x, its derivative $D'(x)$ is the sum of the determinants formed by differentiating the elements of one row only and leaving the others unchanged. Prove this for 2-rowed and 3-rowed determinants.

13. If $f(x) = 1/(1 + e^{1/x})$, show that $f(x)$ and $f'(x)$ are continuous when $x \neq 0$; and that

$$f(0+) = 0, \quad f(0-) = 1; \quad f'(0+) = f'(0-) = 0.$$

Does $f'(0)$ exist? Is the function

$$g(x) = \begin{cases} f'(x) & x \neq 0 \\ 0 & x = 0 \end{cases} \quad \text{continuous at } x = 0?$$

Describe the behavior of $f(x)$ and $f'(x)$ as $x \to \pm\infty$.

14. If $\tanh x = \sinh x/\cosh x$, show that

$$\tanh^{-1} x = \frac{1}{2}\log\frac{1+x}{1-x}, \qquad |x| < 1;$$

and $\tanh^{-1}(-x) = -\tanh^{-1} x$.

15. If $\coth x = \cosh x/\sinh x$, show that

$$\coth^{-1} x = \frac{1}{2}\log\frac{x+1}{x-1}, \qquad |x| > 1;$$

and $\coth^{-1}(-x) = -\coth x$.

16. If $f'_{+}(a)$ and $f'_{-}(a)$ both exist but have different values, show that $f(x)$ is continuous at a.

54. Higher Derivatives. Let the function $y = f(x)$ admit a derivative

$$(1) \qquad\qquad y' = \frac{dy}{dx} = f'(x)$$

in some neighborhood (§ 12) of the point x. If the derived function $f'(x)$ admits a derivative at x, we denote this *second derivative* by

$$(2) \qquad\qquad y'' = \frac{d^2y}{dx^2} = f''(x).$$

Thus, by successive differentiation we may obtain a sequence of derivatives $f'(x), f''(x), f'''(x), \cdots$, the nth derivative being denoted by

$$(3) \qquad\qquad y^{(n)} = \frac{d^ny}{dx^n} = f^{(n)}(x).$$

Note that the existence of $f^{(n)}(x)$ at the point $x = a$ implies that:

(i) $f^{(n-1)}(x)$ exists in some neighborhood of a: $a - \varepsilon < x < a + \varepsilon$;

(ii) $f^{(n-1)}(x)$ is continuous at $x = a$ (§ 50);

(iii) all lower derivatives and also $f(x)$ are continuous throughout this neighborhood.†

When $y = f(x)$, we have defined $dy = f'(x)\,\Delta x$, a function of *two* variables x and $\Delta x = dx$. If we wish to define the *second differential* $d^2y = d(dy)$, we must give dx an arbitrary but *constant* value; then $f'(x)\,dx$ is a function of x alone, and

$$d^2y = d\{f'(x)\,dx\}\,dx = f''(x)\,dx \cdot dx = f''(x)\,dx^2.$$

Always holding dx constant, we have in the same way

$$(4) \qquad\qquad d^ny = f^{(n)}(x)\,dx^n.$$

Division by dx^n now leads to (3) in which the left member is an actual quotient. Except for the case $n = 1$, x must be an independent variable in (4) so that we may impose the condition: $dx = $ const.

---

† If the right-hand derivative $f_{+}^{(n)}(a)$ exists, the corresponding neighborhood for $f^{(n-1)}(x)$ lies to the right of a: $a \leqq x < a + \varepsilon$.

Example. Let $y = x^3$, $x = t^2$; then $y = t^6$, and

$$dy = 6t^5 \, dt, \qquad d^2y = 30t^4 \, dt^2, \qquad \frac{d^2y}{dt^2} = 30t^4.$$

But, if we compute

$$dy = 3x^2 \, dx, \qquad d^2y = 6x \, dx^2, \qquad dx = 2t \, dt,$$

we have

$$\frac{d^2y}{dt^2} = \frac{6x \cdot 4t^2 \, dt^2}{dt^2} = 24t^4,$$

a false result. The error stems from $d^2y = 6x \, dx^2$ in which x is *not* an independent variable. *Such errors cannot arise if we restrict our use of differentials to those of the first order.* Except in §§ 91–94, we shall do so in this book.

A few functions admit simple expressions for the *n*th derivative. Thus, if

$$y = e^x, \qquad\qquad y^{(n)} = e^x;$$

$$y = \log x, \qquad\quad y^{(n)} = (-1)^{n-1} \frac{(n-1)!}{x^n};$$

$$y = \sin x, \qquad\quad y^{(n)} = \sin\left(x + n\frac{\pi}{2}\right);$$

$$y = \cos x, \qquad\quad y^{(n)} = \cos\left(x + n\frac{\pi}{2}\right).$$

Prove these results.

For the product of two functions we have

$$(uv)' = uv' + u'v,$$
$$(uv)'' = uv'' + 2u'v' + u''v,$$
$$(uv)''' = uv''' + 3u'v'' + 3u''v' + u''',$$

and, in general,

$$(5) \qquad (uv)^{(n)} = uv^{(n)} + c_1 u'v^{(n-1)} + c_2 u''v^{(n-2)} + \cdots$$

to $n + 1$ terms. Since the coefficients c_k are constants independent of u and v, we may take special functions to determine them; thus, if

$$u = e^{rx}, \quad v = e^x; \quad \text{then} \quad uv = e^{(1+r)x},$$
$$(uv)^{(n)} = (1 + r)^n e^{(1+r)x}, \quad u^{(k)} v^{(n-k)} = r^k e^{(1+r)x}.$$

On substituting these values in (5) and dividing out the common factor $e^{(1+r)x}$, we obtain

$$(1 + r)^n = 1 + c_1 r + c_2 r^2 + \cdots + c_n r^n.$$

Hence the coefficients c_k in (5) are the binomial coefficients $\binom{n}{k}$ of § 39, a theorem due to Leibnitz.

If D denotes the operator d/dx while D_1 and D_2 operate only on the first or second of two factors, respectively, Leibnitz's theorem may be put in the symbolic form:

(6) $$D^n(uv) = (D_1 + D_2)^n uv.$$

In particular, if $u = e^{ax}$, $D_1 e^{ax} = a e^{ax}$; we may then replace D_1 by a in (6) and obtain the important *shift formula*:

(7) $$D^n(e^{ax}v) = (a + D_2)^n e^{ax}v = e^{ax}(D + a)^n v.$$

PROBLEMS

1. When $y = A \cos nx + B \sin nx$, show that $D^2 y + n^2 y = 0$.

2. When $y = A e^{nx} + B e^{-nx}$ or $y = A \cosh nx + B \sinh nx$, show that $D^2 y - n^2 y = 0$.

3. If $f(x) = \tan^{-1} x$, show that

$$(1 + x^2) f''(x) + 2x f'(x) = 0.$$

Knowing that $f'(0) = 1$, compute $f''(0), f'''(0), f^{iv}(0)$.

4. Find the nth derivative of

$$\frac{1}{x + a}, \quad \frac{x}{x - 1}, \quad \frac{1}{x^2 - 1}, \quad \frac{x + 2}{x^2 - 1}.$$

5. Generalize (54.6) to give

$$D^n(uvw) = (D_1 + D_2 + D_3)^n uvw.$$

Verify this result when $u = x$, $v = x^2$, $w = x^3$, $n = 2$.

6. If $P(D)$ is a polynomial in D, prove the *shift formula*

$$P(D)e^{ax} v = e^{ax} P(D + a)v.$$

Show that (54.7) is a special case of this result.

7. If $f(x) = x^3/(x^2 - 1)$, prove that $f'(0) = 0$, while $f^{(n)}(0) = 0$ or $-n!$ according as n is even or odd (> 1).

8. If $y = f(x)$ has the derivatives $y', y'', \cdots$, show that the inverse function $x = \varphi(y)$ has the derivatives

$$\varphi'(y) = \frac{1}{y'}, \quad \varphi''(y) = -\frac{y''}{y'^3}, \quad \varphi'''(y) = \frac{3y''^2 - y'y'''}{y'^5}.$$

9. Prove that $D^n(e^x/x) = (-1)^n n! \, P_n(x) \, e^x x^{-n-1}$ where $P_n(x)$ is the polynomial formed by the first n terms in the Maclaurin expansion of e^{-x}.

10. If $y = f(x)$, $x = e^t$, and D denotes d/dt, show that

$$y'(x) = e^{-t} Dy, \quad y''(x) = e^{-2t} D(D - 1)y, \quad y'''(x) = e^{-3t} D(D - 1)(D - 2)y,$$

and generalize by finite induction. [Use (54.7).]

55. Rolle's Theorem. *If a function $f(x)$*

(i)   *is continuous in the closed interval $a \leqq x \leqq b$,*

(ii)   *has a derivative at every interior point, and*

(iii) $f(a) = f(b)$;

then there is a least one interior point ξ such that

(1) $f'(\xi) = 0$, $a < \xi < b$.

Proof. Under hypothesis (i), $f(x)$ must assume its minimum and maximum values, m and M, at least once in the interval (§ 47).

If $m = M$, $f(x)$ has the constant value m, $f'(x) = 0$, and (1) holds for any choice of ξ between a and b.

If $m < M$, either m or M will differ from the equal end values of $f(x)$. If $M > f(a)$, at some interior point ξ, $f(\xi) = M$. Now $f'(\xi)$ cannot be positive or negative; for then $f(x) > M$ for values near ξ and on one side or the other (§ 51), and $f(x)$ would not have a maximum M at ξ. Hence $f'(\xi) = 0$.

If $M = f(a)$, then $m < f(a)$, and, at some interior point ξ, $f(\xi) = m$. Just as before we must have $f'(\xi) = 0$.

Corollary. *If a and b are two roots of the equation $f(x) = 0$, then the equation $f'(x) = 0$ will have at least one root between a and b, provided*

(a) $f(x)$ is continuous in $a \leqq x \leqq b$,

(b) $f'(x)$ exists in $a < x < b$.

If $f(x)$ is a *polynomial*, the conditions (a) and (b) are evidently fulfilled.

56. Theorem of Darboux. In § 48 we proved that, if $f(x)$ is continuous in $a \leqq x \leqq b$ and $f(a) \neq f(b)$, then $f(x)$ will assume every value between $f(a)$ and $f(b)$ at some interior point. This property does not characterize continuous functions because it is also shared by the class of derivatives $f'(x)$, whether continuous or not.

THEOREM OF DARBOUX. *Let $f'(x)$ exist at all points of $a \leqq x \leqq b$ and $f'(a) \neq f'(b)$; then, if γ lies between $f'(a)$ and $f'(b)$, then $f'(c) = \gamma$ for some point c between a and b.*

Proof. Suppose that $f'(a) < \gamma < f'(b)$, and put

$$F(x) = f(x) - \gamma(x - a), \qquad F'(x) = f'(x) - \gamma.$$

Since $F(x)$ is differentiable in (a, b), it is also continuous there and must attain its minimum value at some point c of the interval. But, since

$$F'(a) = f'(a) - \gamma < 0, \qquad F'(b) = f'(b) - \gamma > 0,$$

c is not a or b (§ 51). Hence, at some point c between a and b, $F(x)$ has a minimum, and $F'(c) = 0$. Hence,

$$f'(c) = \gamma, \qquad a < c < b.$$

57. Mean-Value Theorem.   *If the function* $f(x)$

(i)   *is continuous in the closed interval* $a \leqq x \leqq b$,
(ii) *has a derivative at every interior point;*

then at some interior point ξ *of the interval*

(1) $$\frac{f(b) - f(a)}{b - a} = f'(\xi), \qquad a < \xi < b.$$

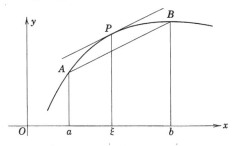

FIG. 57.   Interpretation of the mean-value theorem

Proof.   If $f(a) = f(b)$, the theorem reduces to Rolle's theorem.   We therefore assume $f(a) \neq f(b)$ and construct the function

$$F(x) = f(x) - kx$$

so that $F(a) = F(b)$; then the constant

(2) $$k = \frac{f(b) - f(a)}{b - a}.$$

Since $F(x)$ satisfies all the hypotheses of Rolle's theorem,

$$F'(\xi) = f'(\xi) - k = 0, \qquad a < \xi < b.$$

But this is precisely the desired result (1).

The mean-value theorem has a simple geometric interpretation.   If $y = f(x)$ is represented by the curve AB in Fig. 57, the value of k in (2) is slope of the chord AB.   The theorem $f'(\xi) = k$ now states that at some point P ($x = \xi$) of the curve between A and B, the tangent to the curve is parallel to the chord AB.

If $b = a + h$, we can write

$$\xi = a + \theta h \quad \text{where} \quad 0 < \theta < 1;$$

the mean-value theorem (1) now takes the form

(3) $$f(a + h) - f(a) = h f'(a + \theta h), \qquad 0 < \theta < 1.$$

The mean-value theorem is one of the basic formulas of the differential calculus. Before applying it we must be certain that both hypotheses (a) and (b) are fulfilled. Thus the function $f(x) = |x|$ is everywhere continuous, but the mean-value theorem fails in every interval (a, b) having the origin in its interior. Show this and give the reason for the failure.

We now draw a number of important conclusions from the mean-value theorem.

Corollary 1. If $f'(x) = 0$ in the interval $a < x < b$, then $f(x)$ is constant in this interval.

Proof. Let $a < x_1 < x_2 < b$; since $f(x)$ is continuous in $x_1 \leq x \leq x_2$, we have $f(x_2) - f(x_1) = 0$ from (1). Thus $f(x)$ has the same value at *all* points between a and b.

Corollary 2. If $f'(x) = g'(x)$ in the interval $a < x < b$, then

$$f(x) = g(x) + \text{const}, \qquad (a < x < b).$$

Proof. Apply Corollary 1 to $F(x) = f(x) - g(x)$.

Corollary 3. If $f(x)$ is continuous in $a \leq x \leq b$, and $f'(x) > 0$ in $a < x < b$, then $f(x)$ is an increasing function in $a \leq x \leq b$.

Proof. If x_1 and x_2 are chosen so that $a \leq x_1 < x_2 \leq b$,

$$f(x_2) - f(x_1) = (x_2 - x_1) f'(\xi) > 0, \qquad x_1 < \xi < x_2.$$

Note that we may take $x_1 = a$ or $x_2 = b$.

Corollary 4. If $f'(x)$ exists for $a < x \leq b$ and $|f(x)| \to \infty$ as $x \to a$, then $f'(x)$ is unbounded in the neighborhood of a.

Proof. From the mean-value theorem

$$|f(b) - f(x)| = (b - x)|f'(\xi)|, \qquad a < x < \xi < b.$$

If $f'(x)$ were bounded in $a < x \leq b$, say $|f'(x)| < M$, we would have

$$|f(b) - f(x)| < (b - a)M,$$

and $f(x)$ would remain bounded as $x \to a$. But this contradicts the hypothesis.

PROBLEMS

1. When $f(x) = (x - a)^m(x - b)^n$, where m and n are positive integers, show that ξ in Rolle's theorem divides the segment $a \leq x \leq b$ in the ratio m/n.

2. If $f(x) = Ax^2 + Bx + C$, show that $\theta = \frac{1}{2}$ in (57.3).

3. Prove the theorem: If $f'(x) \geq 0$ (but not identically zero) in $a \leq x \leq b$, then $f(b) > f(a)$.
[Show that, for any x in $a < x < b, f(a) \leq f(x) \leq f(b)$ but not $f(a) = f(x) = f(b)$.]

4. Prove (a): $x/(1 + x) < \log(1 + x) < x, x > 0$; (b): $x < -\log(1 - x) < x/(1 - x)$, $0 < x < 1$.

5. In the neighborhood of the point ξ, $(a)\, f(x)$ is continuous, $(b)\, f'(x)$ exists when $x \neq \xi$, $(c)\, \lim\limits_{x \to \xi} f'(x) = k$. Prove that $f'(\xi) = k$.

6. A twice differentiable function $f(x)$ is such that $f(a) = f(b) = 0$, and $f(c) > 0$ where $a < c < b$. Prove that there is at least one value ξ between a and b for which $f''(\xi) < 0$.

7. If the functions $f(x)$, $g(x)$ satisfy the conditions (i) and (ii) of the mean-value theorem, prove that

$$\begin{vmatrix} f(a) & f(b) \\ g(a) & g(b) \end{vmatrix} = (b - a) \begin{vmatrix} f(a) & f'(\xi) \\ g(a) & g'(\xi) \end{vmatrix}, \qquad a < \xi < b.$$

Consider the case $g(x) = 1$.

8. If $f(x)$, $g(x)$ $h(x)$ satisfy the conditions (i) and (ii) of the mean-value theorem, prove that

$$\begin{vmatrix} f(a) & f(b) & f'(\xi) \\ g(a) & g(b) & g'(\xi) \\ h(a) & h(b) & h'(\xi) \end{vmatrix} = 0, \qquad a < \xi < b.$$

9. If $f'(x) \to k$ as $x \to \infty$, show that $f(x)/x \to k$. When $f'(x) \to \infty$ as $x \to \infty$, show that $f(x) \to \infty$.

10. If $f(0) = f'(0) = 0$ and $f''(x)$ exists in $0 \leq x \leq h$, prove that
$$f(h) = \tfrac{1}{2}h^2 f''(\xi), \qquad 0 < \xi < h.$$
[Consider the function $\varphi(x) = f(x) - (x/h)^2 f(h)$.]

11. If $f''(x)$ exists in $a - h \leq x \leq a + h$, prove that
$$f(a + h) - 2f(a) + f(a - h) = h^2 f''(\xi), \qquad a - h < \xi < a + h.$$
[Apply Prob. 10 to $\varphi(x) = f(a + x) - 2f(a) + f(a - x)$, and use Darboux's theorem.]

12. Prove the theorem: In any interval in which the functions $u(x)$, $v(x)$, $u'(x)$, $v'(x)$ are continuous and $uv' - vu' \neq 0$, the roots of $u(x)$ and $v(x)$ separate each other. Verify this when $u = \sin x$, $v = \cos x$.
[Let a and b be *consecutive* roots of $u(x) = 0$. Then, if $v(x) \neq 0$ when $a < x < b$, u/v is continuous in (a, b) and vanishes at a and b; hence $(u/v)'$ must vanish at an intermediate point: a contradiction. Nor can v vanish *twice* in (a, b), for then $u(x)$ would have a root between a and b.]

13. Prove that $\left(1 + \dfrac{1}{x}\right)^x$ and $\left(1 - \dfrac{1}{x}\right)^x$ $(x > 0)$ are increasing functions

[Take logarithms and show that
$$f(x) = x \log\left(1 + \frac{1}{x}\right) \quad \text{and} \quad g(x) = x \log\left(1 - \frac{1}{x}\right)$$
have positive derivatives by using the mean-value theorem.]

14. Prove that a continuous function which takes on no value more than twice must take on some value exactly once.
[*Am. Math. Monthly*, vol. 61, 1954, p. 425.]

15. If $f'(x)$ exists in $a \leq x \leq b$, prove that $f'(x)$ can have no finite jumps in (a, b).

16. If the polynomial $P(x)$ and its derivative $P'(x)$ have no zeros in common and $P'(x)$ has the distinct zeros $x_1 < x_2 < \cdots < x_k$, show that the number of real zeros of $P(x)$ is equal to the number of variations in sign in the sequence

$$P(-\infty),\ P(x_1),\ P(x_2),\ \cdots,\ P(x_k),\ P(\infty).$$

17. Prove that the equation

$$1 - x + \frac{x^2}{2} - \frac{x^3}{3} + \cdots + (-1)^n \frac{x^n}{n} = 0$$

has one real root if n is odd, no real root if n is even (*Math. Tripos*, 1948).

58. The Iterative Solution of Equations. As an example of the use of the mean-value theorem consider the problem of finding a root α of the equation

$$(1) \qquad\qquad x = f(x).$$

This root may be computed by successive approximations if $f(x)$ fulfils the conditions of the

THEOREM. *Let α be a root of the equation $x = f(x)$. Then, if*

(a) x_0 lies in the interval I: $\alpha - h \leq x \leq \alpha + h$,

(b) $f(x)$ is differentiable throughout I,

(c) $|f'(x)| < M < 1$ in I,

we can find better and better approximations to α by the iterative process,

$$(2) \qquad x_1 = f(x_0),\quad x_2 = f(x_1),\quad \cdots,\quad x_n = f(x_{n-1}),$$

and, in fact,

$$(3) \qquad\qquad \lim_{n \to \infty} x_n = \alpha.$$

Proof. First let us prove that, if x_{n-1} lies in I, x_n is closer to α than x_{n-1} and therefore also lies in I. Since $x_n = f(x_{n-1})$, $\alpha = f(\alpha)$, we have on subtraction

$$x_n - \alpha = f(x_{n-1}) - f(\alpha) = f'(\xi)(x_{n-1} - \alpha),$$

where ξ lies between x_{n-1} and α. Hence, in view of (c),

$$|x_n - \alpha| < M|x_{n-1} - \alpha| \qquad (M < 1).$$

By applying this inequality repeatedly, we have

$$|x_n - \alpha| < M^2|x_{n-2} - \alpha| < \cdots < M^n|x_0 - \alpha|,$$

and from hypothesis (a),

$$(4) \qquad\qquad |x_n - \alpha| < M^n h.$$

Since $0 < M < 1$, $M^n \to 0$ as $n \to \infty$ (15.7); hence $x_n \to \alpha$ and $M^n h$ is an upper limit on the error of the nth approximation.

The formation of the sequence $\{x_n\}$ is shown graphically in Figs. 58a–b. Note that in both cases $|f'(\alpha)| < 1$.

Let us now consider the equation $\varphi(x) = 0$ which has a real root near x_0, say at $\alpha = x_0 + h$. Now from the mean-value theorem (57.3)

$$0 = \varphi(x_0 + h) = \varphi(x_0) + h\varphi'(x_0 + \theta h);$$

and, if $\varphi'(x)$ is continuous and not zero at x_0, $h = -\varphi(x_0)/\varphi'(x_0)$ approximately. Thus, if x_0 is sufficiently near the root α, we may presume that

(5)
$$x_1 = x_0 - \frac{\varphi(x_0)}{\varphi'(x_0)}$$

is even nearer.

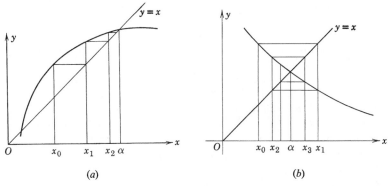

(a)   (b)

Fig. 58. Successive approximations to a root α of $f(x) = x$

This result suggests that we replace $\varphi(x) = 0$ by the equivalent equation,

(6)
$$x = f(x) \quad \text{where} \quad f(x) = x - \frac{\varphi(x)}{\varphi'(x)}.$$

Since

(7)
$$f'(x) = \frac{\varphi(x)\,\varphi''(x)}{\{\varphi'(x)\}^2},$$

the method of iteration can be applied to find a root α of (6) in whose neighborhood $|f'(x)| < 1$. The successive approximations are computed after the pattern of (5), and their accuracy may be estimated by (4). Equation (7) shows that we may expect rapid convergence when $\varphi'(x_0)$ is large, $\varphi(x_0)$ and $\varphi''(x_0)$ small.

This is *Newton's method* for finding the roots of an equation. It has a simple geometric interpretation. If the curve $y = \varphi(x)$ cuts the x-axis near x_0, the tangent to the curve at the point $[x_0, \varphi(x_0)]$, namely,

(8)
$$y - \varphi(x_0) = \varphi'(x_0)(x - x_0),$$

will cut the x-axis at the point $(x_1, 0)$ given by (5).

Example. To find the real root α of the cubic equation $x^3 - 2x - 5 = 0$, put

$$\varphi(x) = x^3 - 2x - 5, \qquad \varphi'(x) = 3x^2 - 2, \qquad \varphi'' = 6x.$$

Since $\varphi(2) = -1$, $\varphi(3) = 16$, the root is near 2. With $x_0 = 2$ we have from (5)

$$x_1 = 2 - \frac{-1}{10} = 2.1, \qquad x_2 = 2.1 - \frac{0.061}{11.23} = 2.09457.$$

The root lies in the interval (2.0, 2.1) and in this interval an upper bound of $|f'(x)|$ is given by

$$M = \frac{|\varphi(2)| |\varphi''(2.1)|}{\{\varphi'(2)\}^2} = \frac{1 \times 12.6}{100} = 0.126.$$

From (4) the error

$$|x_2 - \alpha| < (0.126)^2 \times 0.1 < 0.0016.$$

PROBLEMS

1. If condition (c) in Theorem 58 is replaced by $0 < f'(x) < M < 1$, show that $x_n \to \alpha$, from below if $x_0 < \alpha$, from above if $x_0 > \alpha$.

2. If condition (c) in Theorem 58 is replaced by $-1 < f'(x) < -M < 0$, show that the approximations x_n are alternately greater and smaller than α.

3. Find the smallest positive root of the equation $x^3 - 6x + 1 = 0$. Take $x_0 = 0$, and find x_3 by iteration on $x = \frac{1}{6}(x^3 + 1)$. Then find x_3 by Newton's method.

4. Find a root of $x^3 - 3x^2 - 1 = 0$ between 3 and 4 by iteration on $x = 3 + 1/x^2$ up to x_3, starting with $x_0 = 3$. Solve again starting with $x_0 = 4$.

5. Find the smallest positive root of $x = \tan x$. [Write $x = \pi + \tan^{-1} x$.]

6. Let us attempt to solve (1) by the iterative process (2), but without regard to the properties of $f(x)$. Then, if $x_n \to \alpha$ and $f(x)$ is continuous at α, show that $\alpha = f(\alpha)$.

$$[\,|\alpha - f(\alpha)| \le |\alpha - x_n| + |f(x_{n-1}) - f(\alpha)|\,].$$

59. Second Mean-Value Theorem (Cauchy). *If the functions $f(x)$ and $g(x)$*

(i)   *are continuous in $a \le x \le b$,*

(ii) *have derivatives at every interior point, and*

(iii) *$f'(x)$ and $g'(x)$ do not both vanish at any interior point,*

(iv) *$g(a) \ne g(b)$;*

then

(1) $$\frac{f(b) - f(a)}{g(b) - g(a)} = \frac{f'(\xi)}{g'(\xi)}, \qquad a < \xi < b.$$

Proof. Construct the function

$$F(x) = f(x) - kg(x)$$

so that $F(a) = F(b)$; then the constant

(2) $$k = \frac{f(b) - f(a)}{g(b) - g(a)}$$

in view of (iv). Since $F(x)$ satisfies all the hypotheses of Rolle's theorem,

$$F'(\xi) = f'(\xi) - kg'(\xi) = 0, \qquad a < \xi < b.$$

But this gives precisely the result (1).   Note that $g'(\xi) \neq 0$; for $g'(\xi) = 0$ would imply $f'(\xi) = 0$ and contradict (iii).

Hypotheses (iii) and (iv), which serve to rule out division by zero, may be replaced by the single condition

(iii)′ $g'(x) \neq 0$ $(a < x < b)$,

Both (iii) and (iv) are now fulfilled; for, if $g(a) = g(b)$, $g'(x)$ would vanish between a and b by Rolle's theorem.

If we put $b = a + h$, we can write (1) in the form

(3) $\dfrac{f(a + h) - f(a)}{g(a + h) - g(a)} = \dfrac{f'(a + \theta h)}{g'(a + \theta h)},$ $0 < \theta < 1.$

When $g(x) = x$, $g'(x) = 1$, and (1) becomes the mean-value theorem (57.1).

60. l'Hospital's Rule.   The ratio of two functions $f(x)/g(x)$ is not defined for $x = a$ when $g(a) = 0$.   If both $f(a)$ and $g(a)$ vanish, the ratio has the "indeterminate form" $0/0$ for $x = a$; in this case $f(x)/g(x)$ may approach a definite limit when $x \to a$.

Whenever we compute a derivative

$$\varphi'(a) = \lim_{x \to a} \frac{\varphi(x) - \varphi(a)}{x - a},$$

we have a problem of this sort; for $\varphi(x) - \varphi(a)$ and $x - a$ are functions of x which vanish for $x = a$.   Thus we might expect that a knowledge of derivatives would prove useful in dealing with the form $0/0$.

A basic theorem in this connection is given by

L'HOSPITAL'S RULE.   *If the functions $f(x)$ and $g(x)$ are*

(i)   *continuous for $a \leq x \leq a + h$,*
(ii)   *differentiable for $a < x \leq a + h$, and*
(iii) $f(a) = g(a) = 0,$
(iv) $\lim\limits_{x \to a+} \dfrac{f'(x)}{g'(x)} = A$ (or $\pm\infty$),

then

(1) $\lim\limits_{x \to a+} \dfrac{f(x)}{g(x)} = A$ (or $\pm\infty$).

Proof.   Since the limit in (iv) exists, $g'(x) \neq 0$ in some interval $a < x \leq a + h$ when h is chosen suitably small; otherwise f'/g' would not be defined for an infinite number of values between a and $a + h$ (no matter how small h is chosen) and the limit A would not exist. Thus all the conditions of Cauchy's formula (59.1) are met in the interval $(a, a + h)$ and, in view of (iii),

$$\frac{f(x)}{g(x)} = \frac{f(x) - f(a)}{g(x) - g(a)} = \frac{f'(\xi)}{g'(\xi)}, \qquad a < \xi < x;$$

$$\lim_{x \to a+} \frac{f(x)}{g(x)} = \lim_{\xi \to a+} \frac{f'(\xi)}{g'(\xi)} = A.$$

Note that, no matter how $x \to a+$, $\xi \to a+$ in some way (in general unknown) and hence $f'(\xi)/g'(\xi) \to A$ by (iv).

However, if $f'(\xi)/g'(\xi)$ does *not* approach a limit, we cannot conclude that $f(x)/g(x)$ also has no limit. For *all* sequences $x_n \to a+$ may correspond to only *special* sequences $\xi_n \to a+$, and for *these* $f'(\xi)/g'(\xi)$ may have a limit; whereas, for sequences other than these special ones, $f'(\xi)/g'(\xi)$ may not have a limit. Consider, for example,

$$f(x) = x^2 \sin \frac{1}{x}, \quad f(0) = 0; \qquad g(x) = \sin x;$$

then, as $x \to 0$,

$$\frac{f(x)}{g(x)} = \frac{x}{\sin x} \cdot x \sin \frac{1}{x} \to 0, \quad \text{but} \quad \frac{f'(x)}{g'(x)} = \frac{2x \sin \frac{1}{x} - \cos \frac{1}{x}}{\cos x}$$

approaches no limit whatever. Evidently the conditions of the theorem are sufficient but not necessary for the existence of $\lim f(x)/g(x)$.

Corollary 1. If the conditions hold for the interval $(a - h, a)$ or $(a - h, a + h)$, the theorem holds for the limits $x \to a-$ and $x \to a$, respectively.

Corollary 2. If conditions (i) and (ii) hold when x is sufficiently large (say $x > k$) and

(iii) $$\lim_{x \to \infty} f(x) = \lim_{x \to \infty} g(x) = 0,$$

then

(2) $$\lim_{x \to \infty} \frac{f(x)}{g(x)} = \lim_{t \to \infty} \frac{f'(x)}{g'(x)},$$

when the last limit exists.

Proof. Put $x = 1/t$; then

$$\lim_{x \to \infty} \frac{f(x)}{g(x)} = \lim_{t \to 0} \frac{f(1/t)}{g(1/t)} = \lim_{t \to 0} \frac{-t^{-2} f'(1/t)}{-t^{-2} g'(1/t)} = \lim_{x \to \infty} \frac{f'(x)}{g'(x)}.$$

Thus l'Hospital's rule holds when $a = \infty$.

Corollary 3.   If $f'(a) = g'(a) = 0$, l'Hospital's rule may be applied to find $\lim f'(x)/g'(x)$; then, as $x \to a+$,

$$\lim \frac{f(x)}{g(x)} = \lim \frac{f'(x)}{g'(x)} = \lim \frac{f''(x)}{g''(x)},$$

if this last limit exists.   We may continue in this fashion until at least one derivative in the ratio is not zero for $x = a$.

Example 1. When $x = 0$, $\log (1 + x) = 0$; hence,

$$\lim_{x \to 0} \frac{\log (1 + x)}{x} = \lim_{x \to 0} \frac{1}{1 + x} = 1.$$

Example 2. If m and n are positive integers,

$$\lim_{x \to 1} \frac{x^m - 1}{x^n - 1} = \lim_{x \to 1} \frac{m \, x^{m-1}}{n x^{n-1}} = \frac{m}{n}.$$

Example 3. As $x \to 0$

$$\lim \frac{\tan x - x}{x - \sin x} = \lim \frac{\sec^2 x - 1}{1 - \cos x} = \lim \frac{2 \sec^2 x \tan x}{\sin x},$$

on using the rule twice.   Since the last fraction equals $2 \sec^3 x$ its limiting value is 2; or, if we apply the rule again, we find

$$\lim_{x \to 0} \frac{4 \sec^2 x \tan^2 x + 2 \sec^4 x}{\cos x} = 2.$$

Example 4.   If $f'(x)$ exists in the neighborhood of a

$$\lim_{h \to 0} \frac{f(a + h) - f(a - h)}{2h} = \lim_{h \to 0} \frac{f'(a + h) + f'(a - h)}{2} = f'(a).$$

If $f''(x)$ exists in the neighborhood of a.

$$\lim_{h \to 0} \frac{f(a + h) + f(a - h) - 2f(a)}{h^2} = \lim_{h \to 0} \frac{f'(a + h) - f'(a - h)}{2h} = f''(a).$$

Example 5. As $x \to \infty$, the fraction

$$\frac{f(x)}{g(x)} = \frac{e^{-2x}(\cos x + 2 \sin x)}{e^{-x}(\cos x + \sin x)}$$

has the form 0/0.   Now

$$\frac{f'(x)}{g'(x)} = \frac{-5e^{-2x} \sin x}{-2e^{-x} \sin x}.$$

When $\sin x \neq 0$, $f'/g' = \frac{5}{2} e^{-x} \to 0$ as $x \to \infty$.   Nevertheless f'/g' has no limit as $\to \infty$, for, as x becomes infinite along the sequence $n\pi$, f'/g' is never defined.   (Remember that

$$\lim_{x \to \infty} \frac{f'(x)}{g'(x)} = A \quad \text{means that} \quad \lim \frac{f'(x)}{g'(x)} = A$$

for *all* sequences $\{x_n\} \to \infty$.   Cf. § 44.)

If we cancel $\sin x$ in f'/g' and take $\lim f'/g' = 0$, we arrive at the *false conclusion* that $\lim f/g = 0$; for, as $x \to \infty$,

$$\frac{f}{g} = e^{-x} \frac{1 + 2\tan x}{1 + \tan x}$$

approaches no limit whatever, since it takes on all real values in every neighborhood of ∞.

61. The Form ∞/∞. l'Hospital's rule may also be applied to find $\lim_{x \to a} f(x)/g(x)$ when $f(x)$ and $g(x)$ become infinite as $x \to a$. The proof, however, is more difficult. If we try to reduce this case to the form $0/0$, we have

$$\lim \frac{f(x)}{g(x)} = \lim \frac{1/g(x)}{1/f(x)} = \lim \frac{-g'(x)/\{g(x)\}^2}{-f'(x)/\{f(x)\}^2} = \lim \left\{\frac{f(x)}{g(x)}\right\}^2 \frac{g'(x)}{f'(x)}.$$

Hence, if we know in advance that

$$\lim \frac{f(x)}{g(x)} = L \neq 0, \qquad \lim \frac{f'(x)}{g'(x)} = A \neq 0,$$

we have $L = L^2/A$ and $L = A$. But, since the existence of L is usually not known in advance, this reasoning is not applicable. What we need is a rule that tells us that, when A exists, L also exists and equals A.

THEOREM. *If the functions $f(x)$ and $g(x)$ are*

(i) *continuous for $a \leq x \leq a + h$,*

(ii) *differentiable for $a < x \leq a + h$, and*

(iii) $\lim\limits_{x \to a+} f(x) = \lim\limits_{x \to a+} g(x) = \infty$,

(iv) $\lim\limits_{x \to a+} \dfrac{f'(x)}{g'(x)} = A$ *(or $\pm \infty$),*

then

(1) $$\lim_{x \to a+} \frac{f(x)}{g(x)} = A \quad \text{(or $\pm\infty$)}.$$

Proof. Since the limit in (iv) exists, $g'(x) \neq 0$ in some interval $a < x \leq a + h$ (cf. § 60). Hence we may apply Cauchy's formula (59.1) in the interval $(a, a + h)$; and, if $a < x < b < a + h$,

$$\frac{f(x) - f(b)}{g(x) - g(b)} = \frac{f(x)}{g(x)} \frac{1 - f(b)/f(x)}{1 - g(b)/g(x)} = \frac{f'(\xi)}{g'(\xi)},$$

where $a < x < \xi < b$. Hence,

(2) $$\frac{f(x)}{g(x)} = \frac{f'(\xi)}{g'(\xi)} \frac{1 - g(b)/g(x)}{1 - f(b)/f(x)}.$$

Case 1. *A finite.* Choose a number ε, arbitrarily small, between 0 and 1. By taking b (and hence ξ) close enough to a we can, by virtue of (iv), make $f'(\xi)/g'(\xi)$ differ from A by less than ε:

$$\frac{f'(\xi)}{g'(\xi)} = A + \eta_1, \qquad |\eta_1| < \varepsilon < 1.$$

When b is fixed and $x \to a$, (iii) shows that the last fraction in (2) approaches 1. Hence, with b fixed, we can choose x close enough to a so that

$$\frac{1 - g(b)/g(x)}{1 - f(b)/f(x)} = 1 + \eta_2 \text{ where } |\eta_2| < \begin{cases} \varepsilon & \text{if } |A| \leq 1, \\ \dfrac{\varepsilon}{|A|} & \text{if } |A| > 1. \end{cases}$$

Then

$$\frac{f(x)}{g(x)} = (A + \eta_1)(1 + \eta_2) = A + \eta_1 + A\eta_2 + \eta_1\eta_2,$$

$$\left| \frac{f(x)}{g(x)} - A \right| \leq |\eta_1| + |A|\,|\eta_2| + |\eta_1|\,|\eta_2| < 3\varepsilon.$$

Since 3ε can be taken arbitrarily small, $f(x)/g(x) \to A$.

Case 2: $A = +\infty$. By choosing b close enough to a we can make

$$\frac{f'(x)}{g'(x)} > M, \quad M \text{ arbitrarily large.}$$

Keeping b fixed, we can choose x close enough to a so that

$$\frac{1 - g(b)/g(x)}{1 - f(b)/f(x)} > 1 - \frac{1}{M}.$$

Then, from (2),

$$\frac{f(x)}{g(x)} > M\left(1 - \frac{1}{M}\right) = M - 1,$$

and, hence, $f(x)/g(x) \to \infty$.

Case 3: $A = -\infty$. Since $-f'(x)/g'(x) \to \infty$, by Case 2, $-f(x)/g(x) \to \infty$. The proof is now complete.

Note that Conditions (iii) imply that $f'(a)$ and $g'(a)$ are also infinite (Cor. 57.4); hence $f'(x)/g'(x)$ is again indeterminate for $x = a$. Nevertheless, the fraction f'/g' may be amenable to certain reductions to which f/g is not; for example,

$$\lim_{x \to 0} \frac{x^{-1/2}}{\log x} = \lim_{x \to 0} \frac{-\frac{1}{2}x^{-3/2}}{x^{-1}} = \lim_{x \to 0} \left(-\tfrac{1}{2}x^{-1/2}\right) = \infty.$$

Corollary 1.   If conditions (i) and (ii) hold when x is sufficiently large (say $x > k$) and

(iii) $$\lim_{x \to \infty} f(x) = \lim_{x \to \infty} g(x) = \infty,$$

then

$$\lim_{x \to \infty} \frac{f(x)}{g(x)} = \lim_{x \to \infty} \frac{f'(x)}{g'(x)},$$

when the last limit exists.   The proof is the same as for the form 0/0 (§ 60, Cor. 2).

If $f'(x)/g'(x)$ does not approach a limit we cannot conclude that $f(x)/g(x)$ also has no limit (cf. § 60).   Consider, for example,

$$f(x) = \frac{1}{x}, \qquad g(x) = \frac{1}{x} + \sin \frac{1}{x} \, ;$$

as $x \to 0$, f/g takes the form ∞/∞.   Now

$$\frac{f'(x)}{g'(x)} = \frac{-x^{-2}}{-x^{-2}(1 + \cos x^{-1})} = \frac{1}{1 + \cos x^{-1}}$$

approaches no limit as $x \to 0$; but

$$\frac{f(x)}{g(x)} = \frac{1}{1 + x \sin x^{-1}} \to 1.$$

Example 1.   Using the rule twice, we have

$$\lim_{x \to \infty} \frac{2x^2 - 1}{x^2 + x + 1} = \lim_{x \to \infty} \frac{4x}{2x + 1} = \frac{4}{2} = 2.$$

In general, if

$$a_0 x^n + a_1 x^{n-1} + \cdots, \qquad b_0 x^n + b_1 x^{n-1} + \cdots$$

are two polynomials of degree n, we have on using the rule n times:

(3) $$\lim_{x \to \infty} \frac{a_0 x^n + \cdots}{b_0 x^n + \cdots} = \frac{n! \, a_0}{n! \, b_0} = \frac{a_0}{b_0}.$$

Example 2.   If n is a positive integer, we find after n applications of the rule

$$\lim_{x \to \infty} \frac{x^n}{e^x} = \lim_{x \to \infty} \frac{n!}{e^x} = 0.$$

If α is a positive number between the integers $n - 1$ and n, we can use the rule n times;

(4) $$\lim_{x \to \infty} \frac{x^\alpha}{e^x} = \lim_{x \to \infty} \frac{\alpha(\alpha - 1) \cdots (\alpha - n + 1)x^{\alpha-n}}{e^x} = 0,$$

since $\alpha - n < 0$.†

Equation (4) shows that

The exponential e^x becomes infinite more rapidly than x^α ($\alpha > 0$), no matter how large α is taken.

---

† When $\alpha \leqq 0$, x^α/e^x is not indeterminate as $x \to \infty$.   But now we see directly that $x^\alpha/e^x \to 0$.   Hence (4) holds for all real values of α.

Example 3. If $\alpha > 0$,

(5) $$\lim_{x \to \infty} \frac{\log x}{x^\alpha} = \lim_{x \to \infty} \frac{x^{-1}}{\alpha x^{\alpha - 1}} = \lim_{x \to \infty} \frac{1}{\alpha x^\alpha} = 0.$$

The power x^α ($\alpha > 0$) becomes infinite more rapidly than $\log x$, no matter how small α is taken.

62. Other Indeterminate Forms. The form $0 \cdot \infty$ is obviously reducible to either $0/0$ or ∞/∞. The form $\infty - \infty$ is readily altered to $\infty \cdot 0$; for, if $f(x)$ and $g(x) \to \infty$ as $x \to a$, we can set

$$f - g = fg \left(\frac{1}{g} - \frac{1}{f} \right).$$

The three exponential forms 0^0, ∞^0, 1^∞ are dealt with by taking their logarithm; in all cases this leads to the form $0 \cdot \infty$.

Example 1. ($\infty \cdot 0$). As $x \to 0$,

$$\cot x \log \frac{1 + x}{1 - x} = \frac{\log (1 + x) - \log (1 - x)}{\tan x} \to \frac{\dfrac{1}{1 + x} + \dfrac{1}{1 - x}}{\sec^2 x} \to 2.$$

Example 2. ($\infty - \infty$). As $x \to 0$.

$$\frac{1}{x} - \frac{1}{\sin x} = \frac{\sin x - x}{x \sin x} \to \frac{\cos x - 1}{x \cos x + \sin x} \to \frac{-\sin x}{2 \cos x - x \sin x} \to 0.$$

Example 3. (0^0). As $x \to 0$,

$$\log x^x = x \log x = \frac{\log x}{x^{-1}} \to \frac{-x^{-1}}{x^{-2}} = -x \to 0;$$

hence, $x^x \to e^0 = 1$.

One must not infer from this example that the form 0^0 always leads to 1 as a limit. As a simple counterexample consider

$$\lim_{x \to 0+} x^{a/\log x} = e^a; \quad \text{for} \quad \log x^{a/\log x} = a.$$

Example 4. (∞^0). As $x \to 0$,

$$\log \left(1 + \frac{1}{x} \right)^x = x \log \left(1 + \frac{1}{x} \right) = \frac{\log (x + 1) - \log x}{x^{-1}} \to \frac{(x + 1)^{-1} - x^{-1}}{-x^{-2}}$$

$$= x - \frac{x^2}{x + 1} \to 0;$$

hence, $\left(1 + \dfrac{1}{x} \right)^x \to e^0 = 1$.

Example 5. (1^∞). As $x \to 0$.

$$\log (\cos x)^{1/x^2} = \frac{\log \cos x}{x^2} \to \frac{-\tan x}{2x} \to \frac{-\sec^2 x}{2} \to -\frac{1}{2};$$

hence, $(\cos x)^{1/x^2} \to e^{-\frac{1}{2}} = 1/\sqrt{e}$.

PROBLEMS

1. As $x \to 0$, find the limit of

$$\frac{\sin x - x}{x - \tan x}; \qquad \frac{\tan x - \sin x}{x^3}; \qquad \frac{a^x - b^x}{x}; \qquad \frac{\sin x - x \cos x}{\sin x - x}.$$

2. As $x \to 1$, find the limit of

$$\frac{\sin \pi x}{1 - x}; \qquad \frac{2x^4 - 3x^3 + x}{(x - 1)^2}; \qquad \frac{1 - 4 \sin^2 \frac{1}{6}\pi x}{1 - x^2}.$$

3. As $x \to \pi/2$, find the limit of

$$\frac{\sec 3x}{\sec x}; \qquad \frac{\tan 3x}{\tan x}; \qquad \frac{\log (x - \pi/2)}{\tan x}.$$

4. As $x \to \pi/2$ find the limit of $(\sin x)^{\tan x};$ $(\tan x)^{\sin 2x}$.

5. Find $\lim (1 + ax)^{b/x}$ as $x \to 0$; as $x \to \infty$.

6. Find $\lim (\cot x - 1/x)$ as $x \to 0$.

7. Find $\lim x\{\sqrt{x^2 + a^2} - x\}$ as $x \to \infty$.

8. If $f(x)$ has a continuous second derivative in the neighborhood of $x = a$, prove that

$$\lim_{x \to a} \left\{ \frac{1}{f(x) - f(a)} - \frac{1}{(x - a) f'(a)} \right\} = -\frac{f''(a)}{2f'(a)^2}.$$

9. Find $\displaystyle\lim_{x \to 0} \frac{e^x - e^{\tan x}}{x - \tan x}$.

10. If $f(0) = 0$, $f'(x)$ is continuous in the neighborhood of 0, and $f'(0) \neq 0$, prove that $\displaystyle\lim_{x \to 0} x^{f(x)} = 1$.

11. If $f(x) = \dfrac{x \log x}{1 - x}$ $(x \neq 0, 1)$, and $f(0) = 0, f(1) = -1$, prove that $f(x)$ is a continuous, decreasing function in $(0, 1)$ and that $f'(1) = -1/2$.

12. A body falling from rest through x ft attains a speed of v ft/sec in t sec. If V denotes the limiting or "terminal" speed of the body in a medium whose resistance is proportional to v^2, we have

$$v = V \tanh \frac{gt}{V}, \qquad x = \frac{V^2}{g} \log \cosh \frac{gt}{V}, \qquad v^2 = V^2(1 - e^{-2gx/V^2}).$$

By letting $V \to \infty$, obtain the corresponding results for fall in a vacuum:

$$v = gt, \qquad x = \tfrac{1}{2}gt^2, \qquad v^2 = 2gx.$$

63. Approximating Polynomial. Let the function $f(x)$ be defined at the point $x = a$ and have derivatives there up to the nth order. We propose to find a polynomial of degree n,

$$(1) \qquad P_n(x) = c_0 + c_1(x - a) + c_2(x - a)^2 + \cdots + c_n(x - a)^n,$$

such that

$$(2) \qquad P_n(a) = f(a), \qquad P_n^{(k)}(a) = f^{(k)}(a) \qquad (k = 1, 2, \cdots, n);$$

that is, at the point $x = a$, $P_n(x)$ and its first n derivatives coincide in value with $f(x)$ and its first n derivatives.   Since

(3) $P_n(a) = c_0,$ $P_n^{(k)}(a) = c_k k!$ $(k = 1, 2, \cdots, n),$

we have from (2)

(4) $c_k = \dfrac{f^{(k)}(a)}{k!}.$

Thus the desired polynomial is

(5) $P_n(x) = f(a) + \dfrac{f'(a)}{1!}(x - a) + \cdots + \dfrac{f^{(n)}(a)}{n!}(x - a)^n.$

When $f(x)$ is itself a polynomial of degree n, $f(x) - P_n(x) = F(x)$ is a polynomial for which $F(a) = F'(a) = \cdots = F^{(n)}(a) = 0$.   Consequently, if $F(x)$ is put in the form (1), all of its coefficients vanish by (4); hence $F(x) = 0$, and $P_n(x) = f(x)$.

Example 1.   If $f(x) = x^3 + 2x^2 - x + 3$, $a = 1$, we have
$$f(1) = 5, \quad f'(1) = 6, \quad f''(1) = 10, \quad f'''(1) = 6;$$
$$f(x) = 5 + 6(x - 1) + 5(x - 1)^2 + (x - 1)^3.$$

Example 2. If $f(x) = e^x$, $a = 0$, we have
$$f(0) = f'(0) = f''(0) = \cdots = 1,$$
and
$$P_n(x) = 1 + \frac{x}{1!} + \frac{x^2}{2!} + \cdots + \frac{x^n}{n!}.$$

64. The Remainder.   We now consider the difference between the function $f(x)$ and its approximating polynomial, the *remainder*
$$R_n(x) = f(x) - P_n(x).$$

If we assume that the derivative $f_+^{(n+1)}(a)$ exists, $f^{(n)}(x)$ must exist in some closed interval $a \leq x \leq b$ and be continuous when $x = a$ (§ 54); then
$$f^{(n-1)}(x), \quad f^{(n-2)}(x), \quad \cdots, \quad f'(x), \quad f(x)$$

must all be continuous in this interval.   Since the polynomial $P_n(x)$ also has these properties, the same is true of the remainder $R_n(x)$.   If we drop the subscript n, the two functions,

(1) $R(x) = f(x) - P(x),$ $Q(x) = (x - a)^{n+1},$

and their first n derivatives vanish at $x = a$; moreover, Q and its derivatives vanish *only* at a.   We may therefore apply the second mean-value theorem (§ 59) n times to R/Q:

(2) $\dfrac{R(x)}{Q(x)} = \dfrac{R(x) - R(a)}{Q(x) - Q(a)} = \dfrac{R'(\xi_1)}{Q'(\xi_1)} = \dfrac{R''(\xi_2)}{Q''(\xi_2)} = \cdots = \dfrac{R^{(n)}(\xi_n)}{Q^n(\xi_n)},$

where $a < \xi_n < \xi_{n-1} < \cdots < \xi_2 < \xi_1 < x < b$. Now from (63.1) and (63.4)

$$P^{(n)}(x) = n!c_n = f^{(n)}(a),$$

and from (1)

$$R^{(n)}(x) = f^{(n)}(x) - f^{(n)}(a), \qquad Q^{(n)}(x) = (n+1)!(x-a);$$

hence (2) becomes

(3)
$$\frac{R(x)}{(x-a)^{n+1}} = \frac{1}{(n+1)!} \frac{f^{(n)}(\xi_n) - f^{(n)}(a)}{\xi_n - a}.$$

Since, by definition,

$$\lim_{x \to a+} \frac{f^{(n)}(x) - f^{(n)}(a)}{x - a} = f_+^{(n+1)}(a),$$

we have from (3)

(4)
$$\lim_{x \to a+} \frac{R_n(x)}{(x-a)^{n+1}} = \frac{f_+^{(n+1)}(a)}{(n+1)!}.$$

Hence, given an arbitrary positive ε, we can take $x - a$ sufficiently small so that

$$\left| \frac{R_n(x)}{(x-a)^{n+1}} - \frac{f_+^{(n+1)}(a)}{(n+1)!} \right| < \varepsilon.$$

Thus, if we put $R_n(x) = r_n(x)(x-a)^{n+1}$, the function $r_n(x)$ remains bounded and approaches $f_+^{(n+1)}(a)/(n+1)!$ as $x \to a$. Thus from (1) we have

$$f(x) = P(x) + r_n(x)(x-a)^{n+1},$$

or, on writing $x = a + h$, $r_n(h)$ for $r_n(a+h)$,

(5)　　$$f(a+h) = f(a) + \frac{f'(a)}{1!} h + \cdots + \frac{f^{(n)}(a)}{n!} h^n + r_n(h) h^{n+1}.$$

This development is called the *Taylor expansion* of $f(x)$ about the point $x = a$. In particular, when $a = 0$ (and hence $x = h$), we have the *Maclaurin expansion*,

(6)　　$$f(x) = f(0) + \frac{f'(0)}{1!} x + \cdots + \frac{f^{(n)}(0)}{n!} x^n + r_n(x) x^{n+1},$$

in which

(7)
$$\lim_{x \to 0} r_n(x) = \frac{f^{(n+1)}(0)}{(n+1)!}.$$

The Taylor expansion of a function is unique; more specifically we have the

THEOREM. *If we can express*

(8)　　$$f(a+h) = c_0 + c_1 h + \cdots + c_n h^n + \overline{r}_n(h) h^{n+1},$$

where the coefficients c_k are independent of h and $\bar{r}_n(h)$ is a function that remains bounded as $h \to 0$, we have a Taylor expansion; that is,

$$(9) \qquad c_k = \frac{f^{(k)}(0)}{k!} \quad (k = 0, 1, \cdots, n), \qquad \bar{r}_n(h) = r_n(h).$$

Proof. From the identity in h,

$$c_0 + c_1 h + \cdots + \bar{r}_n h^{n+1} = f(a) + \frac{f'(a)}{1!} h + \cdots + r_n h^{n+1},$$

we deduce $c_0 = f(a)$ by letting $h \to 0$ (note that $\bar{r}_n$ and r_n remain bounded). On dropping these equal terms, dividing by h, and again letting $h \to 0$, we get $c_1 = f'(a)$. Proceeding in this manner, we obtain all the equations (9).

We take this occasion to mention two notations that have gained wide currency in mathematics. If φ is a function involved in a limiting process, then:

(i) $O(\varphi)$ denotes the product of φ by any function that remains bounded,

(ii) $o(\varphi)$ denotes the product of φ by any function that approaches zero in the limiting process in question.

Therefore

$$(10) \qquad f = O(g) \quad \text{implies} \quad \left|\frac{f}{g}\right| < M;$$

$$(11) \qquad f = o(g) \quad \text{implies} \quad \frac{f}{g} \to 0.$$

Thus, if $f = O(1)$, f itself is a bounded function; and, if $f = o(1)$, $f \to 0$.

With respect to the limiting process $x \to 0$, the remainder in (6) may now be written $O(x^{n+1})$ or $o(x^n)$. Of course neither notation conveys the specific information given in (7).

Example 1. The sum of $n + 1$ terms of the geometric series $1 + x + x^2 + \cdots$ is given by (22.2):

$$1 + x + x^2 + \cdots + x^n = \frac{1 - x^{n+1}}{1 - x}, \qquad (x \neq 1).$$

Hence,

$$(12) \qquad \frac{1}{1 - x} = 1 + x + x^2 + \cdots + x^n + \frac{x^{n+1}}{1 - x}.$$

This is a Maclaurin expansion for $(1 - x)^{-1}$; for $r_n = (1 - x)^{-1}$ remains bounded as $x \to 0$. Moreover, when $|x| < 1$, the remainder $r_n x^{n+1} \to 0$ as $n \to \infty$; we thus obtain the *Maclaurin series*: for $(1 - x)^{-1}$, valid when and only when $|x| < 1$:

$$(13) \qquad \frac{1}{1 - x} = 1 + x + x^2 + \cdots x^n + \cdots, \qquad (-1 < x < 1).$$

Example 2. If we differentiate both members of (12), we have

(14)
$$\frac{1}{(1-x)^2} = 1 + 2x + 3x^2 + \cdots + nx^{n-1} + \frac{1+n-nx}{(1-x)^2} x^n.$$

This is the Maclaurin expansion for $(1-x)^{-2}$; for r_{n-1}, the fraction in the last term, remains bounded as $x \to 0$; in fact,

$$\lim_{x \to 0} r_{n-1} = 1 + n = \frac{f^{(n)}(0)}{n!}$$

in agreement with (7).

When $|x| < 1$, both $x^n \to 0$ and $nx^n \to 0$ as $n \to \infty$ (26.6); consequently $R_{n-1} \to 0$ as $n \to \infty$, and we obtain the Maclaurin series for $(1-x)^{-2}$:

(15)
$$\frac{1}{(1-x)^2} = 1 + 2x + 3x^2 + \cdots + (n+1)x^n + \cdots \qquad (-1 < x < 1).$$

Example 3. *Indeterminate Forms.* We can now give another version of the rule for evaluating the form 0/0.

If the functions $f(x)$, $g(x)$, *along with their first n derivatives, all vanish at* $x = a$, *while* $f^{(n+1)}(a)$ *exists and* $g^{(n+1)}(a) \neq 0$, *then*

(16)
$$\lim_{x \to a} \frac{f(x)}{g(x)} = \frac{f^{(n+1)}(a)}{g^{(n+1)}(a)} \ .$$

The proof follows at once from the Taylor expansions,

$$f(x) = r_n(x) (x-a)^{n+1}, \qquad g(x) = \bar{r}_n(x) (x-a)^{n+1},$$

and the limit (4); thus as $x \to a$,

$$\frac{f(x)}{g(x)} = \frac{r_n(x)}{\bar{r}_n(x)} \to \frac{f^{(n+1)}(a)}{g^{(n+1)}(a)}.$$

65. Taylor's Theorem. Instead of merely assuming the existence of $f_+^{(n+1)}(a)$ we now make the stronger assumption that $f^{(n+1)}(x)$ exists at all points of an interval $a \leq x \leq b$. Then $f^{(n)}(x)$ is continuous in this interval, and, from the mean-value theorem (§ 57),

$$\frac{f^{(n)}(x) - f^{(n)}(a)}{x - a} = f^{(n+1)}(\xi), \qquad a < \xi < x.$$

Hence, from (64.3) we can now deduce that

$$\frac{R_n(x)}{(x-a)^{n+1}} = \frac{f^{(n+1)}(\xi)}{(n+1)!}, \qquad a < \xi < x.$$

This gives the Lagrange form of the remainder,

(1)
$$R_n(x) = \frac{f^{(n+1)}(\xi)}{(n+1)!} (x-a)^{n+1}, \qquad a < \xi < x.$$

Since $f(x) = P_n(x) + R_n(x)$, we have proved

TAYLOR'S THEOREM. *If $f^{(n+1)}(x)$ exists at all points of an interval $a \leq x \leq b$, then for any x in the interval*

$$(2) \qquad f(x) = f(a) + \frac{f'(a)}{1!}(x-a) + \cdots + \frac{f^{(n)}(a)}{n!}(x-a)^n$$
$$+ \frac{f^{(n+1)}(\xi)}{(n+1)!}(x-a)^{n+1},$$

where ξ is a number between a and x.†

The case $a = 0$ gives *Maclaurin's theorem*:

$$(3) \qquad f(x) = f(0) + \frac{f'(0)}{1!}x + \cdots + \frac{f^{(n)}(0)}{n!}x^n + \frac{f^{(n+1)}(\xi)}{(n+1)!}x^{n+1}.$$

If we put

$$x = a + h, \qquad \xi = a + \theta h \qquad (0 < \theta < 1),$$

Taylor's theorem takes the form

$$(4) \qquad f(a+h) = f(a) + \frac{f'(a)}{1!}h + \cdots + \frac{f^{(n)}(a)}{n!}h^n$$
$$+ \frac{f^{(n+1)}(a + \theta h)}{(n+1)!}h^{n+1}.$$

When $f^{(n)}(x)$ exists for any n however large and

$$(5) \qquad R_n(h) = f^{(n+1)}(a + \theta h)\frac{h^{n+1}}{(n+1)!} \to 0$$

as $n \to \infty$, we obtain from (4) the expansion of $f(a+h)$ in a convergent infinite series in powers of h:

$$(6) \qquad f(a+h) = f(a) + \frac{f'(a)}{1!}h + \frac{f''(a)}{2!}h^2 + \cdots.$$

In particular, this series holds whenever $f^{(n)}(x)$ is bounded as $n \to \infty$; for, if $|f^{(n)}(x)| < M$,

$$|R_n| < M\frac{|h|^{n+1}}{(n+1)!} \qquad \text{and} \qquad \lim_{n \to \infty} R_n = 0,$$

since $h^n/n!$ is the general term of a convergent series (Ex. 29.3).

The series (6) is called the *Taylor series* for $f(x)$ about the point $x = a$. We have seen that, when $R_n \to 0$, the series converges to $f(a+h)$; but the mere convergence of the series (6) does not imply that it converges to $f(a+h)$. Consider, for example, Cauchy's function (Ex. 66):

$$f(x) = e^{-1/x^2}, \qquad f(0) = 0;$$

---

† If $f(x)$ is not defined to the left of a the derivatives $f^k(a)$ are necessarily right-hand derivatives.

in the next article we shall prove that $f^{(n)}(0) = 0$ for all values of n. The approximating polynomial in (3) is now identically zero and hence $f(x) = R_n(x)$ for all values of n.   Although the Maclaurin expansion for $f(x)$, namely $0 + 0 + \cdots$, converges to zero, it does not represent $f(x)$; $R_n(x) = f(x)$ is independent of n and does not vanish at any point $x \neq 0$.

Consider now the function $F(x) = f(x) + g(x)$ where $f(x)$ is Cauchy's function and $g(x)$ any function whose Maclaurin expansion converges to $g(x)$.   The Maclaurin series for $F(x)$, being identical with that of $g(x)$, converges but not to $F(x)$.

66. Exponential Series.   If $f(x) = e^x$,

$$f^{(n)}(x) = e^x, \qquad f^{(n)}(0) = 1, \qquad R_n(x) = \frac{e^{\theta x}}{(n+1)!} x^{n+1},$$

where $0 < \theta < 1$; hence,

(1) $$e^x = 1 + x + \frac{x^2}{2!} + \cdots + \frac{x^n}{n!} + \frac{e^{\theta x}}{(n+1)!} x^{n+1}.$$

Since $e^{\theta x}$ lies between $e^0 = 1$ and e^x, $e^{\theta x}$ is bounded as $n \to \infty$ and $R_n(x) \to 0$. Hence e^x is represented by the Maclaurin series,

(2) $$e^x = 1 + x + \frac{x^2}{2!} + \cdots + \frac{x^n}{n!} + \cdots,$$

which converges for all values of x.   When $x = 1$,

(3) $$e = 1 + 1 + \frac{1}{2!} + \cdots + \frac{1}{n!} + \frac{e^\theta}{(n+1)!},$$

where $R_n < 3/(n+1)!$.   Thus, when $n = 9$,

$$R_n < \frac{3}{10!} = \frac{3}{3,628,800} < 0.000,001,$$

and e is given correctly to the sixth decimal place by

$$1 + 1 + \frac{1}{2!} + \cdots + \frac{1}{9!} = 2.718\ 281 \cdots.$$

If we multiply (3) by $n!$, we have

$$n!e = \text{an integer} + \frac{e^\theta}{n+1}.$$

This equation shows that e is an irrational number; for, if $e = p/q$ (the quotient of two integers), the left member is an integer when $n \geq q$, whereas the right member is fractional when $n \geq 2$.

As $x \to \infty$, e^x becomes infinite more rapidly than any positive power of x;

(4) $$\lim_{x \to \infty} \frac{e^x}{x^n} = \infty.$$

Since e^x is a positive function which increases monotonely with x (§ 51), it will suffice to prove (4) when n is a positive integer; and in this case it follows at once from (1).

Example. Cauchy's Function:

$$f(x) = e^{-1/x^2}, \qquad f(0) = 0,$$

has the remarkable property that its derivatives of all orders vanish at $x = 0$. In the graph of $y = f(x)$ (Fig. 66) this property accounts for the extreme flattening in the neighborhood of the origin.

Proof. If we put $u = 1/x$,

$$f'(0) = \lim_{x \to 0} \frac{f(x)}{x} = \lim_{u \to \infty} \frac{u}{e^{u^2}} = 0$$

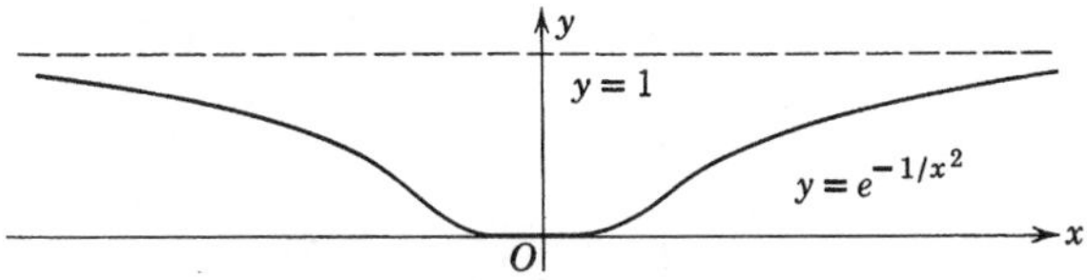

FIG. 66. Graph of Cauchy's function whose derivatives of all orders vanish at the origin

from (4). Since $f(x) = e^{-u^2}$, $u' = -u^2$, we have from the chain rule

$$f'(x) = e^{-u^2}(-2uu') = 2u^3 e^{-u^2}, \qquad x \neq 0.$$

Now

$$f''(0) = \lim_{x \to 0} \frac{f'(x)}{x} = \lim_{u \to \infty} \frac{2u^4}{e^{u^2}} = 0,$$

$$f''(x) = e^{-u^2}(6u^2 u' - 4u^4 u') = e^{-u^2}(4u^6 - 6u^4).$$

Again

$$f'''(0) = \lim_{x \to 0} \frac{f''(x)}{x} = \lim_{u \to \infty} \frac{4u^7 - 6u^5}{e^{u^2}} = 0$$

and so on. For any n we find that

$$f^{(n)}(0) = \lim_{u \to \infty} \frac{P_n(u)}{e^{u^2}} = 0,$$

where $P_n(u)$ is a polynomial in u.

If $f^{(n)}(x) = e^{-u^2} P_n(u)$, we find the recursion formula

$$P_{n+1}(u) = 2u^3 P_n(u) - u^2 P'_n(u).$$

Thus with $P_0(u) = 1$, we have

$$P_1(u) = 2u^3, \quad P_2(u) = 4u^6 - 6u^4, \quad \ldots$$

67. Sine and Cosine Series.   Both $\sin x$ and $\cos x$ are periodic functions with the period 2π and are defined for all values of x.

When $f(x) = \sin x$,

$$f'(x) = \cos x = \sin\left(x + \frac{\pi}{2}\right),$$

$$f^{(n)}(x) = \sin\left(x + n\frac{\pi}{2}\right), \qquad f^{(n)}(0) = \sin n\frac{\pi}{2}.$$

Thus $f(0), f'(0), f''(0), f'''(0), \cdots$ give the set of values $0, 1, 0, -1$ indefinitely repeated. Since the Maclaurin expansion contains only odd powers, put $n = 2k + 1$; then

$$(1) \qquad \sin x = x - \frac{x^3}{3!} + \cdots + (-1)^k \frac{x^{2k+1}}{(2k+1)!} + R_n(x);$$

and, since $\left| f^{(n)}(x) \right| \leq 1$,

$$(2) \qquad \left| R_n \right| \leq \frac{\left| x \right|^{2k+3}}{(2k+3)!}.$$

As $n \to \infty$, $R_n \to 0$, and we have

$$(3) \qquad \sin x = x - \frac{x^3}{3!} + \frac{x^5}{5!} - \cdots + (-1)^k \frac{x^{2k+1}}{(2k+1)!} + \cdots$$

for all values of x.

When $f(x) = \cos x$,

$$f'(x) = -\sin x = \cos\left(x + \frac{\pi}{2}\right),$$

$$f^{(n)}(x) = \cos\left(x + n\frac{\pi}{2}\right), \qquad f^{(n)}(0) = \cos n\frac{\pi}{2}.$$

Now $f(0), f'(0), f''(0), f'''(0), \cdots$ give the set of values $1, 0, -1, 0$ indefinitely repeated. Since the Maclaurin expansion contains only even powers of x, put $n = 2k$; then

$$(4) \qquad \cos x = 1 - \frac{x^2}{2!} + \cdots + (-1)^k \frac{x^{2k}}{(2k)!} + R_n(x),$$

where

$$\left| R_n \right| \leq \frac{\left| x \right|^{2k+2}}{(2k+2)!}, \qquad \lim_{n \to \infty} R_n = 0.$$

Therefore,

$$(5) \qquad \cos x = 1 - \frac{x^2}{2!} + \frac{x^4}{4!} - \cdots + (-1)^k \frac{x^{2k}}{(2k)!} + \cdots$$

for all values of x.

68. Even and Odd Functions. Every function $f(x)$ defined over an interval $-a < x < a$ may be expressed in just one way as the sum of an even function and an odd function (§ 43). For, if we *assume* that

$$(1) \qquad f(x) = g(x) + h(x)$$

where $g(x)$ is even, $h(x)$ odd, we have, on replacing x by $-x$,

$$f(-x) = g(x) - h(x);$$

hence, if the decomposition is possible,

$$(2) \qquad g(x) = \frac{f(x) + f(-x)}{2}, \qquad h(x) = \frac{f(x) - f(-x)}{2}.$$

This is actually the unique solution of our problem; for $g(x)$ is even, $h(x)$ is odd, and their sum is $f(x)$.

Now suppose that $f(x)$ admits the Maclaurin series,

$$(3) \qquad f(x) = c_0 + c_1 x + c_2 x^2 + c_3 x^3 + \cdots,$$

in the interval $0 \leq x < a$. Since this power series converges *absolutely* in this interval (§ 38), the series of even powers and the series of odd powers in (3) also converge in $0 \leq x < a$; and we may write

$$(4) \qquad g(x) = c_0 + c_2 x^2 + c_4 x^4 + \cdots,$$

$$(5) \qquad h(x) = c_1 x + c_3 x^3 + c_5 x^5 + \cdots;$$

for the functions defined by these series are visibly the even and odd parts of $f(x)$. Now the series (4) and (5) obviously converge to $g(x)$ and $h(x)$ in the wider interval $-a < x < a$. Hence their sum also converges to $g(x) + h(x) = f(x)$ in the wider interval $-a < x < a$ (§ 35). In brief, if $f(x)$ admits the Maclaurin series (3) in the interval $0 \leq x < a$, it also admits this expansion in the wider interval $-a < x < a$. Thus, when the Maclaurin remainder $R_n(x) \to 0$ for $0 \leq x < a$, we can conclude that $R_n(x) \to 0$ for $-a < x < a$.

Example. The even and odd parts of the exponential function,

$$e^x = 1 + x + \frac{x^2}{2!} + \frac{x^3}{3!} + \cdots,$$

are, respectively, the hyperbolic sine and cosine:

$$(6) \qquad \cosh x = \frac{e^x + e^{-x}}{2} = 1 + \frac{x^2}{2!} + \frac{x^4}{4!} + \cdots,$$

$$(7) \qquad \sinh x = \frac{e^x - e^{-x}}{2} = x + \frac{x^3}{3!} + \frac{x^5}{5!} + \cdots.$$

These expansions are valid for all values of x (§ 66).

69. Logarithmic Series. The function $f(x) = \log(1 + x)$ is defined when $x > -1$; and

$$\lim_{x \to -1} \log(1 + x) = -\infty, \qquad \lim_{x \to \infty} \log(1 + x) = +\infty.$$

In the Maclaurin expansion the constant term $f(0) = \log 1 = 0$; and since

$$f'(x) = \frac{1}{1 + x}, \qquad f^{(n)}(x) = (-1)^{n-1} \frac{(n - 1)!}{(1 + x)^n},$$

the term in x^n is

$$\frac{f^{(n)}(0)}{n!} x^n = (-1)^{n-1} \frac{x^n}{n},$$

and the remainder

$$R_n(x) = \frac{f^{(n+1)}(\theta x)}{(n+1)!} x^{n+1} = \frac{(-1)^n}{n+1} \left(\frac{x}{1+\theta x}\right)^{n+1}.$$

When $0 < x < 1$,

(1) $$|R_n| < \frac{x^{n+1}}{n+1} \quad \text{and} \quad \lim_{n \to \infty} R_n = 0;$$

hence, in this interval

(2) $$\log(1+x) = x - \frac{x^2}{2} + \frac{x^3}{3} - \frac{x^4}{4} + \cdots.$$

But, since $\log(1+x)$ is defined in the symmetric interval $-1 < x < 1$, we conclude (§ 68) that (2) is valid throughout this wider interval, and even at the end point $x = 1$ (since $R_n(1) \to 0$). For $x = 1$ we have

(3) $$\log 2 = 1 - \tfrac{1}{2} + \tfrac{1}{3} - \tfrac{1}{4} + \cdots,$$

a series that converges too slowly for practical use.

Replacing x by $-x$ in (2), we have

(4) $$\log(1-x) = -x - \frac{x^2}{2} - \frac{x^3}{3} - \frac{x^4}{4} + \cdots, \qquad -1 \leq x < 1.$$

From (68.2), the odd part of $\log(1+x)$ is one-half the difference of (2) and (4):

(5) $$\frac{1}{2} \log \frac{1+x}{1-x} = x + \frac{x^3}{3} + \frac{x^5}{5} + \cdots, \qquad -1 < x < 1.$$

If we put

$$t = \frac{1+x}{1-x}, \quad \text{then} \quad \frac{dt}{dx} = \frac{2}{(1-x)^2} > 0;$$

hence t is an increasing function of x (§ 51), and, as x increases from -1 to 1, t increases from 0 to ∞ through all positive real values. Thus, on putting $x = (t-1)/(t+1)$ in (5), we obtain a convergent series for $\log t$ valid throughout $0 < t < \infty$. If, for example, we wish to compute $\log 2$ we put $x = 1/3$ in (5); thus

(6) $$\tfrac{1}{2} \log 2 = \tfrac{1}{3} + \tfrac{1}{3}(\tfrac{1}{3})^3 + \tfrac{1}{5}(\tfrac{1}{3})^5 + \cdots,$$

a series that converges much more rapidly than (3).

70. Binomial Series. The function $f(x) = (1+x)^k$ is defined for all values of x except $x = -1$ when k is negative. If k is a positive integer, the Maclaurin series ends with the term in x^k; otherwise we obtain an infinite series. Since

$$f^{(n)}(x) = k(k-1)(k-2) \cdots (k-n+1)(1+x)^{k-n},$$

the term in x^n is

(1) $$\frac{f^{(n)}(0)}{n!}\, x^n = \frac{k(k-1)\cdots(k-n+1)}{n!}\, x^n = \binom{k}{n}\, x^n,$$

where $\binom{k}{n}$ is a binomial coefficient (§ 39); moreover,

(2) $$R_n(x) = \frac{k(k-1)\cdots(k-n)}{(n+1)!}\,(1+\theta x)^{k-n-1}x^{n+1}$$

$$= \binom{k}{n+1}\left(\frac{x}{1+\theta x}\right)^{n+1}(1+\theta x)^k.$$

When $0 < x < 1$,

(3) $$|\,R_n(x)\,| < \left|\binom{k}{n+1}\, x^{n+1}\right|\, 2^k \quad\text{and}\quad \lim_{n\to\infty} R_n(x) = 0,$$

since $\binom{k}{n} x^n$ is the general term of the binomial series (39.1) which converges (absolutely) when $|\,x\,| < 1$. Hence when $0 < x < 1$, and also in the wider interval $-1 < x < 1$ (§ 68), we have

(4) $$(1+x)^k = 1 + \binom{k}{1} x + \binom{k}{2} x^2 + \cdots + \binom{k}{n} x^n + \cdots.$$

When $x = 1$,

$$|\,R_n(1)\,| < \left|\binom{k}{n+1}\right|\, 2^k.$$

In § 39 we proved that, when $x = 1$, the binomial series converges, provided $k > -1$; hence in this case also

$$\lim_{n\to\infty} R_n(1) = 0, \qquad k > -1.$$

Therefore, on putting $x = 1$ in (4), we have

(5) $$2^k = 1 + \binom{k}{1} + \binom{k}{2} + \cdots, \qquad k > -1.$$

When $x = -1$, the binomial series converges when $k > 0$ (§ 39); we shall prove later on that (4) holds in this case also:

(6) $$0 = 1 - \binom{k}{1} + \binom{k}{2} - \binom{k}{3} + \cdots, \qquad k > 0.$$

When k is a positive integer, $\binom{k}{r}$ is the number of combinations of k different things taken r at a time. Interpret (5) and (6) in this case.

Example. To expand $(1+x)^{-\frac{1}{2}}$ we have

$$\binom{-\frac12}{n} = \frac{-\frac12\cdot(-\frac12-1)\cdots(-\frac12-n+1)}{1\cdot2\cdots n} = (-1)^n\frac{1\cdot3\cdots(2n-1)}{2\cdot4\cdots2n}.$$

Thus, from (4) and (5),

(7) $(1 + x)^{-\frac{1}{2}} = 1 - \frac{1}{2} x + \frac{1 \cdot 3}{2 \cdot 4} x^2 - \frac{1 \cdot 3 \cdot 5}{2 \cdot 4 \cdot 6} x^3 + \cdots,$ $-1 < x \leqq 1.$

On putting $x = -y$ this becomes

(8) $(1 - y)^{-\frac{1}{2}} = 1 + \frac{1}{2} y + \frac{1 \cdot 3}{2 \cdot 4} y^2 + \frac{1 \cdot 3 \cdot 5}{2 \cdot 4 \cdot 6} y^3 + \cdots,$ $-1 \leqq y < 1.$

PROBLEMS

1. Verify the Maclaurin series as far as given:

(a) $\tan x = x + \frac{1}{3} x^3 + \frac{2}{15} x^5 + \frac{17}{315} x^7 + \cdots$;

(b) $\tan^{-1} x = x - \frac{1}{3} x^3 + \frac{1}{5} x^5 - \frac{1}{7} x^7 + \cdots$;

(c) $x \cot x = 1 - \frac{1}{3} x^2 - \frac{1}{45} x^4 - \cdots$;

(d) $\sec x = 1 + \frac{1}{2} x^2 + \frac{5}{24} x^4 + \cdots$;

(e) $\log \cos x = -\frac{1}{2} x^2 - \frac{1}{12} x^4 - \frac{1}{45} x^6 - \cdots$;

(f) $x/\sin x = 1 + \frac{1}{6} x^2 + \frac{7}{360} x^4 + \cdots$;

(g) $e^{\sin x} = 1 + x + \dfrac{x^2}{2!} - \dfrac{3x^4}{4!} - \cdots.$

2. Find the functions that form the even and odd parts of $(1 - x)^{-1}$ and their Maclaurin expansions.

3. In the sine series (67.1) show that

$$R_{2k+1}(x) = (-1)^{k+1} \frac{x^{2k+3}}{(2k + 3)!} \cos \theta x.$$

4. In the cosine series (67.5) show that

$$R_{2k}(x) = (-1)^{k+1} \frac{x^{2k+2}}{(2k + 2)!} \cos \theta x.$$

5. Find the Maclaurin expansion to three terms plus remainder for (a) e^{-x^2}; (b) $\log (1 - x)$.

6. Expand the function

$$f(x) = \frac{x \log x}{1 - x} \ (x \neq 0, 1), \quad f(0) = 0, \quad f(1) = -1.$$

in a Taylor series about $x = 1$ and determine its precise interval of convergence. [Cf. Prob. 62.11.]

7. From the Maclaurin series for e^{a+x} prove that $e^{a+x} = e^a e^x$.

71. Extremes of $f(x)$. A function $f(x)$ is said to have a (relative) *maximum* at $x = c$ if $f(x) < f(c)$ in some suitably small neighborhood of c. If we write $x = c + h$, where h may be positive or negative, $f(x)$ will have a maximum at $x = c$ if

(1) $f(c + h) - f(c) < 0,$

when $|h|$ is sufficiently small. Similarly $f(x)$ is said to have a (relative) *minimum* at $x = c$ if $f(x) > f(c)$ in some suitably small neighborhood of c; that is,

(2) $f(c + h) - f(c) > 0,$

when $|h|$ is sufficiently small. In either case the function has an *extreme* $f(c)$ at $x = c$.

At a maximum or minimum point the ordinate of the curve $y = f(x)$ is greater or less than adjacent ordinates on either side. In Fig. 71 the curve $y = f(x)$, defined in the interval $a \leq x \leq b$, has maxima at x_1 and x_3, minima at x_2 and x_4. In the interval $a \leq x \leq b$, the greatest value of $f(x)$ is assumed at x_3, the least value at the end point a: thus $f(x_3)$ is the absolute maximum, $f(a)$ the absolute minimum of $f(x)$ in the interval. Note that $f(x_1)$ is also a relative maximum; and that the relative maximum $f(x_1)$ is less than the relative minimum $f(x_4)$.

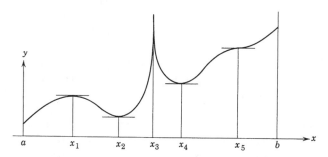

FIG. 71.   Graph of a function showing its extremes

THEOREM 1.   *The extreme values of $f(x)$ can only occur at those points where $f'(x) = 0$ or where $f'(x)$ does not exist.*

Proof.   At any point c where $f'(c)$ exists, $f(x)$ can only have an extreme when $f'(c) = 0$; for, if $f'(c) > 0$, or $f'(c) < 0$, $f(x)$ is increasing or decreasing at $x = c$ (§ 51) and in any neighborhood of c takes on values both smaller and greater than $f(c)$.

Note that the extremes for which $f'(x) = 0$ correspond to points where the tangent to the curve $y = f(x)$ is parallel to the x-axis. In Fig. 71 the curve has zero slope at x_1, x_2, x_4, x_5; $f(x)$ has a maximum at x_1, minima at x_2 and x_4, but no extreme at x_5 where the curve has a point of inflection since it crosses its tangent. Thus, if $f'(c)$ exists, $f'(c) = 0$ is a necessary, but not a sufficient, condition that $f(x)$ have an extreme at c.

THEOREM 2.   *If, at the point c,*

(i) $f'(c) = f''(c) = \cdots = f^{(n)}(c) = 0$,

(ii) $f^{(n+1)}(c) \neq 0$,

then, at $x = c$, $f(x)$ has

(a) *a point of inflection if n is even,*

(b) *an extreme if n is odd:*

   a maximum if $f^{(n+1)}(c) < 0$,

   a minimum if $f^{(n+1)}(c) > 0$.

Proof.   In view of the hypotheses (i) and (ii), (64.5) becomes

(3) $$f(c + h) - f(c) = r_n(h) h^{n+1},$$

where

$$\lim_{h \to 0} r_n(h) = \frac{f^{(n+1)}(c)}{(n + 1)!}.$$

Hence, when $|h|$ is chosen sufficiently small, $r_n(h)$ will always have the same sign as $f^{(n+1)}(c)$.

(a) If n is even, the odd power h^{n+1} will change sign with h; hence $f(c + h) - f(c)$ also changes sign with h and the curve crosses its horizontal tangent at c. But, by definition, a curve has a *point of inflection* where it crosses its tangent.

(b) If n is odd, the even power h^{n+1} is always positive, and (3) shows that $f(c + h) - f(c)$ and $f^{(n+1)}(c)$ have the same sign. Hence, from (1) and (2), $f(x)$ has a maximum or minimum at c according as $f^{(n+1)}(c)$ is negative or positive.

The application of Theorem 2 requires the computation of $f''(x)$ and possibly higher derivatives. In many cases, therefore, the following theorem is easier to apply; moreover, it *must* be used at points where $f'(x)$ does not exist.

THEOREM 3.   *Suppose that*

(i) $f(x)$ *is defined at* $x = c$,
(ii) $f'(x)$ *exists in a suitably small neighborhood of* c, $0 < |x - c| < \varepsilon$; ($f'(c)$ *need not exist*);
(iii) $f'(x)$ *has a fixed sign when* $x < c$ *and also when* $x > c$.

Then, as x increases through c,

(a) $f(x)$ *has no extreme at* c *if the sign of* $f'(x)$ *does not change*;
(b) $f(x)$ *has a maximum at* c *if the sign of* $f'(x)$ *changes from* $+$ *to* $-$;
(c) $f(x)$ *has a minimum at* c *if the sign of* $f'(x)$ *changes from* $-$ *to* $+$.

Proof.   From § 51, $f(x)$ is increasing or decreasing at x according as $f'(x)$ is positive or negative.   In case

(a) $f(x)$ increases or decreases as x passes through c: $f(x)$ has no extreme at c;
(b) $f(x)$ increases as x approaches c, decreases as x leaves c: $f(x)$ has a maximum at c;
(c) $f(x)$ decreases as x approaches c, increases as x leaves c: $f(x)$ has a minimum at c.

Example 1. When $f(x) = |x|$,
$$f'(x) = -1 \quad (x < 0), \quad f'(x) = 1 \quad (x > 0),$$
while $f'(0)$ is not defined.   As x passes through 0, the sign of $f'(x)$ changes from $-$ to $+$; hence, $y = f(x)$ has the minimum point $(0\ 0)$.

Example 2. When $f(x) = 1 - x^{\frac{2}{3}}$,

$$f'(x) = -\tfrac{2}{3}x^{-\frac{1}{3}} \text{ is not defined for } x = 0.$$

As x passes through 0, the sign of $f'(x)$ changes from $+$ to $-$; hence $y = 1 - x^{\frac{2}{3}}$ the maximum point $(0, 1)$, a *cusp* with vertical tengent.

Example 3. When $f(x) = x^{\frac{1}{3}}$

$$f'(x) = \tfrac{1}{3}x^{-\frac{2}{3}} \text{ is not defined for } x = 0.$$

As x passes through 0, the sign of $f'(x)$ remains positive; hence the curve $y = x^{\frac{1}{3}}$ has no extreme at the origin. But, since the curve crosses its vertical tangent at the origin, it has a point of inflection there.

Example 4. The polynomial

$$f(x) = x^2(x - 1)^3$$

is everywhere continuous. To find its extremes, we have

$$f'(x) = x(x - 1)^2(5x - 2),$$
$$f''(x) = 2(x - 1)(10x^2 - 8x + 1),$$
$$f'''(x) = 60x^2 - 72x + 18.$$

Since $f'(x)$ always exists, the extremes of $f(x)$ can only occur where $f'(x) = 0$: namely, $x = 0, \, ^2/_5, \, 1$.

Using Theorem 2:

$$f''(0) = -2: \quad (0, 0) \text{ is a maximum point;}$$

$$f''\!\left(\frac{2}{5}\right) = \frac{18}{25}: \quad \left(\frac{2}{5}, -\frac{108}{3125}\right) \text{is a minimum point;}$$

$$f''(1) = 0, \quad f'''(1) = 6: \quad (1, 0) \text{ is a point of inflection.}$$

Using Theorem 3: The term of highest degree in $f'(x)$, $5x^4$, shows that $f'(\pm\infty) > 0$. Moreover, $f'(x)$ can change sign only at a root of odd multiplicity, in this case at $x = 0$ and $^2/_5$. Thus $f'(x)$ has the following sign pattern:

$$-\infty \, (+) \, 0 \, (-) \, \tfrac{2}{5} \, (+) \, 1 \, (+) \, \infty;$$

hence, $f(x)$ has a maximum at $x = 0$, a minimum at $x = {}^2/_5$, and a point of inflection at $x = 1$.

Example 5. The *rational function* $f(x) = P/Q$, where $P(x)$, $Q(x)$ are polynomials without common factor, has the derivative

$$f'(x) = \frac{QP' - PQ'}{Q^2}.$$

Since $Q^2 \geqq 0$, $f'(x)$ has the same sign as the polynomial $QP' - PQ'$. The function $f(x)$ becomes infinite at all roots of $Q(x) = 0$, but $f'(x)$ only changes sign at the roots of Q of *even* multiplicity, for these are roots of $QP' - PQ'$ of odd multiplicity (why?). If we regard a point where $f(x) = \pm\infty$ and $f'(x)$ changes sign as a maximum $(+\infty)$ or a minimum $(-\infty)$ in an extended sense, we can read off the extremes of $f(x)$ from the sign pattern of the polynomial $QP' - PQ'$.

For example, if

$$P = (x - 1)^2 (3x^2 - 2x - 37) \qquad Q = (x + 5)^2 (3x^2 - 14x - 1),$$
$$QP' - PQ' = 72(x + 5) (x + 2) (x + 1) (x - 1) (x - 3) (x - 7).$$

This polynomial changes sign at each of its simple roots $-5, -2, -1, 1, 3, 7$; its sign pattern,

$$-\infty \, (+) -5 \, (-) -2 \, (+) -1 \, (-) \, 1 \, (+) \, 3 \, (-) \, 7 \, (+) \, \infty,$$

discloses that $f(x) = P/Q$ has maxima at $-5, -1, 3$; minima at $-2, 1, 7$. The "maximum" of $f(x)$ at $x = -5$ (a double zero of Q) is ∞.

PROBLEMS

1. Find the greatest and least values of $y = x \log x$ in the interval $0 < x \leq 1$. Sketch the curve.

2. Find the points of the following curves where the tangent is horizontal and state their nature. Sketch the curve.

   (a) $y = (2x - 3)^2 (3x - 2)^3$;
   (b) $y = (x^2 - 3) (x - 2)^3$;
   (c) $y = (x - 2)^3 (x^2 + 2x + 3)^2$;
   (d) $y = 3x^5 - 5x^3 + 2$.

3. Find the extremes of the curves:

   (a) $y = (3x + 4)/(x^2 + 1)$;
   (b) $y = x + \log (1 + x^2)$;
   (c) $y = a^2 x/(a^2 + x^2)$;
   (d) $y = x + \sqrt{1 - x}$.

4. Find the maximum value of

$$f(x) = \tan^{-1} x - \tan^{-1} (x/4);$$

the inverse functions denote the principal branch (Ex. 49.3).

5. Show that $f(x) = a \sin x + b \cos x$ has $\pm \sqrt{a^2 + b^2}$ as extreme values. Verify by putting $f(x)$ in the form $k \sin (x + \alpha)$.

6. Show that the greatest value of $x^m(a - x)^n$ is $m^m n^n a^{m+n}/(m + n)^{m+n}$.

7. Show that the least value of $a^2 \sec^2 x + b^2 \csc^2 x$ is $(a + b)^2$.

8. The points $P_1(x_1, y_1)$, $P_2(x_2, y_2)$ are above and below the x-axis, regarded as a line of separation between two media. A ray of light from P_1 to P_2 travels along a broken line P_1PP_2, where $P(x, 0)$ is the point of refraction. If the velocity of light is v_1 above the x-axis, v_2 below, show that the path for which the time of passage,

$$T = (P_1P)/v_1 + (PP_2)/v_2,$$

is a minimum obeys *Snell's law*,

$$\sin \theta_1/\sin \theta_2 = v_1/v_2,$$

where θ_1 and θ_2 are the angles of incidence and refraction. (If N_1PN_2 is the normal to the x-axis at P, $\theta_1 = $ angle P_1PN_1, $\theta_2 = $ angle P_2PN_2.)

9. Show that, when m and n are positive integers,

$$f(x) = (x - a)^m (x - c)^n \qquad (a < c)$$

has the extremes given in the following table, where b divides the segment ac in the ratio m/n:

(i) m and n even; minima at a and c, maximum at $b = (na + mc)/(m + n)$;
(ii) m and n odd: minimum at b;
(iii) m even, n odd: maximum at a, minimum at b;
(iv) m odd, n even: maximum at b, minimum at c.

  10. Find the extremes of the curve

$$y = ax^3 + 3bx^2 + 3cx + d,$$

when $D = b^2 - ac > 0$, $D = 0$, $D < 0$. When $D > 0$, prove that $|y''|$ (and hence the curvature) has the same value at the extreme points.

  11. Find the greatest value of k for which $e^x \geq kx^n$ ($x > 0$).

  12. Show that the values of $y = (x^2 + 2x + 11)/(x^2 + 4x + 10)$ are confined to the interval $5/6 \leq y \leq 2$.

  13. Show that the extreme values of

(i) $$y = \frac{ax^2 + 2bx + c}{px^2 + 2qx + r}, \qquad pr - q^2 > 0$$

are roots of the quadratic

(ii) $$(pr - q^2)y^2 - (ar - 2bq + cp)y + (ac - b^2) = 0.$$

Verify from Prob. 12.

  [The condition $y' = 0$ gives $y = (ax + b)/(px + q)$; from this and (i) show that $y = (bx + c)/(qx + r)$. Eliminate x from these equations to get (ii).]

  14. Show that $e^x < (1 - x)^{-1}$ when $x < 1$, $x \neq 0$. [Find maximum of $(1 - x)e^x$.]

  15. Show that $\log x < x - 1$ when $x > 0$, $x \neq 1$. [Find maximum of $\log x - x + 1$.]

  72. Summary: Functions of a Real Variable. The functional relation $y = f(x)$ means that a definite value of y corresponds to every value of x in a certain range. The function $f(x)$ is continuous at $x = c$ if

(1) $$\lim_{x \to c} f(x) = f(c);$$

this definition implies that $f(c)$ is defined and that the limit exists.

  A function $f(x)$ continuous in a *closed* interval $a \leq x \leq b$, has the important properties:

(i) $f(x)$ is *uniformly continuous* in (a, b): for any x, x' in (a, b),

$$|f(x) - f(x')| < \varepsilon \quad \text{when} \quad |x - x'| < \delta;$$

(ii) $f(x)$ is *bounded* in (a, b): $|f(x)| \leq M$;

(ii) $f(x)$ takes on a least and a greatest value at least once in (a, b);

(iv) if $f(a) \neq f(b)$ and μ is an intermediate value, $f(x) = \mu$ for at least one x in (a, b).

  The *derivative* of $f(x)$ at $x = c$ is

(2) $$f'(c) = \lim_{h \to 0} \frac{f(c + h) - f(c)}{h}.$$

When $f'(c)$ exists,

(i) $f(x)$ must be defined in some two-sided neighborhood of c;

(ii) $f(x)$ must be continuous at c.

If $y = f(x)$, we write $f'(x) = dy/dx$; here the *differentials* dy and dx are defined by

$$dy = f'(x)\,\Delta x, \qquad dx = \Delta x.$$

The relation $dy = f'(x)\,dx$ is still true when x is a function of another independent variable.

If $y = f(u)$, $u = g(x)$ we have the *chain rule*,

$$(3) \qquad \frac{dy}{dx} = \frac{dy}{du}\cdot\frac{du}{dx} = f'(u)\,g'(x),$$

when these derivatives exist.

The function $f(x)$ is *increasing* or *decreasing* at $x = c$ according as $f'(c) > 0$ or $f'(c) < 0$. If $y = f(x)$ is an increasing or decreasing function throughout $a < x < b$, then the inverse function $x = \varphi(y)$ exists when y lies between $f(a)$ and $f(b)$ and

$$(4) \qquad \frac{dx}{dy}\cdot\frac{dy}{dx} = 1.$$

Important monotonic functions $f(x)$ over an interval (a, b) and their inverse functions $\varphi(x)$ are tabulated below: while x in $\varphi(x)$ ranges over the interval $(f(a), f(b))$, $\varphi(x)$ is confined to values in (a, b):

$f(x)$	(a, b)	$\varphi(x)$	$(f(a), f(b))$
x^n	$0 \leq x < \infty$	$x^{1/n}$	$0 \leq x < \infty$
e^x	$-\infty < x < \infty$	$\log x$	$0 < x < \infty$
$\sin x$	$-\dfrac{\pi}{2} \leq x \leq \dfrac{\pi}{2}$	$\sin^{-1} x$	$-1 \leq x \leq 1$
$\cos x$	$0 \leq x \leq \pi$	$\cos^{-1} x$	$-1 \leq x \leq 1$
$\tan x$	$-\dfrac{\pi}{2} < x < \dfrac{\pi}{2}$	$\tan^{-1} x$	$-\infty < x < \infty$
$\sinh x$	$-\infty < x < \infty$	$\sinh^{-1} x = \log(x + \sqrt{x^2 + 1})$	$-\infty < x < \infty$
$\cosh x$	$0 \leq x < \infty$	$\cosh^{-1} x = \log(x + \sqrt{x^2 - 1})$	$1 \leq x < \infty$
$\tanh x$	$-\infty < x < \infty$	$\tanh^{-1} x = \dfrac{1}{2}\log\dfrac{1 + x}{1 - x}$	$-1 < x < 1$

If the functions $f(x)$, $g(x)$ are

(i)   continuous in $a \leq x \leq b$,
(ii)   differentiable in $a < x < b$, and
(iii) $g'(x) \neq 0$ in $a < x < b$, then

$$(5) \qquad \frac{f(b) - f(a)}{g(b) - g(a)} = \frac{f'(\xi)}{g'(\xi)}, \qquad a < \xi < b.$$

This theorem of Cauchy reduces to the *mean-value theorem* when $g(x) = x$:

$$(6) \qquad \frac{f(b) - f(a)}{b - a} = f'(\xi), \qquad a < \xi < b.$$

This in turn becomes *Rolle's theorem* when $f(a) = f(b)$.

If $f(a) = g(a) = 0$ in (5) and we let $b \to a$, we obtain *l'Hospital's rule* for evaluating the "indeterminate form 0/0":

$$\lim_{x \to a} \frac{f(x)}{g(x)} = \lim_{x \to a} \frac{f'(x)}{g'(x)} \qquad (a \text{ finite or } \infty)$$

when the second limit exists. l'Hospital's rule also applies to the indeterminate form ∞/∞. The forms $0 \cdot \infty$, $\infty - \infty$, 0^0, ∞^0, 1^∞ are reducible to $0/0$ or ∞/∞.

A function $f(x)$ having derivatives up to the nth order at $x = a$ has the approximating polynomial:

$$P_n(x) = f(a) + \frac{f'(a)}{1!}(x - a) + \cdots + \frac{f^{(n)}(a)}{n!}(x - a)^n.$$

The *remainder*

$$R_n(x) = f(x) - P_n(x) \quad \text{and} \quad (x - a)^{n+1}$$

have this common property: Both functions and their first n derivatives vanish at $x = a$. If we apply Cauchy's theorem (5) n times to these functions, we find that

$$(7) \qquad \lim_{x \to a} \frac{R_n(x)}{(x - a)^{n+1}} = \frac{f^{(n+1)}(a)}{(n + 1)!},$$

when $f^{(n+1)}(a)$ exists. Hence the function on the left (denoted by $r_n(x)$) remains bounded as $x \to a$, and we have the *Taylor expansion* of $f(x)$ about $x = a$:

$$(8) \qquad f(x) = P_n(x) + R_n(x) = P_n(x) + r_n(x)(x - a)^{n+1}.$$

Expansions of this type (with $r_n(x)$ *bounded* as $x \to a$) are unique.

When $f^{(n+1)}(x)$ exists in some neighborhood of a, the remainder can be put in the *Lagrange form*:

$$(9) \qquad R_n(x) = \frac{f^{(n+1)}(\xi)}{(n + 1)!}(x - a)^{n+1}, \qquad a < \xi < x.$$

When $a = 0$, we have the *Maclaurin expansion*:

$$(10) \quad f(x) = f(0) + \frac{f'(0)}{1!} x + \cdots + \frac{f^{(n)}(0)}{n!} x^n + \frac{f^{(n)}(\xi)}{(n+1)!} x^{n+1},$$

where $0 < \xi < x$. When $R_n(x) \to 0$ as $n \to \infty$, we obtain from (8) or (10) an infinite series; thus (10) gives the *Maclaurin series* for $f(x)$:

$$(11) \qquad f(x) = f(0) + \frac{f'(0)}{1!} x + \frac{f''(0)}{2!} x^2 + \cdots .$$

If this series represents $f(x)$ when $0 < x < a$, it also represents $f(x)$ over the wider interval $-a < x < a$ symmetric about the origin. The following series are valid everywhere:

$$e^x = 1 + x + \frac{x^2}{2!} + \frac{x^3}{3!} + \cdots ,$$

$$\sin x = x - \frac{x^3}{3!} + \frac{x^5}{5!} - \frac{x^7}{7!} + \cdots ,$$

$$\sinh x = x + \frac{x^3}{3!} + \frac{x^5}{5!} + \frac{x^7}{7!} + \cdots ,$$

$$\cos x = 1 - \frac{x^2}{2!} + \frac{x^4}{4!} - \frac{x^6}{6!} + \cdots ,$$

$$\cosh x = 1 + \frac{x^2}{2!} + \frac{x^4}{4!} + \frac{x^6}{6!} + \cdots ,$$

The *logarithmic series*,

$$\log (1 + x) = x - \frac{x^2}{2} + \frac{x^3}{3} - \frac{x^4}{4} + \cdots \qquad (-1 < x \leqq 1),$$

and the *binomial series*,

$$(1 + x)^k = 1 + \binom{k}{1} x + \binom{k}{2} x^2 + \cdots \qquad (-1 < x < 1),$$

both diverge when $|x| > 1$. The latter converges at the end point $x = 1$ when $k > -1$, at $x = -1$ when $k > 0$.

The *extremes* of $f(x)$ can only occur where $f'(x) = 0$ or where $f'(x)$ does not exist. If at $x = c$ all derivatives of $f(x)$ up to the nth vanish while $f^{(n+1)}(c) \neq 0$, then $f(x)$ has a point of inflection if n is even, an extreme if n is odd (a maximum or minimum according as $f^{(n+1)}(c) < 0$ or > 0).

If $f'(x)$ exists when $0 < |x - c| < \varepsilon$ and changes sign as x increases through c, the change $+$ to $-$ denotes a maximum at c, $-$ to $+$ a minimum. If $f'(x)$ does not change sign, there is no extreme at c.

CHAPTER 4

Functions of Several Variables

73. Functions of Two Variables. A *region* of the xy-plane is a set of points (x, y), any two of which can be joined by a broken line whose points belong to the set. A point (a, b) is an *interior point* of a set S if it is the center of a square region,

$$(1) \qquad |x - a| < \delta, \qquad |y - b| < \delta,$$

whose points belong to S. A point set is *open* if it consists entirely of interior points. Thus the points (x, y) which satisfy the inequalities (1) form an *open square* of side 2δ; while

$$(2) \qquad (x - a)^2 + (y - b)^2 < \delta^2$$

defines an *open circle* of radius δ about (a, b).

A point set is *closed* if it includes all of its cluster points (§ 75); and the cluster points not interior to the set form its *boundary*. Thus the circumference of the circle,

$$(3) \qquad (x - a)^2 + (y - b)^2 = \delta^2,$$

is the boundary of the open circle (2); its points are all cluster points of the open set (2). When these boundary points are added to the open set (2), we have a closed circular region.

An *open region* is often called a *domain*. The points (x, y) for which $y > 0$ form an open region, the upper half-plane; similarly $y < 0$ defines the lower half-plane. The points in both half-planes form a new set S which contains all points of the plane except those on the x-axis; note that S is not a region (why?).

The variable $u = f(x, y)$ is called a *function* of two independent variables x, y if a single value of u corresponds to each pair of values (x, y) belonging to a certain specified set, say some region of the xy-plane. Thus the function $w = \sqrt{x + y - 1}$ is defined over the closed region $x + y \geqq 1$; while $w = 1/\sqrt{x + y - 1}$ is defined over the open region $x + y > 1$, consisting of all points to the right of the line $x + y = 1$.

The function $u = f(x, y)$ may be defined for *all* real points (x, y); this is always the case for polynomials, for example,

$$u = ax + by + c, \qquad u = ax^2 + bxy + cy^2 + dx + ey + f,$$

and also for functions such as e^{x+y}, $\sin(x - y)$.

A *rational function*, the quotient of two polynomials $f(x, y)/g(x, y)$, is defined at all points where $g(x, y) \neq 0$. Thus,

$$u = \frac{ax + by + c}{\alpha x + \beta y + \gamma}$$

is defined at all points not lying on the line $\alpha x + \beta y + \gamma = 0$.

The *algebraic functions*,

$$u = \sqrt{(x - 1)(y + 2)}, \qquad v = \sqrt{x^2 + 2y^2 - 4},$$

are only defined where the radicand is not negative. Thus u is defined in the closed region formed by the first and third quadrants into which the lines $x - 1 = 0$, $y + 2 = 0$ divide the plane; and v is defined where $x^2 + 2y^2 \geq 4$, that is, for points on or outside of the ellipse $x^2/4 + y^2/2 = 1$.

Consider a function $f(x, y)$ defined in the neighborhood of the point (ξ, η). Let the sequence of points (x_n, y_n) approach (ξ, η) as a limit; then the distance between (x_n, y_n) and (ξ, η) approaches zero:

$$(x_n - \xi)^2 + (y_n - \eta)^2 \to 0.$$

Now, if $f(x_n, y_n) \to A$ for *every* such point sequence that has (ξ, η) as a limit, we say that $f(x, y)$ tends to the limit A as $(x, y) \to (\xi, \eta)$; and we write

(4)
$$\lim_{\substack{x \to \xi \\ y \to \eta}} f(x, y) = A.$$

In much the same manner as in § 44 we may now prove theorems analogous to Theorems 44.1–2–3.

THEOREM 1. *In order that $f(x, y) \to A$ as $(x, y) \to (\xi, \eta)$, it is necessary and sufficient that, for every $\varepsilon > 0$, we can find a number $\delta > 0$ such that*

(5)
$$|f(x, y) - A| < \varepsilon \quad \text{when} \quad 0 < \begin{vmatrix} x - \xi \\ y - \eta \end{vmatrix} < \delta.$$

THEOREM 2. *If $f(x, y) \to A \neq 0$ as $(x, y) \to (\xi, \eta)$, then $f(x, y)$ will have the same sign as A in some circle about (ξ, η) with (ξ, η) excluded.*

THEOREM 3. *If $f(x, y) \to A$, $g(x, y) \to B$ as $(x, y) \to (\xi, \eta)$, then*

$$f(x, y) + g(x, y) \to A + B,$$
$$f(x, y) \cdot g(x, y) \to AB,$$
$$f(x, y)/g(x, y) \to A/B \quad \text{if} \quad B \neq 0.$$

Example 1. The *double limit*,

$$\lim_{\substack{x \to 0 \\ y \to 0}} \frac{x + y}{x - y},$$

does not exist; for, along the line

$$y = mx, \qquad f(x, y) = \frac{1 + m}{1 - m}.$$

Thus, by letting $(x, y) \to (0, 0)$ along a suitable line, $f(x, y)$ will approach any value k specified at pleasure. Note that both *repeated* (or iterated) *limits*

$$\lim_{y \to 0} \lim_{x \to 0} \frac{x + y}{x - y} = -1, \qquad \lim_{x \to 0} \lim_{y \to 0} \frac{x + y}{x - y} = 1$$

exist, although they have different values,

Example 2. Even when both repeated limits exist and have the same value, the double limit may not exist. Thus,

$$f(x, y) = \frac{x^2 y^2}{x^2 y^2 + (x - y)^2}$$

has no limit as $(x, y) \to (0, 0)$ over an arbitrary path; for $f(x, y) = 1$ on the line $y = x$ but $f(x, y) = 0$ on the lines $x = 0$ or $y = 0$. Nevertheless, both repeated limits exist and equal zero.

PROBLEMS

1. Determine the point set in which each of the following functions is defined, and state whether the set forms a region (open or closed) or not.

(*a*) $\{|x| + |y| - 2\}^{1/2}$.

(*b*) $\{(x + y - 1)(x - y + 1)\}^{-1/2}$.

(*c*) $\left(\dfrac{1 - x^2 - y^2}{xy}\right)^{1/2}$.

(*d*) $\sin^{-1}(x^2 + y^2)$.

(*e*) $\{(x^2 + y^2 - 1)(4 - x^2 - y^2)\}^{1/2}$.

(*f*) $\log(x + y)$.

2. Show that the function

$$f(x, y) = \frac{2xy}{x^2 + y^2} \ (x, y \neq 0), \qquad f(0, 0) = 0$$

is not continuous at $(0, 0)$; and prove that, on any circle $x^2 + y^2 = \delta^2$, no matter how small δ is chosen, $f(x, y)$ assumes all values between -1 and 1.

[Write $x = \delta \cos \theta, y = \delta \sin \theta$.]

74. Continuity at a Point. The *neighborhood*† of a point (a, b) is any open region of the xy-plane containing (a, b). Thus,

$$(x - a)^2 + (y - b)^2 < \delta^2$$

defines a circular neighborhood of (a, b); and

$$|x - a| < \delta, \qquad |y - b| < \delta,$$

a square neighborhood of (a, b).

---

† The neighborhood of a real number r is any open interval containing r (§ 12); for example, $|x - r| < \delta$.

A function $f(x, y)$ is said to be continuous at a point (a, b) if

(i) $f(x, y)$ is defined in some neighborhood of (a, b),
(ii) $f(x, y) \to A$ as $(x, y) \to (a, b)$,
(iii) $A = f(a, b)$.

These three requirements for continuity at a point are all comprised in the single equation,

$$(1) \qquad \lim_{\substack{x \to a \\ y \to b}} f(x, y) = f(a, b).$$

In view of Theorem 73.1 we may state

THEOREM 1. *In order that $f(x, y)$ be continuous at the point (a, b), it is necessary and sufficient that, for every $\varepsilon > 0$, we can find a number $\delta > 0$ such that*

$$(2) \qquad |f(x, y) - f(a, b)| < \varepsilon \quad \text{when} \quad |x - a| < \delta, \quad |y - b| < \delta.$$

The requirement $x \neq a$, $y \neq b$ now has no significance for $f(a, b)$ must exist and (2) is obviously fulfilled when $x = a$, $y = b$.

If $f(x, y)$ and $g(x, y)$ are continuous functions, $f + g$ and fg are also continuous; and f/g is continuous at all points where $g \neq 0$ (Theorem 73.2). *Polynomials* are everywhere continuous since they are sums of terms $c_{mn} x^m y^n$. If $f(x, y)$ and $g(x, y)$ are polynomials, the *rational function f/g* is continuous at all points where $g \neq 0$. Finally continuous functions of continuous functions are continuous; thus $\sin xy$ and $e^{-(x^2 + y^2)}$ are continuous.

When $f(x, y)$ is continuous at an interior point (a, b) of a region R and $f(a, b) \neq 0$, we can always find a square region about (a, b) in which $f(x, y)$ has the same sign as $f(a, b)$. We need only take $\varepsilon = |f(a, b)|$ in (2) and determine the corresponding square of side 2δ. Within this square $f(x, y)$ and $f(a, b)$ have the same sign; for otherwise

$$|f(x, y) - f(a, b)| = |f(x, y)| + |f(a, b)| \geq \varepsilon.$$

When $f(x, y)$ is continuous at (a, b), the functions of one variable $f(x, b), f(a, y)$ are continuous at $x = a$ and $y = b$; this follows at once from (2). But the converse need not be true, as we see from the following example.

Example 1. The function

$$f(x, y) = \frac{2xy}{x^2 + y^2}, \qquad f(0, 0) = 0$$

is a continuous function of either variable when the other is given a fixed value. However $f(x, y)$ is not a continuous function of both variables. For, along any line $y = x \tan v$,

$$f(x, y) = \frac{2 \tan v}{1 + \tan^2 v} = 2 \sin v \cos v = \sin 2v,$$

except at the origin where $f(0, 0) = 0$.   The surface $z = f(x, y)$ is a *conoid* whose
parametric equations are

$$x = u \cos v, \qquad y = u \sin v, \qquad z = \sin 2v;$$

it is generated by a line, perpendicular to z-axis, which revolves about and slides along
this axis so that its height above the xy-plane is $\sin 2v$ when it makes an angle v with the
x-axis.   The aspect of this surface near the z-axis shows the nature of the remarkable
discontinuity of $f(x, y)$ at $(0, 0)$.   On any circle $x^2 + y^2 = \delta^2$ the function assumes all
values between -1 and 1, no matter how small δ is chosen.

Example 2. The function

$$f(x, y) = xy \frac{x^2 - y^2}{x^2 + y^2}, \qquad f(0, 0) = 0$$

is continuous at $(0, 0)$; for

$$\left| f(x, y) - f(0, 0) \right| = \left| xy \right| \frac{\left| x^2 - y^2 \right|}{x^2 + y^2} < \left| x \right| \left| y \right| < \varepsilon,$$

when $x^2 + y^2 < \varepsilon$; thus we may take $\delta = \sqrt{\varepsilon}$ in (2).

PROBLEMS

1. Can $f(0, 0)$ be defined so that $f(x, y)$ is continuous at $(0, 0)$ when

$$(a)\ f(x, y) = \frac{\sin (x^2 + y)}{x + y}\ ; \qquad (b)\ f(x, y) = \frac{x^2y}{x^4 + y^2}\ ?$$

2. If $f(x, y) \to 0$ as $(x, y) \to (0, 0)$ along *all* straight lines toward the origin, can we
conclude that $f(x, y) \to 0$ as $(x, y) \to (0, 0)$ along any path? [Consider Prob. 1*b*.]

75. Continuity in a Region.   A function $f(x, y)$ is said to be continuous
in a region (open or closed) if it is continuous at every point of the region.
A function $f(x, y)$ continuous in a *closed* region R has properties analogous
to those of functions of a single variable (§ 44).   Before considering these
properties we first extend the concept of a *cluster point* (§ 12) to a set of
points in a plane.

A set S of points (x, y) in a plane is said to have a *cluster point* (ξ, η)
if every neighborhood of (ξ, η), however small, contains an infinite
number of points of S (cf. § 12).   We may readily extend the Bolzano-
Weierstrass theorem of § 13 to plane sets.

THEOREM 1.   *Every bounded, infinite set of points in a plane has at least
one cluster point.*

Proof.   Since the set S consists of an infinite number of points (x, y),
if there are but a finite number of different x's, one $x = \xi$ must correspond
to an infinite number of y's; since these are bounded, they must cluster
at least about one value $y = \eta$ (§ 13); then (ξ, η) is a cluster point of S.
Similarly, if the points of S contain but a finite number of different y's,
we can infer the existence of a cluster point.

Assume then that the points of S have an infinite number of different
x's *and* y's.   The x's must then cluster at least about one value $x = \xi$;
and any interval $\xi - \varepsilon < x < \xi + \varepsilon$ about ξ must contain an infinite

number of x's belonging to points of S; these x's correspond to an infinite number of y's belonging to points of S, and these y's must also cluster at least about one value $y = \eta$. Then (ξ, η) is a cluster point of S.

We may now prove three theorems, for functions $f(x, y)$ continuous in a *bounded, closed* region which are analogous to theorems of §§ 46, 47.

A function $f(x, y)$ is said to be *uniformly continuous* in a region R, open or closed, when, for any $\varepsilon > 0$, we can choose a $\delta > 0$ so that, if (x, y), (x', y') are any points of R for which $|x - x'| < \delta$ and $|y - y'| < \delta$, then $|f(x, y) - f(x', y')| < \varepsilon$.

We now state in turn the analogues of Theorems 46.1, 46.2 and 47.1.

Theorem 2. *A function $f(x, y)$ which is continuous in a bounded, closed region R is uniformly continuous there; that is, if $(x, y), (x', y')$ are points of R,*

(1) $$|f(x, y) - f(x', y')| < \varepsilon \quad when \quad |x - x'|, |y - y'| < \delta,$$

where ε is arbitrary and δ depends on ε alone.

Theorem 3. *A function $f(x, y)$ which is continuous in a bounded, closed region is bounded there.*

Theorem 4. *A function $f(x, y)$, continuous in a bounded, closed region, takes on a least and a greatest value at least once in the region.*

The proofs are similar to those already given for functions of a single variable and will be omitted.

76. Partial Derivatives. Let $u = f(x, y)$ be a function of two independent variables x, y in a region R. If y is held constant, $f(x, y)$ becomes a function of x alone; and its derivative (if it exists) is called the *partial derivative of $f(x, y)$ with respect to x.* Similarly, if x is held constant, $f(x, y)$ becomes a function of y alone, whose derivative is called the *partial derivative of $f(x, y)$ with respect to y.* Therefore these derivatives, variously denoted by

$$f_x(x, y), \frac{\partial f}{\partial x}, u_x, \frac{\partial u}{\partial x} \quad and \quad f_y(x, y), \frac{\partial f}{\partial y}, u_y, \frac{\partial u}{\partial y},$$

have the defining equations

(1) $$f_x(x, y) = \lim_{h \to 0} \frac{f(x + h, y) - f(x, y)}{h},$$

(2) $$f_y(x, y) = \lim_{k \to 0} \frac{f(x, y + k) - f(x, y)}{k}.$$

When these limits exist at an interior point (x_0, y_0) of R, they are denoted by $f_x(x_0, y_0), f_y(x_0, y_0)$. If they exist at all points of a region, $f_x(x, y)$ and $f_y(x, y)$ are new functions of x, y which may again be differentiated with respect to x and y.

The notations $\partial f/\partial x$, $\partial f/\partial y$ are suggested by the notation df/dx for an ordinary derivative. But, unlike df/dx, $\partial f/\partial x$ is not interpreted as a quotient, for ∂f and ∂x are not given independent meanings.

The function $f(x, y, z)$ of three independent variables x, y, z has three partial derivatives of the first order: $\partial f/\partial x$, $\partial f/\partial y$, $\partial f/\partial z$. These are derivatives of functions of one variable when two others are held constant; thus,

$$f_x(x, y, z) = \frac{\partial f}{\partial x} = \lim_{h \to 0} \frac{f(x + h, y, z) - f(x, y, z)}{h}.$$

Example 1. The function $r = \sqrt{x^2 + y^2}$ is continuous everywhere in the xy-plane. We have

$$\frac{\partial r}{\partial x} = \frac{1}{2} \frac{2x}{\sqrt{x^2 + y^2}} = \frac{x}{r}, \qquad \frac{\partial r}{\partial y} = \frac{y}{r}.$$

Example 2. The partial derivatives of

$$f(x, y) = \frac{x + y - 1}{x - y + 1} \quad \text{at} \quad (2, 1)$$

are given by

$$f_x(2, 1) = \lim_{h \to 0} \frac{f(2 + h, 1) - f(2, 1)}{h} = \lim_{h \to 0} \frac{1 - 1}{h} = 0,$$

$$f_y(2, 1) = \lim_{k \to 0} \frac{f(2, 1 + k) - f(2, 1)}{k} = \lim_{k \to 0} \frac{\dfrac{2 + k}{2 - k} - 1}{k}$$

$$= \lim_{k \to 0} \frac{2}{2 - k} = 1.$$

Example 3. The function of Ex. 74.1,

$$f(x, y) = \frac{2xy}{x^2 + y^2}, \qquad f(0, 0) = 0,$$

is discontinuous at $(0, 0)$; but both partial derivatives exist at $(0, 0)$:

$$f_x(0, 0) = f_y(0, 0) = 0.$$

A function $f(x)$ of one variable is continuous at x_0 when $f'(x_0)$ exists (§ 50). The example shows that $f(x, y)$ is not necessarily continuous at (x_0, y_0) when $f_x(x_0, y_0)$ and $f_y(x_0, y_0)$ exist.

77. Total Differential.

When $f'(x)$ exists,

$$\frac{f(x + h) - f(x)}{h} = f'(x) + \eta,$$

where $\eta \to 0$ as $h \to 0$. Since

$$\Delta f = f(x + h) - f(x), \qquad df = hf'(x),$$

we then have

$$(1) \qquad\qquad\qquad \Delta f = df + \eta h.$$

Then $\Delta f - df$ vanishes to a higher order than h since $\eta h/h \to 0$. Thus $\eta h = o(h)$ in the notation of (64.11), and we have

(2) $$\Delta f = df + o(h).$$

We say that $f(x)$ is *differentiable at* x when (2) holds; and this is *always* the case when $f'(x)$ exists.

The concept of *differentiability* may be extended to functions of several variables. Thus the function $f(x, y)$ is said to be *differentiable* at (x, y) when we can express Δf in the form

(3) $$\Delta f = f(x + h, y + k) - f(x, y) = Ah + Bk + \eta_1 h + \eta_2 k,$$

where A, B are independent of h, k and η_1 and $\eta_2 \to 0$ as h and $k \to 0$.

If we put $k = 0$ in (3), divide by h, and let $h \to 0$, we get $f_x(x, y) = A$; and, if we put $h = 0$, divide by k and let $k \to 0$, we get $f_y(x, y) = B$. Thus, when $f(x, y)$ is differentiable at (x, y), both f_x and f_y exist at (x, y) and

(4) $$\Delta f = hf_x + kf_y + \eta_1 h + \eta_2 k.$$

At the point (x, y), f_x and f_y are ordinary derivatives of functions of *one* variable obtained from $f(x, y)$ by holding the other variable constant. The differentials (§ 50) of these functions, namely, hf_x and kf_y, are called the *partial differentials* of the function $f(x, y)$ of *two* variables; and their sum $hf_x + kf_y$, is called the *total differential df*. Thus, when f_x and f_y exist at (x, y), the total differential of $f(x, y)$ is defined as

(5) $$df = hf_x + kf_y = \frac{\partial f}{\partial x} \Delta x + \frac{\partial f}{\partial y} \Delta y.$$

The total differential of $f(x, y)$ *at a point* (x, y) *is that part of* Δf *which is linear in* h *and* k. In particular, when $f = x$, we have $dx = \Delta x$; and, when $f = y$, $dy = \Delta y$. *The differentials of the independent variables are the same as their increments.* Thus we may write the defining equation (5) as

(6) $$df = f_x dx + f_y dy.$$

When $f(x, y)$ is differentiable at (x, y), we have, from (3),

(7) $$\Delta f = df + \eta_1 h + \eta_2 k,$$

where $\eta_1, \eta_2 \to 0$ as $h, k \to 0$. Then $\Delta f - df$ vanishes to a higher order than either h or k; for *both* factors in $\eta_1 h$, $\eta_2 k$ approach zero. In the notation of (64.11) we may write $\eta_1 h + \eta_2 k = o(\sqrt{h^2 + k^2})$; for

$$\frac{|\eta_1 h + \eta_2 k|}{\sqrt{h^2 + k^2}} < |\eta_1| + |\eta_2| \to 0.$$

At points where $f(x, y)$ is differentiable, df is a close approximation of

(8) $$\Delta f = df + o(\sqrt{h^2 + k^2}),$$

when $\sqrt{h^2 + k^2}$ is sufficiently small. Then $f(x + h, y + k)$ is given approximately by $f(x, y) + hf_x + kf_y$, a *linear* function in h and k.

A function is continuous at any point where it is differentiable; for, if h and $k \to 0$ in any manner in (4), $\Delta f \to 0$; that is,

$$f(x + h, y + k) \to f(x, y).$$

For a function $f(x)$ of *one* variable, the existence of $f'(a)$ implies that the $f(x)$ is differentiable at a. But, for functions $f(x, y)$ of *two* variables, the existence of $f_x(a, b)$ and $f_y(a, b)$ does *not* imply that $f(x, y)$ is differentiable at (a, b), and (8) may not hold (cf. Ex. 2).

Example 1. If $f(x, y) = xy$,
$$\Delta f = (x + h)(y + k) - xy = hy + kx + hk, \qquad df = hy + kx.$$
Obviously xy is differentiable at all points; and $f_x = y$, $f_y = x$. [Let the reader show that $hk = o(\sqrt{h^2 + k^2})$.]

Example 2. At the point $(0, 0)$ the function $f(x, y) = \sqrt{|xy|}$ has the partial derivatives $f_x = f_y = 0$; and
$$df = 0, \qquad \Delta f = \sqrt{|hk|}.$$
If we put $h = r \cos \theta$, $k = r \sin \theta$,
$$\frac{\Delta f - df}{\sqrt{h^2 + k^2}} = \frac{\sqrt{|hk|}}{\sqrt{h^2 + k^2}} = \sqrt{|\cos \theta \sin \theta|},$$
which, in general, does not approach zero as $\sqrt{h^2 + k^2} = r \to 0$. Although f_x and f_y exist at $(0, 0)$, $f(x, y)$ is not differentiable there. However, $f(x, y)$ is continuous at $(0, 0)$.

78. Differentiable Functions. We now give a sufficient (but not necessary) condition that $f(x, y)$ be differentiable at a point (x, y).

THEOREM. *If f_x and f_y exist at a point (a, b) and f_x is continuous there, $f(x, y)$ is differentiable at (a, b).*

Proof. We have

(1) $\Delta f = f(a + h, b + k) - f(a, b)$
$$= \{f(a + h, b + k) - f(a, b + k)\} + \{f(a, b + k) - f(a, b)\}$$

Since $f_x(x, y)$ is continuous at (a, b), $f_x(x, y)$ exists in some neighborhood R of (a, b); and, in R, $f(x, y)$ is a continuous function of x when y is held constant. Hence we can apply the mean-value theorem (57.3) to the first brace in (1):

$$f(a + h, b + k) - f(a, b + k) = hf_x(a + \theta_1 h, b + k), \qquad 0 < \theta_1 < 1.$$

Since f_x is continuous at (a, b) we can write

$$f_x(a + \theta_1 h, b + k) = f_x(a, b) + \eta_1,$$

where $\eta_1 \to 0$ as h and $k \to 0$.

Since f_y exists at (a, b),

$$\frac{f(a, b + k) - f(a, b)}{k} = f_y(a, b) + \eta_2,$$

where $\eta_2 \to 0$ as $k \to 0$; hence,

$$f(a, b + k) - f(a, b) = k f_y(a, b) + k \eta_2.$$

We thus obtain

$$\Delta f = h f_x(a, b) + k f_y(a, b) + h \eta_1 + k \eta_2;$$

that is, $f(x, y)$ is differentiable at (a, b).

79. Composite Functions. If $f(u)$ and $u = u(x)$ are differentiable functions, the chain rule (52.1) gives

$$\frac{df}{dx} = \frac{df}{du} \frac{du}{dx}.$$

Suppose now that $f(u, v)$ is a *differentiable* function of u, v, which in turn are differentiable functions of an independent variable x. Let the increment Δx produce the increments Δu, Δv, while these produce Δf. Then

(1) $$\Delta f = \frac{\partial f}{\partial u} \Delta u + \frac{\partial f}{\partial v} \Delta v + \eta_1 \Delta u + \eta_2 \Delta v,$$

where η_1, $\eta_2 \to 0$ as Δu, $\Delta v \to 0$; and, on division by Δx,

(2) $$\frac{\Delta f}{\Delta x} = \frac{\partial f}{\partial u} \frac{\Delta u}{\Delta x} + \frac{\partial f}{\partial v} \frac{\Delta v}{\Delta x} + \eta_1 \frac{\Delta u}{\Delta x} + \eta_2 \frac{\Delta v}{\Delta x}.$$

As $\Delta x \to 0$, Δu, $\Delta v \to 0$ and η_1, $\eta_2 \to 0$; moreover, $\Delta u / \Delta x$ and $\Delta v / \Delta x$ approach finite limits du/dx and dv/dx. Hence the right member of (2) approaches a limit, namely,

(3) $$\frac{df}{dx} = \frac{\partial f}{\partial u} \frac{du}{dx} + \frac{\partial f}{\partial v} \frac{dv}{dx}.$$

This formula is an instance of the extended *chain rule*.

Equation (3) comprises several important differentiation formulas as special cases. Denoting x-derivatives by primes, we have

$$(u + v)' = u' + v',$$
$$(uv)' = vu' + uv',$$
$$\left(\frac{u}{v}\right)' = \frac{vu' - uv'}{v^2},$$
$$(u^v)' = vu^{v-1}\, u' + u^v \log u\, v'.$$

In each case the coefficients of u' and v' are f_u and f_v.

Next consider the function $f(u, v)$ when u and v are differentiable functions of *two* independent variables x, y. Then (77.4)

$$\Delta u = hu_x + ku_y + \eta_3 h + \eta_4 k, \qquad \Delta v = hv_x + kv_y + \eta_5 h + \eta_6 k,$$

where the η's all approach zero as $h, k \to 0$. On substituting these values of $\Delta u, \Delta v$ in (1), we get

$$\Delta f = (f_u + \eta_1) \{(u_x + \eta_3)h + (u_y + \eta_4)k\} +$$
$$(f_v + \eta_2) \{(v_x + \eta_5)h + (v_y + \eta_6)k\},$$

or, on collecting terms in h and k,

$$(4) \qquad \Delta f = (f_u u_x + f_v v_x)h + (f_u u_y + f_v v_y)k + \mu_1 h + \mu_2 k,$$

where μ_1 and μ_2 consist of sums of terms with at least one η as a factor and hence approach zero as $h, k \to 0$. The consequences of (4) are stated in the following important

THEOREM. *If $f(u, v)$ is a differentiable function of u, v and u, v are differentiable functions of x, y, the composite function,*

$$f \{u(x, y), v(x, y)\} = F(x, y),$$

is a differentiable function of x, y, whose partial derivatives are given by

$$(5) \qquad \frac{\partial f}{\partial x} = \frac{\partial f}{\partial u}\frac{\partial u}{\partial x} + \frac{\partial f}{\partial v}\frac{\partial v}{\partial x},$$

$$(6) \qquad \frac{\partial f}{\partial y} = \frac{\partial f}{\partial u}\frac{\partial u}{\partial y} + \frac{\partial f}{\partial v}\frac{\partial v}{\partial y},$$

and whose total differential,

$$(7) \qquad df = \frac{\partial f}{\partial u}\, du + \frac{\partial f}{\partial v}\, dv,$$

has the same form as if u, v were independent variables.

Proof. The differentiability of $F(x, y)$ follows from (4). If we put $k = 0$ in (4), divide by h, and then let $h \to 0$, we get (5); and (6) follows in similar fashion.

Since x, y are *independent* variables,

$$df = f_x \, dx + f_y \, dy \qquad\qquad (77.6)$$
$$= (f_u u_x + f_v v_x)h + (f_u u_y + f_v v_y)k$$
$$= f_u(u_x h + u_y k) + f_v(v_x h + v_y k)$$
$$= f_u \, du + f_v \, dv,$$

which is (7).

The fact that the total differential of a function has the same form whether the variables are dependent or independent is one of the chief advantages of the differential notation.

When the variables are independent, their differentials are also their increments and may be given *arbitrary* values. From this fact we have the useful rule:

If $f(x, y)$ is a function of independent variables, the equation

(8) $\qquad\qquad df = P \, dx + Q \, dy$ implies $f_x = P, \quad f_y = Q.$

Proof. In the identity

$$f_x \, dx + f_y \, dy = P \, dx + Q \, dy$$

put $dx = 1$, $dy = 0$; then $dx = 0$, $dy = 1$.

Example 1. Consider the function $f(u, v) = \sqrt{|uv|}$ when $u = x$, $v = x$. Then $f = |x|$ and $df/dx = 1$ or -1 according as x is positive or negative; at $x = 0$, df/dx does not exist.

If we apply the chain rule (3),

$$\frac{df}{dx} = f_u \frac{du}{dx} + f_v \frac{dv}{dx} = f_u + f_v.$$

When $x = 0$, $u = v = 0$ and $f_u = f_v = 0$ (Ex. 77.2); thus we get $df/dx = 0$ when $x = 0$. This false result is due to the fact that $\sqrt{|uv|}$ is not differentiable when $u = v = 0$.

Example 2. The plane-polar coordinates r, φ are given in terms of rectangular coordinates x, y by the equations

$$r = \sqrt{x^2 + y^2}, \qquad \varphi = \tan^{-1}\frac{y}{x} \qquad (x > 0).$$

Taking differentials, we have

$$dr = \frac{x \, dx + y \, dy}{\sqrt{x^2 + y^2}}, \qquad d\varphi = \frac{x \, dy - y \, dx}{x^2 + y^2}$$

and, since x and y are independent variables,

$$\frac{\partial r}{\partial x} = \frac{x}{r}, \qquad \frac{\partial r}{\partial y} = \frac{y}{r}; \qquad \frac{\partial \varphi}{\partial x} = -\frac{y}{r^2}, \qquad \frac{\partial \varphi}{\partial y} = \frac{x}{r^2}.$$

80. Functions of Three Variables. Let $f(x, y, z)$ be a function of three independent variables. When y and z are held fast, $f(x, y, z)$ becomes a function of x whose derivative f_x is called the *partial derivative of* $f(x, y, z)$ *with respect to* x: thus,

$$(1) \qquad f_x(x, y, z) = \lim_{h \to 0} \frac{f(x + h, y, z) - f(x, y, z)}{h} \; ;$$

f_y and f_z are defined in similar fashion.

The function $f(x, y, z)$ is said to be *differentiable* at (x, y, z) if the increments Δx, Δy, Δz produce an increment

$$\Delta f = f(x + \Delta x, y + \Delta y, z + \Delta z) - f(x, y, z)$$

having the form

$$(2) \qquad \Delta f = A \, \Delta x + B \, \Delta y + C \, \Delta z + \eta_1 \, \Delta x + \eta_2 \, \Delta y + \eta_3 \, \Delta z,$$

where A, B, C are independent of Δx, Δy, Δz, and η_1, η_2, η_3 tend to zero with these increments. The last condition is equivalent to

$$\eta_1 \, \Delta x + \eta_2 \, \Delta y + \eta_3 \, \Delta z = o(\rho) \quad \text{where} \quad \rho^2 = (\Delta x)^2 + (\Delta y)^2 + (\Delta z)^2.$$

If we put $\Delta y = \Delta z = 0$ in (2), divide by Δx and pass to the limit $\Delta x \to 0$, we get $f_x = A$; similarly, $f_y = B$, $f_z = C$. Thus for a differentiable function

$$(3) \qquad \Delta f = f_x \, \Delta x + f_y \, \Delta y + f_z \, \Delta z + o(\rho).$$

The part of Δf that is linear in the increments is called the *total differential* of f and written

$$(4) \qquad df = f_x \, \Delta x + f_y \, \Delta y + f_z \, \Delta z.$$

In particular, when $f = x$, $f_x = 1$, $f_y = f_z = 0$ and we have $dx = \Delta x$; similarly, $dy = \Delta y$, $dz = \Delta z$. *Thus the differentials of the independent variables are equal to their increments; and* (4) may be written

$$(5) \qquad df = f_x \, dx + f_y \, dy + f_z \, dz.$$

When $f(u, v, w)$ is differentiable and u, v, w are functions of x, we have the analogue of (79.3):

$$(6) \qquad \frac{df}{dx} = \frac{\partial f}{\partial u} \frac{du}{dx} + \frac{\partial f}{\partial v} \frac{dv}{dx} + \frac{\partial f}{\partial w} \frac{dw}{dx} \, .$$

For example, if $f = uvw$,

$$(uvw)' = vwu' + wuv' + uvw'.$$

When u, v, w are functions of two more independent variables x, y, $\cdots$, the composite function $f(u, v, w)$ is differentiable in x, y, $\cdots$, and f_x, f_y, $\cdots$ are computed as in § 79. In the case of three variables x, y, z, we have

$$f_x = f_u u_x + f_v v_x + f_w w_x,$$

(7)
$$f_y = f_u u_y + f_v v_y + f_w w_y,$$

$$f_z = f_u u_z + f_v v_z + f_w w_z,$$

"chain" formulas entirely analogous to (79.5–6). On multiplying these equations by dx, dy, dz, respectively, and adding, we obtain the total differential,

(8) $$df = f_x\, dx + f_y\, dy + f_z\, dz = f_u\, du + f_v\, dv + f_w\, dw.$$

This proves the *invariance of the total differential:*

The form of df is the same for dependent as for independent variables.

PROBLEMS

1. Prove that the function

$$f(x, y) = \frac{xy}{\sqrt{x^2 + y^2}} \qquad (x, y \neq 0), \qquad f(0, 0) = 0$$

is continuous but not differentiable at $(0, 0)$.

[At $(0, 0)$, $f_x = f_y = 0$ and $\Delta f - df = hk/\sqrt{h^2 + k^2}$.]

2. When $x = y = t$ in $f(x, y)$ of Prob. 1, $f(x, y) = t/\sqrt{2}$, and $df/dt = 1/\sqrt{2}$; but, from the chain rule (3),

$$df/dt = f_x \cdot 1 + f_y \cdot 1 = 0 \quad \text{when} \quad t = 0.$$

Explain this contradiction.

3. Find df, and hence f_x, f_y, when $f(x, y)$ is

(a) $(x^2 + y^2)^{-1/2}$;   (b) $\dfrac{x - y}{x + y}$;   (c) $e^{-x}\sin y$;

(d) $\log \sqrt{x^2 + y^2}$;   (e) x^y;   (f) a^{xy}.

4. If $f(x, y, z) = x^2y + y^2z + z^2x$, show that $f_x + f_y + f_z = (x + y + z)^2$.

5. Find du/dt if $u = \tan^{-1}\dfrac{y}{x}$ and $y = \cos t$, $x = \sin t$, using the chain rule. Explain the answer.

6. The equations connecting spherical and rectangular coordinates are given in (87.12–14). Compute the nine partial derivatives of x, y, z with respect to r, θ, φ and vice versa. Show that $x_r r_x + x_\theta \theta_x + x_\varphi \varphi_x = 1$.

81. Homogeneous Functions. A function $f(x, y, z)$ is said to be *homogeneous of degree n* when

$$(1) \qquad f(tx, ty, tz) = t^n f(x, y, z) \qquad (t > 0)$$

for every positive value of t. For example, the functions

$$x^2 + 2yz, \qquad \tan^{-1} \frac{z}{\sqrt{x^2 + y^2}}, \qquad (x^2 + y^2 + z^2)^{-1/2}$$

are homogeneous of degree 2, 0, -1, respectively.

Euler proved that any homogeneous function of degree n satisfies the identity

$$(2) \qquad xf_x + yf_y + zf_z = nf.$$

This partial differential equation is characteristic of homogeneous functions.†

THEOREM. *Any differentiable function that satisfies* (1) *also satisfies* (2) *and conversely.*

Proof. Consider the function

$$(3) \qquad F(x, y, z, t) = t^{-n} f(tx, ty, tz).$$

Put $u = tx$, $v = ty$, $w = tz$, and differentiate F with respect to t:

$$\frac{\partial F}{\partial t} = -n t^{-n-1} f(u, v, w) + t^{-n}(xf_u + yf_v + zf_w),$$

$$(4) \qquad t^{n+1} \frac{\partial F}{\partial t} = uf_u + vf_v + wf_w - nf(u, v, w).$$

When f is homogeneous, $F = f(x, y, z)$ from (1) and $\partial F/\partial t = 0$; hence the right side of (4) vanishes identically. This proves the identity (2).

Conversely, when the identity (2) is satisfied, (4) shows that $\partial F/\partial t = 0$; then F is a function of x, y, z alone, and this function can be found by putting $t = 1$ in (3). Thus,

$$t^{-n} f(tx, ty, tz) = f(x, y, z),$$

and we have the identity (1).

---

† The usual method of solving a linear partial differential equation of the first order gives the solution of (2) in the form

$$f(x, y, z) = x^n \varphi \left(\frac{y}{x}, \frac{z}{x} \right),$$

where φ is an arbitrary differentiable function of its arguments. This function satisfies (1).

PROBLEMS

1. Which of the following functions are homogenous and to what degree:

(a) $x^{2/3}$;   (b) $e^y - e^x$;   (c) $\log y - \log x$;   (d) $xe^{y/x} + y$;

(e) $\dfrac{x^2}{y} + \dfrac{y^2}{x} + z$;   (f) $\dfrac{1+x}{y}$;   (g) $\dfrac{x}{y} + \tan^{-1}\dfrac{y}{x}$;

(h) $x + \sqrt{xy}$;   (i) $ze^{y^2/x}$;   (j) $\cos^{-1}\dfrac{z}{\sqrt{x^2 + y^2 + z^2}}$?

2. When the functions of Prob. 1 are homogeneous, show that they satisfy (2).

3. If $f(x, y, z)$ is homogeneous of degree n, show that when $x > 0$,
$$f(x, y, z) = x^n f(1, y/x, z/x).$$
Illustrate with the fifth and tenth functions of Prob. 1.

4. If $f(x, y, z)$ is homogeneous of degree n, show that f_x, f_y, f_z are homogeneous of degree $n - 1$.

5. If $f(x, y, z)$ has the property
$$f(tx, t^j y, t^k z) = t^n f(x, y, z),$$
prove that
$$f(x, y, z) = x^n f(1, y/x^j, z/x^k), \qquad xf_x + jyf_y + kzf_z = nf.$$

6. If u, v, w are functions of x, show that their *Wronskian* $W(u, v, w) = \begin{vmatrix} u & v & w \\ u' & v' & w' \\ u'' & v'' & w'' \end{vmatrix}$

satisfies $W(u, v, w) = u^3 W(1, v/u, w/u)$.

82. Higher Derivatives. The first partial derivatives $f_x(x, y)$ and $f_y(x, y)$ are functions of x and y that may admit partial derivatives. The derivatives of f_x are denoted by
$$f_{xx}, f_{xy} \quad \text{or} \quad \frac{\partial^2 f}{\partial x^2}, \frac{\partial^2 f}{\partial y\, \partial x},$$
the derivatives of f_y by †
$$f_{yx}, f_{yy} \quad \text{or} \quad \frac{\partial^2 f}{\partial x\, \partial y}, \frac{\partial^2 f}{\partial y^2}.$$

Thus $f(x, y)$ may have four second derivatives. These in turn yield eight third derivatives,
$$f_{xxx}, f_{xxy}; \quad f_{xyx}, f_{xyy}; \quad f_{yxx}, f_{yxy}; \quad f_{yyx}, f_{yyy};$$
and so on. The higher derivatives with respect to both x and y are called *mixed*. Under certain conditions mixed derivatives that differ only in the *order* in which the differentions are performed have the same value. Thus, when the requisite conditions are fulfilled, $f(x, y)$ has but *three* second derivatives,
$$f_{xx}, \quad f_{xy} = f_{yx}, \quad f_{yy},$$

---

† In the subscript notation the order of differentiation is from left to right, in the ∂ notation from right to left. This usage, however, is not always followed by European authors.

and *four* third derivatives, and, in the general case, $n + 1$ partial derivatives of the nth order (for there can be $0, 1, 2, \cdots, n$ derivations with respect to x).

As to f_{xy} and f_{yx} we have by definition

$$f_x(a, y) = \lim_{h \to 0} \frac{f(a + h, y) - f(a, y)}{h},$$

$$f_{xy}(a, b) = \lim_{k \to 0} \lim_{h \to 0} \frac{f(a + h, b + k) - f(a, b + k) - f(a + h, b) + f(a, b)}{kh}$$

If we write

(1) $F(h, k) = f(a + h, b + k) - f(a, b + k) - f(a + h, b) + f(a, b),$

we have

(2) $f_{xy}(a, b) = \lim_{k \to 0} \lim_{h \to 0} \frac{F(h, k)}{kh},$

and, similarly,

(3) $f_{yx}(a, b) = \lim_{h \to 0} \lim_{k \to 0} \frac{F(h, k)}{hk}.$

We have seen that such iterated limits may differ in value (cf. Ex. 73.1). Sufficient conditions for their equality are given in the following theorems.

THEOREM 1 (Schwarz). *If the derivatives*

(i) f_x *and* f_y *exist in some neighborhood of* (a, b),

(ii) f_{xy} *is continuous at* (a, b),
then $f_{yx}(a, b)$ *exists and* $f_{yx}(a, b) = f_{xy}(a, b)$.

Proof. If we put $\varphi(x) = f(x, b + k) - f(x, b)$,
we have

     $F(h, k) = \varphi(a + h) - \varphi(a)$

         $= h \varphi'(a + \theta h), \quad\quad 0 < \theta < 1,$

(a) $= h[f_x(a + \theta h, b + k) - f_x(a + \theta h, b)],$

(b) $= hk f_{xy}(a + \theta h, b + \theta'k), \quad\quad 0 < \theta' < 1,$

(c) $= hk[f_{xy}(a, b) + \eta],$

where $\eta \to 0$ as h and k approach zero *in any manner whatever*. In step (a) the use of the mean-value theorem is justified by (i); in step (b) its use is justified by (ii), for the continuity of f_{xy} at (a, b) requires the existence of f_{xy} in the neighborhood of (a, b). Step (c) is again a consequence of (ii); and this expression for $F(h, k)$ shows that the limits in (2) and (3) are the same.

THEOREM 2 (Young). *If the derivatives*

(i) f_x *and* f_y *exist in some neighborhood of* (a, b),

(iii) f_x *and* f_y *are differentiable at* (a, b),

then

$$f_{xy}(a, b) = f_{yx}(a, b).$$

Proof. After step (a) in the above proof, hypothesis (iii) shows that

$$f_x(a + \theta h, b + k) - f_x(a, b) = \theta h f_{xx}(a, b) + k f_{xy}(a, b) + \eta_1 \theta h + \eta_2 k,$$
$$f_x(a + \theta h, b) - f_x(a, b) \quad = \theta h f_{xx}(a, b) + \eta_3 \theta h,$$

where all $\eta_i \to 0$ as h and $k \to 0$. If we now put $k = h$, we find

$$\lim_{h \to 0} \frac{F(h, h)}{h^2} = \lim_{h \to 0} [f_{xy}(a, b) + (\eta_1 - \eta_3)\theta + \eta_2] = f_{xy}(a, b).$$

Since the hypotheses on f_x and f_y are exactly the same, it is clear that we must also have

$$\lim_{h \to 0} \frac{F(h, h)}{h^2} = f_{yx}(a, b).$$

We may verify this fact by putting

$$\psi(y) = f(a + h, y) - f(a, y)$$

and carrying through the above calculation. Thus the theorem is proved.

In any mixed derivative of order n we can make a succession of interchanges of x and y, provided all the derivatives in question are continuous. We can, for example, write all such mixed derivatives so that the differentiations with respect to x come first, as in f_{xyy} or f_{xxy}.

Example 1. As an instance where $f_{xy} \neq f_{yx}$ consider the function

$$f(x, y) = xy \frac{x^2 - y^2}{x^2 + y^2}, \qquad f(0, 0) = 0.$$

We have

$$f_x(0, y) = \lim_{h \to 0} \frac{f(h, y) - f(0, y)}{h} = \lim_{h \to 0} y \frac{h^2 - y^2}{h^2 + y^2} = -y,$$

$$f_{xy}(0, 0) = \lim_{k \to 0} \frac{f_x(0, k) - f_x(0, 0)}{k} = \lim_{k \to 0} \frac{-k}{k} = -1;$$

$$f_y(x, 0) = \lim_{k \to 0} \frac{f(x, k) - f(x, 0)}{k} = \lim_{k \to 0} x \frac{x^2 - k^2}{x^2 + k^2} = x,$$

$$f_{yx}(0, 0) = \lim_{h \to 0} \frac{f_y(h, 0) - f_y(0, 0)}{h} = \lim_{h \to 0} \frac{h}{h} = 1.$$

Example 2. Consider the function

$$f(x, y) = x^2 \tan^{-1} \frac{y}{x} - y^2 \tan^{-1} \frac{x}{y}, \qquad f(0, y) = f(x, 0) = 0.$$

When $x, y \neq 0$, we find that

$$f_x = 2x \tan^{-1}\frac{y}{x} - y, \qquad f_{xy} = \frac{x^2 - y^2}{x^2 + y^2};$$

$$f_y = x - 2y \tan^{-1}\frac{x}{y}, \qquad f_{yx} = \frac{x^2 - y^2}{x^2 + y^2}.$$

Thus $f_{xy} = f_{yx}$ when $x, y \neq 0$; of course this follows from Theorem 1 since f_{xy} is continuous at a point where $x \neq 0, y \neq 0$.

However, we cannot conclude that $f_{xy} = f_{yx}$ at $(0, 0)$ since $x^2 + y^2 = 0$ there. We must therefore resort to the definition of partial derivatives. Thus we find

$$f_x(0, y) = -y, \qquad f_{xy}(0, 0) = -1,$$

$$f_y(x, 0) = x, \qquad f_{yx}(0, 0) = 1.$$

83. Implicit Functions. If the equation $F(x, y) = 0$, defines y as a function of x, say $y = f(x)$, so that

$$F(x, f(x)) = 0,$$

we say that $y = f(x)$ is defined *implicitly* by $F(x, y) = 0$. Thus in § 49 we found that under certain conditions the equation $x - f(y) = 0$ defined a unique inverse function $y = \varphi(x)$. Not every equation $F(x, y) = 0$ can be "solved" for y as a function of x; for example, the only real values that satisfy $x^2 + y^2 = 0$ are $x = y = 0$. Therefore a first consideration is to give conditions under which the equation $F(x, y) = 0$ defines a unique implicit function. This is done in the following

EXISTENCE THEOREM. *Let the region R contain the point (x_0, y_0) in its interior. Then, if*

(i) $F(x_0, y_0) = 0,$

(ii) F_x, F_y *are continuous in R,*

(iii) $F_y(x_0, y_0) \neq 0;$

there is some interval I_0 about x_0 in which there exists a unique, differentiable function $y = f(x)$ such that

(1) $y_0 = f(x_0),$

(2) $F(x, f(x)) = 0,$

(3) $f'(x) = -F_x/F_y.$

Proof. Assume that $F_y(x_0, y_0) > 0$†; then, since F_y is continuous in R, we can find a rectangle in R and having (x_0, y_0) as center in which

$$F_y(x, y) > 0, \qquad |x - x_0| \leqq a, \qquad |y - y_0| \leqq b.$$

Then $F(x_0, y)$ steadily increases from $y_0 - b$ to $y_0 + b$, passing through zero when $y = y_0$; hence,

$$F(x_0, y_0 - b) < 0, \qquad F(x_0, y_0 + b) > 0.$$

---

† This is no restriction; for, if $F_y(x_0, y_0) < 0$, consider the equation $-F(x, y) = 0$.

But, since $F_x(x, y)$ is continuous in R, the functions of x, $F(x, y_0 - b)$ and $F(x, y_0 + b)$, are also continuous; therefore, we can find an interval

$$I_0: \quad x_0 - \delta_0 \leqq x \leqq x_0 + \delta_0 \qquad (\delta_0 \leqq a)$$

in which

$$F(x, y_0 - b) < 0, \qquad F(x, y_0 + b) > 0.$$

Now in the rectangle $ABCD$ (Fig. 83a) the function $F(x, y)$ is negative along AB, positive along CD. Hence, for each value x_1 in the interval I_0, the increasing function $F(x_1, y)$ must vanish for exactly one value y_1

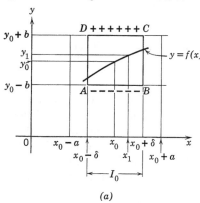

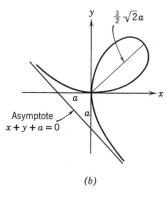

FIG. 83a.   FIG. 83b.

Existence of an implicit function.   Folium: $x^3 + y^3 - 3axy = 0$.

between $y_0 - b$ and $y_0 + b$. Since unique values of y correspond to all values of x in I_0, y is a function of x over this interval, say $y = f(x)$. From (i) we see that y_0 corresponds to x_0, that is, $y_0 = f(x_0)$. Thus we have proved parts (1) and (2) of the theorem.

We next show that $y = f(x)$ is *continuous* at any point x_1 of I_0: that is,

$$|y - y_1| < \varepsilon \quad \text{when} \quad |x - x_1| < \delta_1.$$

Draw horizontal lines $y = y_1 \pm \varepsilon$, and find an x-interval I_1 about x_1 so that

$$F(x, y_1 - \varepsilon) < 0, \qquad F(x, y_1 + \varepsilon) > 0.$$

For any x in this interval the corresponding y lies between $y_1 - \varepsilon$ and $y_1 + \varepsilon$ and therefore differs from y_1 by less than ε. Thus $y = f(x)$ is continuous throughout I_0.

Finally to prove part (3) of the theorem, we note that $F(x, y)$ is differentiable in R since F_x, F_y are continuous in R (§ 78). Now, if x_1 is any point of I_0 and

$$y_1 = f(x_1), \qquad y_1 + k = f(x_1 + h),$$

we have

$$\Delta F = F(x_1 + h, y_1 + k) - F(x_1, y_1) = 0 - 0 = 0,$$

and, hence from (77.3),

$$hF_x(x_1, y_1) + kF_y(x_1, y_1) + \eta_1 h + \eta_2 k = 0,$$

where $\eta_1, \eta_2 \to 0$ as $h, k \to 0$. If we divide this equation by hF_y ($\neq 0$ by hypothesis), we have

$$\frac{F_x(x_1, y_1) + \eta_1}{F_y(x_1, y_1)} + \frac{k}{h}\left(1 + \frac{\eta_2}{F_y}\right) = 0.$$

If we pass to the limit $h \to 0$, the continuity of $f(x)$ requires $k \to 0$ also, and

$$\frac{k}{h} = \frac{f(x_1 + h) - f(x_1)}{h} \to f'(x_1);$$

but, as $h, k \to 0$, $\eta_1, \eta_2 \to 0$ also, and

$$f'(x_1) = - \frac{F_x(x_1, y_1)}{F_y(x_1, y_1)}.$$

This completes the proof.

When the existence of the implicit function is assured, we may find dy/dx from $F(x, y) = 0$ by the chain rule (79.3),

$$F_x \frac{dx}{dx} + F_y \frac{dy}{dx} = 0;$$

or we may take differentials:

$$dF = F_x \, dx + F_y \, dy = 0,$$

remembering that *form* of dF is not altered by the dependence of y upon x.

From $y' = -F_x/F_y$ we can find y'' by using the chain rule and then substituting for y' its value:

$$y'' = - \left(\frac{F_x}{F_y}\right)_x - \left(\frac{F_x}{F_y}\right)_y y'$$

$$= - \frac{F_y F_{xx} - F_x F_{yx}}{F_y{}^2} + \frac{F_y F_{xy} - F_x F_{yy}}{F_y{}^2} \frac{F_x}{F_y},$$

$$y'' = - \frac{F_y{}^2 F_{xx} - 2F_x F_y F_{xy} + F_x{}^2 F_{yy}}{F_y{}^3}.$$

Example 1. From the equation of the *ellipse*

(i) $$b^2 x^2 + a^2 y^2 = a^2 b^2,$$

we have (if $y \neq 0$)

$$2b^2 x + 2a^2 yy' = 0, \qquad y' = - \frac{b^2 x}{a^2 y}.$$

We can also obtain this result from the explicit functions,

(ii)
$$y = \pm \frac{b}{a} \sqrt{a^2 - x^2},$$

representing the upper and lower halves of the ellipse. But implicit differentiation avoids radicals and gives a single formula for y' valid wherever $y \neq 0$. The existence theorem fails at the points $(\pm a, 0)$; at these points equation (i) defines *two* functions (ii). We have also

$$y'' = -\frac{b^2}{a^2} \frac{y - xy'}{y^2} = -\frac{b^4}{a^2 y^3}.$$

As (i) is dimensionally correct, we note that y' and y'' have the dimensions 0 and -1 in length.

Example 2. The equation of the *folium of Descartes* (Fig. 83b),

(iii)
$$x^3 + y^3 - 3axy = 0,$$

has an explicit solution that can be found by solving a cubic equation. To avoid this nuisance we can use the existence theorem at all points where

$$F_y = 3(y^2 - ax) \neq 0.$$

Thus we find by the chain rule:

$$y' = -\frac{x^2 - ay}{y^2 - ax}, \qquad y'' = -\frac{2a^3 xy}{(y^2 - ax)^3}.$$

It is easy to show that the curve has a horizontal tangent at $(2^{1/3}a, 2^{2/3}a)$, a vertical tangent at $(2^{2/3}a, 2^{1/3}a)$. The last point and the origin (where the curve crosses itself) are the points excluded by $F_y \neq 0$. Observe that y' and y'' have the dimensions 0 and -1 in length; for (iii) shows that a has the dimension 1.

84. One Equation. The existence theorem can be generalized to equations having three or more variables. If we wish to solve

(1)
$$F(x, y, z) = 0$$

for z we have the

THEOREM. *Let the region R contain the point (x_0, y_0, z_0) in its interior. Then, if*

(i) $F(x_0, y_0, z_0) = 0$,

(ii) F_x, F_y, F_z *are continuous in R,*

(iii) $F_z(x_0, y_0, z_0) \neq 0$,

there is a neighborhood I_0 of the point (x_0, y_0) in which there exists a unique, differentiable function $z = f(x, y)$ such that

(a)
$$z_0 = f(x_0, y_0),$$

(b)
$$F(x, y, f(x, y)) = 0,$$

(c)
$$z_x = -F_x/F_z, \qquad z_y = -F_y/F_z.$$

The proof of this theorem follows the same pattern as the fundamental theorem of § 83. In the 2-dimensional neighborhood I_0 over which $z = f(x, y)$ is uniquely defined we have $F_z \neq 0$; hence, from

(2) $$dF = F_x \, dx + F_y \, dy + F_z \, dz = 0,$$
we have

$$dz = -\frac{F_x}{F_z} \, dx - \frac{F_y}{F_z} \, dy,$$

and, from (79.8), equations (c). The hypothesis (ii) shows that z_x and z_y are continuous in I_0 so that $z = f(x, y)$ is a differentiable function.

If $F_x \neq 0$, we may solve (1) for x and (2) for dx. Regarding x as a function of the independent variables y, z, we then obtain from (79.9)

$$x_y = -F_y/F_x, \qquad x_z = -F_z/F_x.$$

If $F_y \neq 0$, we may solve (1) for y and (2) for dy. Regarding x and z as independent variables, we now have

$$y_x = -F_x/F_y, \qquad y_z = -F_z/F_y.$$

This theorem admits of further generalization to equations such as

(3) $$F(x, y, z, u) = 0.$$

If $F_u \neq 0$ we can solve (3) for u and $dF = 0$ for du; we thus obtain

$$u_x = -F_x/F_u, \qquad u_y = -F_y/F_u, \qquad u_z = -F_z/F_u$$

Example. The equation
(i) $$F = xyz + x + y - z = 0$$
can be solved for z near $(0, 0, 0)$; for $F_z(0, 0, 0) = -1$. We have, in fact,

(ii) $$z = \frac{x + y}{1 - xy}.$$
From this explicit solution we have

(iii) $$z_x = \frac{1 + y^2}{(1 - xy)^2}, \qquad z_y = \frac{1 + x^2}{(1 - xy)^2};$$

or we may apply the chain rule to (i):

$$yz + 1 + (xy - 1)z_x = 0, \qquad z_x = \frac{yz + 1}{1 - xy};$$

$$xz + 1 + (xy - 1)z_y = 0, \qquad z_y = \frac{xz + 1}{1 - xy}.$$

These results reduce to (iii) when the value (ii) is substituted for z.

PROBLEMS

1. If $f(x, y) = (x^2 + y^2) \tan^{-1}(y/x)$, $f(0, y) = \pi y^2/2$, show that $f_{xy} = f_{yx} = (x^2 - y^2)/(x^2 + y^2)$ when $x, y \neq 0$. But from the definition of partial derivatives,

$$f_x(0, y) = -y, \quad f_y(x, 0) = x; \qquad f_{xy}(0, 0) = -1, \quad f_{yx}(0, 0) = 1.$$

2. Find y' and y'' for the implicit functions $y(x)$ defined by the equations

$$(a)\ x^2 - y^2 = a^2;\qquad (b)\ y^2 - 4ax\ = 0;$$
$$(c)\ x^n + y^n = a^n;\qquad (d)\ y^3 - ax^2\ = 0.$$
$$(e)\ e^{ax+y^2}\ = 1;\qquad (f)\ x^3/y^3 + y^3/x^3 = 2.$$

If we regard x, y, a as having the dimension 1 in length, each equation except (e) is dimensionally correct. Verify in each case that y' and y'' have the dimensions 0 and -1.

3. If $F(x, y, z) = 0$, show that

$$\frac{\partial x}{\partial y}\frac{\partial y}{\partial z}\frac{\partial z}{\partial x} = -1,$$

where each partial derivative is computed by holding the remaining variable constant.

4. Prove that y'' for the implicit function defined by $f(x, y) = 0$ is given by

$$f_y{}^3 y'' = \begin{vmatrix} f_{xx} & f_{xy} & f_x \\ f_{xy} & f_{yy} & f_y \\ f_x & f_y & 0 \end{vmatrix} \quad \text{if}\ f_y \neq 0.$$

5. When $f(x, y) = Ax^2 + 2Bxy + Cy^2 + 2Dx + 2Ey + F$ in Prob. 4, show that

$$y'' = \frac{\Delta}{(Bx + Cy + E)^3} \quad \text{where}\quad \Delta = \begin{vmatrix} A & B & D \\ B & C & E \\ D & E & F \end{vmatrix}.$$

Verify this result in Ex. 83.1.

6. In Prob. 5 prove that the conic $f(x, y) = 0$ degenerates into two straight lines when and only when $\Delta = 0$.

7. If $u = f(x - y, y - z, z - x)$, show that $u_x + u_y + u_z = 0$.

8. If $f(x, y) = \log(x^2 + y^2) + \tan^{-1} y/x$, compute $f_{xy}, f_{yx},$ and $f_{xx} + f_{yy}$.

9. If $u = f(x + at, y + bt)$, show that

$$\frac{\partial^n u}{\partial t^n} = (aD_x + bD_y)^n\, u,$$

where $D_x = \partial/\partial x$, $D_y = \partial/\partial y$, and powers of D in the binomial expansion denote repeated derivations.

85. Two Equations. We next consider the problem of solving two equations,

$$(1) \qquad F(x, y, u, v) = 0, \qquad G(x, y, u, v) = 0,$$

for u and v. We shall speak of a set of values (x_0, y_0, u_0, v_0) as a *point* in four dimensions. Regions having this point in the interior are the open 4-dimensional cube,

$$|x - x_0| < k, \qquad |y - y_0| < k, \qquad |u - u_0| < k, \qquad |v - v_0| < k,$$

or the open 4-dimensional sphere,

$$(x - x_0)^2 + (y - y_0)^2 + (u - u_0)^2 + (v - v_0)^2 < r^2.$$

EXISTENCE THEOREM. *Let the region R contain the point* (x_0, y_0, u_0, v_0) *in its interior. Then, if*

(i) $F(x_0, y_0, u_0, v_0) = 0,$ $G(x_0, y_0, u_0, v_0) = 0,$

(ii) $F_x, F_y, F_u, F_v, G_x, G_y, G_u, G_v$ *are continuous in R,*

(iii) $J = \begin{vmatrix} F_u & F_v \\ G_u & G_v \end{vmatrix} \neq 0$ *at* $(x_0, y_0, u_0, v_0),$

there is some 2-dimensional neighborhood I_0 of (x_0, y_0) in which there exist two unique differentiable functions $u = f(x, y)$, $v = g(x, y)$ such that

(a) $u_0 = f(x_0, y_0),$ $v_0 = g(x_0, y_0);$

(b) $F(x, y, f, g) = 0$ $G(x, y, f, g) = 0;$

(c)   *the partial derivatives of $u = f(x, y)$, $v = g(x, y)$ are continuous functions found by solving $dF = 0$, $dG = 0$ for du and dv in terms of dx and dy.*

Proof. Condition (iii) shows that at least one of the functions F_u, F_v is not zero at (x_0, y_0, u_0, v_0); hence we may assume $F_v(x_0, y_0, u_0, v_0) \neq 0$. By the generalized theorem of § 84 we can solve $F(x, y, u, v) = 0$ for

(1) $v = \varphi(x, y, u)$

in some neighborhood at (x_0, y_0, u_0, v_0) so that $v_0 = \varphi(x_0, y_0, u_0)$. If we differentiate

(2) $F(x, y, u, \varphi) = 0$

with respect to u we have

$$F_u + F_v \varphi_u = 0, \qquad \varphi_u = -F_u/F_v.$$

We next substitute (1) in $G(x, y, u, v)$ and solve

(3) $G(x, y, u, \varphi) = \psi(x, y, u) = 0$

for u in the neighborhood of (x_0, y_0, u_0). This is possible since

$$\psi_u = G_u + G_v \varphi_u = \frac{G_u F_v - G_v F_u}{F_v} = -\frac{J}{F_v}$$

and $\psi_u(x_0, y_0, u_0) \neq 0$ from (iii). Hence there exists a unique function $u = f(x, y)$ such that $u_0 = f(x_0, y_0)$, which makes equation (3) an identity in some neighborhood I_0 of (x_0, y_0).

If we now put

$$v = \varphi(x, y, f(x, y)) = g(x, y),$$

we have

$$g(x_0, y_0) = \varphi(x_0, y_0, u_0) = v_0.$$

The pair of functions,

(4) $u = f(x, y), \qquad v = g(x, y),$

now fulfill all the requirements of the theorem; for from (2) and (3)

$$F(x, y, f, \varphi(x, y, f)) = 0, \qquad G(x, y, f, \varphi(x, y, f)) = 0.$$

As to part (c) of the theorem, we have

(5) $dF = F_x \, dx + F_y \, dy + F_u \, du + F_v \, dv = 0,$
$$dG = G_x \, dx + G_y \, dy + G_u \, du + G_v \, dv = 0.$$

When (x, y) is in the neighborhood I_0, and u, v are given by (4), $J \neq 0$. We may then solve equations (5) for du, dv:

$$du = -\frac{1}{J} \begin{vmatrix} F_x & F_v \\ G_x & G_v \end{vmatrix} dx - \frac{1}{J} \begin{vmatrix} F_y & F_v \\ G_y & G_v \end{vmatrix} dy,$$

$$dv = -\frac{1}{J} \begin{vmatrix} F_u & F_x \\ G_u & G_x \end{vmatrix} dx - \frac{1}{J} \begin{vmatrix} F_u & F_y \\ G_u & G_y \end{vmatrix} dy.$$

The coefficients of dx and dy give u_x, v_x and u_y, v_y, respectively (79.8).

The determinants above (including J) are called *Jacobians*. They are often written in a more compact notation: for example,

(6) $$J = \begin{vmatrix} F_u & F_v \\ G_u & G_v \end{vmatrix} = \frac{\partial(F, G)}{\partial(u, v)}.$$

In this notation

(7) $$u_x = -\frac{1}{J} \frac{\partial(F, G)}{\partial(x, v)}, \qquad u_y = -\frac{1}{J} \frac{\partial(F, G)}{\partial(y, v)},$$

(8) $$v_x = -\frac{1}{J} \frac{\partial(F, G)}{\partial(u, x)}, \qquad v_y = -\frac{1}{J} \frac{\partial(F, G)}{\partial(u, y)}.$$

In each formula the Jacobian written out is formed from J by replacing u or v by x or y as indicated by the letters on the left.

If we wish to solve equations (1) for x and y, part (iii) of the hypothesis must be replaced by

(iii)′ $$J' = \frac{\partial(F, G)}{\partial(x, y)} \neq 0 \quad \text{at} \quad (x_0, y_0, u_0, v_0).$$

The derivatives x_u, x_v, y_u, y_v are found by solving equations (5) for dx, dy. The results may be written down by interchanging x and u, y and v, in equations (7) and (8).† Note that $x_u \neq 1/u_x$.

---

† This changes J into J'.

When there are more than three variables, the partial-derivative notation is ambiguous unless the independent variables are specified. Thus suppose we wish to find $\partial u/\partial x$ from the equations

$$u = f(x, z), \qquad z = g(x, y).$$

There are now two possible choices for the independent variables: (i) x and y, (ii) x and z. Note that y and z cannot be independent if $\partial u/\partial x$ is to have a meaning. Now, in case (i),

$$\left(\frac{\partial u}{\partial x}\right)_y = \frac{\partial f}{\partial x} + \frac{\partial f}{\partial z}\frac{\partial z}{\partial x} = \frac{\partial f}{\partial x} + \frac{\partial f}{\partial z}\frac{\partial g}{\partial x},$$

where the subscript y indicates the other independent variable. In case (ii)

$$\left(\frac{\partial u}{\partial x}\right)_z = \frac{\partial f}{\partial x},$$

where the subscript z again indicates the other independent variable. This subscript notation is useful in cases of ambiguity. In any event before computing a partial derivative it is essential to ask the question: *What are the independent variables?*

Example. The equations

(i)
$$F = u^2 + v^2 - x^2 - y = 0,$$

$$G = u + v - x^2 + y = 0,$$

are satisfied by $x_0 = 2$, $y_0 = 1$, $u_0 = 1$, $v_0 = 2$.
Since

$$J = \begin{vmatrix} F_u & F_v \\ G_u & G_v \end{vmatrix} = \begin{vmatrix} 2u & 2v \\ 1 & 1 \end{vmatrix} = 2(u - v),$$

and $J_0 = -2$, equations (i) can be solved uniquely for u, v in the neighborhood of $x_0 = 2$, $y_0 = 1$. On solving

(ii)
$$dF = 2u\,du + 2v\,dv - 2x\,dx - dy = 0,$$

$$dG = du + dv - 2x\,dx + dy = 0,$$

for du, dv, we find

$$u_x = \frac{x(1 - 2v)}{u - v}, \qquad u_y = \frac{1 + 2v}{2(u - v)},$$

$$v_x = \frac{x(2u - 1)}{u - v}, \qquad v_y = \frac{-2u - 1}{2(u - v)}.$$

Since

$$J' = \begin{vmatrix} F_x & G_x \\ F_y & G_y \end{vmatrix} = \begin{vmatrix} -2x & -2x \\ -1 & 1 \end{vmatrix} = -4x,$$

and $J_0' = -8$, equations (i) can also be solved uniquely for x, y in the neighborhood of $u_0 = 1$, $v_0 = 2$. On solving (ii) for dx, dy, we now find

$$x_u = \frac{2u+1}{4x}, \qquad x_v = \frac{2v+1}{4x},$$

$$y_u = \frac{2u-1}{2}, \qquad y_v = \frac{2v-1}{2}.$$

As a check on the computation we may verify that the Jacobians $\partial(u, v)/\partial(x, y)$ and $\partial(x, y)/\partial(u, v)$ are reciprocals. See § 87.

86. Three Equations. The problem of solving three equations,

(1) $F(x, y, z, u, v, w) = 0, \qquad G = 0, \qquad H = 0,$

for u, v, w depends upon our previous solutions of one and two equations. In the present case we have the

EXISTENCE THEOREM. *Let the region R contain the point* $(x_0, y_0, z_0, u_0, v_0, w_0)$ *in its interior. Then, if*

(i) $F = 0, \quad G = 0, \quad H = 0$ *at* $(x_0, y_0, z_0, u_0, v_0, w_0)$,

(ii) *all first partial derivatives of F, G, H are continuous in R,*

(iii) $J = \begin{vmatrix} F_u & F_v & F_w \\ G_u & G_v & G_w \\ H_u & H_v & H_w \end{vmatrix} \neq 0$ *at* $(x_0, y_0, z_0, u_0, v_0, w_0)$,

there is some 3-dimensional neighborhood I_0 *of* (x_0, y_0, z_0) *in which there exist three unique differentiable functions,*

(2) $u = f(x, y, z), \qquad v = g(x, y, z), \qquad w = h(x, y, z),$

such that

(a) $u_0 = f(x_0, y_0, z_0), \qquad v_0 = g(x_0, y_0, z_0), \qquad w_0 = h(x_0, y_0, z_0),$

(b)   *the functions* (2) *reduce equations* (1) *to identities in* x, y, z,

(c)   *the partial derivatives of f, g, h are continuous functions found by solving*

(3) $dF = 0, \qquad dG = 0, \qquad dH = 0$

for du, dv, dw in terms of dx, dy, dz.

Proof (in outline). Condition (iii) shows that at least one of the three minors in the first two rows of J is not zero at $(x_0, y_0, z_0, u_0, v_0, w_0)$; hence, we may assume

(4) $\begin{vmatrix} F_v & F_w \\ G_v & G_w \end{vmatrix}_0 \neq 0.$

We may now solve equations $F = 0$, $G = 0$ uniquely for v, w (§ 85),

(5) $\qquad\qquad v = \varphi(x, y, z, u), \qquad w = \psi(x, y, z, u),$

and

$$v_0 = \varphi(x_0, y_0, z_0, u_0), \qquad w_0 = \psi(x_0, y_0, z_0, u_0).$$

When the solutions (5) are substituted in $F = 0$, $G = 0$, these equations become identities in x, y, z, u; and, on differentiating these identities with respect to u, we get

(6) $\qquad\qquad F_u + F_v \varphi_u + F_w \psi_u = 0,$

(7) $\qquad\qquad G_u + G_v \varphi_u + G_w \psi_u = 0.$

When the solutions (5) are substituted in H, we obtain a function

$$\chi(x, y, z, u) = H(x, y, z, u, v, w)$$

whose derivative with respect to u is given by

(8) $\qquad\qquad H_u + H_v \varphi_u + H_w \psi_u = \chi_u.$

At the point $(x_0, y_0, z_0, u_0, v_0, w_0)$, the linear equations (6), (7), (8) have the solution 1, φ_u, ψ_u; and by Cramer's rule

(9) $\qquad 1 = \dfrac{1}{J_0} \begin{vmatrix} 0 & F_v & F_w \\ 0 & G_v & G_w \\ \chi_u & H_v & H_w \end{vmatrix}_0 = \dfrac{(\chi_u)_0}{J_0} \begin{vmatrix} F_v & F_w \\ G_v & G_w \end{vmatrix}_0 .$

Hence, from (4), $\chi_u \neq 0$ at (x_0, y_0, z_0, u_0), and the equation $\chi = 0$ may be solved for

(10) $\qquad\qquad u = f(x, y, z)$

in some neighborhood I_0 of (x_0, y_0, z_0) and $u_0 = f(x_0, y_0, z_0)$. If we substitute (10) in equations (5), we get

$$v = g(x, y, z), \qquad w = h(x, y, z),$$

and

$$v_0 = g(x_0, y_0, z_0), \qquad w_0 = h(x_0, y_0, z_0).$$

As to part (c) of the theorem, equations (3) may be solved for du, dv, dw when (x, y, z) is in I_0; for $J \neq 0$ in I_0.

The Jacobian J is also written

(11) $\qquad\qquad J = \dfrac{\partial(F, G, H)}{\partial(u, v, w)} .$

87. Coordinate Transformations. If we change from rectangular coordinates (x, y) to curvilinear coordinates (u, v) through the equations

(1) $\qquad\qquad x = F(u, v), \qquad y = G(u, v),$

the theorem of § 85 shows that we may solve them uniquely for u, v in the vicinity of any point where the *Jacobian of the transformation,*

$$\text{(2)} \qquad \frac{\partial(x, y)}{\partial(u, v)} = \frac{\partial(F, G)}{\partial(u, v)} \neq 0;$$

for this determinant is also the Jacobian of the left members of the equations

$$F(u, v) - x = 0, \qquad G(u, v) - y = 0.$$

We thus obtain the inverse transformation,

$$\text{(3)} \qquad u = f(x, y), \qquad v = g(x, y),$$

whose Jacobian is the reciprocal of (2):

$$\text{(4)} \qquad \frac{\partial(u, v)}{\partial(x, y)} \cdot \frac{\partial(x, y)}{\partial(u, v)} = 1$$

The proof of (4) follows at once from the chain-rule:

$$u_x x_u + u_y y_u = u_u = 1, \qquad u_x x_v + u_y y_v = u_v = 0,$$
$$v_x x_u + v_y y_u = v_u = 0, \qquad v_x x_v + v_y y_v = v_v = 1.$$

In three dimensions the transformation takes the form

$$\text{(5)} \qquad x = F(u, v, w), \qquad y = G(u, v, w), \qquad z = H(u, v, w).$$

The theorem of § 86 shows that we may solve equations (5) uniquely for u, v, w in the vicinity of any point where the Jacobian

$$\text{(6)} \qquad \frac{\partial(x, y, z)}{\partial(u, v, w)} \neq 0.$$

We thus obtain the inverse transformation,

$$\text{(7)} \qquad u = f(x, y, z), \qquad v = g(x, y, z), \qquad w = h(x, y, z)$$

whose Jacobian is the reciprocal of (6):

$$\text{(8)} \qquad \frac{\partial(u, v, w)}{\partial(x, y, z)} \frac{\partial(x, y, z)}{\partial(u, v, w)} = 1.$$

This is proved in the same fashion as (4).

The Jacobians $\partial(u, v)/\partial(x, y)$ and $\partial(u, v, w)/\partial(x, y, z)$ are determinants of the *Jacobian matrices*:

$$\begin{pmatrix} u_x & u_y \\ v_x & v_y \end{pmatrix}, \qquad \begin{pmatrix} u_x & u_y & u_z \\ v_x & v_y & v_z \\ w_x & w_y & w_z \end{pmatrix}.$$

From the row-column rule of matrix multiplication combined with the chain rule, we find that the product of the corresponding Jacobian matrices in (4) and (8) is the *unit matrix*:

(4)′
$$\begin{pmatrix} u_x & u_y \\ v_x & v_y \end{pmatrix} \begin{pmatrix} x_u & x_v \\ y_u & y_v \end{pmatrix} = \begin{pmatrix} 1 & 0 \\ 0 & 1 \end{pmatrix},$$

(8)′
$$\begin{pmatrix} u_x & u_y & u_z \\ v_x & v_y & v_z \\ w_x & w_y & w_z \end{pmatrix} \begin{pmatrix} x_u & x_v & x_w \\ y_u & y_v & y_w \\ z_u & z_v & z_w \end{pmatrix} = \begin{pmatrix} 1 & 0 & 0 \\ 0 & 1 & 0 \\ 0 & 0 & 1 \end{pmatrix}.$$

Consequently, if one of the Jacobian matrices in (4)′ or (8)′ is known, the other is its *inverse*; for example,

$$\begin{pmatrix} u_x & u_y \\ v_x & v_y \end{pmatrix} = \begin{pmatrix} x_u & x_v \\ y_u & y_v \end{pmatrix}^{-1} = \frac{1}{J} \begin{pmatrix} y_v & -x_v \\ -y_u & x_u \end{pmatrix},$$

where $J = \partial(x, y)/\partial(u, v)$.† Since the equality of matrices implies that all corresponding elements are equal, the partial derivatives on the left are all known; thus $u_x = y_v/J$, etc.

Example 1. Rectangular coordinates (x, y) are given in terms of polar coordinates (r, φ) by the equations

(9) $$x = r \cos \varphi, \qquad y = r \sin \varphi.$$

Since the Jacobian of the transformation is

(10) $$\frac{\partial(x, y)}{\partial(r, \varphi)} = \begin{vmatrix} \cos \varphi & -r \sin \varphi \\ \sin \varphi & r \cos \varphi \end{vmatrix} = r$$

we can solve equations (9) uniquely for r, φ at all points except the origin. Their explicit solution is

(11) $$r = (x^2 + y^2)^{1/2}, \qquad \varphi = \begin{cases} \tan^{-1} \dfrac{y}{x} & (x > 0), \\ \pi + \tan^{-1} \dfrac{y}{x} & (x < 0). \end{cases}$$

---

† The inverse of a square matrix whose determinant $D \neq 0$ may be found in three steps: (1) replace each element by its cofactor in the corresponding determinant; (ii) transpose this matrix (interchange rows and columns); (iii) multiply the matrix (i.e. each element) so obtained by $1/D$. Thus $\begin{pmatrix} a & b \\ c & d \end{pmatrix}$ has the cofactor matrix $\begin{pmatrix} d & -c \\ -b & a \end{pmatrix}$; and, if $D = ad - bc \neq 0$.

$$\begin{pmatrix} a & b \\ c & d \end{pmatrix}^{-1} = \frac{1}{D} \begin{pmatrix} d & -b \\ -c & a \end{pmatrix}.$$

The cofactor of an element in row r, column s, of a determinant is the minor obtained by striking out this row and column and multiplied by $(-1)^{r+s}$.

where $\tan^{-1}(y/x)$ denotes the (principal) branch of the inverse tangent between $-\pi/2$ and $\pi/2$ (Ex. 53.3).

The inverse of the matrix

$$\begin{pmatrix} x_r & x_\varphi \\ y_r & y_\varphi \end{pmatrix} = \begin{pmatrix} \cos \varphi & -r \sin \varphi \\ \sin \varphi & r \cos \varphi \end{pmatrix}, \quad \text{namely} \quad \begin{pmatrix} r_x & r_y \\ \varphi_x & \varphi_y \end{pmatrix} = \begin{pmatrix} \cos \varphi & \sin \varphi \\ \dfrac{-\sin \varphi}{r} & \dfrac{\cos \varphi}{r} \end{pmatrix},$$

gives the values of r_x, r_y, φ_x, φ_y.

Example 2. The rectangular coordinates (x, y, z) of a point are given in terms of its spherical coordinates (r, θ, φ) by the equations

$$(12) \qquad x = r \sin \theta \cos \varphi, \qquad y = r \sin \theta \sin \varphi, \qquad z = r \cos \theta.$$

Since the Jacobian of the transformation,

$$(13) \qquad \frac{\partial(x, y, z)}{\partial(r, \theta, \varphi)} = \begin{vmatrix} \sin \theta \cos \varphi & r \cos \theta \cos \varphi & -r \sin \theta \sin \varphi \\ \sin \theta \sin \varphi & r \cos \theta \sin \varphi & r \sin \theta \cos \varphi \\ \cos \theta & -r \sin \theta & 0 \end{vmatrix} = r^2 \sin \theta,$$

we can solve equations (12) uniquely for r, θ, φ at all points where $r^2 \sin \theta \neq 0$, that is, at all points not lying on the z-axis. The explicit solution when $x > 0$ is

$$(14) \qquad r = (x^2 + y^2 + z^2)^{1/2} \qquad \theta = \cos^{-1} z/r, \qquad \varphi = \tan^{-1} y/x.$$

The matrix corresponding to $\partial(x, y, z)/\partial(r, \theta, \varphi)$ has the inverse

$$\begin{pmatrix} r_x & r_y & r_z \\ \theta_x & \theta_y & \theta_z \\ \varphi_x & \varphi_y & \varphi_z \end{pmatrix} = \begin{pmatrix} \sin \theta \cos \varphi & \sin \theta \sin \varphi & \cos \theta \\ \dfrac{\cos \theta \cos \varphi}{r} & \dfrac{\cos \theta \sin \varphi}{r} & -\dfrac{\sin \theta}{r} \\ -\dfrac{\sin \varphi}{r \sin \theta} & \dfrac{\cos \varphi}{r \sin \theta} & 0 \end{pmatrix}$$

from which the values of the partial derivatives r_x, r_y, $\cdots$ can be read off.

PROBLEMS

1. Find du/dx and dv/dx if

$$u(v + x) = u - x, \qquad v(v + x) = u + x.$$

2. If $f(u, x, y) = 0$, $g(v, x, y) = 0$, find $(\partial u/\partial x)_y$ and $(\partial u/\partial x)_v$.

3. If $x^3 + y^3 + z^3 - 3xyz = 0$, find $\partial z/\partial x$ and $\partial z/\partial y$.

4. If $x = uv$, $y = \frac{1}{2}(v^2 - u^2)$, find u_x, u_y, v_x, v_y. Check by (87.4).

5. If $u^2 - v = x^2 - y$, $u + v^2 = x + y^2$, find u_x, u_y, v_x, v_y. Check by (87.4).

6. If $u^2 - v^2 - x^3 + 3y = 0$, $u + v - 2x - y^2 = 0$, find u_x, u_y, v_x, v_y. Check by (87.4).

7. If $u^3 + xv = y$, $v^3 + yu = x$, find u_x, u_y, v_x, v_y. Check by (87.4).

8. Find u_x, u_y, u_z, v_x, v_y, v_z, w_x, w_y, w_z, when

$$x = u + v + w, \qquad y = uv + vw + wu, \qquad z = uvw.$$

Check by (87.8).

9. Find u_x and u_y when

$$x = u + v + w, \qquad y = u^2 + v^2 + w^2, \qquad z = u^3 + v^3 + w^3.$$

10. When $u = x^2 + y^2 + z^2$, $z = xyt$, find $\partial u/\partial x$ when the other independent variables are

(i) y and z;   (ii) y and t;   (iii) z and t.

11. Show that

$$y = f(x - at) + g(x + at)$$

satisfies the equation $y_{tt} = a^2 y_{xx}$.

12. If $z = f(x, y)$, express $z_{xx} + z_{yy}$ in terms of the new variables,

$$u = ax + by, \qquad v = ax - by.$$

13. Express $z_{xx} + z_{yy}$ in plane-polar coordinates r, φ by making the change of variable $x = r \cos \varphi$, $y = r \sin \varphi$.

14. Find du/dx if

$$u = f(v, w, x), \qquad v = g(w, u, x), \qquad w = h(u, v, x).$$

15. If $u(x, y)$ and $v(x, y)$ satisfy the equations

$$\frac{\partial u}{\partial x} = \frac{\partial v}{\partial y}, \qquad \frac{\partial u}{\partial y} = -\frac{\partial v}{\partial x},$$

show that, on introducing polar coordinates $x = r \cos \varphi$, $y = r \sin \varphi$, we have

(i) $$\frac{\partial u}{\partial r} = \frac{1}{r} \frac{\partial v}{\partial \varphi}, \qquad \frac{1}{r} \frac{\partial u}{\partial \varphi} = -\frac{\partial v}{\partial r};$$

(ii) $$\frac{\partial^2 u}{\partial r^2} + \frac{1}{r} \frac{\partial u}{\partial r} + \frac{1}{r^2} \frac{\partial^2 u}{\partial \varphi^2} = 0.$$

16. If x, y are independent variables, and $w = uv$, $u^2 - v^2 + 2x = 0$, $u + v - y = 0$, find w_x and w_y.

88. Jacobians. If $F(u, v)$, $G(u, v)$ are functions of u, v and we put

(1) $$u = f(x, y), \qquad v = g(x, y),$$

we obtain two functions of x and y. These functions will be differentiable if F, G, f, g are differentiable (§ 79); and their Jacobian

(2) $$\begin{vmatrix} F_x & F_y \\ G_x & G_y \end{vmatrix} = \begin{vmatrix} F_u u_x + F_v v_x & F_u u_y + F_v v_y \\ G_u u_x + G_v v_x & G_u u_y + G_v v_y \end{vmatrix} = \begin{vmatrix} F_u & F_v \\ G_u & G_v \end{vmatrix} \begin{vmatrix} u_x & u_y \\ v_x & v_y \end{vmatrix}$$

on using the row-column method of multiplying determinants.† This

---

† The element in the ith row and jth column of the product determinant is found by multiplying the elements in the ith row of the first factor by the corresponding elements in the jth column of the second and adding the results. This is also the rule for multiplying *matrices*.

equation is easily remembered when written in the form

$$(3) \qquad \frac{\partial(F, G)}{\partial(x, y)} = \frac{\partial(F, G)}{\partial(u, v)} \frac{\partial(u, v)}{\partial(x, y)}$$

suggestive of the chain rule.

When $\partial(u, v)/\partial(x, y) \neq 0$, we may solve equations (1) for x and y, say

$$(4) \qquad\qquad x = F(u, v) \qquad y = G(u, v).$$

When (3) is applied to these particular functions we have

$$(4) \qquad\qquad 1 = \frac{\partial(x, y)}{\partial(u, v)} \frac{\partial(u, v)}{\partial(x, y)},$$

as already shown in § 87.

The row-column rule for multiplying determinants leads to the analogous equations:

$$(5) \qquad \frac{\partial(F, G, H)}{\partial(x, y, z)} = \frac{\partial(F, G, H)}{\partial(u, v, w)} \frac{\partial(u, v, w)}{\partial(x, y, z)},$$

$$(6) \qquad\qquad 1 = \frac{\partial(x, y, z)}{\partial(u, v, w)} \frac{\partial(u, v, w)}{\partial(x, y, z)},$$

These results can obviously be extended to the Jacobians of n functions of n variables.

89. Functional Dependence.

THEOREM 1. *If the functions*

$$(1) \qquad\qquad u = f(x, y), \qquad v = g(x, y)$$

have continuous first partial derivatives in a region R, a necessary and sufficient condition that they satisfy a functional relation,

$$(2) \qquad\qquad F(u, v) = 0,$$

is that their Jacobian

$$(3) \qquad\qquad J = \frac{\partial(u, v)}{\partial(x, y)} = 0.$$

Proof. The condition is necessary. For, if we assume (2) and differentiate partially with respect to x and y, we have

$$(4) \qquad F_u u_x + F_v v_x = 0, \qquad F_u u_y + F_v v_y = 0.$$

If $J(x_0, y_0) \neq 0$, $J(x, y) \neq 0$ in some neighborhood I_0 of (x_0, y_0) by reason of its continuity. In I_0 the equations (4) admit only the solution $F_u = 0$, $F_v = 0$; and this implies that $F(u, v)$ involves neither u nor v, so that the relation (2) is illusory.

The condition $J = 0$ is sufficient to ensure (2). If both $f_x = f_y = 0$, u is a constant c and $u - c = 0$ is a functional relation of the form (2). The theorem is then trivial.

Assume therefore that $f_x(x_0, y_0) \neq 0$. Then (§ 84) we may solve the equation

$$u - f(x, y) = 0 \quad \text{for} \quad x = \varphi(u, y)$$

in some neighborhood of (x_0, y_0). Since

$$u = f(\varphi, y)$$

is an *identity* in u and y, $f(\varphi, y)$ is independent of y; hence,

$$f_x \varphi_y + f_y = 0, \qquad \varphi_y = -f_y/f_x.$$

Now

$$v = g(\varphi, y) = G(u, y),$$

$$G_y = g_x \varphi_y + g_y = \frac{f_x g_y - f_y g_x}{f_x} = \frac{J}{f_x} = 0,$$

Hence G is independent of y and a function of u alone; that is,

$$v = G(u).$$

Note that this proof is *constructive*; it shows how to find the functional relation between u and v when $J = 0$. We need only eliminate one of the variables x, y from the equations (1); then the other will automatically disappear, and the result is the desired relation.

Example. Computation shows that the functions

$$u = \frac{x + y}{1 - xy}, \qquad v = \frac{(x + y)(1 - xy)}{(1 + x^2)(1 + y^2)}$$

have a vanishing Jacobian; hence there is a functional relation between u and v. To find it solve the first equation for y and substitute its value in the second; thus,

$$y = \frac{u - x}{1 + xu}, \qquad x + y = \frac{u(1 + x^2)}{1 + xu},$$

$$1 - xy = \frac{1 + x^2}{1 + xu}, \qquad 1 + y^2 = \frac{(1 + u^2)(1 + x^2)}{(1 + xu)^2},$$

and hence

$$v = \frac{u}{1 + u^2}.$$

THEOREM 2.   *If the functions*

$$u = f(x, y, z), \qquad v = g(x, y, z)$$

have continuous first partial derivatives in a region R, a necessary and sufficient condition that they satisfy a functional relation,

$$F(u, v) = 0,$$

is that all two-rowed determinants in the Jacobian matrix

$$\begin{pmatrix} u_x & u_y & u_z \\ v_x & v_y & v_z \end{pmatrix} \quad \text{vanish.}$$

Proof.   The condition is necessary.   For, if we differentiate $F(u, v) = 0$ partially with respect to x, y, z, we have

$$F_u u_x + F_v v_x = 0, \qquad F_u u_y + F_v v_y = 0, \qquad F_u u_z + F_v v_z = 0.$$

If any determinant of the matrix is not zero, say $\partial(u, v)/\partial(x, y) \neq 0$ the first two of these equations show that $F_u = F_v = 0$. Thus F is independent of both u and v, and the relation is illusory.

The condition is sufficient.   If $f_x = f_y = f_z = 0$, $u = c$ and the theorem is trivial.

Assume therefore that $f_x(x_0, y_0, z_0) \neq 0$.   Then (§ 84) we may solve the equation

$$u - f(x, y, z) = 0 \quad \text{for} \quad x = \varphi(u, y, z)$$

in some neighborhood of $(x_0, y_0\ z_0)$.   Since

$$u = f(\varphi,\, y, z)$$

is an identity in u, y, z, $f(\varphi, y, z)$ is independent of y and z;   hence

$$f_x\, \varphi_y + f_y = 0, \qquad \varphi_y = -f_y/f_x,$$
$$f_x\, \varphi_z + f_z = 0, \qquad \varphi_z = -f_z/f_x.$$

Now

$$v = g(\varphi, y, z) = G(u, y, z),$$

and

$$G_y = g_x\, \varphi_y + g_y = \frac{f_x\, g_y - f_y\, g_x}{f_x} = 0,$$

$$G_z = g_x\, \varphi_z + g_z = \frac{f_x\, g_z - f_z\, g_x}{f_x} = 0.$$

Hence G is independent of both y and z and a function of u alone;   that is, $v = G(u)$.

THEOREM 3.   *If the functions*

(5) $u = f(x, y, z), \qquad v = g(x, y, z), \qquad w = h(x, y, z)$

have continuous first partial derivatives in a region R, a necessary and sufficient condition that they satisfy a functional relation,

$$F(u, v, w) = 0,$$

is that their Jacobian

$$J = \frac{\partial(u, v, w)}{\partial(x, y, z)} = 0.$$

Proof.   The necessity of the condition is proved as before.   To prove the condition sufficient, we observe that, if all minors of J involving u and v

vanish, we have a relation $F(u, v) = 0$, and the theorem follows. Assume therefore that $\partial(f, g)/\partial(x, y) \neq 0$ at (x_0, y_0, z_0). Then (§ 85) we can solve the first two equations of (5) for x and y:

(6) $$x = \varphi(u, v, z), \qquad y = \psi(u, v, z).$$

in some neighborhood of (x_0, y_0, z_0). Since

$$u = f(\varphi, \psi, z), \qquad v = g(\varphi, \psi, z)$$

are identities in u, v, z, the right-hand members are independent of z; hence

$$f_x \varphi_z + f_y \psi_z + f_z = 0;$$
$$g_x \varphi_z + g_y \psi_z + g_z = 0.$$

If we substitute x and y from (6) in $w = h(x, y, z)$, we get

$$w = h(\varphi, \psi, z) = H(u, v, z),$$

and, on differentiating with respect to z,

$$h_x \varphi_z + h_y \psi_z + h_z = H_z$$

Now the Jacobian

$$J = \begin{vmatrix} f_x & f_y & f_z \\ g_x & g_y & g_z \\ h_x & h_y & h_z \end{vmatrix} = \begin{vmatrix} f_x & f_y & 0 \\ g_x & g_y & 0 \\ h_x & h_y & H_z \end{vmatrix} = H_z \begin{vmatrix} f_x & f_y \\ g_x & g_y \end{vmatrix},$$

where the second determinant is obtained from J by multiplying the first column by φ_z, the second column by ψ_z, and adding both to the third. Since $J = 0$, $\partial(f, g)/\partial(x, y) \neq 0$, we conclude that $H_z = 0$. Thus H is independent of z and a function of u and v alone; that is, $w = H(u, v)$.

PROBLEMS

1. Show that the following sets of functions are functionally dependent and find how they are related:

(a) $u = \tan^{-1} x - \tan^{-1} y$, $v = \dfrac{x - y}{1 + xy}$;

(b) $u = \log x - \log y$, $v = \dfrac{x^2 + y^2}{xy}$;

(c) $u = x + y + z$, $v = xy + xz$, $w = x^2 + y^2 + z^2 + 2yz$.

2. If x, y, z are functions of the independent variables u, v, and $J(x, y) = \partial(x, y)/\partial(u, v)$, prove that

$$J(x, J(y, z)) + J(y, J(z, x)) + J(z, J(x, y)) = 0,$$

[Differentiate $x_u J(y, z) + y_u J(z, x) + z_u J(x, y) = 0$ with respect to v,

$x_v J(y, z) + y_v J(z, x) + z_v J(x, y) = 0$ with respect to u and subtract.]

3. If x, y, z, w are functions of the independent variables u, v, prove that, in the notation of Prob. 2,

$$J(y, z) J(x, w) + J(z, x) J(y, w) + J(x, y) J(z, w) = 0.$$

4. If x, y, z are functions of the independent variables u, v, prove that

$$J(y, z) \, dx + J(z, x) \, dy + J(x, y) \, dz = 0.$$

5. If $u = f(x, y, z)$, $v = g(x, y, z)$, and x, y, z are functions of s, t, show that

$$\frac{\partial(u, v)}{\partial(s, t)} = \frac{\partial(u, v)}{\partial(x, y)} \frac{\partial(x, y)}{\partial(s, t)} + \frac{\partial(u, v)}{\partial(y, z)} \frac{\partial(y, z)}{\partial(s, t)} + \frac{\partial(u, v)}{\partial(z, x)} \frac{\partial(z, x)}{\partial(s, t)}.$$

90. Mean-Value Theorem for $f(x, y)$. For differentiable functions of a single variable we have the mean-value theorem (57.3):

$$(1) \qquad F(a + h) - F(a) = h \, F'(a + \theta h), \qquad 0 < \theta < 1.$$

The analogous theorem for a *differentiable* function $f(x, y)$ of two variables is readily obtained from (1) by the simple device of writing

$$x = a + ht, \qquad y = b + kt.$$

With these values of x and y, define

$$(2) \qquad F(t) = f(x, y) = f(a + ht, b + kt);$$

hence from (1)

$$(3) \qquad F(1) - F(0) = F'(\theta), \qquad 0 < \theta < 1.$$

From the chain rule applied to (2) we have

$$F'(t) = h f_x(a + ht, \ b + kt) + k f_y(a + ht, \ b + kt,)$$
$$F'(\theta) = h f_x(a + \theta h, b + \theta k) + k f_y(a + \theta h, b + \theta k.)$$

Equation (3) now becomes

$$(4) \quad f(a + h, b + k) - f(a, b)$$
$$= h f_x(a + \theta h, b + \theta k) + k f_y(a + \theta h, b + \theta k),$$

where $0 < \theta < 1$. This is the *mean-value theorem* for $f(x, y)$. Note that $(a + \theta h, b + \theta k)$ is an intermediate point on the line joining (a, b) to $(a + h, b + k)$.

If the first partial derivatives of $f(x, y)$ vanish throughout a region R, then f is a constant in R; for from (4)

$$f(a + h, b + k) = f(a, b),$$

whenever $(a + h, b + k)$ is a point of R.

The formula (4) admits of an obvious generalization to functions of three or more variables.

91. Taylor's Theorem for $f(x, y)$. Taylor's expansion with remainder (65.4) may also be extended to functions of two variables by the same device of defining

$$F(t) = f(a + ht, b + kt).$$

We assume that $f(x, y)$ has continuous partial derivatives up to order $n + 1$. Then from (65.3) we have

$$(1) \qquad F(1) = F(0) + F'(0) + \frac{F''(0)}{2!} + \cdots + \frac{F^{(n)}(0)}{n!} + \frac{F^{(n+1)}(\theta)}{(n+1)!},$$

where $0 < \theta < 1$; and the successive derivatives of $F(t)$ are computed by the chain rule:

$$F'(t) = hf_x + kf_y = (hD_x + kD_y)f,$$
$$(2) \qquad F^{(n)}(t) = (hD_x + kD_y)^n f, \qquad n = 1, 2, \cdots.$$

Here D_x, D_y denote the operations $\partial/\partial x$ and $\partial/\partial y$, $hD_x + kD_y$ is a linear operator with constant coefficients h, k, and $(hD_x + kD_y)^n$ denotes n successive applications of this operator. Under the hypotheses on $f(x, y)$, all mixed derivatives of f that differ only in the *order* of operations are equal (§ 82); hence $(hD_x + kD_y)^n$ may be expanded by the binomial theorem,

$$(3) \quad (hD_x + kD_y)^n$$
$$= h^n D_x^n + \binom{n}{1} h^{n-1}k \, D_x^{n-1}D_y + \binom{n}{2} h^{n-2}k^2 \, D_x^{n-2}D_y^2 + \cdots;$$

for this theorem depends only on the commutative, associative and distributive laws. For example,

$$(hD_x + kD_y)^2 f = h^2 f_{xx} + 2hk f_{xy} + k^2 f_{yy}.$$

Since $x = a$, $y = b$ when $t = 0$, we write

$$F^{(n)}(0) = (hD_x + kD_y)^n f(a, b), \qquad n = 1, 2, \cdots,$$

with the understanding that we *first* apply the operator (3) to $f(x, y)$ and *then* put $x = a, y, = b$. Since $x = a + \theta h$, $y = b + \theta k$ when $t = \theta$, we use the analogous notation for

$$F^{(n+1)}(\theta) = (hD_x + kD_y)^{n+1} f(a + \theta h, b + \theta k).$$

Equation (1) now gives Taylor's theorem:

$$(3) \quad f(a + h, b + k) = f(a, b) + (hD_x + kD_y)f(a, b) + \cdots$$
$$+ \frac{1}{n!}(hD_x + kD_y)^n f(a, b)$$
$$+ \frac{1}{(n+1)!}(hD_x + kD_y)^{n+1} f(a + \theta h, b + \theta k).$$

For $n = 0$ this reduces to the mean-value theorem (90.4).

The generalization of Taylor's theorem to functions $f(x, y, z)$ is immediate.

If we write

$$a = x, \quad b = y; \quad a + \theta h = \xi, \quad b + \theta k = \eta,$$

Taylor's theorem takes the form

$$\Delta f = f(x + h, y + k) - f(x, y)$$

$$= \sum_{k=1}^{n} \frac{1}{k!} (hD_x + kD_y)^k f(x, y) + \frac{1}{(n+1)!} (hD_x + kD_y)^{n+1} f(\xi, \eta).$$

The first term in the sum

$$(4) \qquad df = hf_x + kf_y = (hD_x + kD_y)f$$

is a function of four variables x, y, h, k. We now define the higher differentials by this same formula *with the proviso that h and k be held constant*. Thus, if f_x, f_y are differentiable (so that $f_{xy} = f_{yx}$),

$$(5) \qquad d^2f = (hD_x + kD_y) \, df = (hD_x + kD_y)^2 f$$

$$= h^2 f_{xx} + 2hk f_{xy} + k^2 f_{yy};$$

and, in general,

$$(6) \qquad d^n f = (hD_x + kD_y)^n f(x, y).$$

Taylor's theorem can now be written

$$(7) \qquad \Delta f = df + \frac{1}{2!} d^2 f + \cdots + \frac{1}{n!} d^n f + \frac{1}{(n+1)!} d^{n+1} f(\xi, \eta),$$

where the last differential is computed at some intermediate point (ξ, η) of the line joining (x, y) to $(x + h, y + k)$.

When all partial derivatives involved are bounded functions, $d^{k+1}f$ is an infinitesimal of higher order than $d^k f$: thus (64.11)

$$(8) \qquad d^{k+1}f = o(d^k f), \qquad k = 1, 2, \cdots.$$

92. Extremes of $f(x, y)$. The function $f(x, y)$ is said to have a *maximum* at the point (a, b) if $f(x, y) < f(a, b)$ in some suitably small neighborhood of (a, b). Then

$$(1) \qquad \Delta f = f(a + h, b + k) - f(a, b) < 0,$$

when $h^2 + k^2$ is sufficiently small. Similarly, $f(x, y)$ is said to have a *minimum* at (a, b) when

$$(2) \qquad \Delta f = f(a + h, b + k) - f(a, b) > 0,$$

when $h^2 + k^2$ is sufficiently small.

Since x and y are independent variables, if $f(x, y)$ has an extreme at (a, b), the same will be true of the functions $f(x, b)$ and $f(a, y)$ of a single variable. Hence from Theorem 1, § 71, we have

THEOREM 1. *The extreme values of $f(x, y)$ can only occur at the points where both $f_x = f_y = 0$, or where these derivatives do not exist.*

If $f(x, y)$ is differentiable, we have the necessary conditions

$$(3) \qquad f_x(a, b) = 0, \qquad f_y(a, b) = 0$$

(or simply $df = 0$) for an extreme at (a, b).

To find sufficient conditions we assume that f_x and f_y are also differentiable at (a, b). Then from the mean-value theorem

$$(4) \qquad \Delta f = h\, f_x(a + \theta h, b + \theta k) + k\, f_y(a + \theta h, b + \theta k)$$

$$= h\theta\, [h f_{xx}(a, b) + k\, f_{xy}(a, b) + \eta_1 h + \eta_2 k]$$

$$+ k\theta\, [h\, f_{yx}(a, b) + k\, f_{yy}(a, b) + \eta_3 h + \eta_4 k],$$

where all $\eta_i \to 0$ as $h, k \to 0$. Hence,

$$\eta_1 h^2 + (\eta_2 + \eta_3)kh + \eta_4 k^2 = o(h^2 + k^2); \dagger$$

and, if we write

$$A = f_{xx}(a, b), \qquad B = f_{xy}(a, b), \qquad C = f_{yy}(a, b),$$

$$(5) \qquad \frac{\Delta f}{\theta} = Ah^2 + 2Bhk + Ck^2 + o(h^2 + k^2).$$

When $h^2 + k^2$ is sufficiently small, the sign of Δf depends on the quadratic form,

$$(6) \qquad d^2 f = Ah^2 + 2Bhk + Ck^2,$$

provided A, B, C are not all zero. The *discriminant* of this form is

$$(7) \qquad H = AC - B^2.$$

We have now three cases to consider.

I. $H > 0$. Then $AC > 0$ and A and C have the same sign. Since

$$(8) \qquad A\, d^2 f = (Ah + Bk)^2 + (AC - B^2)k^2$$

is positive when $h^2 + k^2 \neq 0$, $d^2 f$ always has the sign of A. We therefore have a maximum or a minimum at (a, b) according as $A < 0$ or $A > 0$.

II. $H < 0$. (*a*) If $A \neq 0$, we see from (8) that

$$A\, d^2 f > 0 \quad \text{when} \quad h \neq 0, \qquad k = 0;$$

$$A\, d^2 f < 0 \quad \text{when} \quad Ah + Bk = 0, \quad k \neq 0.$$

---

† This is obvious if we write $h = r \cos \varphi$, $k = r \sin \varphi$.

(b)   If $C \neq 0$, we can draw similar conclusions by merely interchanging A and C, h and k.

(c)   If $A = C = 0$, $B \neq 0$, $d^2f = 2Bhk$.   Then

$$B\,d^2f > 0 \quad \text{when} \quad hk > 0,$$

$$B\,d^2f < 0 \quad \text{when} \quad hk < 0.$$

In all cases d^2f changes sign in every neighborhood of (a, b), however small, and $f(x, y)$ has no extreme at (a, b), but a *saddle point*.   This apt term should recall that a saddle, cut lengthwise, shows a curve with a minimum, but, cut across, a curve with a maximum.

III. $H = 0$.   When $A \neq 0$, we have from (8)

$$A\,d^2f = (Ah + Bk)^2, \qquad A\,d^2f \geq 0.$$

Now $d^2f = 0$ when $Ah + Bk = 0$;   otherwise d^2f has the same sign as A.   When $C \neq 0$, we have from (6)

$$C\,d^2f = (Bh + Ck)^2, \qquad C\,d^2f \geq 0,$$

and we can draw similar conclusions.   The sign of Δf now depends on d^3f or possibly higher differentials.   The discussion is intricate and will be omitted.[†]

The quadratic form (6) in case III is said to be *singular* for it vanishes for values other than $h = k = 0$.   This is the *doubtful case*.

In cases I and II the form is *nonsingular* for it vanishes only when $h = k = 0$.   In case I the form is said to be (positive or negative) *definite* for it maintains a constant sign when $h^2 + k^2 \neq 0$.   In case II the form is said to be *indefinite* for its sign varies in every neighborhood of $h = k = 0$.

We summarize our results in

THEOREM 2.   *At a point for which $f_x = f_y = 0$, the function $f(x, y)$ will have an extreme when*

$$H = f_{xx}f_{yy} - f^2{}_{xy} > 0,$$

maximum if f_{xx} and $f_{yy} < 0$,

minimum if f_{xx} and $f_{yy} > 0$.

When $H < 0$, we have a saddle point.   The case $H = 0$ is undecided.

Example 1. $f(x, y) = x^3 + y^3 - 3axy$.   Extremes may occur where

$$f_x = 3(x^2 - ay) = 0, \qquad f_y = 3(y^2 - ax) = 0,$$

that is, at $(0, 0)$ or (a, a).   Now

$$f_{xx} = 6x, \quad f_{yy} = 6y, \quad f_{xy} = -3a; \qquad H = 9(4xy - a^2).$$

[†] Cf. Goursat-Hedrick, *A Course in Mathematical Analysis*, vol. 1, Boston, 1904, §§ 57–58.

Since $H(0, 0) = -9a^2 < 0$ we have a saddle point at $(0, 0)$.
Since $H(a, a) = 27a^2 > 0$, f has a relative minimum,

$$f(a, a) = -a^3 \quad \text{at} \quad (a, a).$$

Example 2. $f(x, y) = (x + y)^2 + x^4$. Extremes may occur where

$$f_x = 2(x + y) + 4x^3 = 0, \qquad f_y = 2(x + y) = 0,$$

that is, only at $(0, 0)$. At this point

$$f_{xx} = 2 + 12x^2 = 2, \quad f_{yy} = 2, \quad f_{xy} = 2, \quad \text{and} \quad H = 0.$$

The form d^2f is singular: the doubtful case. Nevertheless, since $f(x, y) \geqq 0$, f has an absolute minimum of 0 at $(0, 0)$.

Example 3. $f(x, y) = (y - x^2)(y - 2x^2)$. Extremes may occur where

$$f_x = -6xy + 8x^3 = 0, \qquad f_y = 2y - 3x^2 = 0,$$

that is, only at $(0, 0)$. At this point

$$f_{xx} = -6y + 24x^2 = 0, \quad f_{yy} = 2, \quad f_{xy} = -6x = 0, \quad \text{and} \quad H = 0.$$

Again the doubtful case, but this time we have no extreme; for

$$f < 0 \quad \text{when} \quad x^2 < y < 2x^2,$$
$$f > 0 \quad \text{when} \quad y < x^2 \quad \text{or} \quad y > 2x^2.$$

The sign pattern of $f(x, y)$ is shown in Fig. 92.

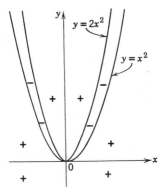

FIG. 92. Sign pattern for $(y - x^2)(y - 2x^2)$.

Example 4. If the function

(9) $f(x, y) = Ax^2 + 2Bxy + Cy^2 + 2Dx + 2Ey + F$ $(A > 0)$

has an extreme at (x_1, y_1), the point will satisfy $f_x = f_y = 0$; namely,

(10) $Ax + By + D = 0,$

(11) $Bx + Cy + E = 0.$

If we multiply (10) by x, (11) by y, and subtract their sum from (9) we have also

$$(12) \qquad\qquad Dx + Ey + F = f(x, y).$$

Since $f_{xx} = 2A > 0$, $H = 4(AC - B^2)$, $f(x, y)$ will have a minimum at the point (x_1, y_1) where the lines (10) and (11) meet, provided $AC - B^2 > 0$. The minimum value of $f(x, y)$ is readily found from (10), (11), (12) regarded as three linear equations in the unknowns $x, y, f(x, y)$. Thus we find

$$(13) \qquad\qquad f(x_1, y_1) = \frac{\Delta}{AC - B^2} \quad \text{where} \quad \Delta = \begin{vmatrix} A & B & D \\ B & C & E \\ D & E & F \end{vmatrix}.$$

When $AC - B^2 > 0$, the *equation* $f(x, y) = 0$ represents an ellipse with its center at (x_1, y_1). Then $f(x, y)$ assumes its minimum value (13) at the center of the ellipse. When the determinant $\Delta = 0$, the center lies *on* the curve; the ellipse must then degenerate into the point (x_1, y_1).

When $AC - B^3 < 0$, the equation $f(x, y) = 0$ represents an hyperbola with its center at (x_1, y_1). Since $H < 0$, $f(x, y)$ has a saddle point at (x_1, y_1). When $\Delta = 0$ the center lies *on* the curve; the hyperbola must then degenerate into a pair of lines through (x_1, y_1).

What can be said when $AC - B^2 = 0$?

Example 5. The Mean Center. Given a set of $n(>1)$ points $P_1(x_1, y_1), \cdots, P_n(x_n, y_n)$, to find the point $P(x, y)$ for which

$$(P_1P)^2 + (P_2P)^2 + \cdots + (P_nP)^2$$

is a minimum.

The function to be minimized is

$$(14) \qquad\qquad f(x, y) = \sum_{i=1}^{n} \{(x - x_i)^2 + (y - y_i)^2\};$$

and we have

$$f_x = 2 \sum_1^n (x - x_i), \qquad f_y = 2 \sum_1^n (y - y_i),$$

$$f_{xx} = 2n, \qquad f_{xy} = 0, \qquad f_{yy} = 2n.$$

The equations $f_x = 0, f_y = 0$ give

$$(15) \qquad\qquad \bar{x} = \frac{1}{n} \sum_1^n x_i, \qquad \bar{y} = \frac{1}{n} \sum_1^n y_i,$$

namely, the coordinates of the *mean center*† of the given points. Since $H > 0, f_{xx} > 0$, a minimum is assured.

If we make n observations to determine a given point, the mean center of these points is the "best" point in the sense of least squares.

The minimum value of $f(x, y)$ is obtained by putting $x = \bar{x}, y = \bar{y}$ in (14), expanding the squares, and making use of equations (15); thus we find

$$(16) \qquad\qquad f(\bar{x}, \bar{y}) = \sum_1^n (x_i^2 + y_i^2) - n(\bar{x}^2 + \bar{y}^2).$$

---

† Cf. Brand, L., *Vector and Tensor Analysis*, Wiley, 1948, p. 20.

Since $f(\bar{x}, \bar{y}) > 0$, we have the inequality

(17) $$n \Sigma(x_i{}^2 + y_i{}^2) > (\Sigma x_i)^2 + (\Sigma y_i)^2, \qquad n > 1.$$

Example 6. Least-Squares Line. Suppose that n points $P_i(x_i, y_i)$ obtained by observation or experiment are theoretically points of a straight line, but due to slight errors in measurement fail to have this property. What is the "best" straight line through the given points in the sense of least squares?

Let the line $y = ax + b$ be the line for which the sum of the squares of the deviations is a minimum. Our problem is to find a and b so that

(18) $$f(a, b) = \sum_{i=1}^{n} (ax_i + b - y_i)^2$$

is a minimum. Now

$$f_a = 2 \sum_{1}^{n} x_i(ax_i + b - y_i), \qquad f_b = 2 \sum_{1}^{n} (ax_i + b - y_i),$$

$$f_{aa} = 2 \sum_{1}^{n} x_i{}^2, \qquad f_{ab} = 2 \sum_{1}^{n} x_i, \qquad f_{bb} = 2n.$$

The equations $f_a = 0, f_b = 0$ gives

(19) $$a \Sigma x_i{}^2 + b \Sigma x_i = \Sigma x_i y_i,$$

(20) $$a \Sigma x_i + bn = \Sigma y_i.$$

These equations may always be solved uniquely for a and b since their determinant

(21) $$n \Sigma x_i{}^2 - (\Sigma x_i)^2 > 0;$$

for this inequality follows from (17) if we put all $y_i = 0$. This also shows that $H > 0$; and, since $f_{aa} > 0$, a minimum is assured.

If we divide equation (20) by n and use the notation (15), we obtain

$$a\bar{x} + b = \bar{y};$$

the mean center of the given points lies on the least-squares line.

The minimum value of $f(a, b)$ is also readily found. Since

$$af_a + bf_b - 2f = 2 \Sigma y_i(ax_i + b - y_i),$$

we see that, when a and b satisfy the equations $f_a = 0, f_b = 0$, we have $-f = \Sigma y_i(ax_i + b - y_i)$ or

(22) $$f(a, b) + a \Sigma x_i y_i + b \Sigma y_i = \Sigma y_i{}^2.$$

where a, b now denote the values of a, b that give $f(a, b)$ its minimum value. If we regard (22), (19), (20) as three linear equations in the unknowns $f(a, b), a, b$, and solve for $f(a, b)$ by Cramer's rule, we find

(23) $$\text{Min } f(a, b) = \begin{vmatrix} \Sigma y_i{}^2 & \Sigma x_i y_i & \Sigma y_i \\ \Sigma x_i y_i & \Sigma x_i{}^2 & \Sigma x_i \\ \Sigma y_i & \Sigma x_i & n \end{vmatrix} : \begin{vmatrix} \Sigma x_i{}^2 & \Sigma x_i \\ \Sigma x_i & n \end{vmatrix}.$$

Equations (18) and (21) show that the 3-rowed determinant D in (23) is nonnegative: $D \geq 0$. The smaller the value of D, the better the least-squares line fits the given points. Evidently $D = 0$ is a necessary and sufficient condition that all the points lie on a line.

PROBLEMS

Test the functions numbered **1** to **8** for extremes and saddle points:

1. $x^2 + xy + y^2 - 6x - 3y$.

2. $x^3 + y^2 - 3x$.

3. $x^4 + y^4 - 2x^2 + 4xy - 2y^2$.

4. $x^4 + y^4 - x^2 + xy - y^2$.

5. $x^3 - 3xy + y^3$.

6. $x^3 y^2 (a - x - y)$.

7. $\sin x + \sin y + \cos (x + y)$.

8. $2x^2 + 2xy + y^2 + 6x + 6y + 3$.

9. Find the least distance between the line $x + y = 4$ and the ellipse $x^2/4 + y^2 = 1$.

10. Find the least distance between the lines

$$4x = 3y = -z \quad \text{and} \quad 3x - 3 = -y - 2 = -4z + 2$$

11. Find the plane $x/a + y/b + z/c = 1$ through the point (x_1, y_1, z_1) in the first octant that cuts off the least volume from this octant. $[V = abc/6.]$

12. What values of x, y, z give $x^a y^b z^c$ its maximum value when $x + y + z = k$?

93. Constrained Extremes. The extremes of $f(x, y)$ previously considered were "free" in the sense that both x and y were independent variables. But problems frequently arise in which the extremes of $f(x, y)$ are required subject to the condition

(1) $$g(x, y) = 0.$$

If $g_y \neq 0$, we can solve (1) for $y = \varphi(x)$, and the problem is then reduced to finding the extremes of

(2) $$F(x) = f(x, \varphi).$$

The necessary condition to be fulfilled is

(3) $$F'(x) = f_x + f_y \varphi'(x) = 0,$$

where from (1)

(4) $$g_x + g_y \varphi'(x) = 0.$$

If we eliminate $\varphi'(x)$ from (3) and (4), we get

(5) $$\begin{vmatrix} f_x & f_y \\ g_x & g_y \end{vmatrix} = \frac{\partial(f, g)}{\partial(x, y)} = 0.$$

The values (x, y) which give extremes must now be sought among the solutions of equations (5) and (1). The further discussion is carried out as in § 92.

We next find the extremes of $f(x, y, z)$ subject to the condition

(6) $$g(x, y, z) = 0.$$

If $g_z \neq 0$, we can solve (6) for $z = \varphi(x, y)$ and the problem is reduced to finding the free extremes of

(7) $$F(x, y) = f(x, y, \varphi).$$

The necessary conditions are now

(8) $$F_x = f_x + f_z \varphi_x = 0, \qquad F_y = f_y + f_z \varphi_y = 0,$$

where from (6)

(9) $$g_x + g_z \varphi_x = 0, \qquad\qquad g_y + g_z \varphi_y = 0.$$

If we eliminate φ_x and φ_y from (8) and (9), we obtain the equations

(10) $$\frac{\partial(f, g)}{\partial(x, z)} = 0, \qquad \frac{\partial(f, g)}{\partial(y, z)} = 0.$$

The values (x, y, z) which give extremes must now be sought among the solutions of equations (10) and (6). Further discussion then depends on d^2F (§ 92).

Example. To find the shortest distance from the point (a, b, c) to the plane

(i) $$Ax + By + Cz + D = 0.$$

The square of the distance from (a, b, c) to any point (x, y, z) of the plane is given by

(ii) $$f(x, y) = (x - a)^2 + (y - b)^2 + (z - c)^2,$$

where z is given by (i):

$$z = (-D - Ax - By)/C.$$

At the point (x, y, z) for which $f(x, y)$ is a minimum,

$$f_x = 2(x - a) - 2(z - c)\frac{A}{C} = 0,$$

$$f_y = 2(y - b) - 2(z - c)\frac{B}{C} = 0;$$

hence,

(iii) $$\frac{x - a}{A} = \frac{y - b}{B} = \frac{z - c}{C} = -\frac{Aa + Bb + Cc + D}{A^2 + B^2 + C^2},$$

where the common value of the fractions is found by multiplying the fractions by A/A, B/B, C/C, respectively, and adding numerators and denominators. Thus an extreme can only occur at the single point (x, y, z) given by (iii).

To show that we actually have a minimum, we compute

$$f_{xx} = 2\left(1 + \frac{A^2}{C^2}\right), \qquad f_{yy} = 2\left(1 + \frac{B^2}{C^2}\right), \qquad f_{xy} = \frac{2AB}{C^2};$$

$$H = f_{xx}f_{yy} - f_{xy}^2 = 4\left(1 + \frac{A^2 + B^2}{C^2}\right).$$

Since H, f_{xx}, f_{vv} are all positive, a minimum is assured. The squared minimum distance $f(x, y)$ is given by substituting from (iii) in (ii); thus we find

$$\text{Min} f(x, y) = \frac{(Aa + Bb + Cc + D)^2}{A^2 + B^2 + C^2}.$$

94. Lagrangian Multipliers. Let us find the extremes of $f(x, y, u, v)$ subject to the conditions

(1) $$g(x, y, u, v) = 0, \qquad h(x, y, u, v) = 0.$$

If the Jacobian $\partial(g, h)/\partial(u, v) \neq 0$, we can solve the equations (1) for u and v in terms of x and y (§ 85):

(2) $$u = \varphi(x, y), \qquad v = \psi(x, y).$$

If we substitute these values in $f(x, y, u, v)$, the problem reduces to finding the free extremes of the function

(3) $$F(x, y) = f(x, y, \varphi, \psi).$$

The necessary conditions are now

(4) $$F_x = f_x + f_u \varphi_x + f_v \psi_x = 0,$$
(5) $$F_y = f_y + f_u \varphi_y + f_v \psi_y = 0.$$

where $\varphi_x, \varphi_y, \psi_x, \psi_y$ are found from

$$g(x, y, \varphi, \psi) = 0, \qquad h(x, y, \varphi, \psi) = 0.$$

Differentiating these equations, we get

(6) $$g_x + g_u \varphi_x + g_v \psi_x = 0, \qquad h_x + h_u \varphi_x + h_v \psi_x = 0$$
(7) $$g_y + g_u \varphi_y + g_v \psi_y = 0, \qquad h_y + h_u \varphi_y + h_v \psi_y = 0.$$

Since $\partial(g, h)/\partial(u, v) \neq 0$, we can solve equations (6) for φ_x, ψ_x, equations (7) for φ_y, ψ_y, and, on substituting in (4) and (5), we get the equivalent necessary conditions:

(8) $$\frac{\partial(f, g, h)}{\partial(x, u, v)} = 0, \qquad \frac{\partial(f, g, h)}{\partial(y, u, v)} = 0.$$

The values of x, y, u, v that give extremes must now be sought among the solutions of equations (1) and (8).

This problem may be formulated in a more symmetric fashion by use of *Lagrangian multipliers*. The necessary condition for an extreme of $f(x, y, u, v)$ is

$$df = f_x \, dx + f_y \, dy + f_u \, du + f_v \, dv = 0.$$

Moreover, from (1) and (2) we have

$$dg = g_x \, dx + g_y \, dy + g_u \, du + g_v \, dv = 0,$$
$$dh = h_x \, dx + h_y \, dy + h_u \, du + h_v \, dv = 0.$$

Multiply these equations by the constants 1, α, β, respectively, and add them together; we thus obtain

$$(9) \qquad (f_x + \alpha g_x + \beta h_x)\, dx + (f_y + \alpha g_y + \beta h_y)\, dy +$$
$$(f_u + \alpha g_u + \beta h_u)\, du + (f_v + \alpha g_v + \beta h_v)\, dv = 0.$$

If $\partial(g, h)/\partial(u, v) \neq 0$, we can remove the last two terms of (9) by choosing α and β so that

$$(10) \qquad f_u + \alpha g_u + \beta h_u = 0, \qquad f_v + \alpha g_v + \beta h_v = 0.$$

But now the first two terms of (9) also vanish, for dx and dy are *independent* increments:

$$(11) \qquad f_x + \alpha g_x + \beta h_x = 0, \qquad f_y + \alpha g_y + \beta h_y = 0.$$

We have now six equations (1), (10), (11) for the unknowns x, y, u, v and multipliers α, β. When equations (10) are solved for α and β and these values substituted in (11), we again obtain equations (8).

Equations (10) and (11) may be derived from the Lagrangian function,

$$(12) \qquad L(x, y, u, v) = f + \alpha g + \beta h,$$

by writing down the necessary conditions for a free extreme in the four variables:

$$(13) \qquad L_x = 0, \qquad L_y = 0, \qquad L_u = 0, \qquad L_v = 0.$$

When u and v are replaced by their values (2),

$$L(x, y, \varphi, \psi) = F(x, y);$$

and evidently $dF = dL = 0$. From

$$dL = L_x\, dx + L_y\, dy + L_u\, du + L_v\, dv$$

we may compute d^2F as $d(dL)$ if we remember that $d^2x = d^2y = 0$ as x and y are independent variables, but that d^2u and d^2v do not in general vanish. Thus we find

$$d^2F = (dL_x)\, dx + (dL_y)\, dy + (dL_u)\, du + (dL_v)\, dv$$

plus the terms $L_u\, d^2u$ and $L_v\, d^2v$. But, since $L_u = L_v = 0$ by equations (13), we see that we may compute

$$(14) \qquad d^2F(x, y) = d^2L(x, y, u, v)$$

as if all variables were independent. This fact may facilitate the discussion of the nature of our solutions. This discussion, however, is apt to be difficult, and one usually relies on the nature of the problem to disclose the existence of maxima and minima.

The simpler problems treated in § 93 may also be treated by use of the Lagrangian function $L = f + \alpha g$. This method is applicable to all problems in extremes in which we have a function of $n + m$ variables

$$f(x_1, \cdots, x_n, u_1, \cdots, u_m)$$

subject to m equations of conditions $g_1 = 0, \cdots, g_m = 0$. If the Jacobian

$$\frac{\partial(g_1, g_2, \cdots, g_m)}{\partial(u_1, u_2, \cdots, u_m)} \neq 0,$$

we form the Lagrangian function,

$$L = f + \alpha_1 g_1 + \cdots + \alpha_m g_m,$$

and write down the necessary conditions for a free extreme:

$$L_{x_1} = 0, \quad \cdots, \quad L_{x_n} = 0, \quad L_{u_1} = 0, \quad \cdots, \quad L_{u_m} = 0.$$

These $n + m$ equations together with the m equations of condition give $n + 2m$ equations for the $n + m$ variables x_i and the m multipliers α_j.

Example. Maximum Value of a Determinant. Consider the determinant

(i)
$$D = \begin{vmatrix} x_1 & x_2 & x_3 \\ y_1 & y_2 & y_3 \\ z_1 & z_2 & z_3 \end{vmatrix}$$

as a function of its nine elements. What is the maximum value of D when the squares of the elements in any row have a prescribed sum?

D is a function of nine variables subject to three equations of condition:

(ii)
$$f(x_1, x_2, x_3) = x_1{}^2 + x_2{}^2 + x_3{}^3 - a^2 = 0,$$
$$g(y_1, y_2, y_3) = y_1{}^2 + y_2{}^2 + y_3{}^2 - b^2 = 0,$$
$$h(z_1, z_2, z_3) = z_1{}^2 + z_2{}^2 + z_3{}^2 - c^2 = 0,$$

where a, b, c are arbitrary positive numbers. The Lagrangian function is now

(iii)
$$L = D + \alpha f + \beta g + \gamma h,$$

and we have the nine necessary conditions for an extreme:

(iv)
$$L_{x_1} = 0, \quad L_{x_2} = 0, \quad \cdots, \quad L_{z_3} = 0.$$

If we expand D according to the elements of its first row,

$$D = x_1 X_1 + x_2 X_2 + x_3 X_3,$$

where X_i denotes the cofactor of x_i. Hence the first three equations of (iv) are

$$X_1 + 2\alpha x_1 = 0, \qquad X_2 + 2\alpha x_2 = 0, \qquad X_3 + 2\alpha x_3 = 0,$$

and hence

$$\frac{x_1}{X_1} = \frac{x_2}{X_2} = \frac{x_3}{X_3} = -\frac{1}{2\alpha}.$$

The remaining equations of (iv) give two analogous equations. From these we see that D can only assume an extreme value when the elements of any row are proportional to their cofactors. Now, if the cofactors in any row are multiplied by the corresponding elements of *another* row and the products added, the result is zero. Hence we see that, when D has an extreme value, it has the property

(v)
$$
\begin{aligned}
x_1 y_1 + x_2 y_2 + x_3 y_3 &= 0, \\
y_1 z_1 + y_2 z_2 + y_3 z_3 &= 0, \\
z_1 x_1 + z_2 x_2 + z_3 x_3 &= 0.
\end{aligned}
$$

Such a determinant is called *orthogonal*.

If we compute D^2 by the row-row method when conditions (ii) and (v) are fulfilled, we get

(vi)
$$
D^2 = \begin{vmatrix} a^2 & 0 & 0 \\ 0 & b^2 & 0 \\ 0 & 0 & c^2 \end{vmatrix} = a^2 b^2 c^2.
$$

Since D, a polynomial, is a continuous function of its arguments, and

$$
|x_i| \leq a, \qquad |y_i| \leq b, \qquad |z_i| \leq c \qquad (i = 1, 2, 3)
$$

by equations (ii), its absolute maximum and minimum are evidently abc and $-abc$:

(vii)
$$
-abc \leq D \leq abc.
$$

Now let D_3 be any third-order determinant whatever whose elements do not exceed M in absolute value. Then (vii) shows that

$$
|D_3| \leq (x_1^2 + x_2^2 + x_3^2)^{1/2} (y_1^2 + y_2^2 + y_3^2)^{1/2} (z_1^2 + z_2^2 + z_3^2)^{1/2},
$$

or, since each factor $\leq (3M^2)^{1/2}$,

(viii)
$$
|D_3| \leq \sqrt{3^3}\, M^3.
$$

Essentially the same argument shows that, for any nth-order determinant D_n whose elements do not exceed M in absolute value,

(ix)
$$
|D_n| \leq \sqrt{n^n}\, M^n.
$$

This is the famous *determinant theorem of Hadamard*.

PROBLEMS

1. Of all the rectangles inscribed in the ellipse $x^2/a^2 + y^2/b^2 = 1$ which has (*a*) the greatest area, (*b*) the greatest perimeter? What are these maxima?

2. The points P and Q lie on nonintersecting curves $f(x, y) = 0$, $g(X, Y) = 0$, respectively. Show that, when the distance PQ has a relative minimum or maximum, the line PQ is normal to both curves.

$$
[L = (x - X)^2 + (y - Y)^2 + \alpha f(x, y) + \beta g(X, Y).]
$$

3. Obtain the axes of the central conic $ax^2 + 2bxy + cy^2 = k$ by finding the extreme distances λ of the curve from the origin; and show that

$$
\left(a - \frac{k}{\lambda^2}\right)\left(c - \frac{k}{\lambda^2}\right) = b^2.
$$

4. Obtain the axes of the central quadric,

$$\varphi(x, y, z) = ax^2 + by^2 + cz^2 + 2dxy + 2eyz + 2fzx = k,$$

by finding the extreme distances λ of the surface from the origin; and show that

$$\begin{vmatrix} a - k/\lambda^2 & d & f \\ d & b - k/\lambda^2 & e \\ f & e & c - k/\lambda^2 \end{vmatrix} = 0.$$

[From the Lagrangian function

$$L = x^2 + y^2 + z^2 + \alpha(\varphi(x, y, z) - k)$$

we have $L_x = L_y = L_z = 0$; whence

$$\frac{ax + dy + fz}{x} = \frac{dx + by + ez}{y} = \frac{fx + ey + cz}{z} = \frac{k}{\lambda^2}.$$

On equating each fraction to k/λ^2, we obtain three linear homogeneous equations in x, y, z which admit solutions other than $x = y = z = 0$ only when their determinant vanishes.]

5. Find the semiaxes of the quadric

$$5x^2 - 2y^2 + 11z^2 + 12xy + 12yz = 14$$

in magnitude and direction. What is the quadric?

6. A triangle of area K has sides a, b, c. From a point inside the triangle three perpendiculars x, y, z are drawn to its sides. Find x, y, z when $x^2 + y^2 + z^2$ assumes its minimum value.

[$ax + by + cz = 2K$.]

7. Find the dimensions of the rectangular parallelepiped inscribed in the ellipsoid $x^2/a^2 + y^2/b^2 + z^2/c^2 = 1$ that has the greatest volume.

95. Summary: Functions of Several Variables. The function $f(x, y)$ of the independent variables x, y is said to be *continuous* at (a, b) when

$$\lim_{\substack{x \to a \\ y \to b}} f(x, y) = f(a, b)$$

for any mode of approach of (x, y) to (a, b). A function $f(x, y)$ continuous at every point of a bounded, closed region R has the properties:

(i) $f(x, y)$ is *uniformly continuous* in R: for any points (x, y), (x', y') of R,

$$|f(x, y) - f(x', y')| < \varepsilon \quad \text{when} \quad (x - x')^2 + (y - y')^2 < \delta^2;$$

(ii) $f(x, y)$ is bounded in R: $|f(x, y)| \leqq M$;

(iii) $f(x, y)$ takes on a least and a greatest value at least once in R.

The *partial derivatives* of $f(x, y)$ at (a, b) are defined as

$$f_x(a, b) = \lim_{h \to 0} \frac{f(a + h, b) - f(a, b)}{h},$$

$$f_y(a, b) = \lim_{k \to 0} \frac{f(a, b + k) - f(a, b)}{k}.$$

The function $f(x, y)$ is said to be *differentiable* at (a, b) when $f = f(x + h, y + k) - f(x, y)$ can be expressed in the form

$$\Delta f = Ah + Bk + \eta_1 h + \eta_2 k,$$

where A, B are independent of h, k and η_1, $\eta_2 \to 0$ as h, $k \to 0$. When $f(x, y)$ is differentiable at (a, b), we find that $A = f_x(a, b)$, $B = f_y(a, b)$; and we define

$$df = f_x(a, b)h + f_y(a, b)k$$

as the *total differential* of $f(x, y)$. In particular, $dx = h$, $dy = k$: *the differentials of the independent variables are equal to their increments.* Thus, when $f(x, y)$ is differentiable at (a, b),

$$\Delta f = df + \text{(a term vanishing to a higher order than } h \text{ or } k).$$

The mere existence of $f_x(a, b)$ and $f_y(a, b)$ does not imply that $f(x, y)$ is continuous at (a, b); but, if $f(x, y)$ is differentiable at (a, b), it is also continuous there.

If $f(u, v)$ is a differentiable function of u, v, which in turn are differentiable functions of x, y, we have the chain rule

$$f_x = f_u u_x + f_v v_x, \qquad f_y = f_u u_y + f_v v_y;$$

and the total differential

$$df = f_x \, dx + f_y \, dy = f_u \, du + f_v \, dv.$$

This result is general: *The total differential of a function of several dependent variables has the same form as if the variables were independent:*

$$df(u, v, w) = f_u \, du + f_v \, dv + f_w \, dw.$$

The *mixed derivatives* $f_{xy}(a, b)$, $f_{yx}(ab)$ are not necessarily equal: but, when

(i) f_x, f_y exist in some neighborhood of (a, b);
(ii) f_x, f_y are differentiable at (a, b); *or*
(ii)' f_{xy} is continuous at (a, b);
then $f_{xy}(a, b) = f_{yx}(a, b)$.

The conditions under which an equation $F(x, y) = 0$ defines a unique *implicit function* $y = f(x)$ are as follows. If (a, b) is an interior point of a region R, and

(i) $F(a, b) = 0$,   (ii) F_y, F_y are cont. in R,   (iii) $F_y(a, b) \neq 0$,

then in some x-interval about a there exists a unique differentiable function $y = f(x)$ such that

(1) $b = f(a)$,   (2) $F(x, f(x)) = 0$,   (3) $f'(x) = -F_x/F_y$.

The derivative $dy/dx = f'(x)$ is obtained from

$$dF = F_x \, dx + F_y \, dy = 0.$$

Two equations $F(x, y, u, v) = 0$, $G(x, y, u, v) = 0$ define unique, differentiable functions $u(x, y)$, $v(x, y)$ in the neighborhood of $x = a$, $y = b$, $u = \alpha$, $v = \beta$ under analogous conditions; (iii) is now replaced by

$$\frac{\partial(F, G)}{\partial(u, v)} \neq 0 \quad \text{at} \quad (a, b, \alpha, \beta).$$

The partial derivatives of u and v are found by solving $dF = 0$, $dG = 0$ for du and dv.

Three or more equations are dealt with in a similar fashion.

If f and g are functions of u, v, and u, v are functions of x, y, we have

$$\frac{\partial(f, g)}{\partial(x, y)} = \frac{\partial(f, g)}{\partial(u, v)} \frac{\partial(u, v)}{\partial(x, y)},$$

an identity that suggests the chain rule. If the functions u and v satisfy the equations

$$x = f(u, v), \qquad y = g(u, v),$$

the equation becomes

$$1 = \frac{\partial(x, y)}{\partial(u, v)} \frac{\partial(u, v)}{\partial(x, y)}.$$

Both Jacobian identities admit of obvious generalizations.

The functions $u = f(x, y)$, $v = g(x, y)$ with continuous first partial derivatives, satisfy a functional relation $F(u, v) = 0$ when and only when their Jacobian $\partial(u, v)/\partial(x, y) = 0$. When the Jacobian is zero, we can find the relation between u and v by eliminating x from the given equations; in the process, y will disappear.

The *mean-value theorem* for a differentiable function $f(x, y)$ is obtained from the corresponding one-variable theorem for

$$F(t) = f(a + ht, b + kt); \quad \text{namely} \quad F(1) - F(0) = F'(\theta), \quad 0 < \theta < 1.$$

If we apply Maclaurin's theorem to find $F(1)$, we obtain

$$f(a + h, b + k) = f(a, b) + (hD_x + kD_y)f(a, b) + \cdots$$

$$+ \frac{1}{n!}(hD_x + kD_y)^n f(a, b)$$

$$+ \frac{1}{(n + 1)!}(hD_x + kD_y)^{n+1} f(a + \theta h, b + \theta k), \qquad 0 < \theta < 1.$$

This is *Taylor's theorem* for $f(x, y)$; for $n = 1$ it reduces to the mean-value theorem.

The *extremes* of $f(x, y)$ can only occur where $f_x = f_y = 0$ or where these derivatives do not exist. At a point for which $f_x = f_y = 0$, $f(x, y)$ will have an extreme when

$$H = f_{xx}f_{yy} - f_{xy}^2 > 0,$$

max if $f_{xx}, f_{yy} < 0$,

min if $f_{xx}, f_{yy} > 0$.

When $H < 0$ we have a saddle point. The case $H = 0$ is undecided.

The extremes of a function f of n variables when subject to $m < n$ equations of condition $g = 0$, $h = 0$, $\cdots$ are included among the free extremes of the Lagrangian function $L = f + \alpha g + \beta h + \cdots$, in which the multipliers α, β, $\cdots$ are constants.

PROBLEMS: Theory of Games

1. Show that the function

$$F(x, y) = (x, 1 - x) \begin{pmatrix} a & b \\ c & d \end{pmatrix} \begin{pmatrix} y \\ 1 - y \end{pmatrix},$$

where $s = a + d - b - c \neq 0$, has a saddle point at $x_0 = (d - c)/s$, $y_0 = (d - b)/s$; and $F(x_0, y_0) = (ad - bc)/s$.

2. Let $X = (x_1, x_2, x_3)$, $Y = (y_1, y_2, y_3)$ $J = (1, 1, 1)$, and X', Y', J' be the corresponding column vectors; and let $A = (a_{ij})$ be a 3×3 matrix whose determinant $| A | \neq 0$. When X and Y are subjected to the conditions $XJ' = 1$, $JY' = 1$, show that the function $F(X, Y) = XAY'$ has a saddle point at

$$X_0 = F_0 J A^{-1}, \ \ Y_0 = F_0 A^{-1} J', \text{ where } F_0 = X_0 A Y_0' = \frac{1}{J A^{-1} J'} = \frac{|A|}{\Sigma A_{ij}}$$

and A_{ij} is the cofactor of a_{ij} in $| A |$.

[Form the Lagrangian function

$$L = XAY' - \alpha(XJ' - 1) - \beta(JY' - 1)$$

and show that

$$X_0 A = \beta J, \ AY_0' = \alpha J'; \ \alpha = \beta = X_0 A Y_0' = F_0; \ \ \ F(X_0, Y') = F_0 = F(X_0, Y').]$$

3. Show that the results of Prob. 2 apply when X, Y, J are $1 \times n$ vectors and the matrix A is $n \times n$. Check the solution of Prob. 1.

CHAPTER 5

Vectors

96. Vectors. A vector is a directed line segment. The vector from A to B is denoted by $\overrightarrow{AB}$ or by a single bold-face letter as **u**.

Two vectors are equal when they have the same length and direction. Thus a vector is not altered by a parallel displacement: our vectors are "free" in that their initial point may be chosen at pleasure. In the parallelogram $ABCD$ (Fig. 96a)

$$\overrightarrow{AB} = \overrightarrow{DC} = \mathbf{u}, \qquad \overrightarrow{AD} = \overrightarrow{BC} = \mathbf{v}.$$

The symbol $|\mathbf{u}|$ denotes the length of a vector. A vector of length 1 is called a *unit vector*. When $|\mathbf{u}| \neq \mathbf{0}$, **u** is called a *proper* vector. We extend the term vector to include not only proper vectors, but also the *zero vector* **0**, a segment of zero length but devoid of direction.

Vectors are added by the *triangle construction*; thus in Fig. 96a

$$(1) \quad \mathbf{u} + \mathbf{v} = \overrightarrow{AB} + \overrightarrow{BC} = \overrightarrow{AC}, \qquad \mathbf{v} + \mathbf{u} = \overrightarrow{AD} + \overrightarrow{DC} = \overrightarrow{AC}.$$

Vector addition is commutative and associative; for

$$(2) \qquad\qquad \mathbf{u} + \mathbf{v} = \mathbf{v} + \mathbf{u} \qquad\qquad \text{(Fig. 96a)},$$

$$(3) \qquad\qquad (\mathbf{u} + \mathbf{v}) + \mathbf{w} = \mathbf{u} + (\mathbf{v} + \mathbf{w}) \qquad\qquad \text{(Fig. 96b)}.$$

Since both sums in (3) are the same, they are simply written $\mathbf{u} + \mathbf{v} + \mathbf{w}$. *The sum of any number of vectors is independent of the order in which they are added and of their grouping to form partial sums.*

If we put $\overrightarrow{AA} = \overrightarrow{BB} = \mathbf{0}$, we have

$$\overrightarrow{AA} + \overrightarrow{AB} = \overrightarrow{AB}, \qquad \overrightarrow{AB} + \overrightarrow{BB} = \overrightarrow{AB};$$

$$(4) \qquad\qquad \mathbf{0} + \mathbf{u} = \mathbf{u}, \qquad \mathbf{u} + \mathbf{0} = \mathbf{u}.$$

If $\overrightarrow{AB} = \mathbf{u}$, we write $\overrightarrow{BA} = -\mathbf{u}$; for $\mathbf{u}$ and $-\mathbf{u}$ have the characteristic property of negatives:

(5) $$\mathbf{u} + (-\mathbf{u}) = \mathbf{0}.$$

The negative of a vector is a vector of the same length but opposite direction. Note also that $-(-\mathbf{u}) = \mathbf{u}$.

The difference $\mathbf{u} - \mathbf{v}$ is defined by the equation

$$(\mathbf{u} - \mathbf{v}) + \mathbf{v} = \mathbf{u}.$$

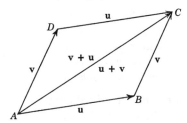

Fig. 96a. Vector addition is commutative.

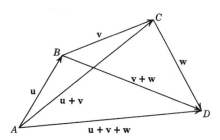

Fig. 96b. Vector addition is associative.

On adding $-\mathbf{v}$ to both sides, we have

(6) $$\mathbf{u} - \mathbf{v} = \mathbf{u} + (-\mathbf{v}):$$

subtracting a vector is the same as adding its negative.

With respect to an origin O, any point P of space is determined by its *position vector* $\overrightarrow{OP}$. Any vector $\overrightarrow{AB}$ may be expressed in terms of position vectors; for

(7) $$\overrightarrow{AB} = \overrightarrow{AO} + \overrightarrow{OB} = \overrightarrow{OB} - \overrightarrow{OA}.$$

The product of a vector $\mathbf{u}$ by a number k (written $k\mathbf{u}$ or $\mathbf{u}k$) is defined as a vector $|k|$ times as long as $\mathbf{u}$ and having the same direction as $\mathbf{u}$, or the opposite, according as k is positive or negative. In accordance with this definition

(8) $(-a)\mathbf{u} = a(-\mathbf{u}) = -a\mathbf{u}, \qquad (-a)(-\mathbf{u}) = a\mathbf{u};$

and

(9) $a\mathbf{u} = \mathbf{u}a, \qquad (ab)\mathbf{u} = a(b\mathbf{u}), \qquad (a+b)\mathbf{u} = a\mathbf{u} + b\mathbf{u}.$

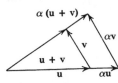

FIG. 96c.
Multiplication by a scalar is distributive.

Moreover, since corresponding sides of similar triangles are proportional, we have

(10) $a(\mathbf{u} + \mathbf{v}) = a\mathbf{u} + a\mathbf{v}$ (Fig. 96c).

As far as addition, subtraction, and multiplication by numbers are concerned, vectors may be treated formally in accordance with the rules of ordinary algebra.

If we adopt a dextral† set of rectangular axes $O\text{–}xyz$ (Fig. 96d) as system of reference, any vector $\mathbf{u}$ may be shifted to become a position vector $\overrightarrow{OP}$ with its initial point at the origin. Then, if P has the rectangular co-ordinates (x, y, z), we say that $\mathbf{u} = \overrightarrow{OP}$ has the *components*

$$u_1 = x, \qquad u_2 = y, \qquad u_3 = z.$$

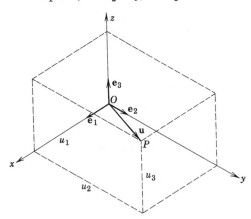

FIG. 96d.   Components of a vector.

If $\mathbf{e}_1$, $\mathbf{e}_2$, $\mathbf{e}_3$ denote unit vectors in the positive directions of x, y, z,‡ we have

(11) $\mathbf{u} = u_1\mathbf{e}_1 + u_2\mathbf{e}_2 + u_3\mathbf{e}_3.$

---

† In a dextral system of axes the rotation of $+x$ towards $+y$ will cause a right-handed screw to advance in the direction $+z$. Such a rotation, viewed from the $+z$ side of the xy-plane, is counterclockwise.

‡ These vectors are often written $\mathbf{i}$, $\mathbf{j}$, $\mathbf{k}$.

Thus **u** is completely determined by its components, a number triple written in a definite order. In fact we may regard number triples as vectors and write

(12) $$\mathbf{u} = [u_1, u_2, u_3].$$

By the Pythagorean theorem,

(13) $$|\mathbf{u}|^2 = u_1^2 + u_2^2 + u_3^2;$$

and the direction of **u** is determined by its direction cosines given by

(14) $$u_1 = |\mathbf{u}| \cos \alpha, \qquad u_2 = |\mathbf{u}| \cos \beta, \qquad u_3 = |\mathbf{u}| \cos \gamma.$$

The zero vector $(0, 0, 0)$ now presents no peculiarity; and the negative

(15) $$-\mathbf{u} = [-u_1, -u_2, -u_3].$$

Addition, subtraction, and multiplication by numbers now conform to the rules:

(16) $$\mathbf{u} + \mathbf{v} = [u_1 + v_1, u_2 + v_2, u_3 + v_3],$$

(17) $$\mathbf{u} - \mathbf{v} = [u_1 - v_1, u_2 - v_2, u_3 - v_3],$$

(18) $$k\mathbf{u} = [ku_1, ku_2, ku_3].$$

The components of a unit vector **e** are its direction cosines

(19) $$\mathbf{e} = [\cos \alpha, \cos \beta, \cos \gamma];$$

and, in particular, the *base vectors*

(20) $$\mathbf{e}_1 = [1, 0, 0], \qquad \mathbf{e}_2 = [0, 1, 0], \qquad \mathbf{e}_3 = [0, 0, 1].$$

This algebra of number triples is entirely equivalent to a vector algebra based on geometric concepts. Moreover, it is free from the vague idea of *direction* and readily lends itself to generalization in space of n dimensions.

Example 1. *Point of Division.* If A, B, C are points of a straight line, C is said to divide the segment AB in the ratio λ if

(21) $$\overrightarrow{AC} = \lambda \overrightarrow{CB}.$$

As C describes the entire line from left to right, λ varies as follows:

$$-1 < \lambda \leqq 0 \, (A) \, 0 \leqq \lambda < \infty \, (B) \, -\infty < \lambda < -1.$$

Thus $\lambda = 0$, $\lambda = \pm \infty$, $\lambda = -1$ correspond to A, B and the infinitely distant "point" at the line. The ratio λ is positive or negative according as C lies within or without the segment AB.

To find the position vector of C relative to an origin O (Fig. 96e), write (21) in the form

$$\overrightarrow{OC} - \overrightarrow{OA} = \lambda(\overrightarrow{OB} - \overrightarrow{OC});$$

then

(22)
$$\overrightarrow{OC} = \frac{\overrightarrow{OA} + \lambda \overrightarrow{OB}}{1 + \lambda}.$$

We shall denote the position vectors of A, B, C by $\mathbf{a}$, $\mathbf{b}$, $\mathbf{c}$; then, if $\lambda = \beta/\alpha$, (22) becomes

(23)
$$\mathbf{c} = \frac{\alpha\mathbf{a} + \beta\mathbf{b}}{\alpha + \beta}.$$

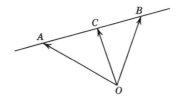

Fig. 96e.   Point of division.

This equation states that C divides AB in the ratio β/α. Thus, if $\mathbf{c} = \frac{1}{2}(\mathbf{a} + \mathbf{b})$, C is the mid-point of AB.

Formula (23) is equivalent to the

THEOREM 1.   *Three distinct points A, B, C will lie on a line when, and only when, there exist three nonzero numbers α, β, γ such that*

(24) $\alpha\mathbf{a} + \beta\mathbf{b} + \gamma\mathbf{c} = 0$, $\alpha + \beta + \gamma = 0$.

Proof. If A, B, C are collinear, C divides AB in some ratio β/α; then on writing $\gamma = -\alpha - \beta$, (23) implies (24).   Conversely, the relations (24) imply (23); hence C lies on the line AB.

The symmetrical relations (24) disclose how each point divides the segment formed by the other two.   Thus (24) states that A, B, C divide BC, CA, AB, respectively, in the ratios γ/β, α/γ, β/α, whose product is 1.

Example 2.   The plane analogue of Theorem 1 is given in

THEOREM 2.   *Four points A, B, C, D, no three of which are collinear, will be in a plane when, and only when, there exist four nonzero numbers α, β, γ, δ, such that*

(25) $\alpha\mathbf{a} + \beta\mathbf{b} + \gamma\mathbf{c} + \delta\mathbf{d} = 0$, $\alpha + \beta + \gamma + \delta = 0$.

Proof. If A, B, C, D, are coplanar, either AB is parallel to CD or AB cuts CD in a point P (not A, B, C, D).   In the respective cases, we have

$$\mathbf{b} - \mathbf{a} = \kappa(\mathbf{d} - \mathbf{c}); \qquad \frac{\mathbf{a} + \lambda\mathbf{b}}{1 + \lambda} = \frac{\mathbf{c} + \lambda'\mathbf{d}}{1 + \lambda'} = \mathbf{p},$$

where λ, λ' are neither 0 nor -1. In both cases, **a**, **b**, **c**, **d** are connected by a linear relation of the form (25). The proof that the relations (25) imply that A, B, C, D are coplanar is left to the reader.

Note that (25) is packed with information. For example, if $ABCD$ is a plane quadrilateral such that $2\mathbf{a} + 3\mathbf{b} - \mathbf{c} - 4\mathbf{d} = 0$, the sides AB, CD meet in a point P given by

$$5\mathbf{p} = 2\mathbf{a} + 3\mathbf{b} = \mathbf{c} + 4\mathbf{d},$$

and P divides AB in the ratio 3/2, CD in the ratio 4/1. Moreover, AC and BD meet in a point Q given by

$$\mathbf{q} = 2\mathbf{a} - \mathbf{c} = 4\mathbf{d} - 3\mathbf{b};$$

and Q divides AC in the ratio $-1/2$, BD in the ratio $-4/3$. If AD and BC meet at R, how does R divide these segments? What can be read from the equation $4\mathbf{d} = 2\mathbf{a} + 2\mathbf{s}$ where $2\mathbf{s} = 3\mathbf{b} - \mathbf{c}$?

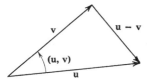

FIG. 97.   Vector subtraction.

97. Scalar Product.   The *scalar* or *inner product* $\mathbf{u} \cdot \mathbf{v}$ of two vectors, referred to a system of rectangular axes, is defined as

(1) $$\mathbf{u} \cdot \mathbf{v} = u_1 v_1 + u_2 v_2 + u_3 v_3.$$

The geometric interpretation of $\mathbf{u} \cdot \mathbf{v}$ is readily found by considering the triangle formed by the vectors, **u**, **v** and $\mathbf{u} - \mathbf{v}$ (Fig. 97). If we denote the smallest angle between **u** and **v** by $(\mathbf{u}, \mathbf{v})$, we have from the cosine law,

(2) $$|\mathbf{u} - \mathbf{v}|^2 = |\mathbf{u}|^2 + |\mathbf{v}|^2 - 2|\mathbf{u}||\mathbf{v}|\cos(\mathbf{u}, \mathbf{v}).$$

If we put

$$|\mathbf{u}|^2 = u_1^2 + u_2^2 + u_3^2, \qquad |\mathbf{v}|^2 = v_1^2 + v_2^2 + v_3^3,$$
$$|\mathbf{u} - \mathbf{v}|^2 = (u_1 - v_1)^2 + (u_2 - v_2)^2 + (u_3 - v_3)^2$$

in (2), we find that

(3) $$\mathbf{u} \cdot \mathbf{v} = |\mathbf{u}||\mathbf{v}|\cos(\mathbf{u}, \mathbf{v}).$$

The scalar product of two vectors is equal to the product of their lengths and the cosine of their included angle.   Consequently $\mathbf{u} \cdot \mathbf{v}$ has the same value in all systems of rectangular axes.   *The scalar product is invariant to translations or rotations of the axes.*

If in (3) we put $\mathbf{v} = \mathbf{e}$, a *unit* vector,

(4) $$\mathbf{u} \cdot \mathbf{e} = |\mathbf{u}| \cos (\mathbf{u}, \mathbf{e})$$

is the *component* of $\mathbf{u}$ on an axis parallel to $\mathbf{e}$. In particular, the components of $\underline{\mathbf{u}}$ on the coordinate axis are

(5) $$u_1 = \mathbf{u} \cdot \mathbf{e}_1, \qquad u_2 = \mathbf{u} \cdot \mathbf{e}_2, \qquad u_3 = \mathbf{u} \cdot \mathbf{e}_3.$$

Thus the vector $\mathbf{u}$ can be written

(6) $$\mathbf{u} = \mathbf{u} \cdot \mathbf{e}_1 \, \mathbf{e}_1 + \mathbf{u} \cdot \mathbf{e}_2 \, \mathbf{e}_2 + \mathbf{u} \cdot \mathbf{e}_3 \, \mathbf{e}_3.$$

From (3) we see that $\mathbf{u} \cdot \mathbf{v} = 0$ not only when $\mathbf{u} = \mathbf{0}$ or $\mathbf{v} = \mathbf{0}$ but also when $\cos (\mathbf{u}, \mathbf{v}) = 0$, that is, when $\mathbf{u}$ and $\mathbf{v}$ are perpendicular. Thus, for *proper* vectors

(7) $$\mathbf{u} \cdot \mathbf{v} = 0 \quad \text{implies} \quad \mathbf{u} \perp \mathbf{v}.$$

This analytic condition for perpendicularity is of great service in geometry.

From the definition (1) we see that scalar multiplication is commutative and distributive:

(8) $$\mathbf{u} \cdot \mathbf{v} = \mathbf{v} \cdot \mathbf{u},$$

(9) $$\mathbf{u} \cdot (\mathbf{v} + \mathbf{w}) = \mathbf{u} \cdot \mathbf{v} + \mathbf{u} \cdot \mathbf{w}.$$

The associative law is meaningless; for $(\mathbf{u} \cdot \mathbf{v}) \cdot \mathbf{w}$, the dot product of a number $\mathbf{u} \cdot \mathbf{v}$ and a vector $\mathbf{w}$, has not been defined.

As to the *base vectors* $\mathbf{e}_1, \mathbf{e}_2, \mathbf{e}_3$ we have

(10) $$\mathbf{e}_i \cdot \mathbf{e}_j = \delta_{ij},$$

where the *Kronecker delta* $\delta_{ij} = 0$ or 1 according as $i \neq j$ or $i = j$. If we expand the product

$$\mathbf{u} \cdot \mathbf{v} = (u_1\mathbf{e}_1 + u_2\mathbf{e}_2 + u_3\mathbf{e}_3) \cdot (v_1\mathbf{e}_1 + v_2\mathbf{e}_2 + v_3\mathbf{e}_3)$$

by the distributive law and use these relations, we revert to the value of $\mathbf{u} \cdot \mathbf{v}$ as originally defined in (1).

Example 1. Find the angle $(\mathbf{u}, \mathbf{v})$ between the vectors $\mathbf{u} = [2, -1, 3]$, $\mathbf{v} = [0, 2, 4]$. From (1), $\mathbf{u} \cdot \mathbf{v} = 0 - 2 + 12 = 10$; and from (3)

$$\cos (\mathbf{u}, \mathbf{v}) = \frac{\mathbf{u} \cdot \mathbf{v}}{|u||v|} = \frac{10}{\sqrt{14}\,\sqrt{20}} = 0\text{·}5979, \qquad \text{angle } (\mathbf{u}, \mathbf{v}) = 53°18'.$$

Example 2. In the triangle ABC let $\overrightarrow{BC} = \mathbf{a}$, $\overrightarrow{CA} = \mathbf{b}$, $\overrightarrow{AB} = \mathbf{c}$; then, since $\mathbf{a} + \mathbf{b} = \mathbf{c}$, we have

$$\mathbf{c} \cdot \mathbf{c} = (\mathbf{a} + \mathbf{b}) \cdot (\mathbf{a} + \mathbf{b}) = \mathbf{a} \cdot \mathbf{a} + 2\mathbf{a} \cdot \mathbf{b} + \mathbf{b} \cdot \mathbf{b}.$$

Hence, if we denote the sides of the triangle by a, b, c and the angles opposite them by A, B, C, we have the *cosine law*,

$$c^2 = a^2 + b^2 - 2ab \cos C.$$

Note that angle $(\mathbf{a}, \mathbf{b}) = \pi - C$.

98. Vector Product. The *vector* or *outer product* of two vectors referred to a dextral system of rectangular axes is defined as

$$(1) \qquad \mathbf{u} \times \mathbf{v} = [u_2 v_3 - u_3 v_2, \, u_3 v_1 - u_1 v_3, \, u_1 v_2 - u_2 v_1].$$

In terms of the base vectors

$$(2) \qquad \mathbf{u} \times \mathbf{v} = \begin{vmatrix} \mathbf{e}_1 & \mathbf{e}_2 & \mathbf{e}_3 \\ u_1 & u_2 & u_3 \\ v_1 & v_2 & v_3 \end{vmatrix}.$$

The geometric interpretation of $\mathbf{u} \times \mathbf{v}$ may be found as follows. When $\mathbf{u} \times \mathbf{v} \neq \mathbf{0}$, it represents a vector perpendicular to both $\mathbf{u}$ and $\mathbf{v}$; for from (97.1)

$$\mathbf{u} \cdot (\mathbf{u} \times \mathbf{v}) = 0, \qquad \mathbf{v} \cdot (\mathbf{u} \times \mathbf{v}) = 0.$$

From the definitions of $\mathbf{u} \times \mathbf{v}$ and $\mathbf{u} \cdot \mathbf{v}$ we have the identity

$$|\mathbf{u} \times \mathbf{v}|^2 = |\mathbf{u}|^2 \, |\mathbf{v}|^2 - (\mathbf{u} \cdot \mathbf{v})^2,$$

and hence from (97.3)

$$(3) \qquad |\mathbf{u} \times \mathbf{v}| = |\mathbf{u}| \, |\mathbf{v}| \sin (\mathbf{u}, \mathbf{v}).$$

To find the direction of $\mathbf{u} \times \mathbf{v}$, shift $\mathbf{u}$, $\mathbf{v}$, and $\mathbf{u} \times \mathbf{v}$ to the origin, and revolve the trihedral $O\text{–}xyz$ so that the x-axis points along $\mathbf{u}$ and the y-axis lies in the **uv**-plane and makes an acute angle with $\mathbf{v}$ (Fig. 98). During this continuous rotation $\mathbf{u} \times \mathbf{v}$ remains *fixed*; for it is a vector of known length perpendicular to the **uv**-plane and continuity forbids a change in direction. Since

$$\mathbf{u} = [|\mathbf{u}|, 0, 0], \qquad \mathbf{v} = [|\mathbf{v}| \cos (\mathbf{u}, \mathbf{v}), |\mathbf{v}| \sin (\mathbf{u}, \mathbf{v}), 0],$$

$$\mathbf{u} \times \mathbf{v} = [0, 0, |\mathbf{u}| \, |\mathbf{v}| \sin (\mathbf{u}, \mathbf{v})].$$

Hence $\mathbf{u} \times \mathbf{v}$ points along the positive z-axis, and, since the axes are dextral, the vectors $\mathbf{u}$, $\mathbf{v}$, $\mathbf{u} \times \mathbf{v}$ also form a dextral set.

The product $\mathbf{u} \times \mathbf{v}$ *is a vector of length* $|\mathbf{u}| |\mathbf{v}| \sin (\mathbf{u}, \mathbf{v})$ *perpendicular to the plane of* $\mathbf{u}$ *and* $\mathbf{v}$ *and pointing in the direction that a right-handed screw will advance when turned from* $\mathbf{u}$ *towards* $\mathbf{v}$.

From (3) we see that $\mathbf{u} \times \mathbf{v} = \mathbf{0}$ not only when $\mathbf{u} = \mathbf{0}$ or $\mathbf{v} = \mathbf{0}$ but also when $\sin (\mathbf{u}, \mathbf{v}) = 0$, that is, when $\mathbf{u}$ and $\mathbf{v}$ are parallel. Thus for *proper* vectors

$$(4) \qquad \mathbf{u} \times \mathbf{v} = \mathbf{0} \quad \text{implies} \quad \mathbf{u} \, || \, \mathbf{v}.$$

The definition (1) shows that vector multiplication is *anticommutative*, that is,

(5) $\mathbf{v} \times \mathbf{u} = -\mathbf{u} \times \mathbf{v}.$

We can readily verify that vector multiplication is distributive:

(6) $\mathbf{u} \times (\mathbf{v} + \mathbf{w}) = \mathbf{u} \times \mathbf{v} + \mathbf{u} \times \mathbf{w}.$

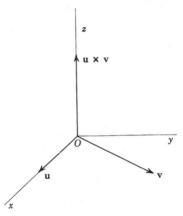

FIG. 98.   Vector product.

The associative law is not in general true; for $(\mathbf{u} \times \mathbf{v}) \times \mathbf{w}$ is parallel to the plane of $\mathbf{u}$ and $\mathbf{v}$, while $\mathbf{u} \times (\mathbf{v} \times \mathbf{w})$ is parallel to the plane of $\mathbf{v}$ and $\mathbf{w}$.

The triple vector product has the important expansion:

(7) $(\mathbf{u} \times \mathbf{v}) \times \mathbf{w} = (\mathbf{u} \cdot \mathbf{w})\,\mathbf{v} - (\mathbf{v} \cdot \mathbf{w})\,\mathbf{u}.$

For, if $\mathbf{p} = (\mathbf{u} \times \mathbf{v}) \times \mathbf{w}$, we have from (1)

$$p_1 = (u_3 v_1 - u_1 v_3) w_3 - (u_1 v_2 - u_2 v_1) w_2$$
$$= (u_3 w_3 + u_2 w_2) v_1 - (v_3 w_3 + v_2 w_2) u_1$$
$$= (u_1 w_1 + u_2 w_2 + u_3 w_3) v_1 - (v_1 w_1 + v_2 w_2 + v_3 w_3) u_1.$$

Thus we have

$$p_1 = (\mathbf{u} \cdot \mathbf{w}) v_1 - (\mathbf{v} \cdot \mathbf{w}) u_1$$

and two similar equations in which the subscript 1 is replaced by 2 or 3. These are the components of $\mathbf{p}$ given by (7).

As to the dextral set of base vectors $\mathbf{e}_1, \mathbf{e}_2, \mathbf{e}_3$, we have the cyclic relations,

(8) $\mathbf{e}_1 \times \mathbf{e}_2 = \mathbf{e}_3,$ $\mathbf{e}_2 \times \mathbf{e}_3 = \mathbf{e}_1,$ $\mathbf{e}_3 \times \mathbf{e}_1 = \mathbf{e}_2;$

Obviously $\mathbf{e}_i \times \mathbf{e}_i = \mathbf{0}.$ If we expand the product

$$\mathbf{u} \times \mathbf{v} = (u_1 \mathbf{e}_1 + u_2 \mathbf{e}_2 + u_3 \mathbf{e}_3) \times (v_1 \mathbf{e}_1 + v_2 \mathbf{e}_2 + v_3 \mathbf{e}_3)$$

by the distributive law and use these relations, we revert to the value of $\mathbf{u} \times \mathbf{v}$ given in (2).

Example 1. Let $\mathbf{u} = \overrightarrow{PQ}$ be a *line vector*: a vector whose mobility is confined to a line, its *line of action*. Then the *moment* of $\mathbf{u}$ *about a point A* is defined as the *vector* $\mathbf{r} \times \mathbf{u}$ where $\mathbf{r}$ is any vector from A to the line of action.

The *moment* of $\mathbf{u}$ *about an axis* through A in the direction of the unit vector $\mathbf{e}$ is defined as the component of $\mathbf{r} \times \mathbf{u}$ on this axis, namely, the *scalar* $\mathbf{e} \cdot \mathbf{r} \times \mathbf{u}$.

For example, let the line of action of the force $\mathbf{f} = [2, -4, 1]$ pass through the point $P(2, 4, -1)$. To find the moment of $\mathbf{f}$ about an axis through $A(3, -1, 2)$ in the direction of the vector $[2, -1, 2]$ we first compute

$$\text{Moment } \mathbf{f} \text{ about } A = \overrightarrow{AP} \times \mathbf{f} = [-1, 5, -3] \times [2, -4, 1] = [-7, -5, -6];$$

and since $\mathbf{e} = \frac{1}{3}[2, -1, 2]$ is a unit vector in the direction of the axis,

$$\text{Moment } \mathbf{f} \text{ about axis} = \frac{1}{3}[2, -1, 2] \cdot [-7, -5, -6] = -7.$$

Example 2. Find the shortest distance d between two lines AB and CD given by the points $A(1, -2, -1)$, $B(4, 0, -3)$; $C(1, 2, -1)$, $D(2, -4, -5)$.

We have

$$\overrightarrow{AB} = [3, 2, -2], \qquad \overrightarrow{CD} = [1, -6, -4];$$

hence the common normal to the lines has the direction

$$\overrightarrow{AB} \times \overrightarrow{CD} = [-20, 10, -20] = 10[-2, 1, -2].$$

A unit vector in this direction is $\mathbf{e} = \frac{1}{3}[-2, 1, -2]$. If $\mathbf{u}$ is any vector from one line to the other (such as $\overrightarrow{AC}$ or $\overrightarrow{BD}$), the component of $\mathbf{u}$ in the direction of $\mathbf{e}$ will give the desired distance d; thus $d = |\mathbf{u} \cdot \mathbf{e}|$ from (97.4). Since $\overrightarrow{AC} = [0, 4, 0]$, we have

$$d = \mathbf{e} \cdot \overrightarrow{AC} = \frac{1}{3}(0 + 4 + 0) = \frac{4}{3}.$$

To check we compute $\mathbf{e} \cdot \overrightarrow{BD}$ or $\mathbf{e} \cdot \overrightarrow{AD}$.

99. Box Product. The triple product $(\mathbf{u} \times \mathbf{v}) \cdot \mathbf{w}$† or

$$\text{(1)} \qquad \mathbf{u} \times \mathbf{v} \cdot \mathbf{w} = \begin{vmatrix} u_1 & u_2 & u_3 \\ v_1 & v_2 & v_3 \\ w_1 & w_2 & w_3 \end{vmatrix},$$

as we see at once on expanding the determinant according to the elements of the last row; for the components of $\mathbf{u} \times \mathbf{v}$ are the cofactors of the respective components of $\mathbf{w}$. From the properties of determinants, $\mathbf{u} \times \mathbf{v} \cdot \mathbf{w}$ is not altered by a cyclical change in the vectors:

$$\text{(2)} \qquad \mathbf{u} \times \mathbf{v} \cdot \mathbf{w} = \mathbf{v} \times \mathbf{w} \cdot \mathbf{u} = \mathbf{w} \times \mathbf{u} \cdot \mathbf{v};$$

but a change in cyclical order produces a change in sign:

$$\text{(3)} \qquad \mathbf{u} \times \mathbf{w} \cdot \mathbf{v} = -\mathbf{u} \times \mathbf{v} \cdot \mathbf{w}.$$

---

† The parentheses may be omitted as $\mathbf{u} \times (\mathbf{v} \cdot \mathbf{w})$ is meaningless.

Moreover, $\mathbf{u} \times \mathbf{v} \cdot \mathbf{w}$ is not altered by an interchange of the dot and cross; for

(3) $\mathbf{u} \times \mathbf{v} \cdot \mathbf{w} = \mathbf{v} \times \mathbf{w} \cdot \mathbf{u} = \mathbf{u} \cdot \mathbf{v} \times \mathbf{w}.$

Let the noncoplanar vectors $\mathbf{u} = \overrightarrow{PA}$, $\mathbf{v} = \overrightarrow{PB}$, $\mathbf{w} = \overrightarrow{PC}$ be drawn from a point P. Then $\mathbf{u}$, $\mathbf{v}$, $\mathbf{w}$ form a dextral or sinistral set according as the circuit ABC, as viewed from P, is clockwise or counterclockwise.

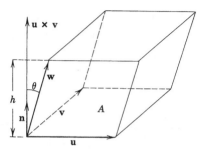

FIG. 99.   Box product.

THEOREM. *The product* $\mathbf{u} \times \mathbf{v} \cdot \mathbf{w}$ *is equal to the volume* V *of a parallelepiped having* $\mathbf{u}$, $\mathbf{v}$, $\mathbf{w}$ *as concurrent edges. Its sign is positive or negative according as* $\mathbf{u}$, $\mathbf{v}$, $\mathbf{w}$ *form a dextral or a sinistral set.*

Proof. We may write the vector product $\mathbf{u} \times \mathbf{v} = A\mathbf{n}$, where

$$A = |\mathbf{u} \times \mathbf{v}| = |\mathbf{u}| |\mathbf{v}| \sin (\mathbf{u}, \mathbf{v})$$

is the area of the parallelogram having $\mathbf{u}$ and $\mathbf{v}$ as adjacent sides and $\mathbf{n}$ is a unit vector perpendicular to $\mathbf{u}$ and $\mathbf{v}$ and pointing as a right-handed screw would advance when turned from $\mathbf{u}$ to $\mathbf{v}$. When the set $\mathbf{u}$, $\mathbf{v}$, $\mathbf{w}$ is dextral (as in Fig. 99) the angle $\theta = (\mathbf{n}, \mathbf{w})$ is acute, and

$$\mathbf{u} \times \mathbf{v} \cdot \mathbf{w} = A\mathbf{n} \cdot \mathbf{w} = A|\mathbf{w}| \cos \theta = Ah = V,$$

where h is the altitude of the box. When the set $\mathbf{u}$, $\mathbf{v}$, $\mathbf{w}$ is sinistral, θ is obtuse, $\cos \theta$ is negative, and $\mathbf{u} \times \mathbf{v} \cdot \mathbf{w} = -V$.

If $\mathbf{u}$, $\mathbf{v}$, $\mathbf{w}$ are proper vectors, $V = 0$ only when the vectors are coplanar. Therefore, for proper vectors

(4) $\mathbf{u} \times \mathbf{v} \cdot \mathbf{w} = 0$ implies $\mathbf{u}$, $\mathbf{v}$, $\mathbf{w}$ coplanar.

Finally we note the useful identity,

(5) $(\mathbf{a} \times \mathbf{b}) \cdot (\mathbf{c} \times \mathbf{d}) = (\mathbf{a} \cdot \mathbf{c})(\mathbf{b} \cdot \mathbf{d}) - (\mathbf{a} \cdot \mathbf{d})(\mathbf{b} \cdot \mathbf{c}).$

To prove it interchange the first cross and dot and expand $\mathbf{b} \times (\mathbf{c} \times \mathbf{d})$ by (98.7).

PROBLEMS

1. Prove that the medians of a triangle ABC meet at a point D given by $\mathbf{d} = \frac{1}{3}(\mathbf{a} + \mathbf{b} + \mathbf{c})$; and that D divides each median in the ratio 2/1.

2. If $\overrightarrow{OA'} = 3\overrightarrow{OA}$, $\overrightarrow{OB'} = 2\overrightarrow{OB}$, in what ratio does the point P in which AB and $A'B'$ intersect divide these segments?

3. In a plane quadrilateral $ABCD$, the diagonals AC, BD meet at P; the sides AB, CD at Q; and the sides BC, DA at R. If P divides AC in the ratio 3/2, BD in the ratio 1/2, find the ratios in which Q divides AB and CD, R divides BC and DA.

4. If $3\mathbf{a} - 2\mathbf{b} + \mathbf{c} - 2\mathbf{d} = 0$, show that the points A, B, C, D are coplanar. Find the point P in which AC and BD meet. In what ratio does P divide AC and BD?

5. If $ABCD$ is a quadrilateral (plane or skew), show that the mid-points of its sides are the vertices of a parallelogram whose center is $(\mathbf{a} + \mathbf{b} + \mathbf{c} + \mathbf{d})/4$.

6. If $\mathbf{u} = [2, 3, -1]$, $\mathbf{v} = [0, 4, 2]$, $\mathbf{w} = [-1, 2, 5]$, compute $\mathbf{u} \cdot \mathbf{v}$, $\mathbf{u} \times \mathbf{v}$, $\mathbf{v} \times \mathbf{w}$, $\mathbf{w} \times \mathbf{u}$, $\mathbf{u} \times \mathbf{v} \cdot \mathbf{w}$, $(\mathbf{u} \times \mathbf{v}) \times \mathbf{w}$, and $\mathbf{u} \times (\mathbf{v} \times \mathbf{w})$.

7. Find the equation of the plane through $A(2, -1, 1)$ normal to the vector $[4, 2, -3]$.

8. Show that the vector $[a, b, c]$ is normal to the plane whose equation is $ax + by + cz + d = 0$.

9. Find the equation of the plane through $A(2, -1, 6)$ and $B(1, -2, 4)$ and perpendicular to the plane $x - 2y - 2z + 9 = 0$.

10. Find the moment of $\overrightarrow{PQ}$ about an axis through $A(3, 1, 0)$ in the direction of $[3, 4, 12]$, given $P(1, 3, 1)$ and $Q(3, 5, 2)$. Check.

11. Show that $\mathbf{a} = [2, 3, -1]$, $\mathbf{b} = [1, 1, 5]$, $\mathbf{c} = [16, -11, -1]$ form a dextral set of mutually orthogonal vectors.

12. Show that the vector $\mathbf{n} = \mathbf{a} \times \mathbf{b} + \mathbf{b} \times \mathbf{c} + \mathbf{c} \times \mathbf{a}$ is normal to the plane through the points A, B, C.

13. Find the perpendicular distance from $P(1, -2, 1)$ to the plane ABC, given the points

$$A(2, 4, 1), \qquad B(-1, 0, 1), \qquad C(-1, 4, 2).$$

14. Find the shortest distance between the lines AB and CD, given the points

$$A(1, 2, 3), \qquad B(-1, 0, 2); \qquad C(0, 1, 7), \qquad D(2, 0, 5).$$

15. Show that the lines AB, CD are coplanar, and find the point P in which they meet, given

$$A(-2, -3, 4), \qquad B(2, 3, 0); \qquad C(-2, 3, 2), \qquad D(2, 0, 1).$$

[Show that $\overrightarrow{AB} \cdot \overrightarrow{AC} \times \overrightarrow{AD} = 0$; let P divide AB in the ratio λ, CD in ratio μ; then $\mathbf{a} + \lambda\mathbf{b} \parallel \mathbf{c} + \mu\mathbf{d}$.]

16. Solve the equation $\mathbf{r} \cdot \mathbf{a} = \alpha$, $\mathbf{r} \times \mathbf{b} = \mathbf{c}$ for $\mathbf{r}$ if $\mathbf{a} \cdot \mathbf{b} \neq 0$ and the vectors $\mathbf{a}$, $\mathbf{b}$, $\mathbf{c}$ and scalar α are given.

17. Prove the formulas

(a) $\mathbf{a} \times (\mathbf{b} \times \mathbf{c}) + \mathbf{b} \times (\mathbf{c} \times \mathbf{a}) + \mathbf{c} \times (\mathbf{a} \times \mathbf{b}) = 0$;

(b) $(\mathbf{a} \times \mathbf{b}) \cdot (\mathbf{b} \times \mathbf{c}) \times (\mathbf{c} \times \mathbf{a}) = (\mathbf{a} \cdot \mathbf{b} \times \mathbf{c})^2$;

(c) $(\mathbf{a} \cdot \mathbf{b} \times \mathbf{c})(\mathbf{u} \cdot \mathbf{v} \times \mathbf{w}) = \begin{vmatrix} \mathbf{a} \cdot \mathbf{u} & \mathbf{a} \cdot \mathbf{v} & \mathbf{a} \cdot \mathbf{w} \\ \mathbf{b} \cdot \mathbf{u} & \mathbf{b} \cdot \mathbf{v} & \mathbf{b} \cdot \mathbf{w} \\ \mathbf{c} \cdot \mathbf{u} & \mathbf{c} \cdot \mathbf{v} & \mathbf{c} \cdot \mathbf{w} \end{vmatrix}$.   [Use (99.1).]

18. If $\mathbf{a} \cdot \mathbf{b} \times \mathbf{c} \neq 0$, show that any vector $\mathbf{u}$ may be expressed as a sum of vectors parallel to $\mathbf{a}$, $\mathbf{b}$, $\mathbf{c}$ by the formula:

$$(\mathbf{a} \cdot \mathbf{b} \times \mathbf{c})\,\mathbf{u} = (\mathbf{u} \cdot \mathbf{b} \times \mathbf{c})\,\mathbf{a} + (\mathbf{u} \cdot \mathbf{c} \times \mathbf{a})\,\mathbf{b} + (\mathbf{u} \cdot \mathbf{a} \times \mathbf{b})\,\mathbf{c}.$$

[Expand $(\mathbf{a} \times \mathbf{b}) \times (\mathbf{c} \times \mathbf{u})$ in two ways.]

19. Prove that the associative law $\mathbf{a} \times (\mathbf{b} \times \mathbf{c}) = (\mathbf{a} \times \mathbf{b}) \times \mathbf{c}$ is valid when and only when $\mathbf{b} \times (\mathbf{c} \times \mathbf{a}) = 0$.

20. If A, B, C, D are any four points, prove that

$$\overrightarrow{AB} \cdot \overrightarrow{CD} + \overrightarrow{BC} \cdot \overrightarrow{AD} + \overrightarrow{CA} \cdot \overrightarrow{BD} = 0,$$

$$\overrightarrow{AB} \times \overrightarrow{CD} + \overrightarrow{BC} \times \overrightarrow{AD} + \overrightarrow{CA} \times \overrightarrow{BD} = 2\,\overrightarrow{AB} \times \overrightarrow{CA}.$$

100. Derivative of a Vector. Let $\mathbf{u}(t)$ denote a vector function of a scalar variable t over the interval $a \leq t \leq b$. Then $\mathbf{u}(t)$ is said to be *continuous* for a value t_0 if

$$(1) \qquad \lim_{t \to t_0} \mathbf{u}(t) = \mathbf{u}(t_0).$$

The derivative of $\mathbf{u}(t)$ is defined as

$$(2) \qquad \frac{d\mathbf{u}}{dt} = \lim_{\Delta t \to 0} \frac{\mathbf{u}(t + \Delta t) - \mathbf{u}(t)}{\Delta t}.$$

If $d\mathbf{u}/dt$ exists when $t = t_0$, then $\mathbf{u}(t)$ is continuous at t_0 (cf. § 50).

Just as in the calculus of scalars we can now prove that

$$(3) \qquad \frac{d\mathbf{c}}{dt} = 0 \quad (\mathbf{c} \text{ const}),$$

$$(4) \qquad \frac{d}{dt}(\mathbf{u} + \mathbf{v}) = \frac{d\mathbf{u}}{dt} + \frac{d\mathbf{v}}{dt},$$

$$(5) \qquad \frac{d}{dt}(f\mathbf{u}) = f\frac{d\mathbf{u}}{dt} + \frac{df}{dt}\mathbf{u},$$

$$(6) \qquad \frac{d}{dt}(\mathbf{u} \cdot \mathbf{v}) = \mathbf{u} \cdot \frac{d\mathbf{v}}{dt} + \frac{d\mathbf{u}}{dt} \cdot \mathbf{v},$$

$$(7) \qquad \frac{d}{dt}(\mathbf{u} \times \mathbf{v}) = \mathbf{u} \times \frac{d\mathbf{v}}{dt} + \frac{d\mathbf{u}}{dt} \times \mathbf{v}.$$

The proofs of (6) and (7) depend essentially on the distributive laws for the dot and cross products. In (7) the *order* of the factors must be maintained.

If the components of

$$\mathbf{u} = u_1 \mathbf{e}_1 + u_2 \mathbf{e}_2 + u_3 \mathbf{e}_3$$

are functions of t, we have from (3) and (5)

$$(8) \qquad \frac{d\mathbf{u}}{dt} = \frac{du_1}{dt}\mathbf{e}_1 + \frac{du_2}{dt}\mathbf{e}_2 + \frac{du_3}{dt}\mathbf{e}_3.$$

THEOREM 1. *A necessary and sufficient condition that a proper vector* **u** *have a constant length is that*

$$(9) \qquad \mathbf{u} \cdot \frac{d\mathbf{u}}{dt} = 0,$$

Proof. Since $|\mathbf{u}|^2 = \mathbf{u} \cdot \mathbf{u}$, we have from (6)

$$\frac{d}{dt} |\mathbf{u}|^2 = 2\mathbf{u} \cdot \frac{d\mathbf{u}}{dt}.$$

Hence $|\mathbf{u}| = $ const implies (9) and conversely.

THEOREM 2. *A necessary and sufficient condition that a proper vector* **u** *always remain parallel to fixed line is that*

$$(10) \qquad \mathbf{u} \times \frac{d\mathbf{u}}{dt} = \mathbf{0}.$$

Proof. Let $\mathbf{u} = u(t)\mathbf{e}$ where **e** is a *unit* vector; then

$$\mathbf{u} \times \frac{d\mathbf{u}}{dt} = u\mathbf{e} \times \left(\frac{du}{dt} \mathbf{e} + u \frac{d\mathbf{e}}{dt} \right) = u^2 \mathbf{e} \times \frac{d\mathbf{e}}{dt}.$$

If **e** is constant, $d\mathbf{e}/dt = \mathbf{0}$, and the condition follows. Conversely, since $\mathbf{u} \neq \mathbf{0}$, the condition implies that $\mathbf{e} \times (d\mathbf{e}/dt) = \mathbf{0}$; but, since $\mathbf{e} \cdot (d\mathbf{e}/dt) = 0$ from Theorem 1, these equations are contradictory unless $d\mathbf{e}/dt = \mathbf{0}$; that is, **e** is constant.

101. Curves. *A curve is an aggregate of points whose coordinates are functions of a single variable.* Thus the equations

$$(1) \qquad x = x(t), \qquad y = y(t), \qquad z = z(t)$$

represent a curve in space. The variable t is called a *parameter*, and each value of the parameter within a certain range $T: a \leq t \leq b$ corresponds to a definite point $P(x, y, z)$ of the curve. If $x(t)$, $y(t)$, $z(t)$ are continuous functions in $a \leq t \leq b$, the curve is said to be *continuous* in this interval.

To avoid having the curve degenerate into a point we exclude the case in which all three functions are constants. We also restrict the interval T so that there is just one value of t corresponding to each point of the curve. Then equations (1) set up a one-to-one correspondence between the points of the curve and the values of t in the interval T.

A parameter value t corresponds to an *ordinary point* of a curve when the three derivatives $x'(t)$, $y'(t)$, $z'(t)$ exist and are continuous at t and at least one is not zero. An arc of the curve which consists entirely of ordinary points is said to be *smooth*.

We can make the change in parameter

$$t = \varphi(u), \quad \text{where} \quad a = \varphi(\alpha), \quad b = \varphi(\beta),$$

provided

(i) $\varphi'(u) > 0$ in U: $\alpha \leqq u \leqq \beta$,

(ii) $\varphi'(u)$ is continuous in U.†

Then the inverse function $u = \psi(t)$ exists in the interval T, and

$$\psi'(t) = \frac{du}{dt} = \frac{1}{\varphi'(u)} \qquad (\S\ 53).$$

Since both $\varphi'(u)$ and $\psi'(t)$ are positive, both $\varphi(u)$ and $\psi(t)$ are increasing functions (Cor. 57.3), and there is one-to-one correspondence between the u-values in U and the t-values in T.   On putting $t = \varphi(u)$ in (1), we obtain another parametric representation of the curve; and, since

$$\frac{dx}{du} = \frac{dx}{dt}\,\varphi'(u), \qquad \frac{dy}{du} = \frac{dy}{dt}\,\varphi'(u), \qquad \frac{dz}{du} = \frac{dz}{dt}\,\varphi'(u),$$

the ordinary points are the same as before.

If P_0 and P are neighboring points of the curve and the line P_0P tends to a limiting position as P approaches P_0 along the curve, the limiting line is called the *tangent* to the curve at the point P_0.   Now, if P_0 and P correspond to t_0 and t and we divide the vector chord,

$$\overrightarrow{P_0P} = [x - x_0, y - y_0, z - z_0] = [\Delta x, \Delta y, \Delta z]$$

by $\Delta t = t - t_0$, the limiting position of the line P_0P is given by the vector

$$(2) \qquad \lim_{\Delta t \to 0} \frac{\overrightarrow{P_0P}}{\Delta t} = \left[\frac{dx}{dt}, \frac{dy}{dt}, \frac{dz}{dt}\right]$$

at every ordinary point.   Hence the vector (2) is a *tangent vector* to the curve that points in the direction of increasing parameter values.

If we denote the position vector $\overrightarrow{OP}$ of the curve by

$$(3) \qquad \mathbf{r} = x\mathbf{e}_1 + y\mathbf{e}_2 + z\mathbf{e}_3,$$

the tangent vector at any ordinary point is

$$(4) \qquad \frac{d\mathbf{r}}{dt} = \frac{dx}{dt}\,\mathbf{e}_1 + \frac{dy}{dt}\,\mathbf{e}_2 + \frac{dz}{dt}\,\mathbf{e}_3.$$

---

† We may replace (i) by $\varphi'(u) < 0$; then $\varphi(u)$ is a decreasing function, and the interval U becomes $\beta \leqq u \leqq \alpha$.

Since the continuity of a function does not ensure its differentiability, a curve may be continuous at a point without having a tangent there. This is the case for the curve

$$x = t, \qquad y = t \sin \frac{1}{t}, \qquad z = 0$$

at the origin ($t = 0$).   Indeed continuous curves are known which nowhere possess a tangent.†

Let the line $P_0 T$ be tangent to the curve (1) at P_0. If T has the coordinates (x, y, z), the vector

$$\overrightarrow{P_0 T} = \overrightarrow{OT} - \overrightarrow{OP_0} = [x - x_0, y - y_0, z - z_0]$$

is parallel to the tangent vector at P_0.   Hence the equations of the tangent line at P_0 are

(5)
$$\frac{x - x_0}{x'(t_0)} = \frac{y - y_0}{y'(t_0)} = \frac{z - z_0}{z'(t_0)},$$

provided no denominators vanish. If $x'(t_0) = 0$, we must have $x - x_0 = 0$, and the tangent lies in the plane $x = x_0$. If $x'(t_0) = y'(t_0) = 0$, the tangent is the line $x = x_0, y = y_0$ parallel to the z-axis.

We shall see in § 129 that a *smooth* curve is *rectifiable*; that is, it has a definite arc length between any two points whose parameter values lie in (a, b). We choose a point $P_0(t_0)$ as the origin of arcs and regard the arc s from P_0 to $P(t)$ as positive or negative according as $t > t_0$ or $t < t_0$. Then $s = \text{arc } P_0 P = f(t)$ is a strictly increasing function of t, and the inverse function $t = \varphi(s)$ is also strictly increasing in an interval $\alpha \leq s \leq \beta$ (§ 49). If we put $t = \varphi(s)$ in (1), we obtain the equations of the curve with s as parameter; and its position vector (3) becomes $\mathbf{r} = \mathbf{r}(s)$.

Example. Find the equations of the tangent to the curve
$$x = 3t, \qquad y = 3t^2, \qquad z = 2t^3$$
at the point $t = 1$.

The point of tangency is $(3, 3, 2)$, and the tangent vector at this point is $[3, 6, 6]$. Hence the required equations are

$$\frac{x - 3}{3} = \frac{y - 3}{6} = \frac{z - 2}{6} \quad \text{or} \quad 2x - 6 = y - 3 = z - 2.$$

102. Unit Tangent Vector.   We now prove the important

THEOREM.   *If* $\mathbf{r} = \mathbf{r}(s)$ *is the vector equation of a smooth curve in terms of* $s = \text{arc } P_0 P$ *as parameter, then*

(1)
$$\frac{d\mathbf{r}}{ds} = \text{T},$$

---

† See Pierpont, J., *The Theory of Functions of Real Variables*, vol. 2, Boston, 1912, pp. 498–500, for the famous example given by Weierstrass. The footnote on p. 96 cites a simpler example.

a unit vector tangent to the curve at P and pointing in the direction of increasing arcs.

Proof. In § 129 we shall show that the ratio of the chord of a smooth curve to the arc it subtends approaches unity as the arc approaches zero (129.9). Hence, if $P(s)$ and $Q(s + \Delta s)$ are points of the curve and **e** denotes a unit vector along $\overrightarrow{PQ}$ (Fig. 102a),

$$\frac{\Delta \mathbf{r}}{\Delta s} = \frac{\overrightarrow{PQ}}{\text{arc } PQ} = \frac{\overrightarrow{PQ}}{\text{chord } PQ} \frac{\text{chord } PQ}{\text{arc } PQ} = \mathbf{e} \frac{\text{chord } PQ}{\text{arc } PQ}.$$

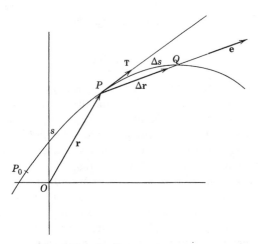

FIG. 102a.   To illustrate $d\mathbf{r}/ds = \mathbf{\tau}$.

Now let $Q \rightarrow P$; then, as $\Delta s \rightarrow 0$, $\mathbf{e} \rightarrow \mathbf{\tau}$ and chord PQ/arc $PQ \rightarrow 1$ on the right; hence, $d\mathbf{r}/ds = \mathbf{\tau}$.

If the position vector of the curve is

$$\mathbf{r} = x\mathbf{e}_1 + y\mathbf{e}_2 + z\mathbf{e}_3,$$

$$\mathbf{\tau} = \frac{dx}{ds}\,\mathbf{e}_1 + \frac{dy}{ds}\,\mathbf{e}_2 + \frac{dz}{ds}\,\mathbf{e}_3.$$

On multiplying this equation by $\mathbf{e}_1\cdot$, $\mathbf{e}_2\cdot$, $\mathbf{e}_3\cdot$ in turn, we get

(2) $\dfrac{dx}{ds} = \cos(\mathbf{e}_1, \mathbf{\tau})$, $\dfrac{dy}{ds} = \cos(\mathbf{e}_2, \mathbf{\tau})$, $\dfrac{dz}{ds} = \cos(\mathbf{e}_3, \mathbf{\tau})$.

An important special case arises when $\mathbf{r} = \mathbf{R}(\theta)$, a variable unit vector in a *plane* making an angle of θ radians with some fixed line as the x-axis.

If $\mathbf{R}(\theta)$ is drawn from the origin, the locus of its end point is a circle of unit radius (Fig. 102b) and $s = \theta$. Now (1) becomes

(3) $$d\mathbf{R}/d\theta = \mathbf{R}(\theta + \tfrac{1}{2}\pi) = \mathbf{P},$$

a unit vector perpendicular to $\mathbf{R}$ in the direction of increasing angles. Using this result we have also

(4) $$d\mathbf{P}/d\theta = \mathbf{R}(\theta + \pi) = -\mathbf{R}.$$

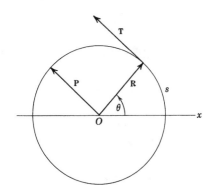

Fig. 102b.   To illustrate $d\mathbf{R}/d\theta = \mathbf{P}$.

If we put $\mathbf{R} = \mathbf{e}_1 \cos\theta + \mathbf{e}_2 \sin\theta$ in (3), we find at one stroke the derivatives of both sine and cosine:

$$\frac{d}{d\theta}\cos\theta = \cos\left(\theta + \frac{\pi}{2}\right), \qquad \frac{d}{d\theta}\sin\theta = \sin\left(\theta + \frac{\pi}{2}\right).$$

If $r = f(\theta)$ is the equation of a plane curve in polar coordinates, its vector equation is

$$\mathbf{r} = r\,\mathbf{R}(\theta);$$

hence, on differentiating with respect to s,

$$\mathbf{T} = \frac{dr}{ds}\mathbf{R} + r\frac{d\mathbf{R}}{d\theta}\frac{d\theta}{ds} = \frac{dr}{ds}\mathbf{R} + r\frac{d\theta}{ds}\mathbf{P}.$$

On multiplying by $\mathbf{R}\cdot$ and $\mathbf{P}\cdot$ we find

(5) $$\frac{dr}{ds} = \cos(\mathbf{R}, \mathbf{T}), \qquad r\frac{d\theta}{ds} = \sin(\mathbf{R}, \mathbf{T}).$$

103. Frenet's Formulas.   For a rectifiable curve

$$\mathbf{r} = \overrightarrow{OP} = \mathbf{r}(s), \qquad \frac{d\mathbf{r}}{ds} = \mathbf{T},$$

the unit tangent at P.   Since the length of $\mathbf{T}$ is constant, $d\mathbf{T}/ds$, if not zero, must be perpendicular to $\mathbf{T}$ (Theorem 100.1).   A directed line through P in the direction of $d\mathbf{T}/ds$ is called the *principal normal* of the curve at P. Let $\mathbf{N}$ denote a unit vector in the direction of the principal normal; then we may write

(1) $$\frac{d\mathbf{T}}{ds} = \kappa \mathbf{N},$$

where κ is a nonnegative scalar called the *curvature* of the curve at P.

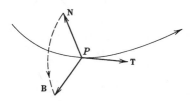

FIG. 103a.   The moving trihedral

Now $\mathbf{B} = \mathbf{T} \times \mathbf{N}$ is third unit vector such that $\mathbf{T}, \mathbf{N}, \mathbf{B}$ form a dextral set of orthogonal unit vectors at each point of the curve where $\kappa \neq 0$.   As P traverses the curve, we speak of the *moving trihedral* $\mathbf{T N B}$.   A directed line through P in the direction of $\mathbf{B}$ is called the *binormal* to the curve at P.

Since $\mathbf{B}$ is a unit vector, $d\mathbf{B}/ds$, if not zero, must be perpendicular to $\mathbf{B}$. Differentiating $\mathbf{B} = \mathbf{T} \times \mathbf{N}$, we have

$$\frac{d\mathbf{B}}{ds} = \frac{d\mathbf{T}}{ds} \times \mathbf{N} + \mathbf{T} \times \frac{d\mathbf{N}}{ds} = \mathbf{T} \times \frac{d\mathbf{N}}{ds}$$

in view of (1).   Hence $d\mathbf{B}/ds$ is perpendicular to $\mathbf{T}$ as well as $\mathbf{B}$ and therefore must be parallel to $\mathbf{N}$.   We therefore may write

(2) $$\frac{d\mathbf{B}}{ds} = -\tau \mathbf{N},$$

where τ is a scalar called the *torsion* of the curve at P.   The minus sign in (2) has this purpose: when $\tau > 0$, $d\mathbf{B}/ds$ has the direction of $-\mathbf{N}$; then, as P moves along the curve in a positive direction, $\mathbf{B}$ revolves about $\mathbf{T}$ in the same sense as a right-handed screw advancing in the direction of $\mathbf{T}$ (Fig. 103a).

We may now compute $d\mathbf{N}/ds$ by differentiating $\mathbf{N} = \mathbf{B} \times \mathbf{T}$, and, using (1) and (2):

$$(3) \qquad \frac{d\mathbf{N}}{ds} = \frac{d\mathbf{B}}{ds} \times \mathbf{T} + \mathbf{B} \times \frac{d\mathbf{T}}{ds} = -\tau\mathbf{N} \times \mathbf{T} + \kappa\mathbf{B} \times \mathbf{N}.$$

Collecting (1), (2), and (3), we have the set of equations,

$$(4) \qquad \begin{aligned} d\mathbf{T}/ds &= \qquad \kappa\mathbf{N}, \\ d\mathbf{N}/ds &= -\kappa\mathbf{T} \qquad +\tau\mathbf{B}, \\ d\mathbf{B}/ds &= \qquad -\tau\mathbf{N}, \end{aligned}$$

known as *Frenet's formulas*. They are fundamental in the theory of space curves.

From (4) we see that both κ and τ have the dimensions of the reciprocal of length; hence,

$$(5)\ (6) \qquad\qquad \rho = 1/\kappa, \qquad \sigma = 1/\tau$$

have the dimensions of length and are called the *radius of curvature* and *radius of torsion*, respectively.

Since $\mathbf{N}$, by definition, has the same direction as $d\mathbf{T}/ds$, the curvature κ is never negative. If $\kappa = 0$ (identically), $d\mathbf{T}/ds = 0$, and $\mathbf{T}$ is a unit vector of constant direction; the curve is therefore a straight line. Conversely, for a straight line $\mathbf{T}$ is constant, $d\mathbf{T}/ds = \mathbf{0}$, and $\kappa = 0$. *The only curves of zero curvature are straight lines.*

The torsion may be positive or negative. As P traverses the curve in a positive direction, the trihedral $\mathbf{T}\,\mathbf{N}\,\mathbf{B}$ will revolve about $\mathbf{T}$ as a right-handed or left-handed screw according as τ is positive or negative. The sign of τ is independent of the choice of positive direction along the curve; for, if we reverse the positive direction, we must replace

$$s, \mathbf{T}, \frac{d\mathbf{T}}{ds}, \mathbf{N}, \mathbf{B}, \frac{d\mathbf{B}}{ds} \qquad \text{by} \qquad -s, -\mathbf{T}, \frac{d\mathbf{T}}{ds}, \mathbf{N}, -\mathbf{B}, \frac{d\mathbf{B}}{ds},$$

and equations (4) maintain their form with unaltered κ and τ.

If $\tau = 0$ (identically), $d\mathbf{B}/ds = \mathbf{0}$, and $\mathbf{B}$ is a constant vector; hence, from

$$\mathbf{B} \cdot \mathbf{T} = \mathbf{B} \cdot \frac{d\mathbf{r}}{ds} = 0, \qquad \mathbf{B} \cdot (\mathbf{r} - \mathbf{r}_0) = 0,$$

and the curve lies in a plane normal to $\mathbf{B}$. Conversely, for a plane curve, $\mathbf{T}$ and $\mathbf{N}$ always lie in a fixed plane while $\mathbf{B}$ is a unit normal to that plane; hence $d\mathbf{B}/ds = \mathbf{0}$ at all points where $\mathbf{N}$ is defined ($\kappa \neq 0$) and $\tau = 0$. *The only curves of zero torsion are plane.*†

---

† For a straight line ($\kappa = 0$), $\mathbf{N}$ is not determined. We then agree to give $\mathbf{N}$ any fixed direction normal to $\mathbf{T}$ and, as before, define $\mathbf{B} = \mathbf{T} \times \mathbf{N}$. Then $\mathbf{B}$ is constant, $d\mathbf{B}/ds = \mathbf{0}$, and $\tau = 0$.

For plane curves Frenet's formulas reduce to

(7) (8) $$d\textsc{t}/ds = \kappa\textsc{n}, \qquad d\textsc{n}/ds = -\kappa\textsc{t}.$$

Since $d\textsc{t}/ds$ is always directed to the concave side of a plane curve, the same is true of $\textsc{n}$. At points of inflection $d\textsc{t}/ds = \mathbf{0}, \kappa = 0$, and $\textsc{n}$ is not defined. When P passes through a point of inflection, $\textsc{n}$ reverses its direction and $\textsc{b} = \textsc{t} \times \textsc{n}$ does the same. To remedy this discontinuous behavior of $\textsc{n}$ and $\textsc{b}$ at points of inflection, the following convention often is adopted in the differential geometry of plane curves. Take $\textsc{b}$ as a *fixed* unit vector normal to the plane of the curve, and define $\textsc{n} = \textsc{b} \times \textsc{t}$. As before, $\textsc{t}, \textsc{n}, \textsc{b}$ form a dextral set of orthogonal unit vectors. The curvature κ, defined by (7), is now positive or negative according as $d\textsc{t}/ds$ has the direction of $\textsc{n}$ or the opposite. Equation (8) still holds good; for

$$d\textsc{n}/ds = \textsc{b} \times (d\textsc{t}/ds) = \textsc{b} \times \kappa\textsc{n} = -\kappa\textsc{t}.$$

Let ψ be the angle from a fixed line in the plane to the tangent at P, taken positive in the sense determined by $\textsc{b}$.† Then, from (102.3),

$$\frac{d\textsc{t}}{ds} = \frac{d\textsc{t}}{d\psi}\frac{d\psi}{ds} = \textsc{b} \times \textsc{t}\,\frac{d\psi}{ds} = \textsc{n}\,\frac{d\psi}{ds},$$

and hence, from (1),

(9) $$\kappa = d\psi/ds, \qquad \rho = ds/d\psi.$$

If a moving particle P has the position vector $\mathbf{r} = \overrightarrow{OP}$ relative to some reference frame, its *velocity* $\mathbf{v}$ and *acceleration* $\mathbf{a}$ relative to this frame are defined as

(10) (11) $$\mathbf{v} = \frac{d\mathbf{r}}{dt}, \qquad \mathbf{a} = \frac{d\mathbf{v}}{dt} = \frac{d^2\mathbf{r}}{dt^2}.$$

If P describes a curve $\mathbf{r} = \mathbf{r}(t)$,

(12) $$\mathbf{v} = \frac{d\mathbf{r}}{ds}\frac{ds}{dt} = \textsc{t}\,\frac{ds}{dt} = v\textsc{t},$$

where $v = ds/dt$ is defined as the *speed* of the particle. *The velocity of P is a vector tangent to the path of P in the direction of motion and of length numerically equal to the speed.*

From (12) we have

$$\mathbf{a} = \frac{dv}{dt}\,\textsc{t} + v\,\frac{d\textsc{t}}{dt};$$

---

† A vector normal to a plane determines a positive sense of rotation by the right-handed screw convention. Thus, if $\textsc{b}$ points *up* from the paper, the positive sense is counterclockwise; then $\textsc{n}$ is always 90° *ahead* of $\textsc{t}$.

or, since τ may be regarded as a function of s,

$$\frac{d\mathbf{T}}{dt} = \frac{d\mathbf{T}}{ds}\frac{ds}{dt} = (\kappa\mathbf{N})v = \frac{v}{\rho}\mathbf{N},$$

(13) $$\mathbf{a} = \frac{dv}{dt}\mathbf{T} + \frac{v^2}{\rho}\mathbf{N}.$$

The acceleration of P is a vector lying in the plane of the tangent and principal normal to the path at P; its tangential and normal components are dv/dt and v^2/ρ.

Example 1. *The Circle.* A circle of radius a about the origin has the position vector $\mathbf{r} = a\,\mathbf{R}(\theta)$ where $\theta = s/a$ radians. On differentiating $\mathbf{r}$ twice with respect to s, we have

$$\mathbf{T} = \mathbf{P}, \qquad \kappa\mathbf{N} = -\mathbf{R}/a;$$

hence,

$$\mathbf{N} = -\mathbf{R}, \qquad \kappa = 1/a, \qquad \rho = a.$$

A circle has a constant curvature equal to the reciprocal of its radius.
    The only plane curves of constant nonzero curvature are circles. For from (8)

$$\frac{d\mathbf{N}}{ds} = -\kappa\frac{d\mathbf{r}}{ds} \quad \text{or} \quad \frac{d}{ds}(\rho\mathbf{N} + \mathbf{r}) = 0;$$

$$\rho\mathbf{N} + \mathbf{r} = \mathbf{c}, \qquad |\mathbf{r} - \mathbf{c}|^2 = \rho^2.$$

This is the equation of a circle of radius ρ whose center has the position vector $\mathbf{c}$.
    Example 2. *The Circular Helix.* The helix (or screw thread) on a circular cylinder of radius a has the vector equation

$$\mathbf{r} = a\mathbf{R}(\theta) + b\theta\mathbf{e} \qquad (a, b > 0),$$

where $\mathbf{e}$ is a unit vector along the axis. Now

$$\frac{d\mathbf{r}}{d\theta} = \frac{d\mathbf{r}}{ds}\frac{ds}{d\theta} = \mathbf{T}\frac{ds}{d\theta} = a\mathbf{P} + b\mathbf{e},$$ (102.3)

and hence $ds/d\theta = \sqrt{a^2 + b^2}$. Differentiating again with respect to θ,

$$\frac{d\mathbf{T}}{ds}\left(\frac{ds}{d\theta}\right)^2 = \kappa\mathbf{N}(a^2 + b^2) = -a\mathbf{R}$$ (102.4)

and, since $\kappa > 0$, $\mathbf{N} = -\mathbf{R}$ and

$$\kappa = a/(a^2 + b^2).$$

For a right-handed screw, $\mathbf{R}$, $\mathbf{P}$, $\mathbf{e}$ form a dextral, orthogonal set of unit vectors; then

$$\mathbf{B} = \mathbf{T} \times \mathbf{N} = \frac{\mathbf{R} \times (a\mathbf{P} + b\mathbf{e})}{\sqrt{a^2 + b^2}} = \frac{a\mathbf{e} - b\mathbf{P}}{\sqrt{a^2 + b^2}},$$

$$\frac{d\mathbf{B}}{d\theta} = \frac{d\mathbf{B}}{ds}\frac{ds}{d\theta} = -\tau\mathbf{N}\sqrt{a^2 + b^2} = \frac{b\mathbf{R}}{\sqrt{a^2 + b^2}},$$

and, since $\mathbf{N} = -\mathbf{R}$,

$$\tau = b/(a^2 + b^2).$$

Example 3. *Evolute.* At any point P of a plane curve where $\kappa \neq 0$, the *center of curvature* P_1 is given by

(14)
$$\mathbf{r}_1 = \mathbf{r} + \rho \mathbf{N}.$$

The locus of P_1 is called the *evolute* of the curve. If $s = AP$ and $s_1 = A_1P_1$ denote corresponding arcs on the curve Γ and its evolute Γ_1 (Fig. 103b), we have, on differentiating (14) with respect to s,

$$\mathbf{T}_1 \frac{ds_1}{ds} = \mathbf{T} + \rho \frac{d\mathbf{N}}{ds} + \frac{d\rho}{ds} \mathbf{N} = \frac{d\rho}{ds} \mathbf{N}.$$

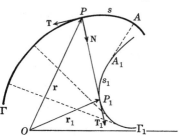

FIG. 103b.   Evolute of a plane curve

Choose the positive direction on Γ_1 so that $\mathbf{T}_1 = \mathbf{N}$; then

$$\frac{ds_1}{ds} = \frac{d\rho}{ds}, \qquad s_1 = \rho + \text{const};$$

and, since $\Delta s_1 = \Delta \rho$, an arc of the evolute is equal to the difference in the values of ρ at its end points. These properties show that a curve Γ may be traced by the end P of a taut string unwound from its evolute Γ_1; the string is always tangent to Γ_1, and its free portion is equal to ρ. From this point of view, Γ is called the *involute* of Γ_1.

104. Curvature and Torsion. By use of Frenet's formulas we may compute the curvature and torsion of a curve from its parametric equations. Thus, on differentiating $\mathbf{r} = \mathbf{r}(t)$ three times and denoting t derivatives by dots, we have

$$\dot{\mathbf{r}} = \frac{d\mathbf{r}}{ds} \frac{ds}{dt} = \dot{s} \mathbf{T},$$

$$\ddot{\mathbf{r}} = \ddot{s} \, \mathbf{T} + \dot{s}^2 \kappa \mathbf{N},$$

$$\dddot{\mathbf{r}} = \dddot{s} \, \mathbf{T} + \dot{s}\ddot{s} \, \kappa \mathbf{N} + (2\dot{s}\ddot{s} \, \kappa + \dot{s}^2 \dot{\kappa})\mathbf{N} + \dot{s}^3 \kappa(- \kappa \mathbf{T} + \tau \mathbf{B})$$

$$= (\dddot{s} - \dot{s}^3 \kappa^2)\mathbf{T} + (3\dot{s}\ddot{s} \, \kappa + \dot{s}^2 \dot{\kappa})\mathbf{N} + \dot{s}^3 \kappa \tau \, \mathbf{B}.$$

Hence,

$$\dot{\mathbf{r}} \times \ddot{\mathbf{r}} = \dot{s}^3 \kappa \mathbf{B}, \qquad \dot{\mathbf{r}} \times \ddot{\mathbf{r}} \cdot \dddot{\mathbf{r}} = \dot{s}^6 \kappa^2 \tau,$$

and, since $|\dot{\mathbf{r}}| = |\dot{s}| \neq 0$ at an ordinary point,

(1) (2)
$$\kappa = \frac{|\dot{\mathbf{r}} \times \ddot{\mathbf{r}}|}{|\dot{\mathbf{r}}|^3}, \qquad \tau = \frac{\dot{\mathbf{r}} \times \ddot{\mathbf{r}} \cdot \dddot{\mathbf{r}}}{|\dot{\mathbf{r}} \times \ddot{\mathbf{r}}|^2}.$$

If the positive direction on the curve is that of increasing t, $\dot{s} = ds/dt > 0$; and the preceding equations show that

(3) $\mathbf{T}$, $\mathbf{B}$, $\mathbf{N}$ have the directions of $\dot{\mathbf{r}}$, $\dot{\mathbf{r}} \times \ddot{\mathbf{r}}$, $(\dot{\mathbf{r}} \times \ddot{\mathbf{r}}) \times \dot{\mathbf{r}}$.

The planes through a point $\mathbf{r}(t)$ of the curve and perpendicular to $\mathbf{T}$, $\mathbf{N}$, or $\mathbf{B}$ are called, respectively, the *normal, rectifying*, and *osculating* planes to the curve.   Their equations are readily obtained from (3).   Thus, if $\mathbf{q}$ is a variable position vector to the osculating plane, its equation is

(4) $(\mathbf{q} - \mathbf{r}) \cdot \dot{\mathbf{r}} \times \ddot{\mathbf{r}} = 0.$

If the parametric equations of a plane curve are $x = x(t)$, $y = y(t)$, we have

$$\mathbf{r} = x\mathbf{e}_1 + y\mathbf{e}_2, \qquad \dot{\mathbf{r}} = \dot{x}\mathbf{e}_1 + \dot{y}\mathbf{e}_2, \qquad \ddot{\mathbf{r}} = \ddot{x}\mathbf{e}_1 + \ddot{y}\mathbf{e}_2;$$

and from (1) and (2)

(5) $\kappa = \dfrac{|\dot{x}\ddot{y} - \dot{y}\ddot{x}|}{(\dot{x}^2 + \dot{y}^2)^{3/2}}$, $\tau = 0.$

If the curve has the Cartesian equation $y = f(x)$, we can regard x as the parameter: $x = t$, $y = f(t)$.   Then κ in (5) becomes

(6) $\kappa = \dfrac{|y''|}{(1 + y'^2)^{3/2}}$,

where the primes denote x derivatives.

If the curve has the polar equation $r = f(\theta)$, we take $t = \theta$ as parameter. Then

$$\mathbf{r} = r\mathbf{R}, \qquad \dot{\mathbf{r}} = \dot{r}\mathbf{R} + r\mathbf{P}, \qquad \ddot{\mathbf{r}} = (\ddot{r} - r)\mathbf{R} + 2\dot{r}\,\mathbf{P} ;$$

and from (1)

(7) $\kappa = \dfrac{|r^2 + 2\dot{r}^2 - r\ddot{r}|}{(r^2 + \dot{r}^2)^{3/2}}.$

Example. At the point $t = 1$ of the twisted cubic

$$x = 2t, \qquad y = t^2, \qquad z = t^3/3,$$

find $\mathbf{T}$, $\mathbf{N}$, $\mathbf{B}$, κ, τ, and the equation of its osculating plane.

Since $\mathbf{r}(t) = [2t,\ t^2,\ t^3/3]$,

$$\dot{\mathbf{r}} = [2,\ 2t,\ t^2], \qquad \ddot{\mathbf{r}} = [0,\ 2,\ 2t], \qquad \dddot{\mathbf{r}} = [0,\ 0,\ 2];$$

hence, when $t = 1$,

$$\dot{\mathbf{r}} = [2, 2, 1], \qquad \ddot{\mathbf{r}} = [0, 2, 2], \qquad \dddot{\mathbf{r}} = [0, 0, 2],$$

$$\dot{\mathbf{r}} \times \ddot{\mathbf{r}} = [2, -4, 4], \qquad \dot{\mathbf{r}} \times \ddot{\mathbf{r}} \cdot \dddot{\mathbf{r}} = 8.$$

From (3)

$$= \tfrac{1}{3}[2, 2, 1], \qquad \mathbf{B} = \tfrac{1}{3}[1, -2, 2], \qquad \mathbf{N} = \tfrac{1}{3}[-2, 1, 2] \ ;$$

the osculating plane is

$$(x - 2) - 2(y - 1) + 2(z - \tfrac{1}{3}) = 0;$$

and, from (1) and (2), $\kappa = 2/9$, $\tau = 2/9$.

Since $\dot{\mathbf{r}} = \dot{s}\mathbf{T}$, the length of the curve is found from the equation

$$ds/dt = | \dot{\mathbf{r}} | = \sqrt{4 + 4t^2 + t^4}.$$

PROBLEMS

1. If r_1 and r_2 are the focal distances of a point P on an ellipse, $r_1 + r_2 = 2a$. Prove that the normal to the ellipse at P bisects the angle between the focal radii.
[Differentiate the equation with respect to s; then $(\mathbf{R}_1 + \mathbf{R}_2) \cdot \mathbf{T} = 0$ from (102.5).]
What analogous property has the hyperbola $r_1 - r_2 = 2a$? the parabola $r = x$?

2. If $\mathbf{r} = \overrightarrow{OP}$ is the position vector to the

$$\text{ellipse:} \qquad x = a \cos t, \quad y = b \sin t, \quad \text{or}$$
$$\text{hyperbola:} \quad x = a \cosh t, \quad y = b \sinh t,$$

prove that $d\mathbf{r}/dt = \mathbf{r}'$ is conjugate to $\mathbf{r}$.

3. If the proper vector $\mathbf{u}(t)$ is not parallel to a fixed line, prove that it will remain parallel to a fixed plane when and only when $\mathbf{u} \times \mathbf{u}' \cdot \mathbf{u}'' = 0$.

4. Two curves are called *parallel* if a plane normal to one at any point P is also normal to the other at a corresponding point P_1. Prove that the distance PP_1 is constant.
[$\mathbf{r}_1 = \mathbf{r} + \alpha\mathbf{N} + \beta\mathbf{B}$ where $\mathbf{r}$, $\mathbf{N}$, $\mathbf{B}$ refer to the first curve and α and β are scalars; show that $d(\alpha^2 + \beta^2)/ds = 0$.]

5. In Prob. 4, prove that $\kappa/\tau = \kappa_1/\tau_1$ at corresponding points of the curves.

6. Prove that Frenet's formulas (103.4) take the form

$$\frac{d\mathbf{T}}{ds} = \boldsymbol{\delta} \times \mathbf{T}, \qquad \frac{d\mathbf{N}}{ds} = \boldsymbol{\delta} \times \mathbf{N}, \qquad \frac{d\mathbf{B}}{ds} = \boldsymbol{\delta} \times \mathbf{B},$$

where $\boldsymbol{\delta} = \tau\mathbf{T} + \kappa\mathbf{B}$ is the *Darboux vector* of the curve.

7. A *helix* is a twisted curve whose tangent makes a constant angle α with a fixed line, its *axis*. If the unit vector $\mathbf{e}$ has the direction of the axis, the defining equation of a helix is $\mathbf{e} \cdot \mathbf{T} = \cos \alpha \ (0 < \alpha < \tfrac{1}{2}\pi)$.
For a helix prove that

(i) the principal normal is always perpendicular to the axis;

(ii) the Darboux vector $\boldsymbol{\delta} = \tau\mathbf{T} + \kappa\mathbf{B}$ is always parallel to the axis;

(iii) the ratio of curvature to torsion is constant: $\kappa/\tau = \tan \alpha$.

8. Prove that a necessary and sufficient condition that a curve $\mathbf{r} = \mathbf{r}(s)$ be a helix is that $\mathbf{r}'' \times \mathbf{r}''' \cdot \mathbf{r}^{iv} = 0$.

9. A particle moves in a circle of radius r with the angular speed ω radians per second. Prove that its speed is $r\omega$ and that its acceleration has the tangential and normal components $r \, d\omega/dt$, $r\omega^2$.

10. A particle moves along the curve

$$x = t^2, \qquad y = \tfrac{2}{3}t^3, \qquad z = t.$$

When $t = 1$, find

(*a*) its velocity **v** and acceleration **a**;

(*b*) **T**, **N**, **B** of the path;

(*c*) κ and τ of the path;

(*d*) the tangential and normal components of **a**.

11. A particle moves along a plane curve whose polar equation is $r = f(\theta)$. Prove that its velocity and acceleration are

$$\mathbf{v} = \dot{r}\mathbf{R} + r\dot{\theta}\mathbf{P}, \qquad \mathbf{a} = (\ddot{r} - r\dot{\theta}^2)\mathbf{R} + (r\ddot{\theta} + 2\dot{r}\dot{\theta})\mathbf{P},$$

where the dots denote time derivatives. [Write $\mathbf{r} = r\mathbf{R}$.]

12. If the acceleration is purely radial in Prob. 11, prove the *law of areas*: $r^2\dot{\theta} = $ const.

105. Surfaces. *A surface is an aggregate of points whose coordinates are functions of two variables.* Thus the equations

$$(1) \qquad x = x(u, v), \qquad y = y(u, v), \qquad z = z(u, v)$$

represent a surface. The variables u, v are called *parameters* or *surface coordinates*; and each pair of values u, v within a prescribed region corresponds to a definite surface point. If x, y, z are functions of $t = \varphi(u, v)$, the equations (1) will represent a *curve*. In order to exclude this case we shall require that the matrix

$$(2) \qquad \begin{pmatrix} x_u & y_u & z_u \\ x_v & y_v & z_v \end{pmatrix}$$

be of *rank two*; then at least one of its two rowed determinants,

$$(3) \qquad A = \frac{\partial(y, z)}{\partial(u, v)}, \qquad B = \frac{\partial(z, x)}{\partial(u, v)}, \qquad C = \frac{\partial(x, y)}{\partial(u, v)},$$

is not identically zero.* However, the three determinants may all vanish for certain surface points. Such points are called *singular* in contrast to the *regular points* where at least one determinant is not zero.

If $\mathbf{r} = [x, y, z]$ is the position vector to the surface, equations (1) may be replaced by a single vector equation,

$$(4) \qquad \mathbf{r} = \mathbf{r}(u, v);$$

and the matrix will be of rank two if

$$(5) \qquad \mathbf{r}_u \times \mathbf{r}_v = [A, B, C] \neq \mathbf{0}.$$

---

*In the excluded case $x_u = x_t t_u$, $x_v = x_t t_v$, etc., and A, B, C vanish identically.

If we put $u = f(t)$, $v = g(t)$ in (1), we obtain a curve on the surface. The curves $v = b$, $u = a$ are called the *parametric curves* of the surface through the point $u = a$, $v = b$; their vector equations are

$$\mathbf{r} = \mathbf{r}(u, b), \qquad \mathbf{r} = \mathbf{r}(a, v).$$

Their tangent vectors are $\mathbf{r}_u$ and $\mathbf{r}_v$ and at any regular point $\mathbf{r}_u \times \mathbf{r}_v \neq \mathbf{0}$; that is, the tangents are not parallel and determine a normal vector $\mathbf{r}_u \times \mathbf{r}_v$ to the surface. Hence the surface has a unique normal at every regular point.

Example 1. The equations

(6) $x = a \sin u \cos v, \qquad y = a \sin u \sin v, \qquad z = a \cos u$

represent a sphere of radius a about the origin. Referring to Ex. 87.2, we see that $u = \theta$, the colatitude measured from the $+z$-pole, while $v = \varphi$ is the longitude measured from the xz-plane. The parametric curves $u = c$, $v = b$ are parallels of latitude and meridians, respectively. The matrix (2) is now

$$\begin{pmatrix} a \cos u \cos v & a \cos u \sin v & -a \sin u \\ -a \sin u \sin v & a \sin u \cos v & 0 \end{pmatrix},$$

and hence

$$\mathbf{r}_u \times \mathbf{r}_v = a^2 \sin u[\sin u \cos v, \sin u \sin v, \cos u].$$

The vector in brackets is normal to the sphere and indeed a *unit normal*:

$$\mathbf{n} = [\sin u \cos v, \sin u \sin v, \cos u].$$

Since $\mathbf{r}_u \times \mathbf{r}_v = 0$ when $\sin u = 0$, the poles $u = 0$ and $u = \pi$ are singular points; these points however are not intrinsic singularities but are due to the parametric representation; in fact, the unit normals at the poles are correctly given by $\mathbf{n}$.

At a regular point u_0, v_0 one of the Jacobians (3) is not zero. If $\partial(x, y)/\partial(u, v) \neq 0$, we may solve the first two equations of (1) for u and v in terms of x and y (§ 85); and, on substituting these values in the third equation, we obtain z as a function of x and y:

(7) $z = f(x, y).$

This equation also represents the surface which now has the parametric form,

(7)' $x = u, \qquad y = v, \qquad z = f(u, v).$

Finally a surface may be given by the equation

(8) $F(x, y, z) = 0.$

This equation may be solved for z in the neighborhood of any point for which $F_z \neq 0$; we then again obtain an equation of the form (7).

Any three functions,

(9) $x = x(t), \qquad y = y(t), \qquad z = z(t),$

which reduce (8) to an identity in t, correspond to a surface curve having the equations (9). If we substitute from (9) in (8) and differentiate the resulting equation with respect to t, we obtain

$$(10) \qquad F_x x_t + F_y y_t + F_z z_t = 0.$$

Now $[x_t, y_t, z_t]$ represents a tangent vector to curve (9) at a point $P(t)$; and, since (10) holds for *all* surface curves through this point, the vector

$$(11) \qquad \nabla F = [F_x, F_y, F_z]$$

is normal to all surface curves through P and hence to the surface $F = 0$ itself. This vector, denoted by ∇F, is called the *gradient* of the function $F(x, y, z)$.

If $Q(x, y, z)$ is any point on the tangent plane to the surface (8) at the point $P_0(x_0, y_0, z_0)$, the vector $\overrightarrow{P_0 Q}$ is perpendicular to ∇F at P_0; hence the equation of the tangent plane to the surface at P_0 is

$$(12) \quad F_x(x_0, y_0, z_0)(x - x_0) + F_y(x_0, y_0, z_0)(y - y_0)$$
$$+ F_z(x_0, y_0, z_0)(z - z_0) = 0.$$

Example 2. The ellipsoid

$$(13) \qquad F = \frac{x^2}{a^2} + \frac{y^2}{b^2} + \frac{z^2}{c^2} - 1 = 0$$

has the gradient

$$\nabla F = 2[x/a^2, y/b^2, z/c^2].$$

The equation of a tangent plane at the point (x_0, y_0, z_0) is therefore

$$\frac{x_0}{a^2}(x - x_0) + \frac{y_0}{b^2}(y - y_0) + \frac{z_0}{c^2}(z - z_0) = 0,$$

or, since (x_0, y_0, z_0) satisfies (13),

$$\frac{x_0 x}{a^2} + \frac{y_0 y}{b^2} + \frac{z_0 z}{c^2} - 1 = 0.$$

Example 3. If a surface is given by equations (1), a functional relation between u and v, say $w(u, v) = 0$, determines a curve on the surface. Along this curve a tangent vector has the components

$$\frac{dx}{du} = x_u + x_v \frac{dv}{du}, \qquad \frac{dy}{du} = y_u + y_v \frac{dv}{du}, \qquad \frac{dz}{du} = z_u + z_v \frac{dv}{du}$$

where dv/du satisfies $dw = w_u du + w_v dv = 0$. Hence,

$$\frac{dv}{du} = -\frac{w_u}{w_v},$$

and with this value we find that

$$(14) \qquad \frac{dx}{du} : \frac{dy}{du} : \frac{dz}{du} = \frac{\partial(x, w)}{\partial(u, v)} : \frac{\partial(y, w)}{\partial(u, v)} : \frac{\partial(z, w)}{\partial(u, v)}$$

Thus the three Jacobians in (14) are the components of the tangent vector to the curve defined by $w(u, v) = 0$. Since the Jacobians A, B, C in (3) are the components of a surface normal, we have the identity

(15) $$\frac{\partial(y, z)}{\partial(u, v)}\frac{\partial(x, w)}{\partial(u, v)} + \frac{\partial(z, x)}{\partial(u, v)}\frac{\partial(y, w)}{\partial(u, v)} + \frac{\partial(x, y)}{\partial(u, v)}\frac{\partial(z, w)}{\partial(u, v)} = 0$$

satisfied by any four differentiable functions x, y, z, w of u and v.

106. Directional Derivative. A scalar or vector which is uniquely defined at every point of a region is said to be a *point function*. If the point $P(x, y, z)$ is given by the position vector $\mathbf{r} = \overrightarrow{OP}$, we denote a scalar point function by $f(\mathbf{r})$ or $f(x, y, z)$, a vector point function by $\mathbf{f}(\mathbf{r})$ or $\mathbf{f}(x, y, z)$.

Consider now a scalar point function $f(x, y, z)$ in neighborhood of a point P_0 where it is continuous and differentiable. A ray (or half-line) through P_0 in the direction of the unit vector,

$$\mathbf{e} = [\cos\alpha, \cos\beta, \cos\gamma],$$

has the vector equation,

$$\mathbf{r} = \mathbf{r}_0 + s\mathbf{e} \quad \text{where} \quad s = |\overrightarrow{P_0P}| > 0.$$

Hence the parametric equations of the ray are

(1) $\quad x = x_0 + s\cos\alpha, \qquad y = y_0 + s\cos\beta, \qquad z = z_0 + s\cos\gamma.$

Along this ray $f(x, y, z)$ is a function of s alone whose right-hand derivative may be computed by the chain rule (80.6):

$$\frac{df}{ds} = \frac{\partial f}{\partial x}\frac{dx}{ds} + \frac{\partial f}{\partial y}\frac{dy}{ds} + \frac{\partial f}{\partial z}\frac{dz}{ds};$$

or, in view of equations (1),

(2) $$\frac{df}{ds} = f_x\cos\alpha + f_y\cos\beta + f_z\cos\gamma.\dagger$$

If f_x, f_y, f_z are computed at the point (x_0, y_0, z_0), formula (2) gives the *directional derivative* of $f(x, y, z)$ at P_0 in the direction of $\mathbf{e}$.

---

† For a function $f(x, y)$ of two variables the direction $\mathbf{e}$ in the xy-plane has the direction cosines,

$$\cos\alpha, \quad \cos(\tfrac{1}{2}\pi - \alpha) = \sin\alpha, \quad \cos\gamma = 0; \quad \text{and}$$

(2)′ $$\frac{df}{ds} = f_x\cos\alpha + f_y\sin\alpha.$$

Since the direction cosines of $\mathbf{e}$ are equal to $\mathbf{e} \cdot \mathbf{e}_1$, $\mathbf{e} \cdot \mathbf{e}_2$, $\mathbf{e} \cdot \mathbf{e}_3$, we may also write (2) in the form

$$(3) \qquad \frac{df}{ds} = \mathbf{e} \cdot (\mathbf{e}_1 f_x + \mathbf{e}_2 f_y + \mathbf{e}_3 f_z),$$

in which the first factor indicates the direction of the derivative, while the second, the *gradient*

$$(4) \qquad \nabla f = \mathbf{e}_1 f_x + \mathbf{e}_2 f_y + \mathbf{e}_3 f_z \qquad\qquad (105.10),$$

depends only on the point where ∇f is computed. Thus the formula

$$(5) \qquad \frac{df}{ds} = \mathbf{e} \cdot \nabla f = |\nabla f| \cos (\nabla f, \mathbf{e})$$

sets up a one-to-one correspondence between directions $\mathbf{e}$ at a point and df/ds, the directional derivatives there. In effect, the *vector* ∇f replaces the infinity of *scalars* df/ds.

From (5) we see that, at a given point P_0, the maximum value of df/ds is assumed when $\mathbf{e} = \mathbf{n}$, a unit vector in the direction of ∇f, and this maximum is $|\nabla f|$. Moreover, ∇f at P_0 is normal to the level surface

$$(6) \qquad f(x, y, z) = \text{const} = f(x_0, y_0, z_0)$$

passing through P_0 (§ 105). Thus the *direction* of ∇f at P_0 is normal to the level surface (6) and pointing in the direction of increasing f; and its magnitude,

$$(7) \qquad |\nabla f| = \mathbf{n} \cdot \nabla f = \frac{df}{dn},$$

is the derivative of f in this direction. Hence, in terms of this *normal derivative* we have an expression for the gradient

$$(8) \qquad \nabla f = \mathbf{n} \frac{df}{dn}$$

entirely independent of the coordinate system. Since

$$(9) \qquad \nabla x = \mathbf{e}_1, \qquad \nabla y = \mathbf{e}_2, \qquad \nabla z = \mathbf{e}_3,$$

the equations

$$(10) \qquad \nabla f = \nabla x f_x + \nabla y f_y + \nabla z f_z,$$

$$(11) \qquad |\nabla f|^2 = f_x^2 + f_y^2 + f_z^2$$

hold in any system of rectangular coordinates.

If we replace x, y, z by functions of other independent variables u, v, w $f(x, y, z)$ becomes $F(u, v, w)$. Along the ray (1)

$$\frac{dF}{ds} = \frac{du}{ds}\frac{\partial F}{\partial u} + \frac{dv}{ds}\frac{\partial F}{\partial v} + \frac{dw}{ds}\frac{\partial F}{\partial w},$$

or in view of (5)

$$\mathbf{e} \cdot \nabla F = \mathbf{e} \cdot (\nabla u\, F_u + \nabla v\, F_v + \nabla w\, F_w).$$

But, since this holds for all vectors $\mathbf{e}$,

(12) $$\nabla F = \nabla u\, F_u + \nabla v\, F_v + \nabla w\, F_w.$$

This expression for the gradient in curvilinear coordinates u, v, w obviously includes (10).

Example 1. The polar distance,

$$OP = r = \sqrt{x^2 + y^2 + z^2},$$

has the gradient

(13) $$\nabla r = [x/r, y/r, z/r] = \mathbf{r}/r,$$

the unit radial vector in the direction OP. This result also follows from the intrinsic definition (8); for the level surfaces of r are spheres about the origin for which

$$\mathbf{n} = \frac{\mathbf{r}}{r}, \qquad \frac{dr}{dn} = \frac{dr}{dr} = 1.$$

Example 2. If the point P has the plane-polar coordinates r, φ, we have

(14) $$\nabla r = \mathbf{R}, \qquad \nabla \varphi = \mathbf{P}/r,$$

where $\mathbf{P}$ is the unit vector 90° ahead of $\mathbf{R}$ (§ 102). These results follow at once from (8) For example, the level curves of φ are rays through the origin for which

$$\mathbf{n} = \mathbf{P}, \qquad \frac{d\varphi}{dn} = \lim_{\Delta\varphi \to 0} \frac{\Delta\varphi}{r \tan\Delta\varphi} = \frac{1}{r}.$$

If $f(r, \varphi) = c$ is the equation of a curve in polar coordinates, a normal to the curve at any point is given by

(15) $$\nabla f(r, \varphi) = f_r\, \nabla r + f_\varphi\, \nabla \varphi = f_r\, \mathbf{R} + \frac{f_\varphi}{r}\, \mathbf{P}.$$

Example 3. *The Law of Refraction.* A ray of light leaves the point P_1 in a medium in which the velocity of light is v_1, penetrates a second medium in which the velocity is v_2, and arrives at a point P_2 (Fig. 106). If the surface of separation between the media is $f(x, y, z) = 0$, what rectilinear path P_1PP_2 (where P is a point on the surface) must the light pursue in order to go from P_1 to P_2 in the least time?

If we write

$$r_i = P_iP = \sqrt{(x - x_i)^2 + (y - y_i)^2 + (z - z_i)^2} \qquad (i = 1, 2),$$

the problem is to minimize the time

$$T = \frac{r_1}{v_1} + \frac{r_2}{v_2} \qquad \text{when} \qquad f(x, y, z) = 0.$$

The necessary conditions for a minimum, found from the Lagrangian function

$$L = T + \alpha f, \quad \text{are} \quad L_x = L_y = L_z = 0$$

or simply

$$\nabla L = \frac{\nabla r_1}{v_1} + \frac{\nabla r_2}{v_2} + \alpha \nabla f = 0.$$

From Ex. 1, $\nabla r_1, = R_1, \nabla r_2 = R_2$ where R_1, R_2 are unit vectors along $P_1 P$ and $P_2 P$; hence,

(i) $$\frac{R_1}{v_1} + \frac{R_2}{v_2} + \alpha n \mid \nabla f \mid = 0,$$

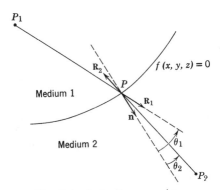

FIG. 106.   Refraction at a surface

where n is a *unit* normal to the surface $f(x, y, z) = 0$ at P. This equation shows that the vectors R_1, R_2, n are coplanar; hence the incident ray, the refracted ray, and the surface normal lie in a plane. Moreover, if we multiply (i) by $n \times$, we get

$$\frac{n \times R_1}{v_1} + \frac{n \times R_2}{v_2} = 0 \quad \text{or} \quad \frac{\sin \theta_1}{v_1} - \frac{\sin \theta_2}{v_2} = 0$$

in view of (98.3). This is *Snell's* law:

$$\frac{\sin \theta_1}{\sin \theta_2} = \frac{v_1}{v_2} = \text{refractive index};$$

θ_1 is the *angle of incidence*, θ_2 the *angle of refraction*.

PROBLEMS

1. If $u = r$, $v = \varphi$ are polar coordinates in the xy-plane (Ex. 87.1), show that the *surface of revolution* obtained by revolving the curve $z = f(x)$, $y = 0$ about the z-axis has the parametric equations:

$$x = u \cos v, \quad y = u \sin v, \quad z = f(u).$$

What are the parametric curves? Find the unit normal vector to the surface.

2. A straight line, which always cuts the z-axis at right angles, is revolved about and moved along this axis. The surface thus generated is called a *conoid*. If $u = r, v = \varphi$ are polar coordinates in the xy-plane, and the height of the line is given in terms of the angle turned, $z = f(v)$, show that the conoid has the parametric equations,

$$x = u \cos v, \qquad y = u \sin v, \qquad z = f(v).$$

In particular when dz/dv is constant the conoid is a *right helicoid* (a spiral ramp):

$$x = u \cos v, \qquad y = u \sin v, \qquad z = av.$$

What are the parametric curves?

3. The parametric equations of a space curve C are

$$x = f(u), \qquad y = g(u), \qquad z = h(u).$$

What surfaces are represented by the equations, and what meaning has v in each case:

(i) $x = f(u) + v \cos \alpha$, $\qquad y = g(u) + v \cos \beta$, $\qquad z = h(u) + v \cos \gamma$;

(ii) $x = vf(u)$, $\qquad\qquad y = vg(u)$, $\qquad\qquad z = vh(u)$?

What are the parametric curves $u = $ const.?

4. Give parametric equations for the ellipsoid (105.13) patterned after equations (105.6) for the sphere.

5. Show that the equations

$$x = u + av, \qquad y = f(u) + bv, \qquad z = cv$$

represent all cylinders whose rulings are parallel to the vector $[a, b, c]$. What are the parametric curves?

6. Show that the equations

$$x = uv + a, \quad y = vf(u) + b, \qquad z = v + c$$

represent all cones whose vertices are at the point (a, b, c). What are the rulings?

7. If $f(x, y) = x^2 - 2xy + y^2$, find

(a) ∇f at $(2, 3)$;

(b) df/ds at $(2, 3)$ along a ray inclined $+45°$ to the x-axis;

(c) the direction and magnitude of df/ds at $(2, 3)$ when its value is greatest.

8. If $f(x, y, z) = xy + yz + zx$, find

(a) ∇f at $(1, 1, 3)$;

(b) df/ds at $(1, 1, 3)$ in the direction of $[1, 1, 1]$;

(c) the normal derivative df/dn at $(1, 1, 3)$;

(d) the equations of the tangent plane and normal line to the surface $xy + yz + zx = 7$ at $(1, 1, 3)$.

9. If r, φ are plane-polar coordinates, show that

(a) $\nabla f(r, \varphi) = f_r(r, \varphi)\, \mathbf{R} + \dfrac{f_\varphi(r, \varphi)}{r}\, \mathbf{P}$;

(b) $(df/dn)^2 = f_r^2 + f_\varphi^2/r^2$. $\qquad$ [Ex. 106.2].

10. If $u(x, y)$ and $v(x, y)$ are functions that satisfy the equations $u_x = v_y$, $u_y = -v_x$, show that

(a) $\nabla u = \nabla v \times \mathbf{e}_3$, $\qquad\qquad \nabla v = \mathbf{e}_3 \times \nabla u$;

(b) at any point (x, y), $du/ds = dv/ds'$ where s' is a ray $90°$ in advance of s.

11. If $r = \sqrt{x^2 + y^2 + z^2}$, show that

$$\nabla r^2 = 2\mathbf{r}, \qquad \nabla (1/r) = -\mathbf{r}/r^3, \qquad \nabla \log r = \mathbf{r}/r^2, \qquad \nabla (\mathbf{a} \cdot \mathbf{r}) = \mathbf{a}.$$

12. Show that the systems of equilateral hyperbolas $x^2 - y^2 = a$, $xy = b$ cut at right angles.

[The normal to the curve $f(x, y) = c$ is parallel to ∇f.]

13. Show that the systems of cardioids $r = a(1 - \cos \varphi)$, $r = b(1 + \cos \varphi)$ cut at right angles. [Cf. (106.15).]

14. Show that the curve $f(r_1, r_2) = c$ in bipolar coordinates has a normal parallel to $f_{r_1}\mathbf{R}_1 + f_{r_2}\mathbf{R}_2$.

Show that the systems of confocal ellipses and hyperbolas $r_1 + r_2 = a$, $r_1 - r_2 = b$ cut at right angles.

15. Prove that the systems of coaxal circles $r_1/r_2 = a$, $\varphi_2 - \varphi_1 = b$ cut at right angles.

107. Gradient of a Vector. A vector point function $\mathbf{f}(\mathbf{r})$ is given by three scalar point functions f_1, f_2, f_3 which form its rectangular components:

$$(1) \qquad\qquad \mathbf{f}(\mathbf{r}) = f_1\mathbf{e}_1 + f_2\mathbf{e}_2 + f_3\mathbf{e}_3.$$

If these scalar functions are differentiable, we say that $\mathbf{f}(\mathbf{r})$ is differentiable.

If $\mathbf{f}(\mathbf{r})$ is differentiable, we may compute its directional derivative at P_0 in the direction $\mathbf{e}$ just as in § 106.

$$(2) \qquad\qquad \frac{d\mathbf{f}}{ds} = \cos \alpha \, \mathbf{f}_x + \cos \beta \, \mathbf{f}_y + \cos \gamma \, \mathbf{f}_z.$$

If we replace the direction cosines by $\mathbf{e} \cdot \mathbf{e}_1$, $\mathbf{e} \cdot \mathbf{e}_2$, $\mathbf{e} \cdot \mathbf{e}_3$, we may write symbolically

$$(3) \qquad\qquad \frac{d\mathbf{f}}{ds} = \mathbf{e} \cdot (\mathbf{e}_1\mathbf{f}_x + \mathbf{e}_2\mathbf{f}_y + \mathbf{e}_3\mathbf{f}_z).$$

Here the quantity in parenthesis has the same form as the gradient of a scalar. But, since $\mathbf{f}$ is a *vector*, the *gradient*

$$(4) \qquad\qquad \nabla \mathbf{f} = \mathbf{e}_1\mathbf{f}_x + \mathbf{e}_2\mathbf{f}_y + \mathbf{e}_3\mathbf{f}_z$$

has as yet no meaning; for it consists of the sum of three pairs of juxtaposed vectors. An expression of this sort is called a *dyadic*, while the ordered vector pairs are called *dyads*. We could now develop an algebra of dyadics by defining equivalence, addition, and multiplication†; but for our purpose this is unnecessary. We shall merely regard $\nabla \mathbf{f}$ defined by (4) as an *operator* that sets up a one-to-one correspondence between directions $\mathbf{e}$ at a point and $d\mathbf{f}/ds$, the directional derivatives there. In effect, the *dyadic* $\nabla \mathbf{f}$ replaces the infinity of *vectors* $d\mathbf{f}/ds$.

---

† See Brand, L., *Vector and Tensor Analysis*, Wiley, 1948, Chapter IV.

We shall regard two dyadic operators as equal when both transform an arbitrary vector in exactly the same way. Thus, if **P** and **Q** are dyadics,

$$\mathbf{P} = \mathbf{Q} \quad \text{when, and only when,} \quad \mathbf{u} \cdot \mathbf{P} = \mathbf{u} \cdot \mathbf{Q}$$

for any vector **u**.

From (3) and (4) we now have

(5)
$$\frac{d\mathbf{f}}{ds} = \mathbf{e} \cdot \nabla \mathbf{f}.$$

When **f** is the position vector,

$$\mathbf{r} = x\mathbf{e}_1 + y\mathbf{e}_2 + z\mathbf{e}_3,$$

$$\mathbf{r}_x = \mathbf{e}_1, \qquad \mathbf{r}_y = \mathbf{e}_2, \qquad \mathbf{r}_z = \mathbf{e}_3 \qquad \text{and}$$

(6)
$$\nabla \mathbf{r} = \mathbf{e}_1 \mathbf{e}_1 + \mathbf{e}_2 \mathbf{e}_2 + \mathbf{e}_3 \mathbf{e}_3 = \mathbf{I}.$$

The dyadic **I** is called the *unit dyadic* (or *idemfactor*) because it transforms any vector **u** into itself:

(7)
$$\mathbf{u} = \mathbf{u} \cdot \mathbf{I} \tag{97.6}.$$

108. Invariants of a Dyadic. By definition the dyadic equation $\mathbf{P} = \mathbf{Q}$ is equivalent to the vector equations,

(1)
$$\mathbf{u} \cdot \mathbf{P} = \mathbf{u} \cdot \mathbf{Q} \quad \text{for every } \mathbf{u},$$

Hence it is also equivalent to the scalar equations,

(2)
$$\mathbf{u} \cdot \mathbf{P} \cdot \mathbf{v} = \mathbf{u} \cdot \mathbf{Q} \cdot \mathbf{v} \quad \text{for any } \mathbf{u}, \mathbf{v};$$

or to the vector equations,

(3)
$$\mathbf{P} \cdot \mathbf{v} = \mathbf{Q} \cdot \mathbf{v} \quad \text{for any } \mathbf{v}.$$

In these equations **u** is used as a *prefactor*, **v** as a *postfactor*.

From (1) and (3) we have the dyadic distributive laws:

(4) (5) $\quad (\mathbf{a} + \mathbf{b})\mathbf{c} = \mathbf{ac} + \mathbf{bc}, \qquad \mathbf{a}(\mathbf{b} + \mathbf{c}) = \mathbf{ab} + \mathbf{ac}.$

Thus (4) follows at once from

$$\mathbf{u} \cdot (\mathbf{a} + \mathbf{b})\mathbf{c} = \mathbf{u} \cdot (\mathbf{ac} + \mathbf{bc}).$$

Using these laws we may express a dyadic $\mathbf{P} = \sum \mathbf{a}_i \mathbf{b}_i$ in various forms by substituting vector sums for $\mathbf{a}_i$, $\mathbf{b}_i$, and expanding or collecting terms. In all these changes, however, there are certain quantities formed from the vectors of the dyadic which remain the same. Among these *invariants* of **P**, the most important are the scalar and the vector

(6) (7) $\quad\quad P_s = \sum \mathbf{a}_i \cdot \mathbf{b}_i, \qquad \mathbf{p} = \sum \mathbf{a}_i \times \mathbf{b}_i,$

obtained by placing a dot or a cross between the vectors of each dyad in P.

THEOREM 1. $\sum \mathbf{a}_i \cdot \mathbf{b}_i$ *and* $\sum \mathbf{a}_i \times \mathbf{b}_i$ *are invariants of the dyadic* $\sum \mathbf{a}_i \mathbf{b}_i$.

Proof.   Let $\mathbf{P} = \mathbf{Q}$.   From (97.1),

$$\mathbf{u} \cdot \mathbf{v} = (\mathbf{e}_1 \cdot \mathbf{u})(\mathbf{v} \cdot \mathbf{e}_1) + (\mathbf{e}_2 \cdot \mathbf{u})(\mathbf{v} \cdot \mathbf{e}_2) + (\mathbf{e}_3 \cdot \mathbf{u})(\mathbf{v} \cdot \mathbf{e}_3);$$

hence

$$P_s = \sum_{j=1}^{3} \mathbf{e}_j \cdot \mathbf{P} \cdot \mathbf{e}_j, \qquad Q_s = \sum_{j=1}^{3} \mathbf{e}_j \cdot \mathbf{Q} \cdot \mathbf{e}_j.$$

Since the corresponding terms in these sums are equal by (2), $P_s = Q_s$. Again, from (98.1),

$$(\mathbf{u} \times \mathbf{v})_1 = u_2 v_3 - u_3 v_2 = \mathbf{e}_2 \cdot (\mathbf{uv}) \cdot \mathbf{e}_3 - \mathbf{e}_3 \cdot (\mathbf{uv}) \cdot \mathbf{e}_2;$$

hence the first components of $\mathbf{p}$ and $\mathbf{q}$ are

$$p_1 = \mathbf{e}_2 \cdot \mathbf{P} \cdot \mathbf{e}_3 - \mathbf{e}_3 \cdot \mathbf{P} \cdot \mathbf{e}_2, \qquad q_1 = \mathbf{e}_2 \cdot \mathbf{Q} \cdot \mathbf{e}_3 - \mathbf{e}_3 \cdot \mathbf{Q} \cdot \mathbf{e}_2.$$

Since the corresponding terms on the right are equal by (2), $p_1 = q_1$; similarly $p_2 = q_2$, $p_3 = q_3$, and hence $\mathbf{p} = \mathbf{q}$.

109. Divergence and Rotation.   If we apply the results of § 108 to the gradient of a vector,

(1) $$\nabla \mathbf{f} = \mathbf{e}_1 \, \mathbf{f}_x + \mathbf{e}_2 \, \mathbf{f}_y + \mathbf{e}_3 \, \mathbf{f}_z,$$

we obtain the invariants

(2) $$\nabla \cdot \mathbf{f} = \mathbf{e}_1 \cdot \mathbf{f}_x + \mathbf{e}_2 \cdot \mathbf{f}_y + \mathbf{e}_3 \cdot \mathbf{f}_z,$$

(3) $$\nabla \times \mathbf{f} = \mathbf{e}_1 \times \mathbf{f}_x + \mathbf{e}_2 \times \mathbf{f}_y + \mathbf{e}_3 \times \mathbf{f}_z,$$

known as the *divergence* and *rotation* (or *curl*) of $\mathbf{f}$.   The notation on the left employs the vector operator *nabla*:

(4) $$\nabla = \mathbf{e}_1 \frac{\partial}{\partial x} + \mathbf{e}_2 \frac{\partial}{\partial y} + \mathbf{e}_3 \frac{\partial}{\partial z} .$$

Other notations in common use are

$$\nabla \cdot \mathbf{f} = \operatorname{div} \mathbf{f}, \qquad \nabla \times \mathbf{f} = \operatorname{rot} \mathbf{f} = \operatorname{curl} \mathbf{f}.$$

If $\mathbf{f}$ is resolved into its rectangular components,

$$\mathbf{f} = f_1 \mathbf{e}_1 + f_2 \mathbf{e}_2 + f_3 \mathbf{e}_3,$$

we have from (2) and (3)

$$\nabla \cdot \mathbf{f} = \mathbf{e}_1 \cdot \left(\frac{\partial f_1}{\partial x} \mathbf{e}_1 + \frac{\partial f_2}{\partial x} \mathbf{e}_2 + \frac{\partial f_3}{\partial x} \mathbf{e}_3 \right) + \cdots ,$$

(5) $$\operatorname{div} \mathbf{f} = \frac{\partial f_1}{\partial x} + \frac{\partial f_2}{\partial y} + \frac{\partial f_3}{\partial z} \, ;$$

$$\nabla \times \mathbf{f} = \mathbf{e}_1 \times \left(\frac{\partial f_1}{\partial x} \mathbf{e}_1 + \frac{\partial f_2}{\partial x} \mathbf{e}_2 + \frac{\partial f_3}{\partial x} \mathbf{e}_3 \right) + \cdots ,$$

(6) $$\operatorname{rot} \mathbf{f} = \left(\frac{\partial f_3}{\partial y} - \frac{\partial f_2}{\partial z} \right) \mathbf{e}_1 + \left(\frac{\partial f_1}{\partial z} - \frac{\partial f_3}{\partial x} \right) \mathbf{e}_2 + \left(\frac{\partial f_2}{\partial x} - \frac{\partial f_1}{\partial y} \right) \mathbf{e}_3;$$

in which terms 2, 3, 1 are cyclic permutations of terms 1, 2, 3. This result is easily remembered by forming the "product" $\nabla \times \mathbf{f}$ or by expanding the symbolic determinant

(7)
$$\operatorname{rot} \mathbf{f} = \begin{vmatrix} \mathbf{e}_1 & \mathbf{e}_2 & \mathbf{e}_3 \\ \partial/\partial_x & \partial/\partial_y & \partial/\partial_z \\ f_1 & f_2 & f_3 \end{vmatrix},$$

according to the elements of its first row.

In particular, when $\mathbf{f} = \mathbf{r} = x\mathbf{e}_1 + y\mathbf{e}_2 + z\mathbf{e}_3$,

(8) (9) $\operatorname{div} \mathbf{r} = 3,$ $\operatorname{rot} \mathbf{r} = \mathbf{0}.$

These results also follow from (107.6).

If in (2) we put $\mathbf{f}$ equal to the gradient of a scalar,

$$\nabla \varphi = \mathbf{e}_1 \varphi_x + \mathbf{e}_2 \varphi_y + \mathbf{e}_3 \varphi_z,$$

we have

(10) $\nabla^2 \varphi = \varphi_{xx} + \varphi_{yy} + \varphi_{zz},$

where ∇^2 denotes the symbolic product $\nabla \cdot \nabla$. This invariant, of fundamental importance in mathematical physics, is called the *Laplacian* of φ; and the equation

(11) $\nabla^2 \varphi = 0$

is known as *Laplace's equation*.

We finally note two important identities. If the scalar $\varphi(x, y, z)$ and vector $\mathbf{f}(x, y, z)$ have continuous second partial derivatives,

(12) $\operatorname{rot} \nabla \varphi = \mathbf{0},$

(13) $\operatorname{div} \operatorname{rot} \mathbf{f} = 0.$

The proofs follow at once from the equality of mixed derivatives (Theorem 82.1).

$$\operatorname{rot} \nabla \varphi = \begin{vmatrix} \mathbf{e}_1 & \mathbf{e}_2 & \mathbf{e}_3 \\ \partial/\partial x & \partial/\partial y & \partial/\partial z \\ \varphi_x & \varphi_y & \varphi_z \end{vmatrix} = \mathbf{0},$$

$$\operatorname{div} \operatorname{rot} \mathbf{f} = \frac{\partial}{\partial x}\left(\frac{\partial f_3}{\partial y} - \frac{\partial f_2}{\partial z}\right) + \cdots = 0.$$

PROBLEMS

1. If λ and $\mathbf{f}$ are point functions, prove that

$$\nabla(\lambda \mathbf{f}) = (\nabla \lambda)\mathbf{f} + \lambda \nabla \mathbf{f},$$
$$\operatorname{div}(\lambda \mathbf{f}) = (\nabla \lambda) \cdot \mathbf{f} + \lambda \operatorname{div} \mathbf{f},$$
$$\operatorname{rot}(\lambda \mathbf{f}) = (\nabla \lambda) \times \mathbf{f} + \lambda \operatorname{rot} \mathbf{f}.$$

2. If **f** and **g** are point functions, prove that

$$\nabla(\mathbf{f} \cdot \mathbf{g}) = \mathbf{g} \cdot \nabla \mathbf{f} + \mathbf{f} \cdot \nabla \mathbf{g},$$
$$\operatorname{div}(\mathbf{f} \times \mathbf{g}) = \mathbf{g} \cdot \operatorname{rot} \mathbf{f} - \mathbf{f} \cdot \operatorname{rot} \mathbf{g},$$
$$\operatorname{rot}(\mathbf{f} \times \mathbf{g}) = \mathbf{g} \cdot \nabla \mathbf{f} - \mathbf{f} \cdot \nabla \mathbf{g} + \mathbf{f}\operatorname{div}\mathbf{g} - \mathbf{g}\operatorname{div}\mathbf{f}.$$

Specialize these results when $\mathbf{f} = \mathbf{a}$ (const), $\mathbf{g} = \mathbf{r}$.

3. If R, P are the unit vectors of § 102 in the xy-plane, prove that

$$\operatorname{div}\mathbf{R} = \frac{1}{r}, \qquad \operatorname{rot}\mathbf{R} = \mathbf{0}; \qquad \operatorname{div}\mathbf{P} = 0, \qquad \operatorname{rot}\mathbf{P} = \frac{\mathbf{e_3}}{r}.$$

4. If $\mathbf{r}/r$ is the unit radial vector in space, prove that

$$\operatorname{div}\frac{\mathbf{r}}{r} = \frac{2}{r}, \qquad \operatorname{rot}\frac{\mathbf{r}}{r} = \mathbf{0}.$$

5. If u, v are scalar point functions, prove that

$$\operatorname{div}(\nabla u \times \nabla v) = 0.$$

6. Find the divergence and rotation of

$$\mathbf{f} = [x - y, \, y - z, \, z - x]; \qquad \mathbf{g} = [x^2 + yz, \, y^2 + zx, \, z^2 + xy].$$

7. If u is a scalar point function, prove that

$$\operatorname{rot}(f(u)\nabla u) = \mathbf{0} \quad \text{and hence} \quad \operatorname{rot}(f(r)\mathbf{r}) = \mathbf{0}.$$

8. If $r = \sqrt{x^2 + y^2 + z^2}$, prove that

$$\nabla^2 f(r) = f''(r) + \frac{2f'(r)}{r}.$$

110. Reciprocal Bases. Three noncoplanar vectors $\mathbf{a}_1$, $\mathbf{a}_2$, $\mathbf{a}_3$ are said to form a *basis*; then $\mathbf{a}_1 \times \mathbf{a}_2 \cdot \mathbf{a}_3 \neq 0$. Two bases, $\mathbf{a}_1, \mathbf{a}_2, \mathbf{a}_3$ and $\mathbf{a}^1, \mathbf{a}^2, \mathbf{a}^3$, are said to be *reciprocal* when they satisfy the nine equations,

$$(1) \qquad\qquad \mathbf{a}_i \cdot \mathbf{a}^j = \delta_i^j \qquad (i, j = 1, 2, 3),$$

where δ_i^j is the Kronecker delta (§ 97).† The three equations that contain $\mathbf{a}^1$ are thus

$$\mathbf{a}_1 \cdot \mathbf{a}^1 = 1, \qquad \mathbf{a}_2 \cdot \mathbf{a}^1 = 0, \qquad \mathbf{a}_3 \cdot \mathbf{a}^1 = 0;$$

the second and third state that $\mathbf{a}^1$ is perpendicular to both $\mathbf{a}_2$ and $\mathbf{a}_3$, that is, parallel to $\mathbf{a}_2 \times \mathbf{a}_3$. Hence $\mathbf{a}^1 = \lambda \mathbf{a}_2 \times \mathbf{a}_3$; and, from the first equation, $1 = \lambda \mathbf{a}_1 \cdot \mathbf{a}_2 \times \mathbf{a}_3$. If $A = \mathbf{a}_1 \cdot \mathbf{a}_2 \times \mathbf{a}_3$, we thus obtain

$$(2) \qquad \mathbf{a}^1 = \frac{\mathbf{a}_2 \times \mathbf{a}_3}{A}, \qquad \mathbf{a}^2 = \frac{\mathbf{a}_3 \times \mathbf{a}_1}{A}, \qquad \mathbf{a}^3 = \frac{\mathbf{a}_1 \times \mathbf{a}_2}{A},$$

$\mathbf{a}^2$ and $\mathbf{a}^3$ being derived from $\mathbf{a}^1$ by cyclical permutation. Now from (98.7)

$$\mathbf{a}^2 \times \mathbf{a}^3 = \frac{A\mathbf{a}_1}{A^2} = \frac{\mathbf{a}_1}{A}, \qquad \mathbf{a}^1 \cdot \mathbf{a}^2 \times \mathbf{a}^3 = \frac{1}{A};$$

---

† The superscripts on the second basis are not exponents but mere tags to denote order; and δ_i^j has the same meaning as δ_{ij}.

hence the box products of reciprocal sets are reciprocal numbers. From the symmetry of equations (5) in the two sets, we also have

$$(3) \qquad \mathbf{a}_1 = \frac{\mathbf{a}^2 \times \mathbf{a}^3}{A^{-1}}, \qquad \mathbf{a}_2 = \frac{\mathbf{a}^3 \times \mathbf{a}^1}{A^{-1}}, \qquad \mathbf{a}_3 = \frac{\mathbf{a}^1 \times \mathbf{a}^2}{A^{-1}}.$$

When a basis and its reciprocal are identical ($\mathbf{a}_i = \mathbf{a}^i$), the basis is called *self-reciprocal*. The equations (1) then become $\mathbf{a}_i \cdot \mathbf{a}_j = \delta_{ij}$; thus $\mathbf{a}_1$ is a unit vector perpendicular to $\mathbf{a}_2$ and $\mathbf{a}_3$, etc. Hence:

The only self-reciprocal bases are triads of mutually orthogonal unit vectors. Since $A = A^{-1}$, $A^2 = 1$ and $A = \pm 1$; the basis is dextral or sinistral according as $A = 1$ or -1. We denote a *dextral* self-reciprocal basis by $\mathbf{e}_1, \mathbf{e}_2, \mathbf{e}_3$.

Since the vectors of a basis are noncoplanar, any vector can be expressed as a linear combination of the base vectors. For the basis $\mathbf{a}_1, \mathbf{a}_2, \mathbf{a}_3$ let

$$\mathbf{u} = u^1 \mathbf{a}_1 + u^2 \mathbf{a}_2 + u^2 \mathbf{a}_3;$$

if we dot multiply this equation in turn by $\mathbf{a}^1, \mathbf{a}^2, \mathbf{a}^3$, we find that $u^i = \mathbf{u} \cdot \mathbf{a}^i$; hence,

$$(4) \qquad \mathbf{u} = \mathbf{u} \cdot (\mathbf{a}^1 \mathbf{a}_1 + \mathbf{a}^2 \mathbf{a}_2 + \mathbf{a}^3 \mathbf{a}_3).$$

Similarly, if we write

$$\mathbf{u} = u_1 \mathbf{a}^1 + u_2 \mathbf{a}^2 + u_3 \mathbf{a}^3,$$

we find $u_i = \mathbf{u} \cdot \mathbf{a}_i$; hence,

$$(5) \qquad \mathbf{u} = \mathbf{u} \cdot (\mathbf{a}_1 \mathbf{a}^1 + \mathbf{a}_2 \mathbf{a}^2 + \mathbf{a}_3 \mathbf{a}^3).$$

Since (4) and (5) hold for any vector $\mathbf{u}$, the definition of dyadic equality shows that

$$(6) \qquad \mathbf{a}^1 \mathbf{a}_1 + \mathbf{a}^2 \mathbf{a}_2 + \mathbf{a}^3 \mathbf{a}_3 = \mathbf{a}_1 \mathbf{a}^1 + \mathbf{a}_2 \mathbf{a}^2 + \mathbf{a}_3 \mathbf{a}^3 = \mathbf{I},$$

the *unit dyadic* (107.7). Conversely, if

$$(7) \qquad \mathbf{a}_1 \mathbf{b}^1 + \mathbf{a}_2 \mathbf{b}^2 + \mathbf{a}_3 \mathbf{b}^3 = \mathbf{I},$$

the triads $\mathbf{a}_i$ and $\mathbf{b}^i$ form reciprocal bases. If $\mathbf{a}_1, \mathbf{a}_2, \mathbf{a}_3$ were coplanar, the left member of (7) would nullify any prefactor $\mathbf{n}$ normal to their plane: an impossibility since $\mathbf{n} \cdot \mathbf{I} = \mathbf{n}$. Hence $\mathbf{a}_1, \mathbf{a}_2, \mathbf{a}_3$ form a basis and have a reciprocal set $\mathbf{a}^1, \mathbf{a}^2, \mathbf{a}^3$; multiplying (7) by these vectors as prefactors we obtain $\mathbf{b}^1 = \mathbf{a}^1, \mathbf{b}^2 = \mathbf{a}^2, \mathbf{b}^3 = \mathbf{a}^3$.

111. Curvilinear Coordinates. Let us change from rectangular coordinates (x, y, z) to curvilinear coordinates (u, v, w) by means of the equations

$$(1) \qquad x = F(u, v, w), \qquad y = G(u, v, w), \qquad z = H(u, v, w),$$

where F, G, H have continuous first partial derivatives in a certain region in which their Jacobian is not zero (§ 87). The functions, therefore, are not connected by any functional relation (Theorem 89.3). Since the Jacobian is continuous, its sign cannot change in the region; and, to be explicit, we shall suppose that

$$(2) \qquad J = \frac{\partial(x, y, z)}{\partial(u, v, w)} = \mathbf{r}_u \times \mathbf{r}_v \cdot \mathbf{r}_w > 0.$$

This involves no loss in generality; for, if the Jacobian were negative, an interchange of v and w (for example) would make it positive.

Under these hypotheses the equations (1) have a unique inverse in some neighborhood I_0 of any point (x_0, y_0, z_0) of the region (§ 86):

$$(3) \qquad u = f(x, y, z), \qquad v = g(x, y, z), \qquad w = h(x, y, z);$$

and f, g, h also have continuous first partial derivatives. In the neighborhood I_0 the correspondence $(x, y, z) \leftrightarrow (u, v, w)$ is one-to-one, so that a point $P(x, y, z)$ may also be specified by the three numbers (u, v, w) given by (3).

When (u_0, v_0, w_0) are given, P_0 is the point of intersection of the three surfaces

$$f(x, y, z) = u_0, \qquad g(x, y, z) = v_0, \qquad h(x, y, z) = w_0.$$

These will intersect in three curves, the *coordinate curves*, along which only one of the quantities u, v, w can vary. For this reason, u, v, w are called *curvilinear coordinates*, in distinction to the rectangular coordinates x, y, z, for which the coordinate curves are straight lines.

Consider now any vector point function $\mathbf{f}(u, v, w)$; its derivative in the direction $\mathbf{e}$ is

$$\frac{d\mathbf{f}}{ds} = \frac{\partial \mathbf{f}}{\partial u} \frac{du}{ds} + \frac{\partial \mathbf{f}}{\partial v} \frac{dv}{ds} + \frac{\partial \mathbf{f}}{\partial w} \frac{dw}{ds}.$$

Since $d\mathbf{f}/ds = \mathbf{e} \cdot \nabla \mathbf{f}$, $du/ds = \mathbf{e} \cdot \nabla u$, we have

$$\mathbf{e} \cdot \nabla \mathbf{f} = \mathbf{e} \cdot (\nabla u \, \mathbf{f}_u + \nabla v \, \mathbf{f}_v + \nabla w \, \mathbf{f}_w)$$

for all vectors $\mathbf{e}$. Hence,

$$(4) \qquad \nabla \mathbf{f} = \nabla u \, \mathbf{f}_u + \nabla v \, \mathbf{f}_v + \nabla w \, \mathbf{f}_w,$$

$$(5) \qquad \nabla \cdot \mathbf{f} = \nabla u \cdot \mathbf{f}_u + \nabla v \cdot \mathbf{f}_v + \nabla w \cdot \mathbf{f}_w,$$

$$(6) \qquad \nabla \times \mathbf{f} = \nabla u \times \mathbf{f}_u + \nabla v \times \mathbf{f}_v + \nabla w \times \mathbf{f}_w$$

give the gradient, divergence, and rotation of a vector point function in general curvilinear coordinates.

When the position vector $\mathbf{r}$ is regarded as a function of x, y, z, we have $\nabla\mathbf{r} = \mathbf{I}$ (107.6); but, when $\mathbf{r}$ is regarded as a function of u, v, w, $\nabla\mathbf{r}$ is given by (4); hence,

$$(7) \qquad \mathbf{I} = \nabla u\,\mathbf{r}_u + \nabla v\,\mathbf{r}_v + \nabla w\,\mathbf{r}_w.$$

We therefore conclude that:

The vector triads ∇u, ∇v, ∇w and $\mathbf{r}_u$, $\mathbf{r}_v$, $\mathbf{r}_w$ form reciprocal sets.†

The vectors $\mathbf{r}_u$ are tangent to the u-curves, the curves along which v and w are constant. Thus at any point $P(u, v, w)$, $\mathbf{r}_u$, $\mathbf{r}_v$, $\mathbf{r}_w$ are tangent to the three coordinate curves meeting there. Moreover, from the properties of reciprocal sets:

$$(8) \qquad \nabla u = \frac{\mathbf{r}_v \times \mathbf{r}_w}{J}, \qquad \nabla v = \frac{\mathbf{r}_w \times \mathbf{r}_u}{J}, \qquad \nabla w = \frac{\mathbf{r}_u \times \mathbf{r}_v}{J},$$

$$(9) \qquad \nabla u \times \nabla v \cdot \nabla w = \frac{\partial(u, v, w)}{\partial(x, y, z)} = \frac{1}{J}.$$

In computing the invariants of $\nabla\mathbf{f}$ it is usually more convenient to eliminate ∇u, ∇v, ∇w from the above formulas by means of equations (8). Thus from (5) we obtain

$$(10) \qquad \nabla \cdot \mathbf{f} = \frac{1}{J}\left\{\mathbf{r}_v \times \mathbf{r}_w \cdot \mathbf{f}_u + \mathbf{r}_w \times \mathbf{r}_u \cdot \mathbf{f}_v + \mathbf{r}_u \times \mathbf{r}_v \cdot \mathbf{f}_w\right\},$$

or, in view of the identity,

$$(\mathbf{r}_v \times \mathbf{r}_w)_u + (\mathbf{r}_w \times \mathbf{r}_u)_v + (\mathbf{r}_u \times \mathbf{r}_v)_w = \mathbf{0},$$

$$(11) \qquad J \operatorname{div} \mathbf{f} = (\mathbf{r}_v \times \mathbf{r}_w \cdot \mathbf{f})_u + (\mathbf{r}_w \times \mathbf{r}_u \cdot \mathbf{f})_v + (\mathbf{r}_u \times \mathbf{r}_v \cdot \mathbf{f})_w.$$

In the important case when all coordinate curves cut at right angles, the coordinates are said to be *orthogonal*; then

$$(12) \qquad \mathbf{r}_v \cdot \mathbf{r}_w = \mathbf{r}_w \cdot \mathbf{r}_u = \mathbf{r}_u \cdot \mathbf{r}_v = 0.$$

We choose the notation so that

$$(13) \qquad \mathbf{r}_u = U\mathbf{a}, \qquad \mathbf{r}_v = V\mathbf{b}, \qquad \mathbf{r}_w = W\mathbf{c},$$

where $\mathbf{a}$, $\mathbf{b}$, $\mathbf{c}$ are a dextral set of (variable) unit vectors and U, V, W are all positive; then $J = UVW$. Now (11) becomes

$$(14) \quad \operatorname{div} \mathbf{f} = \frac{1}{UVW}\left\{\frac{\partial}{\partial u}\left(\frac{VW}{U}\mathbf{r}_u \cdot \mathbf{f}\right) + \frac{\partial}{\partial v}\left(\frac{WU}{V}\mathbf{r}_v \cdot \mathbf{f}\right) + \frac{\partial}{\partial w}\left(\frac{UV}{W}\mathbf{r}_w \cdot \mathbf{f}\right)\right\}.$$

---

† The nine equations. $\nabla u \cdot \mathbf{r}_u = 1$, $\nabla u \cdot \mathbf{r}_v = 0$, $\cdots$, also follow from the chain rule: for example,

$$u_x x_u + u_y y_u + u_z z_u = u_u = 1, \qquad u_x x_v + u_y y_v + u_z z_v = u_v = 0.$$

When we replace $\mathbf{f}$ by the gradient

$$(15) \qquad \nabla g = \nabla u\, g_u + \nabla v\, g_v + \nabla w\, g_w,\dagger$$

we obtain the important formula for the Laplacian of a scalar function in orthogonal coordinates:

$$(16) \quad \nabla^2 g = \frac{1}{UVW}\left\{\frac{\partial}{\partial u}\left(\frac{VW}{U}g_u\right) + \frac{\partial}{\partial v}\left(\frac{WU}{V}g_v\right) + \frac{\partial}{\partial w}\left(\frac{UV}{W}g_w\right)\right\}.$$

When the curvilinear coordinates are given by equations (1), we may compute U, V, W from equations of the type

$$(17) \qquad U = |\mathbf{r}_u| = |x_u\,\mathbf{e}_1 + y_u\,\mathbf{e}_2 + z_u\,\mathbf{e}_3| = \sqrt{x_u{}^2 + y_u{}^2 + z_u{}^2}.$$

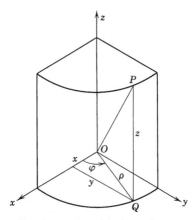

FIG. 111a.   Cylindrical coordinates

Example 1. *Cylindrical Coordinates.*   The point $P(x, y, z)$ projects into the point $Q(x, y, 0)$ in the xy-plane. If ρ, φ are polar coordinates of Q in the xy-plane, $u = \rho$, $v = \varphi$, $w = z$ are called the cylindrical coordinates of P (Fig. 111a). They are related to the rectangular coordinates by the equations.

$$(18) \qquad\qquad x = \rho\cos\varphi, \qquad y = \rho\sin\varphi, \qquad z = z.$$

The level surfaces $\rho = a$, $\varphi = b$, $z = c$ are cylinders about the z-axis, planes through the z-axis, and planes perpendicular to the z-axis. The coordinate curves for ρ are rays perpendicular to the z-axis; for φ, horizontal circles centered on the z-axis; for z, lines parallel to the z-axis.

From $\mathbf{r} = [x, y, z]$ we have

$$\mathbf{r}_\rho = [\cos\varphi, \sin\varphi, 0], \qquad\qquad U = 1;$$
$$\mathbf{r}_\varphi = [-\rho\sin\varphi, \rho\cos\varphi, 0], \qquad V = \rho;$$
$$\mathbf{r}_z = [0, 0, 1], \qquad\qquad\qquad W = 1.$$

---

† This formula is deduced in exactly the same way as (4).

Since these vectors are mutually perpendicular and $J = UVW = \rho > 0$, cylindrical coordinates form an orthogonal system which is dextral in the order ρ, φ, z.

From (16) the Laplacian

(19) $$\nabla^2 g = \frac{1}{\rho} \frac{\partial}{\partial \rho} (\rho g_\rho) + \frac{1}{\rho^2} g_{\varphi\varphi} + g_{zz}.$$

When $g = \log \rho$, $\rho g_\rho = 1$; hence $\log \rho$ is a particular solution of Laplace's equation $\nabla^2 g = 0$.

Example 2. Spherical Coordinates. The spherical coordinates of a point $P(x, y, z)$ are its distance $r = OP$ from the origin, the angle θ between OP and the z-axis, and the dihedral angle φ between the xz-plane and the plane zOP (Fig. 111*b*). They are related to the rectangular coordinates by the equations:

(20) $$x = r \sin \theta \cos \varphi, \qquad y = r \sin \theta \sin \varphi, \qquad z = r \cos \theta.$$

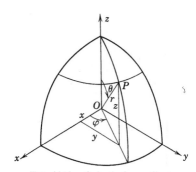

FIG. 111*b*.   Spherical coordinates

The level surfaces $r = a$, $\theta = b$, $\varphi = c$ are spheres about O, cones about the z-axis with vertex at O, and planes through the z-axis. The coordinate curves for r are rays from the origin; for θ, vertical circles centered at the origin (*meridians*); for φ, horizontal circles centered on the z-axis.

From $\mathbf{r} = [x, y, z]$ we have

$$\mathbf{r}_r = [\sin \theta \cos \varphi, \sin \theta \sin \varphi, \cos \theta], \qquad U = 1;$$
$$\mathbf{r}_\theta = [r \cos \theta \cos \varphi, r \cos \theta \sin \varphi, -r \sin \theta], \qquad V = r;$$
$$\mathbf{r}_\varphi = [-r \sin \theta \sin \varphi, r \sin \theta \cos \varphi, 0], \qquad W = r \sin \theta.$$

Since these vectors are mutually perpendicular and $J = UVW = r^2 \sin \theta > 0$, spherical coordinates form an orthogonal system which is dextral in the order r, θ, φ.

From (16) the Laplacian

(21) $$\nabla^2 g = \frac{1}{r^2 \sin \theta} \left[\sin \theta \frac{\partial}{\partial r} (r^2 g_r) + \frac{\partial}{\partial \theta} (\sin \theta \, g_\theta) + \frac{1}{\sin \theta} g_{\varphi\varphi} \right].$$

When $g = 1/r$, $r^2 g_r = -1$; hence $1/r$ satisfies Laplace's equation.

PROBLEMS

1. Find the set of vectors reciprocal to

$$\mathbf{a}_1 = [1, 0, 0], \qquad \mathbf{a}_2 = [1, 1, 0], \qquad \mathbf{a}_3 = [1, 1, 1].$$

Express $\mathbf{u} = [1, 2, 3]$ as a linear combination of $\mathbf{a}_1$, $\mathbf{a}_2$, $\mathbf{a}_3$ and also of $\mathbf{a}^1$, $\mathbf{a}^2$, $\mathbf{a}^3$. Multiply these expressions to find $\mathbf{u} \cdot \mathbf{u}$. Is the result correct?

2. Prove that $\mathbf{a}_1 \times \mathbf{a}^1 + \mathbf{a}_2 \times \mathbf{a}^2 + \mathbf{a}_3 \times \mathbf{a}^3 = \mathbf{0}$.

3. If $\mathbf{a}_1$ and $\mathbf{a}_2$ are vectors in the xy-plane, $\mathbf{a}_1 \times \mathbf{a}_2 = \lambda\mathbf{e}_3$. Prove that the set reciprocal to $\mathbf{a}_1$, $\mathbf{a}_2$, $\mathbf{e}_3$ is $\lambda^{-1}\mathbf{a}_2 \times \mathbf{e}_3$, $\lambda^{-1}\mathbf{e}_3 \times \mathbf{a}_1$, $\mathbf{e}_3$.

4. If the sets $\mathbf{a}_1$, $\mathbf{a}_2$, $\mathbf{a}_3$ and $\mathbf{a}^1$, $\mathbf{a}^2$, $\mathbf{a}^3$ are reciprocal, the sets $\alpha\mathbf{a}_1$, $\beta\mathbf{a}_2$, $\gamma\mathbf{a}_3$ and $\mathbf{a}^1/\alpha$, $\mathbf{a}^2/\beta$, $\mathbf{a}^3/\gamma$ are also.

5. If $\mathbf{a}_1$, $\mathbf{a}_2$, $\mathbf{a}_3$ is a set of mutually orthogonal vectors, the reciprocal set has the form $\mathbf{a}_1/|\mathbf{a}_1|^2$, $\mathbf{a}_2/|\mathbf{a}_2|^2$, $\mathbf{a}_3/|\mathbf{a}_3|^2$.

6. With cylindrical coordinates ρ, φ, z show that

$$\nabla\rho = \mathbf{r}_\rho, \qquad \nabla\varphi = \mathbf{r}_\varphi/\rho^2, \qquad \nabla z = \mathbf{r}_z.$$

7. With spherical coordinates r, θ, φ show that

$$\nabla r = \mathbf{r}_r, \qquad \nabla\theta = \mathbf{r}_\theta/r^2, \qquad \nabla\varphi = \mathbf{r}_\varphi/r^2\sin^2\theta.$$

8. Prove that the equations

$$x = a_1u + a_2v + a_3w, \qquad y = b_1u + b_2v + b_3w, \qquad z = c_1u + c_2v + c_3w$$

represent a change from one set of rectangular coordinates x, y, z into another u, v, w when the vectors $\mathbf{r}_u$, $\mathbf{r}_v$, $\mathbf{r}_w$ form a self-reciprocal set.

[The identity $x^2 + y^2 + z^2 = u^2 + v^2 + w^2$ implies the self-reciprocity of $\mathbf{r}_u$, $\mathbf{r}_v$, $\mathbf{r}_w$.]

112. Vector Algebra and Calculus. Vectors may be added, subtracted, and multiplied by real numbers in conformity with the laws of ordinary algebra. When vectors are given by their rectangular components (referred to the basis $\mathbf{e}_1$, $\mathbf{e}_2$, $\mathbf{e}_3$ of orthogonal unit vectors), these operations given by

$$\mathbf{u} + \mathbf{v} = [u_1 + v_1,\ u_2 + v_2,\ u_3 + v_3],$$

$$\mathbf{u} - \mathbf{v} = [u_1 - v_1,\ u_2 - v_2,\ u_3 - v_3] = \mathbf{u} + (-\mathbf{v}),$$

$$k\mathbf{u} = [ku_1,\ ku_2,\ ku_3].$$

The components of a unit vector $\mathbf{e}$ are its direction cosines:

$$\mathbf{e} = [\cos\alpha,\ \cos\beta,\ \cos\gamma].$$

The *scalar product* is defined by

$$\mathbf{u} \cdot \mathbf{v} = u_1v_1 + u_2v_2 + u_3v_3.$$

Independently of coordinates,

$$\mathbf{u} \cdot \mathbf{v} = |\mathbf{u}||\mathbf{v}|\cos(\mathbf{u}, \mathbf{v}).$$

For proper vectors, $\mathbf{u} \cdot \mathbf{v} = 0$ implies that $\mathbf{u}$ and $\mathbf{v}$ are perpendicular. Scalar multiplication is commutative and distributive:

$$\mathbf{u} \cdot \mathbf{v} = \mathbf{v} \cdot \mathbf{u}, \qquad \mathbf{u} \cdot (\mathbf{v} + \mathbf{w}) = \mathbf{u} \cdot \mathbf{v} + \mathbf{u} \cdot \mathbf{w}.$$

The associative law is meaningless.

When the basis $\mathbf{e}_1$, $\mathbf{e}_2$, $\mathbf{e}_3$ is dextral, the *vector product* is defined by

$$\mathbf{u} \times \mathbf{v} = [u_2v_3 - u_3v_2, \ u_3v_1 - u_1v_3, \ u_1v_2 - u_2v_1].$$

Independently of coordinates, the vector $\mathbf{u} \times \mathbf{v}$ has the magnitude

$$|\mathbf{u} \times \mathbf{v}| = |\mathbf{u}||\mathbf{v}| \sin(\mathbf{u}, \mathbf{v}),$$

and the direction of advance of a right-handed screw turned from $\mathbf{u}$ toward $\mathbf{v}$. For proper vectors, $\mathbf{u} \times \mathbf{v} = 0$ implies that $\mathbf{u}$ and $\mathbf{v}$ are parallel. Vector multiplication is distributive,

$$\mathbf{u} \times (\mathbf{v} + \mathbf{w}) = \mathbf{u} \times \mathbf{v} + \mathbf{u} \times \mathbf{w},$$

but *not* commutative or associative; for

$$\mathbf{v} \times \mathbf{u} = -\mathbf{u} \times \mathbf{v},$$

$$(\mathbf{u} \times \mathbf{v}) \times \mathbf{w} = (\mathbf{u} \cdot \mathbf{w})\mathbf{v} - (\mathbf{v} \cdot \mathbf{w})\mathbf{u}.$$

For the dextral basis $\mathbf{e}_1$, $\mathbf{e}_2$, $\mathbf{e}_3$ we have the cyclic relations,

$$\mathbf{e}_1 \times \mathbf{e}_2 = \mathbf{e}_3, \qquad \mathbf{e}_2 \times \mathbf{e}_3 = \mathbf{e}_1, \qquad \mathbf{e}_3 \times \mathbf{e}_1 = \mathbf{e}_2,$$

The *box product*,

$$\mathbf{u} \times \mathbf{v} \cdot \mathbf{w} = \begin{vmatrix} u_1 & u_2 & u_3 \\ v_1 & v_2 & v_3 \\ w_1 & w_2 & w_3 \end{vmatrix} = \mathbf{v} \times \mathbf{w} \cdot \mathbf{u} = \mathbf{w} \times \mathbf{u} \cdot \mathbf{v},$$

is equal to the volume of the parallelepiped having $\mathbf{u}$, $\mathbf{v}$, $\mathbf{w}$ as concurrent edges; its sign is positive or negative according as $\mathbf{u}$, $\mathbf{v}$, $\mathbf{w}$ form a dextral or a sinistral set. For proper vectors, $\mathbf{u} \times \mathbf{v} \cdot \mathbf{w} = 0$ implies that $\mathbf{u}$, $\mathbf{v}$, $\mathbf{w}$ are coplanar.

A set of three vectors $\mathbf{a}_1$, $\mathbf{a}_2$, $\mathbf{a}_3$ forms a basis if $A = \mathbf{a}_1 \times \mathbf{a}_2 \cdot \mathbf{a}_3 \neq 0$. Two bases $\mathbf{a}_1$, $\mathbf{a}_2$, $\mathbf{a}_3$ and $\mathbf{a}^1$, $\mathbf{a}^2$, $\mathbf{a}^3$ are called *reciprocal* when $\mathbf{a}_i \cdot \mathbf{a}^j = \delta_i^j$. If the indices i, j, k form a cyclical permulation of 1, 2, 3,

$$\mathbf{a}^i = \frac{\mathbf{a}_j \times \mathbf{a}_k}{A}, \qquad \mathbf{a}_i = \frac{\mathbf{a}^j \times \mathbf{a}^k}{A^{-1}}, \qquad A^{-1} = \mathbf{a}^1 \times \mathbf{a}^2 \cdot \mathbf{a}^3.$$

Both bases are dextral if $A > 0$, sinistral if $A < 0$.

The derivative $d\mathbf{u}/dt$ of a vector function $\mathbf{u}(t)$ is defined as the limit of $\Delta\mathbf{u}/\Delta t$ as $\Delta t \to 0$. The derivative of a constant vector is zero. The derivatives of the sum $\mathbf{u} + \mathbf{v}$ and the products $f\mathbf{u}$, $\mathbf{u} \cdot \mathbf{v}$, $\mathbf{u} \times \mathbf{v}$ are found by the familiar rules of calculus; but for $\mathbf{u} \times \mathbf{v}$ the order of the factors must be preserved.

The locus of $\overrightarrow{OP} = \mathbf{r} = \mathbf{r}(t)$ is a curve. When $t = s$ (the arc along the curve), $d\mathbf{r}/ds = \boldsymbol{\tau}$, the unit *tangent* vector to the curve in the direction of increasing s. The unit *principal normal* N to the curve has the direction of $d\mathrm{T}/ds$; and the unit *binormal* $\mathrm{B} = \mathrm{T} \times \mathrm{N}$; then $\mathrm{T} \times \mathrm{N} \cdot \mathrm{B} = 1$. The vectors of the moving trihedral TNB change conformably to *Frenet's formulas*:

$$d\mathrm{T}/ds = \kappa\mathrm{N}, \qquad d\mathrm{N}/ds = -\kappa\mathrm{T} + \tau\mathrm{B}, \qquad d\mathrm{B}/ds = -\tau\mathrm{N};$$

κ is the *curvature*, τ the *torsion* of the curve. For a plane curve $\tau = 0$.

The locus of $\overrightarrow{OP} = \mathrm{r} = \mathrm{r}(u, v)$ is a surface. The curves $\mathrm{r} = r(u, b)$ and $\mathrm{r} = \mathrm{r}(a, v)$ are the *parametric* curves on the surface; their tangent vectors are r_u and r_v, and at any regular point $\mathrm{r}_u \times \mathrm{r}_v \neq 0$ and determines a normal vector to the surface.

If a surface is given in implicit form $f(x, y, z) = $ const, the gradient vector $\nabla f = [f_x, f_y, f_z]$ is normal to the surface. The derivative of $f(x, y, z)$ in the direction e is $df/ds = \mathrm{e} \cdot \nabla f$; it assumes its maximum value $|\nabla f|$ when e is normal to the surface in the direction of increasing f.

The vector function $\mathrm{f}(\mathrm{r}) = \mathrm{e}_1 f_1 + \mathrm{e}_2 f_2 + \mathrm{e}_3 f_3$ has the directional derivative,

$$\frac{d\mathrm{f}}{ds} = \mathrm{e} \cdot \nabla\mathrm{f} = \mathrm{e} \cdot \left(\mathrm{e}_1 \frac{\partial \mathrm{f}}{\partial x} + \mathrm{e}_2 \frac{\partial \mathrm{f}}{\partial y} + \mathrm{e}_3 \frac{\partial \mathrm{f}}{\partial z} \right).$$

The gradient $\nabla\mathrm{f}$ is a *dyadic*. The *divergence* of f is its scalar invariant $\nabla \cdot \mathrm{f}$, the *rotation* (or *curl*) of f is its vector invariant $\nabla \times \mathrm{f}$; and

$$\mathrm{div}\,\mathrm{f} = \frac{\partial f_1}{\partial x} + \frac{\partial f_2}{\partial y} + \frac{\partial f_3}{\partial z}, \qquad \mathrm{rot}\,\mathrm{f} = \begin{vmatrix} \mathrm{e}_1 & \mathrm{e}_2 & \mathrm{e}_3 \\ \partial/\partial x & \partial/\partial y & \partial/\partial z \\ f_1 & f_2 & f_3 \end{vmatrix}.$$

If $g(\mathrm{r})$ is a scalar function, its *Laplacian*

$$\nabla^2 g = \mathrm{div}\,\nabla g = \frac{\partial^2 g}{\partial x^2} + \frac{\partial^2 g}{\partial y^2} + \frac{\partial^2 g}{\partial z^2}.$$

In curvilinear coordinates u, v, w, the gradient

$$\nabla\mathrm{f} = \nabla u\,\mathrm{f}_u + \nabla v\,\mathrm{f}_v + \nabla w\,\mathrm{f}_w,$$

$$\nabla\mathrm{r} = \nabla u\,\mathrm{r}_u + \nabla v\,\mathrm{r}_v + \nabla w\,\mathrm{r}_w = \mathrm{I}.$$

The triads ∇u, ∇v, ∇w and r_u, r_v, r_w form reciprocal sets; and, if $J = \mathrm{r}_u \times \mathrm{r}_v \cdot \mathrm{r}_w$,

$$J\,\mathrm{div}\,\mathrm{f} = (\mathrm{r}_v \times \mathrm{r}_w \cdot \mathrm{f})_u + \mathrm{cycl}, \qquad J\,\mathrm{rot}\,\mathrm{f} = ((\mathrm{r}_v \times \mathrm{r}_w) \times \mathrm{f})_u + \mathrm{cycl}.$$

C H A P T E R   6

The Definite Integral

113. The Riemann Integral. Let $f(x)$ be a *bounded* function over the *finite* interval $a \leq x \leq b$. Then $f(x)$ will have a greatest lower bound (g.l.b.) m and a least upper bound (l.u.b.) M in (a, b). These bounds are the closest numbers for which

$$m \leq f(x) \leq M \quad \text{in} \quad (a, b);$$

and $M - m$ is called the *oscillation* of $f(x)$ in the interval. If $f(x)$ is continuous, $f(x)$ actually equals m and M for values of x in (a, b).†

Divide (a, b) into n subintervals by a *net* of $n - 1$ intermediate points:

$$a = x_0 < x_1 < x_2 < \cdots < x_n = b;$$

and let ξ_i be any point in the ith subinterval,

$$\delta_i = x_i - x_{i-1}; \quad \text{then} \quad x_{i-1} \leq \xi_i \leq x_i.$$

Now consider the sum

$$(1) \qquad \sum_{i=1}^{n} f(\xi_i)\delta_i = f(\xi_1)\delta_1 + \cdots + f(\xi_n)\delta_n,$$

which depends on the net x_i and the choice of the points ξ_i. The *mesh* δ of the net is defined as the greatest of the n numbers δ_i. If the sum (1) approaches a limit as $\delta \to 0$ (and consequently $n \to \infty$) which is independent of the way in which the net and the points ξ_i are chosen, we say that $f(x)$ is *integrable* over (a, b), and we write

$$(2) \qquad \int_a^b f(x)\, dx = \lim_{\delta \to 0} \sum_{i=1}^{n} f(\xi_i)\delta_i, \qquad \delta = \max \delta_i.$$

---

† This is not true for all bounded functions. Thus the function

$$f(x) = 1 - x \quad (0 < x < 1), \qquad f(0) = f(1) = \tfrac{1}{2}$$

has $m = 0$, $M = 1$ in the interval $0 \leq x \leq 1$ but attains neither bound.

The process of obtaining this limit is called *Riemann integration*; and the limit (2) is called the *definite integral* of $f(x)$ between the *limits*† a and b. The function $f(x)$ is called the *integrand*. The integral sign is merely a formalized S, the initial of Latin *summa*. The notations in the left and right members of (2) have an obvious analogy.

Just as the value of the sum (1) does not depend on the letter i chosen as the summation index, the integral (2) is also independent of the letter x chosen as the variable of integration; thus,

$$(3) \qquad \int_a^b f(x)\,dx = \int_a^b f(t)\,dt.$$

There are obviously many different nets in an interval having the same mesh δ, and for any one of these the points ξ_i may be chosen in infinitely many ways. Thus, for a given δ, the numbers $\sum f(\xi_i)\delta_i$ are infinitely numerous, and the limit in (2) is not an ordinary limit. Equation (2) means that, for any $\varepsilon > 0$, there is a positive number Δ such that

$$(4) \qquad \left| \int_a^b f(x)\,dx - \sum f(\xi_i)\delta_i \right| < \varepsilon \quad \text{when} \quad \delta < \Delta.$$

This inequality must hold for all nets over (a, b) whose mesh $\delta < \Delta$, irrespective of how the points ξ_i are chosen.

When $f(x) = 1$, all the sums (1) equal the length $b - a$ of the interval (a, b), and the integral equals this length.

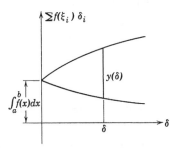

FIG. 113. Limiting process for a Riemann integral

In general, when the integral exists, for any given δ the numbers $\sum f(\xi_i)\delta_i$ lie on a definite interval $y(\delta)$; and $y(\delta) \to 0$ as $\delta \to 0$ (Fig. 113). In the definition (2) we assumed $a < b$. When $a > b$, the net

$$a = x_0 > x_1 > x_2 > \cdots > x_n = b$$

---

† This use of "limit" has nothing to do with its technical meaning in mathematics.

gives negative subintervals $\delta_i = x_i - x_{i-1}$. According to the definition (2), the integral

(5) $$\int_a^b f(x)\,dx = \lim_{\delta \to 0} \sum f(\xi_i)\delta_i = -\lim_{\delta \to 0} \sum f(\xi_i)(-\delta_i) = -\int_b^a f(x)\,dx.$$

Hence an interchange in the limits of a definite integral changes its sign.

When $a = b$ we may regard all $\delta_i = 0$ and put

(6) $$\int_a^a f(x)\,dx = 0.$$

This definition is also suggested by (5).

114. Condition for Integrability. In order to discuss the limit (113.2), consider any net x_i over (a, b), and let m_i and M_i be the g.l.b. and l.u.b. of $f(x)$ in the subinterval δ_i. For this net we define the *lower sum s* and the *upper sum S* by

(1) $$s = \sum_{i=1}^{n} m_i \delta_i, \qquad S = \sum_{i=1}^{n} M_i \delta_i.$$

These sums depend only on the net x_i and not on the points ξ_i; and, since $m_i \leqq f(\xi_i) \leqq M_i$,

(2) $$s \leqq \sum f(\xi_i)\delta_i \leqq S.$$

Hence, if we can show that both s and S approach the same limit L as $\delta \to 0$, then $\sum f(\xi_i)\delta_i$ also approaches L.

When $f(x) = C$, a constant,

$$s = S = \sum C\delta_i = C(b - a); \quad \text{hence}$$

(3) $$\int_a^b C\,dx = C(b - a).$$

In the following we specifically exclude this simple case; then always $s < S$.

We next prove two important properties of s and S.

Lemma 1. If a net of (a, b) having the sums s', S' is refined by the addition of new points of division, the new sums s, S satisfy

(4) $$s' \leqq s < S \leqq S'.$$

Proof. If m and M denote the g.l.b. and l.u.b. of $f(x)$ in (a, b), then $m \leqq m_i$, $M_i \leqq M$, and we have

(5) $$m(b - a) \leqq s < S \leqq M(b - a).$$

Let us now add new points of division to those already present. If a former δ_i now contains several subintervals, and we apply (5) to the interval δ_i,

$$m_i \delta_i \leqq s_i \leqq S_i \leqq M_i \delta_i,$$

where s_i and S_i are the lower and upper sums over δ_i. On adding these equations over (a, b), we obtain (4).† Thus when extra points are added to a subdivision, the lower sum will increase and the upper sum decrease, provided the sums change at all. In other words, a refinement of a net tends to bring the lower and upper sums closer together.

Lemma 2. If we consider all possible nets over an interval, a lower sum is less than any upper sum.

Proof. Let s', S' and s'', S'' denote the sums for any two nets of (a, b); and let s, S denote the sums when both nets are superposed. Since s and S correspond to a refinement of both original nets, we have from (4):

$$s' \leqq s < S \leqq S', \qquad s'' \leqq s < S \leqq S'';$$

hence $s' < S''$ and $s'' < S'$; that is, a lower sum is less than any upper sum.

From (5) we see that s and S lie between fixed bounds. Hence, the set of lower sums s has a least upper bound I, and the set of upper sums S has a greatest lower bound J; and from Lemma 2

$$(6) \qquad\qquad s \leqq I \leqq J \leqq S.$$

I is called the *lower integral*, J the *upper integral* of $f(x)$ over (a, b). They always exist when $f(x)$ is bounded in (a, b).

We next prove the important

THEOREM 1. *If we consider s and S for all possible nets of mesh δ over an interval (a, b),*

$$(7) \qquad\qquad \lim_{\delta \to 0} s = I, \qquad \lim_{\delta \to 0} S = J.$$

Proof. We first consider a function $f(x)$ in (a, b) for which $M > m > 0$. Let N be any net of mesh δ giving a lower sum s. Take $\varepsilon > 0$ arbitrarily small; then, since I is the l.u.b. of s, we can find a net N_1 whose lower sum $s_1 > I - \frac{1}{2}\varepsilon$. Let N_1 have k points between a and b; henceforth we hold N_1 fast so that k is a constant. We must now show that $s > I - \varepsilon$ for *all* nets N whose mesh δ is less than some fixed number.

Denote by N_2 the net formed by points of both N and N_1; then, by Lemma 1, its lower sum

$$(8) \qquad\qquad s_2 \geqq s_1 > I - \tfrac{1}{2}\varepsilon.$$

Now N has two types of subintervals:

(i)   those with *no* points of N_1 inside,

(ii) those having points of N_1 inside.

---

† In at least one subinterval δ_i we must have $s_i < S_i$.

The subintervals of type (i) contribute to s_2 an amount that certainly is less than s. The subintervals of type (ii) number at most k and contain at most $2k$ subintervals of N_2; and, since the mesh of N is δ, their total contribution to s_2 cannot exceed $2kM\delta$; hence,

$$s + 2kM\delta > s_2$$

or, in view of (8),

$$s > I - \tfrac{1}{2}\varepsilon - 2kM\delta.$$

If we now choose $\delta < \varepsilon/4kM$, we have

$$I \geqq s > I - \varepsilon \quad \text{when} \quad \delta < \frac{\varepsilon}{4kM}.$$

Consequently, $s \to I$ as $\delta \to 0$.

If $m \leqq 0$ for a function $f(x)$, replace it by $f(x) + C$ where $m + C > 0$. Then, if the lower sum and integral of $f(x)$ are s, I, the lower sum and integral of $f(x) + C$ are $s + C(b - a)$, $I + C(b - a)$; and, since the former approaches the latter, we again have $s \to I$.

To prove that $S \to J$ we need only consider the function $-f(x)$; the lower sums are now $-S$ and converge to the lower integral $-J$.

THEOREM 2 (Darboux). *A bounded function is integrable over (a, b) when and only when its lower and upper integrals are equal; and then*

$$(9) \qquad \int_a^b f(x)\, dx = I = J.$$

Proof. If $I = J$, both s and S approach I as $\delta \to 0$; then (2) shows that

$$\lim_{\delta = 0} \sum f(\xi_i)\delta_i = I.$$

The condition is therefore sufficient.

To show that the condition is also necessary, assume that $f(x)$ is integrable over (a, b):

$$(10) \qquad \lim_{\delta = 0} \sum f(\xi_i)\delta_i = L.$$

Since m_i is the g.l.b. of $f(x)$ in δ_i, we can choose the points ξ_i so that

$$f(\xi_i) - m_i < \frac{\varepsilon}{2(b - a)};$$

then

$$\sum f(\xi_i)\delta_i - \sum m_i \delta_i < \frac{\varepsilon}{2}.$$

In view of (10) we can choose Δ so small that

$$\left| L - \sum f(\xi_i)\delta_i \right| < \frac{\varepsilon}{2} \quad \text{when} \quad \delta < \Delta.$$

The last two inequalities now show that

$$\left| L - \sum m_i \delta_i \right| < \varepsilon \quad \text{when} \quad \delta < \Delta;$$

that is, $s \to L$ as $\delta \to 0$. Hence $I = L$ from Theorem 1. Applying this result to the function $-f(x)$, we have also $-J = -L$; therefore $I = L = J$.

Example. Consider *Dirichlet's function*:

$$f(x) = \begin{cases} 0 & x \text{ irrational,} \\ 1 & x \text{ rational,} \end{cases}$$

over any interval (a, b). Since both rational and irrational points are *dense*, $m = 0$ and $M = 1$ in any interval. Hence $I = 0$, $J = b - a$, and $f(x)$ is not integrable.

We can now deduce the following useful criterion for integrability.

THEOREM 3.   *In order that a bounded function be integrable over* (a, b) *it is necessary and sufficient that for any* $\varepsilon > 0$ *there is a net over* (a, b) *for which*

(11) $$S - s < \varepsilon.$$

Proof.   When (11) holds, we have from (6)

$$J - I \leqq S - s < \varepsilon.$$

But I and J are definite numbers, and ε can be chosen arbitrarily small; hence $I = J$, and $f(x)$ is integrable.

Conversely, if $f(x)$ is integrable,

$$S - s \to J - I = 0 \quad \text{as} \quad \delta \to 0;$$

hence for any ε we can find a net for which (11) holds.

115. Integrable Functions.   We can show that functions of several important types are integrable.

THEOREM 1.   *A function is integrable in any closed interval in which it is continuous.*

Proof.   The continuity of $f(x)$ in (a, b) implies its *uniform* continuity (Theorem 46.1); hence we can choose Δ so that

$$\left| f(x) - f(x') \right| < \frac{\varepsilon}{b - a} \quad \text{when} \quad \left| x - x' \right| < \Delta.$$

Hence, for any net over (a, b) whose mesh $\delta < \Delta$,

$$M_i - m_i < \frac{\varepsilon}{b - a} \quad \text{in each } \delta_i;$$

for $f(x)$ actually assumes the values m_i and M_i in each δ_i (Theorem 47). Consequently,

$$S - s = \sum (M_i - m_i)\delta_i < \frac{\varepsilon}{b - a} \sum \delta_i = \varepsilon$$

for all nets whose mesh $\delta < \Delta$.

THEOREM 2. *A bounded function is integrable in any interval in which it has but a finite number of discontinuities.*

Proof. Let $f(x)$ be continuous in (a, b) except for a single discontinuity at $x = c$. In any net over (a, b), c is a point of at most two subintervals; *two* when $c = x_i$, a net point. The oscillation in these intervals is $\leq M - m$, and for a net of mesh δ, their contribution to $S - s$ is $\leq 2(M - m)\delta$. We can now choose Δ so small that, when $\delta < \Delta$,

$$2(M - m)\delta < \tfrac{1}{2}\varepsilon,$$

and at the same time

$$M_i - m_i < \frac{\varepsilon}{2(b - a)}$$

in all subintervals in which $f(x)$ is continuous. Then

$$S - s < \frac{1}{2}\varepsilon + \frac{\varepsilon}{2(b - a)}(b - a) = \varepsilon,$$

and $f(x)$ is integrable over (a, b).

The case of any finite number of discontinuities is amenable to a similar argument.

Corollary. *If $f(x)$ is bounded and continuous in $a < x \leq b$, $f(x)$ is integrable over (a, b) when $f(a)$ is assigned any finite value.*

Thus the functions

$$f(x) = \frac{\sin x}{x} \quad \text{and} \quad g(x) = \sin \frac{1}{x}$$

are integrable in $0 \leq x \leq 1$ when $f(0)$ and $g(0)$ are defined at pleasure. If we define $f(0) = 1$, $f(x)$ is continuous at $x = 0$; but, since $g(x)$ approaches no limit as $x \to 0$, the discontinuity of $g(x)$ at $x = 0$ cannot be removed.

THEOREM 3. *A bounded function is integrable in any interval in which it is monotonic.*

Proof. Let $f(x)$ never decrease in (a, b); then its total oscillation over (a, b) is the sum of its oscillations in each subinterval δ_i:

$$\sum(M_i - m_i) = f(b) - f(a).$$

Now, for any net over (a, b) of mesh δ,

$$S - s = \sum(M_i - m_i)\delta_i \leq \delta\{f(b) - f(a)\};$$

hence,

$$S - s < \varepsilon \quad \text{when} \quad \delta < \frac{\varepsilon}{f(b) - f(a)}.$$

A bounded function may be integrable even when it has an infinite number of discontinuities in an interval. In order to state the facts in this case we need the concept of a *set of points of measure zero*. A set of points is said to have zero measure when it is possible to enclose them in a countable set of intervals whose total length $\sum \lambda_k \leq \varepsilon$ where λ_k is the length of the kth interval. Thus a finite set of points is always of measure zero; for n points can be enclosed in n intervals of length ε/n. Moreover, any *countable* infinity of points is of measure zero; for such a set may be placed in one-to-one correspondence with the natural numbers (§ 5) and the nth point enclosed in an interval of length $\varepsilon/2^n$; then

$$\frac{\varepsilon}{2} + \frac{\varepsilon}{2^2} + \frac{\varepsilon}{2^3} + \cdots = \frac{\tfrac{1}{2}\varepsilon}{1 - \tfrac{1}{2}} = \varepsilon.$$

We now state without proof

THEOREM 4 (Lebesgue). *In order that a bounded function with an infinite number of discontinuities be integrable it is necessary and sufficient that its points of discontinuity form a set of zero measure.*[†]

Example. The function

(i) $f(x) = 0$ (x irrational), $f(p/q) = 1/q$,

where p/q is a rational fraction in its lowest terms, is integrable in the interval $0 \leq x \leq 1$. For we have seen in Ex. 45.6 that this function is continuous except at the rational points p/q, and these points form a countable set of measure zero. But the integrability of this function also follows from our basic criterion, Theorem 114.3.

For any $\varepsilon > 0$, there are at most a finite number n of points $x = p/q$ in the interval $(0, 1)$ for which

(ii) $f(p/q) = 1/q \geq \varepsilon/2$;

for there are only a finite number of integers $q \leq 2/\varepsilon$, and consequently only a finite number of proper fractions p/q with these denominators. Let Σ' denote a summation over the intervals δ_i (at most $2n$) containing the n points for which (ii) holds, while Σ'' refers to the remaining intervals. Then, since

$$M_i' \leq 1, \qquad M_i'' < \frac{\varepsilon}{2}, \qquad m_i' = m_i'' = 0,$$

$$S - s < {\sum}' \delta_i + \frac{\varepsilon}{2} {\sum}'' \delta_i;$$

and, for all nets of mesh $\delta < \varepsilon/4n$,

$$S - s < 2n \cdot \frac{\varepsilon}{4n} + \frac{\varepsilon}{2} \cdot 1 = \varepsilon.$$

Since $s = 0$ for all nets, both s and $S \to 0$ as $\delta \to 0$, and, hence,

$$\int_0^1 f(x)\, dx = 0.$$

[†] Franklin, Philip, *A Treatise on Advanced Calculus*, Wiley, 1940, p. 254.

116. Integration by Summation. When $f(x)$ is integrable, we may compute its integral by choosing n subintervals δ_i and the points ξ_i in some specific way. For example, if all δ_i are equal and we choose ξ_i at the right end of each subinterval,

$$\delta_i = h = \frac{b-a}{n}, \qquad \xi_i = a + ih,$$

the sum $\sum f(\xi_i)\delta_i$ is a function of n alone, and we may compute the integral as an ordinary limit as $n \to \infty$ or $h \to 0$:

(1) $$\int_a^b f(x)\, dx = \lim_{n \to \infty} h\{f(a+h) + f(a+2h) + \cdots + f(a+nh)\}.$$

Example 1. We shall prove that

(i) $$\int_a^b e^x\, dx = e^b - e^a.$$

Take equal subintervals $h = (b-a)/n$, and choose $\xi_i = a + (i-1)h$ at the left end of each subinterval. Then the integral (i) is the limit of the sum,

$$\sum_{i=1}^{n} e^{\xi_i}\delta_i = h(e^a + e^{a+h} + e^{a+2h} + \cdots + e^{a+(n-1)h})$$

$$= he^a(1 + e^h + e^{2h} + \cdots + e^{(n-1)h})$$

$$= he^a \frac{e^{nh} - 1}{e^h - 1} = he^a \frac{e^{b-a} - 1}{e^h - 1} \qquad (22.2)$$

$$= (e^b - e^a) \frac{h}{e^h - 1}.$$

The result (i) now follows from l'Hospital's rule (§ 60):

$$\lim_{h \to 0} \frac{h}{e^h - 1} = \lim_{h \to 0} \frac{1}{e^h} = 1.$$

Example 2. If m is an integer $\neq -1$ and a and b are positive,

(ii) $$\int_a^b x^m\, dx = \frac{b^{m+1} - a^{m+1}}{m+1}.$$

Instead of choosing the points of partition in an arithmetic progression, we now take them in a geometric progression:

$$a, ar, ar^2, \cdots, ar^n = b; \quad \text{then}$$

$$\delta_1 = a(r-1), \qquad \delta_2 = ar(r-1), \cdots, \qquad \delta_n = ar^{n-1}(r-1).$$

If we choose $\xi_i = ar^{i-1}$ at the left end of δ_i, the integral (ii) is the limit of the sum

$$\sum_{i=1}^{n} \xi_i^m \delta_i = a^{m+1}(r-1)\,[1 + r^{m+1} + r^{2(m+1)} + \cdots + r^{(n-1)(m+1)}]$$

$$= a^{m+1}(r-1)\frac{r^{n(m+1)} - 1}{r^{m+1} - 1} \qquad (22.2)$$

$$= (b^{m+1} - a^{m+1})\frac{r-1}{r^{m+1} - 1}.$$

The result (ii) now follows from

$$\lim_{r \to 1} \frac{r-1}{r^{m+1}-1} = \lim_{r \to 1} \frac{1}{(m+1)r^m} = \frac{1}{m+1}.$$

Example 3. The case $m = -1$, excluded in Ex. 2, leads to the integral

(iii) $$\int_a^b \frac{dx}{x} = \log b - \log a \qquad (a, b > 0).$$

Since $m + 1 = 0$, the sum in Ex. 2 becomes $(r-1)n$; and, since $ar^n = b$,

$$\sum_{i=1}^n \frac{\delta_i}{\xi_i} = (r-1)n = (\log b - \log a)\frac{r-1}{\log r}.$$

The result (iii) now follows from l'Hospital's rule:

$$\lim_{r \to 1} \frac{r-1}{\log r} = \lim_{r \to 1} \frac{1}{1/r} = 1.$$

Example 4. When the integral is known, we may also use (1) to establish certain limits. Thus, with $h = 1/n$,

$$\int_0^1 \frac{dx}{1+x^2} = \lim_{n \to \infty} \sum_{i=1}^n \frac{1}{n} \frac{1}{1 + (i/n)^2} \; ;$$

or, since the integral equals $\pi/4$,

$$\frac{\pi}{4} = \lim_{n \to \infty} n \left(\frac{1}{n^2 + 1^2} + \frac{1}{n^2 + 2^2} + \cdots + \frac{1}{n^2 + n^2} \right).$$

117. Properties of the Integral.

THEOREM 1. *If $f(x)$ is integrable over (a, b), $f(x)$ is integrable over any subinterval (a', b').*

Proof. Let $a \le a' < b' \le b$ and let $S - s$ refer to any net over (a, b) having a', b' as net points, while $S' - s'$ refers to the portion of this net over (a', b'). Since $S - s$ contains all terms of $S' - s'$ plus other nonnegative terms, $S' - s' \le S - s$; hence,

$$S' - s' < \varepsilon \quad \text{whenever} \quad S - s < \varepsilon.$$

THEOREM 2. *If a, b, c are any three points of an interval in which $f(x)$ is integrable,*

(1) $$\int_a^b f(x)\, dx + \int_b^c f(x)\, dx = \int_a^c f(x)\, dx.$$

Proof. The integrals all exist by Theorem 1. If $a < b < c$, for all nets over (a, c) that have b as net point,

$$\sum_{ab} f(\xi_i)\delta_i + \sum_{bc} f(\xi_i)\delta_i = \sum_{ac} f(\xi_i)\delta_i;$$

and, as their mesh $\delta \to 0$, we obtain (1).

Thus, when $a < c < b$, we have

$$\int_a^c f(x)\, dx + \int_c^b f(x)\, dx = \int_a^b f(x)\, dx.$$

On adding $\int_b^c f(x)\, dx$ to both members, we again obtain (1).

We can also write (1) in the symmetrical form,

(2) $$\int_a^b f(x)\, dx + \int_b^c f(x)\, dx + \int_c^a f(x)\, dx = 0.$$

THEOREM 3. *If $f(x)$ and $g(x)$ are integrable over $a \leq x \leq b$, then*

(3) $$f(x) \leq g(x) \quad implies \quad \int_a^b f(x)\, dx \leq \int_a^b g(x)\, dx.$$

Proof. For any net over (a, b)

$$\sum f(\xi_i)\delta_i \leq \sum g(\xi_i)\delta_i;$$

hence, as the mesh $\delta \to 0$, we obtain (3).

From this theorem we next deduce the important

MEAN-VALUE THEOREM. *If $f(x)$ is integrable in (a, b) and has m, M as closest bounds,*

(4) $$\int_a^b f(x)\, dx = \mu(b - a), \qquad m \leq \mu \leq M.$$

Proof. Since $m \leq f(x) \leq M$, Theorem 3 shows that, when $a < b$,

$$m(b - a) \leq \int_a^b f(x)\, dx \leq M(b - a) \qquad (114.3).$$

Consequently, the integral divided by $b - a$ is some number μ in the interval (m, M).

If $b < a$, we have

$$\int_b^a f(x)\, dx = \mu(a - b);$$

this, multiplied by -1, again gives (4).

Corollary. *When $f(x)$ is continuous in (a, b),*

(5) $$\int_a^b f(x)\, dx = f(\xi)(b - a), \qquad \xi \text{ in } (a, b).$$

For a continuous function must assume its extreme values m, M in (a, b), and consequently all intermediate values (§ 48); hence, at some point ξ of (a, b), $f(\xi) = \mu$.

PROBLEMS

1. Evaluate $\int_0^1 f(x)\, dx$ if $f(x) = \begin{cases} 0, & x = 1, \frac{1}{2}, \frac{1}{3}, \cdots, \\ 1, & x \neq 1/n. \end{cases}$

2. Show that $\int_0^2 x^2(x - [x])\, dx = \dfrac{5}{3}$ (cf. Prob. 44.4).

3. Using the method of Ex. 116.1, prove that

$$\int_0^b \cos x\, dx = \sin b, \qquad \int_0^b \sin x\, dx = 1 - \cos b.$$

[Use the summation formulas (201.2–3).]

4. Show that $\displaystyle\lim_{n \to \infty} \frac{1}{n^2} \sum_{k=1}^n \sqrt{n^2 - k^2} = \int_0^1 \sqrt{1 - x^2}\, dx.$

5. Prove that $\displaystyle\lim_{n \to \infty} \sum_{k=1}^n \frac{1}{n + k - 1} = \log 2.$

6. Prove that $\displaystyle\lim_{n \to \infty} \sum_{k=1}^n \frac{k}{n^2} = \frac{1}{2}.$

7. Discuss the integrability of Dirichlet's function (Ex. 114) by means of Lebesgue's theorem.

8. Show that $\displaystyle\int_0^{2\pi} |\, 1 + 2 \cos x\,|\, dx = \frac{2}{3}\pi + 4\sqrt{3}.$

118. Formation of Integrable Functions. When $|\, f(x)\, |$ is integrable over an interval, we say the $f(x)$ is *absolutely integrable*.

THEOREM 1.   *When $f(x)$ is integrable in (a, b), so is $|\, f(x)\, |$; and*

(1)
$$\left| \int_a^b f(x)\, dx \right| \leq \int_a^b |\, f(x)\, |\, dx, \qquad a < b.$$

Proof. Let $S - s$ and $S' - s'$ refer to $f(x)$ and $|\, f(x)\, |$, respectively. Since the oscillation (cf. § 113) of $|\, f(x)\, |$ in any interval cannot exceed that of $f(x)$, $S' - s' \leq S - s$ for any net over (a, b); hence,

$$S' - s' < \varepsilon \quad \text{when} \quad S - s < \varepsilon.$$

The inequality (1) now follows from

$$\left| \sum f(\xi_i)\delta_i \right| \leq \sum |\, f(\xi_i)\, |\, \delta_i$$

on passing to the limit $\delta \to 0$.

The example,

$$f(x) = \begin{cases} 1 & x \text{ irrational}, \\ -1 & x \text{ rational}, \end{cases}$$

shows that *absolute integrability does not imply integrability*; for $|f(x)| = 1$ is integrable while $f(x)$ is not (Ex. 114).

THEOREM 2. *If $f(x)$ and $g(x)$ are integrable in (a, b), so is their sum; and*

$$(2) \qquad \int_a^b [f(x) + g(x)]\, dx = \int_a^b f(x)\, dx + \int_a^b g(x)\, dx.$$

Proof. Pass to the limit $\delta \to 0$ in

$$\sum [f(\xi_i) + g(\xi_i)]\delta_i = \sum f(\xi_i)\delta_i + \sum g(\xi_i)\delta_i.$$

THEOREM 3. *If $f(x)$ and $g(x)$ are integrable in (a, b), so is their product.*

Proof. Consider any subinterval of (a, b) in which $f(x)$, $g(x)$ and $p(x) = f(x)\, g(x)$ have the closest bounds (g.l.b. and l.u.b.) m_f, M_f; m_g, M_g; m_p, M_p. Then for any $\varepsilon > 0$ there are points ξ, ξ' of the subinterval δ_i such that

$$p(\xi) > M_p - \varepsilon, \qquad p(\xi') < m_p + \varepsilon,$$

and, hence,

$$\begin{aligned} M_p - m_p &< p(\xi) + \varepsilon - p(\xi') + \varepsilon \\ &= f(\xi)\, g(\xi) - f(\xi')\, g(\xi') + 2\varepsilon \\ &= f(\xi)\, [g(\xi) - g(\xi')] + g(\xi')\, [f(\xi) - f(\xi')] + 2\varepsilon. \end{aligned}$$

Then, if K is an upper bound of both $f(x)$ and $g(x)$ in (a, b),

$$(3) \qquad M_p - m_p < K(M_g - m_g) + K(M_f - m_f) + 2\varepsilon.$$

The integrability of fg now follows readily from Theorem 114.3; for

$$S_p - s_p < K(S_g - s_g) + K(S_f - s_f) + 2\varepsilon(b - a).$$

THEOREM 4. *If $f(x)$ is positive and integrable in (a, b), so are*

$$g(x) = 1/f(x) \quad and \quad h(x) = \sqrt{f(x)}.$$

Proof. Let $f(x) > m$ in (a, b). Then in any subinterval in which $f(x)$ has the closest bounds m_f, M_f,

$$M_g - m_g = \frac{1}{m_f} - \frac{1}{M_f} = \frac{M_f - m_f}{m_f M_f} < \frac{M_f - m_f}{m^2},$$

$$M_h - m_h = \sqrt{M_f} - \sqrt{m_f} = \frac{M_f - m_f}{\sqrt{M_f} + \sqrt{m_f}} < \frac{M_f - m_f}{2\sqrt{m}},$$

The integrability of g and h now follows from Theorem 114.3.

119. The Integral as a Function of Its Upper Limit. Let $f(t)$ be integrable over the interval $a \leq t \leq b$, and let x be any number in this interval. Then

$$(1) \qquad \varphi(x) = \int_a^x f(t)\, dt$$

is a function of x alone. The variable of integration is denoted by t to avoid confusion with the upper limit x.

THEOREM 1. *The function* $\int_a^x f(t)\,dt$ *is continuous in any interval* (a, b) *in which* $f(x)$ *is integrable.*

Proof. Let x and $x + h$ be points of (a, b). Then from (117.1)

$$\varphi(x + h) = \int_a^{x+h} f(t)\,dt = \int_a^x f(t)\,dt + \int_x^{x+h} f(t)\,dt,$$

whether h is positive or negative; and

$$(2) \qquad \varphi(x + h) - \varphi(x) = \int_x^{x+h} f(t)\,dt = \mu h, \qquad (117.4)$$

where μ lies between the bounds of $f(t)$ in $(x, x + h)$. As $h \to 0$, $\mu h \to 0$ and $\varphi(x + h) \to \varphi(x)$; hence $\varphi(x)$ is continuous at any point of (a, b).

THEOREM 2. *If* $f(x)$ *is integrable in* (a, b),

$$(3) \qquad \frac{d}{dx}\int_a^x f(t)\,dt = f(x), \qquad a \leqq x \leqq b,$$

at every point x *where* $f(x)$ *is continuous.*

Proof. If $f(t)$ is continuous when $t = x$, we can replace μ in (2) by $f(x) + \eta$ where $\eta \to 0$ as $h \to 0$. Hence,

$$\lim_{h \to 0} \frac{\varphi(x + h) - \varphi(x)}{h} = \lim_{h \to 0} [f(x) + \eta] = f(x);$$

that is, $\varphi'(x) = f(x)$. Of course $\varphi'(a)$ and $\varphi'(b)$ are right-hand and left-hand derivatives, respectively.

Corollary. *At every point where* $f(x)$ *is continuous*

$$(4) \qquad \frac{d}{dx}\int_x^b f(t)\,dt = -f(x).$$

Note that Theorem 2 is essentially an *existence theorem*; for it proves the existence of a function $\varphi(x)$ whose derivative is a given function $f(x)$ at all of its points of continuity.

If $f(x)$ is *continuous* in (a, b), $d\varphi/dx = f(x)$ at all points of (a, b). Hence the differential equation $dF/dx = f(x)$ has the general solution,

$$(5) \qquad F(x) = \varphi(x) + C = \int_a^x f(t)\,dt + C;$$

for any function that has the derivative $f(x)$ over (a, b) can only differ from $\varphi(x)$ by a constant (Theorem 57, Cor. 2). Any of the functions $F(x)$ is called an *indefinite integral* or *primitive* of $f(x)$. The particular solution $\varphi(x)$ is characterized by the fact that $\varphi(a) = 0$.

When $f(x)$ is continuous, the function $\varphi(x)$ is continuous in $a \leq x \leq b$ and differentiable in $a < x < b$.* Hence we may apply the mean-value theorem (57.1):

$$\varphi(b) - \varphi(a) = (b - a)\varphi'(\xi), \qquad a < \xi < b,$$

or, in view of (1) and (3),

(6) $$\int_a^b f(x)\,dx = (b - a)f(\xi), \qquad a < \xi < b.$$

This result, previously given in (117.5), is often called the *mean-value theorem of the integral calculus.* Moreover,

(7) $$f(\xi) = \frac{1}{b - a} \int_a^b f(x)\,dx$$

is defined as the *mean* of the continuous function $f(x)$ over the interval (a, b); for, in a sense, it is a generalization of an arithmetic mean.

If we divide the interval (a, b) into n equal intervals $h = (b - a)/n$, the mean of the n values $f(a + h), f(a + 2h), \cdots, f(a + nh)$ is

$$\frac{1}{n} \sum_{i=1}^n f(a + ih) = \frac{1}{b - a} \sum_{i=1}^n h f(a + ih).$$

If we pass to the limit $n \to \infty$, the right-hand sum approaches the integral in (7); hence,

(8) $$f(\xi) = \lim_{n \to \infty} \frac{1}{n}[f(a + h) + f(a + 2h) + \cdots + f(a + nh)].$$

Example. The function (Ex. 115)

$$f(x) = 0 \quad (x \text{ irrational}), \qquad f(p/q) = 1/q$$

is integrable in $(0, 1)$ and hence in any subinterval; and we find as in § 115

$$\varphi(x) = \int_0^x f(t)\,dt = 0, \qquad 0 \leq x \leq 1.$$

Thus $\varphi'(x) = 0$ throughout $(0, 1)$. Consequently $\varphi'(x) = f(x)$ at the irrational points (where $f(x)$ is continuous); but $\varphi'(x) \neq f(x)$ at the rational points (where $f(x)$ is discontinuous).

120. The Fundamental Theorem of the Integral Calculus. Although a definite integral may sometimes be computed as the limit of a sum (as in the examples of § 116), this direct computation is usually long or beset with difficulties. A systematic plan of computation is given by the following

* $\varphi(x)$ has the right-hand derivative $f(a)$ at a, the left-hand derivative $f(b)$ at b.

FUNDAMENTAL THEOREM. *If $f(x)$ is integrable in (a, b) and $F(x)$ is any function having $f(x)$ as derivative, then*

$$\text{(1)} \qquad \int_a^b f(x)\, dx = F(b) - F(a).\dagger$$

Proof. If we divide (a, b) into n subintervals by the points

$$a = x_0 < x_1 < x_2 < \cdots < x_n = b,$$

we have

$$\text{(2)} \qquad \int_a^b f(x)\, dx = \lim_{\delta \to 0} \sum_{i=1}^n f(\xi_i)\delta_i. \qquad \text{(113.2)}$$

By hypothesis $F'(x) = f(x)$ in the interval (a, b). Since $F(x)$ is differentiable, it is also continuous, and we may apply the mean-value theorem (57.1) in each subinterval $\delta_i = x_i - x_{i-1}$:

$$F(x_i) - F(x_{i-1}) = F'(\xi_i)(x_i - x_{i-1}) = f(\xi_i)\delta_i,$$

where ξ_i is an interior point of δ_i. Consequently, if we choose the intermediate points ξ_i in (2) as these precise points,

$$\int_a^b f(x)\, dx = \lim_{\delta \to 0} \sum_{i=1}^n [F(x_i) - F(x_{i-1})].$$

The "telescopic" sum on the right reduces to

$$F(x_n) - F(x_0) = F(b) - F(a);$$

and, since this reduction is possible for any and all subdivisions of (a, b), it holds good as we approach the limit $\delta \to 0$. Thus (1) is established.

When $f(x)$ is *continuous*, Theorem 119.2 shows that

$$\text{(3)} \qquad F(x) = \int_a^x f(t)\, dx + C$$

comprises all functions for which $F'(x) = f(x)$. In particular, $F(a) = C$, and, hence,

$$\text{(4)} \qquad F(x) - F(a) = \int_a^x f(x)\, dx.$$

In this case we have a simple proof of (1) by putting $x = b$.

A function $F(x)$ having $f(x)$ as derivative is called a *primitive* (or *antiderivative*) of $f(x)$.

---

† The notation $F(x)\Big|_a^b = F(b) - F(a)$ is often used. See Appendix 3.

Let us consider the three examples of § 116 in the light of this theorem. For the continuous functions,

$$f(x) = e^x, \qquad\qquad\qquad F(x) = e^x;$$

$$f(x) = x^m \quad (m \neq -1, x > 0), \qquad F(x) = \frac{x^{m+1}}{m+1};$$

$$f(x) = 1/x \quad (x > 0), \qquad\qquad F(x) = \log x;$$

The integrals (i), (ii), (iii) of § 116 now follow directly from (1). In each case the most general primitive is $F(x)$ plus an arbitrary constant.

When an integrable function $f(x)$ has discontinuities, the function $F(x)$ may not exist. This is the case in Ex. 119. That $F(x)$ may exist in some cases when $f(x)$ is discontinuous is shown by

Example 1. The continuous function,

$$F(x) = x^2 \sin \frac{1}{x} \quad (x \neq 0), \qquad F(0) = 0,$$

has the derivative

$$f(x) = 2x \sin \frac{1}{x} - \cos \frac{1}{x} \ (x \neq 0), \qquad f(0) = 0,$$

which is discontinuous at $x = 0$ (Ex. 50.5). Since $f(x)$ is bounded in $(-1, 1)$ it is integrable; and from (1)

$$\int_{-1}^{1} f(x)\,dx = F(1) - F(-1) = \sin 1 - \sin(-1) = 2 \sin 1.$$

Example 2. Remembering that $\log x$ is real only when $x > 0$, we see that the primitive of $1/x$ is

$$F(x) = \begin{cases} \log x & \text{when} \quad x > 0, \\ \log(-x) & \text{when} \quad x < 0; \end{cases}$$

hence,

$$\int_a^b \frac{dx}{x} = \begin{cases} \log b - \log a & \text{when} \quad 0 < a < b, \\ \log(-b) - \log(-a) & \text{when} \quad a < b < 0. \end{cases}$$

Thus, when a and b have the same sign, we may take $F(x) = \log|x|$; and

(5) $$\int_a^b \frac{dx}{x} = \log|b| - \log|a|, \qquad ab > 0,$$

comprises both cases. But, if the interval (a, b) includes $x = 0$, the integral does not exist for $1/x \to \infty$ as $x \to 0$.

Example 3. The primitive of $1/x^2$ is $-1/x$ when $x \neq 0$; hence,

(6) $$\int_a^b \frac{dx}{x^2} = \frac{1}{a} - \frac{1}{b} \quad \text{when} \quad ab > 0.$$

But, if the interval (a, b) includes $x = 0$, the integral does not exist. Thus, when $a < 0$, $b > 0$, (6) is manifestly absurd; for the left member is positive (why?), the right member negative.

Example 4. This integral

$$(7) \qquad \int_a^b \frac{dx}{1 + x^2} = \tan^{-1} b - \tan^{-1} a,$$

Where $\tan^{-1} x$ denotes the *principal branch* of the inverse tangent:

$$-\pi/2 < \tan^{-1} x < \pi/2 \qquad \text{(Ex. 49.3)}.$$

Thus,

$$\int_{-1}^1 \frac{dx}{1 + x^2} = \frac{\pi}{4} - \left(-\frac{\pi}{4}\right) = \frac{\pi}{2}.$$

If we had chosen the branch of the inverse tangent between $\pi/2$ and $3\pi/2$, the result would have been the same $(5\pi/4 - 3\pi/4)$. But, if $\tan^{-1} a$ and $\tan^{-1} b$ are taken from *different* branches of the multivalued inverse tangent, we get a false result. Although $\tan \pi/4 = 1$, $\tan 3\pi/4 = -1$,

$$\int_{-1}^1 \frac{dx}{1 + x^2} \neq \frac{\pi}{4} - \frac{3\pi}{4} = -\frac{\pi}{2},$$

for $\pi/4$ and $3\pi/4$ do not belong to the same branch. To obviate errors of this kind we shall always use $\tan^{-1} x$ to denote the principal branch. Then, for each integer k, $\tan^{-1} x + k\pi$ represents a definite branch of the inverse tangent.

Example 5. If $-1 < a < b < 1$,

$$(8) \qquad \int_a^b \frac{dx}{\sqrt{1 - x^2}} = \sin^{-1} b - \sin^{-1} a,$$

where $\sin^{-1} x$ denotes the *principal branch* of the inverse sine:

$$-\frac{\pi}{2} \leq \sin^{-1} x \leq \frac{\pi}{2} \qquad \text{(Ex. 49.2)}.$$

We may also take $F(x) = -\cos^{-1} x$, where $\cos^{-1} x$ is the principal branch of the inverse cosine:

$$0 \leq \cos^{-1} x \leq \pi \qquad \text{(Ex. 49.4)}.$$

Then

$$(9) \qquad \int_a^b \frac{dx}{\sqrt{1 - x^2}} = \cos^{-1} a - \cos^{-1} b.$$

This result agrees with (8) since the identity

$$(10) \qquad \cos^{-1} x = \frac{\pi}{2} - \sin^{-1} x$$

connects the principal branches of the inverse sine and cosine.

PROBLEM

1. *Lubin's Problem.*† Consider the integral $G(x) = \int_{-1}^x g(t)\, dt$ where

$$g(t) = \begin{cases} f'(t) & t \neq 0 \\ 0 & t = 0 \end{cases}, \qquad f(t) = \frac{1}{1 + e^{1/t}}.$$

---

† Lubin, C. I., *Am. Math. Monthly*, vol. 53, p. 586.

Show that $g(t)$ is continuous when $t \geq -1$, and hence $G(x)$ is continuous when $x \geq -1$. Prove that

$$G(x) = \begin{cases} f(x) - f(-1) & -1 \leq x < 0, \\ f(1) & x = 0, \\ f(x) + f(1) & x > 0. \end{cases}$$

Find $f'(0+), f'(0-), f'(0), G'(0)$. [Cf. Prob. 53.13.]

121. Mean-Value Theorems. We can generalize the mean-value theorem (117.4) as follows:

THEOREM 1. *If $f(x)$ and $g(x)$ are integrable in (a, b) and $g(x)$ does not change sign, then*

$$(1) \qquad \int_a^b f(x) g(x)\, dx = \mu \int_a^b g(x)\, dx, \qquad m \leq \mu \leq M,$$

where m, M are the closest bounds of $f(x)$ in (a, b).

Proof. Let $g(x) \geq 0$ and $a < b$; then

$$\int_a^b [f(x) - m]g(x)\, dx \geq 0, \qquad \int_a^b [M - f(x)] g(x)\, dx \geq 0,$$

for each integral is the limit of a sum of nonnegative terms. Hence,

$$m \int_a^b g(x)\, dx \leq \int_a^b f(x) g(x)\, dx \leq M \int_a^b g(x)\, dx.$$

When $g(x) \leq 0$ or $a > b$, these inequalities are reversed; but in any case the central integral lies between the given extremes and therefore has the value given in (1).

When $f(x)$ is continuous, $f(x)$ must assume the values m, M, and also all intermediate values; hence, at some point ξ of (a, b), $f(\xi) = \mu$, and

$$(2) \qquad \int_a^b f(x) g(x)\, dx = f(\xi) \int_a^b g(x)\, dx, \qquad a \leq \xi \leq b.$$

THEOREM 2. *If $f(x)$ and $g(x)$ are integrable, $g(x) \geq 0$, and $f(x)$ is monotone in (a, b), then*

$$(3) \qquad \int_a^b f(x) g(x)\, dx = f(a) \int_a^\xi g(x)\, dx + f(b) \int_\xi^b g(x)\, dx, \qquad a < \xi < b.$$

Proof. From Theorem 119.1, the function

$$\varphi(t) = f(a) \int_a^t g(x)\, dx + f(b) \int_t^b g(x)\, dx$$

is continuous in (a, b); moreover it assumes the values

$$\varphi(a) = f(b) \int_a^b g(x)\, dx, \qquad \varphi(b) = f(a) \int_a^b g(x)\, dx.$$

at a and b. Since $f(x)$ is monotone, the number μ in (1) lies between $f(a)$ and $f(b)$; hence for some value $t = \xi$ between a and b, $\varphi(t)$ will assume the value $\mu \int_a^b g(x)\, dx$ (Theorem 48.2). Thus (3) is proved.

Corollary 1. If $f(x)$ is a *positive decreasing* function, the integral (3) is not altered if we define $f(b) = 0$ (Theorem 115.2, Cor.)† Then (3) can be written in *Bonnet's form*:

(4) $$\int_a^b f(x)\, g(x)\, dx = f(a) \int_a^\xi g(x)\, dx, \qquad a < \xi < b.†$$

Corollary 2. If $f(x)$ is a *positive increasing* function, the integral (3) is not altered if we define $f(a) = 0$. Then (3) becomes

(5) $$\int_a^b f(x)\, g(x)\, dx = f(b) \int_\xi^b g(x)\, dx, \qquad a < \xi < b.$$

122. Integration by Parts. This powerful method of integration depends on the

THEOREM. *If $u(x)$ and $v(x)$ are functions whose derivatives $u'(x)$, $v'(x)$ are integrable in (a, b),*

(1) $$\int_a^b uv'\, dx = u(x)\, v(x) \, \Big|_a^b - \int_a^b vu'\, dx.$$

Proof. The functions $u(x)$, $v(x)$ are differentiable and therefore continuous; hence uv', vu' and $uv' + vu'$ are integrable (§ 118). Since

$$uv' + vu' = (uv)',$$

$$\int_a^b (uv' + vu')\, dx = u(b)\, v(b) - u(a)\, v(a)$$

by the fundamental theorem (120.1). This result is equivalent to (1).

Since $u'\, dx = du$, $v'\, dx = dv$, we can put this important rule in the condensed form,

(2) $$\int_a^b u\, dv = uv \, \Big|_a^b - \int_a^b v\, du,$$

where the limits refer to the independent variable x.

Indefinite integrals may also be found by integration by parts. Thus

(3) $$\int_a^x uv'\, dx = u(x)\, v(x) - u(a)\, v(a) - \int_a^x vu'\, dx;$$

---

† Unless $f(b)$ is already zero, the definition $f(b) = 0$ will alter ξ in (3) to another number between a and b.

and, if the indefinite integrals of $f(x)$ are denoted by $\int f(x)\, dx$, we may write

(4) $$\int uv'\, dx = uv - \int vu'\, dx,$$

since $-u(a)\,v(a)$ is a constant.

If $v'(x)$ is discontinuous at a finite number of points where $v'_+(x)$ and $v'_-(x)$ exist but are not equal, (1) is still valid; for we can apply (1) to each subinterval in which $v'(x)$ is continuous and add the resulting equations.

Example 1. With $u = \tan^{-1} x$, $dv = dx$,

$$\int \tan^{-1} x\, dx = x \tan^{-1} x - \int \frac{x}{x^2 + 1}\, dx$$

$$= x \tan^{-1} x - \tfrac{1}{2} \log (x^2 + 1) + C;$$

$$\int_0^x \tan^{-1} t\, dt = x \tan^{-1} x - \tfrac{1}{2} \log (x^2 + 1).$$

Similarly, with $u = \sin^{-1} x$, $dv = dx$,

$$\int \sin^{-1} x\, dx = x \sin^{-1} x + \sqrt{1 - x^2} + C;$$

$$\int_0^x \sin^{-1} t\, dt = x \sin^{-1} x + \sqrt{1 - x^2} - 1.$$

Example 2. To compute

$$I = \int e^{ax} \sin bx\, dx, \qquad J = \int e^{ax} \cos bx\, dx,$$

take

$$u = \sin bx \quad \text{or} \quad \cos bx, \qquad dv = e^{ax}\, dx,$$

$$du = b \cos bx \quad \text{or} \quad -b \sin bx, \qquad v = \frac{1}{a} e^{ax},$$

and integrate by parts. Omitting constants,

$$I = \frac{1}{a} e^{ax} \sin bx - \frac{b}{a} J, \qquad J = \frac{1}{a} e^{ax} \cos bx + \frac{b}{a} I;$$

and, on solving these equations for I and J,

(5) $$I = \int e^{ax} \sin bx\, dx = e^{ax} \frac{a \sin bx - b \cos bx}{a^2 + b^2},$$

(6) $$J = \int e^{ax} \cos bx\, dx = e^{ax} \frac{a \cos bx + b \sin bx}{a^2 + b^2}.$$

Example 3. Integration by parts is often used to set up recursion relations to facilitate the computation of an integral. Thus, if n is a positive integer, let

$$I_n = \int \sin^n x\, dx, \qquad J_n = \int \cos^n x\, dx.$$

Integrate I_n by parts, taking

$$u = \sin^{n-1} x, \qquad dv = \sin x \, dx,$$

$$du = (n-1) \sin^{n-1} x \cos x, \qquad v = -\cos x;$$

then

$$I_n = -\sin^{n-1} x \cos x + (n-1) \int \sin^{n-2} x \cos^2 x \, dx.$$

On putting $\cos^2 x = 1 - \sin^2 x$ in the last integral, we obtain the *reduction formula*:

$$(7) \qquad\qquad n I_n = -\sin^{n-1} x \cos x + (n-1) I_{n-2}.$$

By successive applications of (7) we can eventually express I_n in terms of

$$I_1 = \int \sin x \, dx = -\cos x \quad \text{or} \quad I_0 = \int dx = x,$$

according as n is odd or even; for each step reduces the exponent by 2.

In similar fashion we find

$$(8) \qquad\qquad n J_n = \cos^{n-1} x \sin x + (n-1) J_{n-2}.$$

If we integrate between 0 and $\pi/2$, the part uv vanishes at each limit, and we have the chain of equations,

$$n I_n = (n-1) I_{n-2},$$

$$(n-2) I_{n-2} = (n-3) I_{n-4}, \cdots,$$

ending with

$$3 I_3 = 2 I_1 = 2 \int_0^{\pi/2} \sin x \, dx = 2, \qquad n \text{ odd};$$

or

$$2 I_2 = I_0 = \int_0^{\pi/2} dx = \frac{\pi}{2}, \qquad n \text{ even}.$$

When these equations are multiplied together, all I's cancel except I_n, and we find

$$\int_0^{\pi/2} \sin^n x \, dx = \begin{cases} \dfrac{(n-1)(n-3)\cdots 4 \cdot 2}{n(n-2)\cdots 5 \cdot 3} & (n \text{ odd}), \\[2ex] \dfrac{(n-1)(n-3)\cdots 3 \cdot 1}{n(n-2)\cdots 4 \cdot 2} \dfrac{\pi}{2} & (n \text{ even}). \end{cases}$$

It is easy to see that J_n has precisely the same values. With the notation !! for *semifactorials* in which the factors are all even or all odd, we have

$$(9) \qquad \int_0^{\pi/2} \sin^n x \, dx = \int_0^{\pi/2} \cos^n x \, dx = \frac{(n-1)!!}{n!!} \cdot \begin{cases} 1 & n \text{ odd}, \\[1ex] \dfrac{\pi}{2} & n \text{ even}. \end{cases}$$

PROBLEMS

1. Integrate by parts:

(a) $\displaystyle\int \cos(\log x) \, dx;$ (b) $\displaystyle\int e^{-x} \log(1 + e^x) \, dx;$

(c) $\displaystyle\int x^2 \tan^{-1} x \, dx;$ (d) $\displaystyle\int \frac{x+2}{(x+1)^2} e^{-x} \, dx.$

2. Find a recursion relation between the integrals for n and $n - 1$:

$$(a) \ E_n(x) = \int x^n e^x \, dx, \qquad (b) \ L_n(x) = \int (\log x)^n \, dx.$$

3. Find a recursion relation between S_n and C_{n-1}, C_n and S_{n-1}, when

$$S_n(x) = \int x^n \sin x \, dx, \qquad C_n(x) = \int x^n \cos x \, dx.$$

4. Show that $\displaystyle \lim_{\varepsilon \to 0} \int_\varepsilon^1 (\log x)^n \, dx = (-1)^n \, n!$

5. Show that $\displaystyle \int_a^b xf''(x) \, dx = x f'(x) - f(x) \Big|_a^b .$

6. If $P(x)$ is a polynomial of degree n, prove that

$$\int P(x) e^{ax} \, dx = \frac{e^{ax}}{a} \left\{ P(x) - \frac{P'(x)}{a} + \frac{P''(x)}{a^2} + \cdots + (-1)^n \frac{P^{(n)}(x)}{a^n} \right\}.$$

If $D^{-1} f(x)$ denotes the primitive (antiderivative) of $f(x)$, show that this result may be written

$$D^{-1} [e^{ax} P(x)] = e^{ax} (D + a)^{-1} P(x) = \frac{e^{ax}}{a} \left(1 + \frac{D}{a} \right)^{-1} P(x)$$

when $(1 + D/a)^{-1}$ is expanded into a geometric series.

7. By repeated integration by parts show that

$$\int_0^1 e^{\frac{1}{2}(1-x^2)} \, dx = 1 + \frac{1}{1 \cdot 3} + \frac{1}{1 \cdot 3 \cdot 5} + \cdots + \frac{1}{1 \cdot 3 \cdot 5 \ldots (2n-1)}$$

$$+ \int_0^1 \frac{x^{2n} \, e^{\frac{1}{2}(1-x^2)}}{1 \cdot 3 \cdot 5 \ldots (2n-1)} \, dx.$$

Show that the last integral tends to 0 as $n \to \infty$; hence verify that the given integral equals 1.41068 . . . .

8. If $f^{(n+1)}(x)$ is continuous when $|x - a| < k$, prove that

$$R_n(x) = \frac{1}{n!} \int_a^x f^{(n+1)}(t)(x - t)^n \, dt, \qquad |x - a| < k,$$

is the remainder in Taylor's expansion for $f(x)$ about $x = a$.

$$\left[\text{Show that } R_n(x) = R_{n-1}(x) - \frac{f^{(n)}(a)}{n!} (x - a)^n \text{ and use finite induction.} \right]$$

123. Change of Variable. The evaluation of an integral $\displaystyle \int_a^b f(x) \, dx$ is often simplified by making a change of variable $x = \varphi(t)$. Let $\varphi(t)$ vary continuously from a to b as t varies from α to β; if we then replace $f(x) \, dx$ by $f[\varphi(t)] \, \varphi'(t) \, dt$, we have, at least formally,

$$(1) \qquad \int_a^b f(x) \, dx = \int_\alpha^\beta f[\varphi(t)] \, \varphi'(t) \, dt.$$

Such a substitution is easy to remember for it is just what one would expect. But, since we are dealing with limiting processes, the above is not a *proof*. We therefore state and prove the

THEOREM. *We may transform* $\int_a^b f(x)\,dx$ *by the change of variable* $x = \varphi(t)$ *in accordance with equation* (1) *if*

(a) $a = \varphi(\alpha)$, $b = \varphi(\beta)$,

(b) $\varphi(t)$ *has a continuous derivative* $\varphi'(t)$ *in* (α, β),

(c) $f[\varphi(t)]$ *is continuous in* (α, β).

Proof. Consider the two functions of t,

$$F(t) = \int_a^{x=\varphi(t)} f(x)\,dx, \qquad G(t) = \int_\alpha^t f[\varphi(t)]\,\varphi'(t)\,dt.$$

The conditions show that $f(x)$ is continuous in (a, b), $f[\varphi(t)]\,\varphi'(t)$ in (α, β); hence, from (119.3),

$$\frac{dF}{dt} = \frac{dF}{dx}\frac{dx}{dt} = f(x)\,\varphi'(t), \qquad \frac{dG}{dt} = f[\varphi(t)]\,\varphi'(t).$$

Since $x = \varphi(t)$, these derivatives are equal; hence $F(t)$ and $G(t)$ can only differ by a constant C. But $C = 0$, since both functions vanish for $t = \alpha$; hence $F(t) = G(t)$, and, on putting $t = \beta$, we obtain (1).

This theorem can be proved under more general conditions. Thus, if the bounded function $f[\varphi(t)]$ is continuous except at the point $t = \gamma$, we can divide (α, β) into two intervals (α, γ), (γ, β). If we can show that (1) holds in each subinterval it will also hold in (α, β) by virtue of (117.1). Now in the interval $\alpha \leq t < \gamma$ all the requirements of the theorem are met; hence,

$$\int_a^{\varphi(t)} f(x)\,dx = \int_\alpha^t f[\varphi(t)]\,\varphi'(t)\,dt, \qquad (\alpha \leq t < \gamma).$$

But, since both integrals are continuous in the interval $\alpha \leq t \leq \gamma$ (Theorem 119.1), on passing to the limit $t \to \gamma$, we obtain (1) for the interval (α, γ). Similarly we may prove (1) for the interval (γ, β) and hence for (α, β). The same conclusion subsists if $f[\varphi(t)]$ has a finite number of discontinuities.

Example 1. If $f(x)$ is continuous in $0 \leq x \leq a$, the substitution $x = a - t$ shows that

(2) $$\int_0^a f(x)\,dx = -\int_a^0 f(a - t)\,dt = \int_0^a f(a - x)\,dx.$$

Again, if $\varphi(x, y)$ is a continuous function, we have

(3) $$\int_0^{\pi/2} \varphi(\sin x, \cos x)\,dx = \int_0^{\pi/2} \varphi(\cos x, \sin x)\,dx,$$

on making the substitution $x = \pi/2 - t$. Equation (122.9) is an instance of (3).

Example 2. From (117.1)

$$\int_{-a}^{a} f(x)\,dx = \int_{-a}^{0} + \int_{0}^{a} f(x)\,dx = \int_{0}^{a} f(-t)\,dt + \int_{0}^{a} f(x)\,dx,$$

on putting $x = -t$ in the integral over $(-a, 0)$. Hence,

$$\int_{-a}^{a} f(x)\,dx = 2 \int_{0}^{a} f(x)\,dx \quad \text{or} \quad 0,$$

according as $f(x)$ is an even or an odd function (§ 43).

Example 3. When $f(x + p) = f(x)$ for all values of x, the function $f(x)$ is said to b
periodic with the *period p*. Then we have

(4) $$\int_{0}^{p} f(x)\,dx = \int_{a}^{a+p} f(x)\,dx$$

for any a. For from (117.1)

$$\int_{0}^{p} f(x)\,dx = \int_{0}^{a} + \int_{a}^{a+p} + \int_{a+p}^{p} f(x)\,dx;$$

and, on putting $x = t + p$ in the last integral,

$$\int_{a+p}^{p} f(x)\,dx = \int_{a}^{0} f(t + p)\,dt = - \int_{0}^{a} f(t)\,dt.$$

Example 4. If $R(\sin x, \cos x)$ is a rational function of $\sin x$ and $\cos x$, its integral can
be found by the substitution

$$t = \tan \frac{x}{2}, \qquad x = 2 \tan^{-1} t.$$

Since

$$\sin x = 2 \sin \frac{x}{2} \cos \frac{x}{2} = 2 \tan \frac{x}{2} \cos^2 \frac{x}{2} = \frac{2t}{1 + t^2},$$

$$\cos x = \frac{\sin x}{\tan x} = \frac{2t/(1 + t^2)}{2t/(1 - t^2)} = \frac{1 - t^2}{1 + t^2},$$

$$dx = \frac{2dt}{1 + t^2},$$

we have

(5) $$\int R(\sin x, \cos x)\,dx = \int R\left(\frac{2t}{1 + t^2}, \frac{1 - t^2}{1 + t^2}\right) \frac{2dt}{1 + t^2},$$

in which the integrand is a rational function of t.

As an example of this method we have

(6) $$\int \frac{dx}{\sin x} = \int \frac{dt}{t} = \log |t| = \log \left| \tan \frac{x}{2} \right|.$$

If we put $x = t + \tfrac{1}{2}\pi$ in (6), we obtain

(7) $$\int \frac{dx}{\cos x} = \log \left| \tan \left(\frac{x}{2} + \frac{\pi}{4}\right) \right|$$

Of course, a direct application of (5) will also give this result.

Example 5. If $R(\sin^2 x, \cos^2 x)$ is a rational function of $\sin^2 x$ and $\cos^2 x$, we put $t = \tan x$. Then, since

$$\sin^2 x = \frac{t^2}{1 + t^2}, \qquad \cos^2 x = \frac{1}{1 + t^2}, \qquad dx = \frac{dt}{1 + t^2},$$

(8)
$$\int R(\sin^2 x, \cos^2 x)\, dx = \int R\left(\frac{t^2}{1 + t^2}, \frac{1}{1 + t^2}\right) \frac{dt}{1 + t^2}.$$

For example,

$$\int \frac{dx}{a^2 + b^2 \tan^2 x} = \int \frac{dt}{(1 + t^2)(a^2 + b^2 t^2)}$$

$$= \frac{1}{a^2 - b^2} \left(\int \frac{dt}{1 + t^2} - b^2 \int \frac{dt}{a^2 + b^2 t^2}\right)$$

$$= \frac{1}{a^2 - b^2} \left(\tan^{-1} t - \frac{b}{a} \tan^{-1} \frac{b}{a} t\right)$$

$$= \frac{1}{a^2 - b^2} \left[x - \frac{b}{a} \tan^{-1}\left(\frac{b}{a} \tan x\right)\right];$$

Hence,

$$\int_0^{\pi/2} \frac{dx}{a^2 + b^2 \tan^2 x} = \frac{1}{a^2 - b^2}\left(1 - \frac{b}{a}\right)\frac{\pi}{2} = \frac{\frac{1}{2}\pi}{a(a + b)}.$$

Again, if $a > b$,

$$\int \frac{dx}{a^2 - b^2 \sin^2 x} = \int \frac{dt}{a^2 + (a^2 - b^2)t^2}$$

$$= \frac{1}{a\sqrt{a^2 - b^2}} \tan^{-1} \frac{\sqrt{a^2 - b^2}}{a} t + C,$$

where $t = \tan x$; and

$$\int_0^{\pi/2} \frac{dx}{a^2 - b^2 \sin^2 x} = \frac{\frac{1}{2}\pi}{a\sqrt{a^2 - b^2}}.$$

Example 6. It is well to remember that we can write

$$a \sin x + b \cos x = \sqrt{a^2 + b^2} \sin(x + \alpha),$$

where α is the unique angle between 0 and 2π for which

$$\cos \alpha = a/\sqrt{a^2 + b^2}, \qquad \sin \alpha = b/\sqrt{a^2 + b^2}.$$

Thus we find from (6):

(9)
$$\int \frac{dx}{a \cos x + b \sin x} = \frac{1}{\sqrt{a^2 + b^2}} \int \frac{d(x + \alpha)}{\sin(x + \alpha)} = \frac{1}{\sqrt{a^2 + b^2}} \log\left|\tan \frac{x + \alpha}{2}\right|.$$

Example 7. When the primitive of $f(x)$ cannot be expressed in terms of known functions, the fundamental equation (120.1) is of no avail in computing $\int_a^b f(x)\, dx$. Nevertheless the definite integral may sometimes be found by suitable artifices.

Consider, for example,

$$I = \int_0^\pi \frac{x \sin x}{1 + \cos^2 x}\, dx.$$

Putting $x = \pi - t$, we get

$$I = \int_0^\pi \frac{(\pi - t) \sin t}{1 + \cos^2 t}\, dt;$$

hence, on adding these values,

$$2I = \pi \int_0^\pi \frac{\sin t\, dt}{1 + \cos^2 t} = -\pi \tan^{-1} \cos t \Big|_0^\pi,$$

$$I = -\frac{\pi}{2} [\tan^{-1}(-1) - \tan^{-1} 1] = \frac{\pi^2}{4}.$$

PROBLEMS

1. The substitutions

$$x = a \sin \theta, \qquad x = a \tan \theta, \qquad x = a \sec \theta$$

will remove the respective radicals,

$$\sqrt{a^2 - x^2}, \qquad \sqrt{a^2 + x^2}, \qquad \sqrt{x^2 - a^2},$$

from an integrand. Hence, integrate

$$\int (a^2 - x^2)^r\, dx, \qquad \int (a^2 + x^2)^r\, dx, \qquad \int (x^2 - a^2)^r\, dx,$$

when $r = \frac{1}{2},\ -\frac{1}{2}$ and $-\frac{3}{2}$.

2. When m, n, p, q are integers, show that the substitution $a + bx^n = t^q$ will rationalize the integrand

$$x^m(a + bx^n)^{p/q} \quad \text{when} \quad (m + 1)/n \quad \text{is an integer.}$$

Hence, integrate

$$\int \frac{x\, dx}{(4 - x^4)^{1/2}}, \qquad \int \frac{x^5\, dx}{(x^3 + 1)^{1/3}}, \qquad \int \frac{(x^3 - 4)^{1/2}}{x}\, dx.$$

3. Evaluate $\displaystyle\int_{\pi/4}^{\pi/3} \frac{\log \tan x}{\sin 2x}\, dx.$

4. Find $\displaystyle I = \int \frac{dx}{a + b \cos x}$ when $a + b > 0$ by means of the substitution $t = \tan \frac{1}{2}x$.

Show that

$$I = \frac{2}{\sqrt{a^2 - b^2}} \tan^{-1}\left(t\sqrt{\frac{a - b}{a + b}}\right) \quad \text{when} \quad a^2 > b^2,$$

$$I = \frac{1}{\sqrt{b^2 - a^2}} \log \frac{\sqrt{b + a} + t\sqrt{b - a}}{\sqrt{b + a} - t\sqrt{b - a}} \quad \text{when} \quad b^2 > a^2,$$

and reduces to t/a when $b = a$.

5. Integrate $x^3/(x^2 + 1)^3$ by the substitution $x^2 + 1 = t$; and by $x = \tan \theta$.

6. Integrate $1/x(1 + x^n)$, where n is a positive integer, by the substitution $t = 1 + x^n$.

7. Find $\displaystyle\int_0^1 x(1 - x)^n\, dx$ by the substitution $t = 1 - x$ and also by integration by parts.

124. Algebraic Integrals. An *algebraic function* $y = f(x)$ is one that satisfies a polynomial equation,

$$(1) \qquad\qquad\qquad P(x, y) = 0.$$

If $R(x, y)$ is a rational function of x and y,

$$(2) \qquad\qquad \int R(x, y)\, dx = \int R(x, f(x))\, dx$$

is called an *algebraic integral*.

If the plane curve $P(x, y) = 0$ admits a parametric representation,

$$(3) \qquad\qquad x = x(t), \qquad\qquad y = y(t),$$

where $x(t)$, $y(t)$ are rational functions, the curve is called *unicursal*. In this case the integral (2) becomes

$$(4) \qquad\qquad \int R(x, y)\, dx = \int R[x(t), y(t)]\, x'(t)\, dt.$$

The integrand is now a rational function of t, and the integral is always expressible in terms of the elementary functions by the use of partial fractions.

In particular, if $P(x, y)$ is a polynomial of the second degree, the curve (1) is a conic section. Now *all conics are unicursal curves*. For, if (a, b) is a point of the conic, a line

$$(5) \qquad\qquad\qquad y - b = t(x - a)$$

through this point will cut the conic in just *one* other point (x, y). Since the quadratic

$$P(x, t(x - a) + b) = 0$$

has the factor $x - a$, when this is canceled, we have an equation linear in x which gives x as a rational function of t. Then (5) gives y as a rational function of t.

Example 1. $y = \sqrt{\dfrac{ax + b}{cx + d}}$ (where $ad - bc \neq 0$) satisfies the equation

$$(cx + d)y^2 = ax + b,$$

which represents a cubic curve. The line $y = t$ cuts it in the point

$$x = \frac{dt^2 - b}{a - ct^2}, \qquad y = t.$$

These are rational parametric equations of the curve; and with these functions $\displaystyle\int R(x, y)\, dx$ is given by (4).

In particular, if $y = \sqrt{ax + b}$ (the case $c = 0$, $d = 1$), we have $x = (t^2 - b)/a$, and

$$(6) \qquad\qquad \int R(x, \sqrt{ax + b})\, dx = \frac{2}{a} \int R\left(\frac{t^2 - b}{a}, t\right) t\, dt.$$

Example 2. $y = \sqrt{ax^2 + bx + c}$, where $a > 0$, satisfies the equation

(7) $$y^2 = ax^2 + bx + c,$$

which represents a hyperbola with the lines $y = \pm\sqrt{a}\,x$ as asymptotes. A line parallel to an asymptote, such as

$$y = -\sqrt{a}\,x + t,$$

will cut the curve in just one point, namely,

(8) $$x = \frac{t^2 - c}{2t\sqrt{a} + b}, \qquad y = \frac{(t^2 + c)\sqrt{a} + bt}{2t\sqrt{a} + b};$$

and

$$\frac{dx}{dt} = 2\,\frac{(t^2 + c)\sqrt{a} + bt}{(2t\sqrt{a} + b)^2},$$

For example, the integral

(9) $$\int \frac{dx}{y} = \int \frac{dt}{t\sqrt{a} + \frac{1}{2}b} = \frac{1}{\sqrt{a}}\log\left| t + \frac{b}{2\sqrt{a}} \right|$$

$$= \frac{1}{\sqrt{a}}\log\left| y + \sqrt{a}\,x + \frac{b}{2\sqrt{a}} \right|.$$

PROBLEMS

1. Using the method of Ex. 2, show that

$$\int \frac{dx}{\sqrt{x^2 \pm a^2}} = \log\left(x + \sqrt{x^2 \pm a^2}\right).$$

2. Show that the conic $y^2 = ax^2 + bx + c$ $(c > 0)$ has the parametric equations,

$$x = \frac{b - 2\sqrt{c}\,t}{t^2 - a}, \qquad y = \frac{bt - \sqrt{c}\,(t^2 + a)}{t^2 - a}.$$

[Put $y - \sqrt{c} = tx$; note that the point $(0, \sqrt{c})$ lies on the conic.]

3. If $y = \sqrt{(x - \alpha)(x - \beta)}$, where α and β are real and different, put $y = t(x - \alpha)$, and show that

$$\int R(x, y)\,dx = 2(\beta - \alpha) \int R\left(\frac{\alpha t^2 - \beta}{t^2 - 1}, \frac{(\alpha - \beta)t}{t^2 - 1}\right) \frac{t\,dt}{(t^2 - 1)^2}.$$

This method and that of Ex. 2 give, respectively,

$$\int \frac{dx}{y} = \log\left| \frac{x + y - \alpha}{x - y - \alpha} \right| \quad \text{and} \quad \int \frac{dx}{y} = \log\left| x + y - \frac{\alpha + \beta}{2} \right|.$$

Reconcile these results.

4. Reduce $\int R(x, y)\,dx$ to the integral of a rational function when $y^2(x - y) = x^2$. [Put $y = tx$.]

5. If $y(x - y)^2 = x$, show that

$$\int \frac{dx}{x - 3y} = \frac{1}{2}\log\left| (x - y)^2 - 1 \right|.$$

[Put $x - y = t$.]

125. Duhamel's Theorem. We next prove a theorem which is useful in the applications of integration. Its proof depends on the *uniform* continuity of a continuous function in a closed region.

DUHAMEL'S THEOREM.† *If $\varphi(x, y)$ is a continuous function of x and y in the square $a \leq x \leq b$, $a \leq y \leq b$, and*

$$f(x) = \varphi(x, x), \qquad a \leq x \leq b,$$

the integral

(1)
$$\int_a^b f(x)\, dx = \lim_{\delta \to 0} \sum_{i=1}^{i=n} \varphi(\xi_i, \xi_i')\delta_i,$$

where ξ_i, ξ_i' are any points of the subinterval δ_i.

Proof. By definition

(2)
$$\int_a^b f(x)\, dx = \lim_{\delta \to 0} \sum_{i=1}^{n} \varphi(\xi_i, \xi_i)\delta_i.$$

Hence, in order to prove (1) we must show that the sums in (1) and (2) approach the same limit. Since $\varphi(x, y)$ is a continuous function in a closed square, it is uniformly continuous there (§ 75). Hence, we can find a positive number Δ such that

$$|\varphi(x, y') - \varphi(x, y)| < \frac{\varepsilon}{b - a} \quad \text{when} \quad |y' - y| < \Delta.$$

Hence, when $\delta = \max \delta_i < \Delta$,

$$\left| \sum [\varphi(\xi_i, \xi_i') - \varphi(\xi_i, \xi_i)]\delta_i \right| \leq \sum |\varphi(\xi_i, \xi_i') - \varphi(\xi_i, \xi_i)|\delta_i < \frac{\varepsilon}{b - a} \sum \delta_i = \varepsilon;$$

that is, as $\delta \to 0$, the difference between the sums approaches zero and consequently the sums approach the same limit.

The theorem permits of obvious extensions. Thus if $\varphi(x, y, z)$ is continuous in the cube $a \leq x, y, z \leq b$ and

$$f(x) = \varphi(x, x, x), \qquad a \leq x \leq b,$$

the integral

(3)
$$\int_a^b f(x)\, dx = \lim_{\delta \to 0} \sum_{i=1}^{n} \varphi(\xi_i, \xi_i', \xi_i'')_i,$$

where ξ_i, ξ_i', ξ_i'' are any points of the subinterval δ_i.

---

† This theorem differs in form, but not in substance, from Duhamel's theorem. It includes as a special case

BLISS' THEOREM. *If $f(x)$ and $g(x)$ are continuous in the interval $a \leq x \leq b$,*

$$\int_a^b f(x)g(x)\, dx = \lim_{\delta \to 0} \sum_{i=1}^{n} f(\xi_i)g(\xi_i')\delta_i$$

where ξ_i, ξ_i' are any points of the subinterval δ_i.

126. Areas. Let $f(x)$ be a positive continuous function when $a \leq x \leq b$. The curve $y = f(x)$ then lies above the segment (a, b) of the x-axis. To find the area between the curve and the x-axis we divide (a, b) into n subintervals δ_i as in § 113 and erect ordinates at the points of division. Then the areal strip above δ_i lies between $m_i\delta_i$ and $M_i\delta_i$; and hence the entire area is included between

$$s = \sum m_i\delta_i \quad \text{and} \quad S = \sum M_i\delta_i.$$

Since $f(x)$ is integrable in (a, b), both s and S approach the same limit,

$$(1) \qquad A = \int_a^b y \, dx = \int_a^b f(x) \, dx,$$

as $\delta = \max \delta_i \to 0$. Hence A is defined as the area between the curve $y = f(x)$, the x-axis, and the lines $x = a$ and $x = b$. Since we have assumed $f(x) > 0$, $A > 0$. When $f(x) < 0$, $A < 0$. If $f(x)$ changes sign just once between a and b and $f(c) = 0$,

$$A = \int_a^c f(x) \, dx + \int_c^b f(x) \, dx;$$

then A is the difference between the area above the x-axis and the area below.

If the curve $y = f(x)$ has the parametric equations,

$$x = x(t), \qquad y = f[x(t)] = y(t),$$

and $a = x(\alpha)$, $b = x(\beta)$, the substitution $x = x(t)$ gives

$$(2) \qquad A = \int_\alpha^\beta y(t) \, x'(t) \, dt.$$

If the curve $y = f(x)$ is plotted with equal units on the rectangular axes, the unit of area is the unit square. However, if the units on the axes are unequal, the unit of area is the rectangle having the x-unit and y-unit as sides. When the curve is plotted, a rough check on the computed area may be obtained from the figure.

Example 1. The area under one arch of the *sine curve* $y = \sin x$ is

$$A = \int_0^\pi \sin x \, dx = -\cos x \Big|_0^\pi = 1 + 1 = 2.$$

The area from $x = 0$ to $x = 3\pi/2$ is $0 + 1 = 1$; the area 2 above the x-axis is partly canceled by the area -1 below. The area from $x = 0$ to $x = 2\pi$ is $-1 + 1 = 0$; now the positive and negative areas, 2 and -2, cancel each other.

Example 2. The *cycloid* (Fig. 126a), referred to the central angle $t = ACP$ as parameter, has the equations

$$x = OQ = OA - PR = a(t - \sin t),$$
$$y = QP = AC - RC = a(1 - \cos t).$$

Hence, the area under one arch is

$$A = a^2 \int_0^{2\pi} (1 - \cos t)^2 \, dt$$

$$= a^2 \int_0^{2\pi} \left[1 - 2 \cos t + \frac{1}{2}(1 + \cos 2t) \right] dt$$

$$= a^2 \left[\frac{3}{2}t - 2 \sin t + \frac{1}{4} \sin 2t \right]_0^{2\pi} = 3\pi a^2.$$

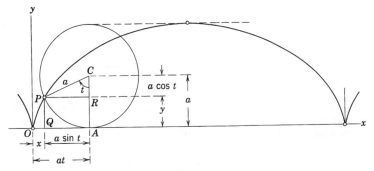

FIG. 126a.   Cycloid

Example 3. The *ellipse* $\dfrac{x^2}{a^2} + \dfrac{y^2}{b^2} = 1$ has the parametric equations

$$x = a \cos t, \qquad y = b \sin t$$

where t is the *eccentric angle AOR* (Fig. 126b). The area S of the elliptic sector OAP, from the vertex $t = 0$ to $t = \varphi$, is the area of the triangle OQP plus the area QAP:

$$S = \frac{1}{2}xy + \int_{a \cos \varphi}^{a} y \, dx$$

$$= \frac{1}{2} ab \cos \varphi \sin \varphi - ab \int_{\varphi}^{0} \sin^2 t \, dt$$

$$= \frac{1}{4} ab \sin 2\varphi + \frac{1}{2} ab \int_0^{\varphi} (1 - \cos 2t) \, dt.$$

Since the integral equals $\varphi - \dfrac{1}{2} \sin 2\varphi$,

(3) $$S = \frac{1}{2} ab\varphi = \frac{1}{2} ab \cos^{-1} \frac{x}{a}.$$

Thus $\varphi = 2S/ab$ gives another geometric meaning to the parameter. At a point P ($t = \varphi$) of the "equilateral ellipse" $x^2 + y^2 = 1$, φ is twice the area of the circular sector OAP.

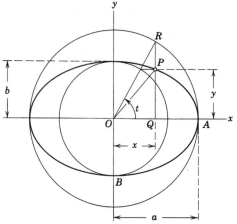

Fig. 126b.   Ellipse

Example 4. The *hyperbola* $x^2/a^2 - y^2/b^2 = 1$ has the parametric equations

$$x = a \cosh t, \qquad y = b \sinh t$$

where t is not an angle but a number whose geometric meaning will soon be clear. The area S of the hyperbolic sector OAP (Fig. 126c), from the vertex $t = 0$ to $t = \varphi$, is the area of the triangle OQP minus the area AQP:

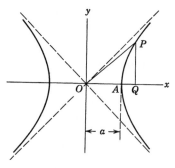

Fig. 126c.   Hyperbola

$$S = \frac{1}{2}xy - ab \int_a^{a \cosh \varphi} y \, dx$$

$$= \frac{1}{2} ab \cosh \varphi \sinh \varphi - ab \int_0^\varphi \sinh^2 t \, dt$$

$$= \frac{1}{4} ab \sinh 2\varphi - \frac{1}{2} ab \int_0^\varphi (\cosh 2t - 1) \, dt.$$

Since the integral equals $\frac{1}{2} \sinh 2\varphi - \varphi$,

$$(4) \qquad S = \frac{1}{2} ab\varphi = \frac{1}{2} ab \cosh^{-1} \frac{x}{a}.$$

Thus $\varphi = 2S/ab$ gives a geometric meaning to the parameter. At a point P ($t = \varphi$) of the equilateral hyperbola $x^2 - y^2 = 1$, φ is twice the area of the hyperbolic sector OAP. From (53.6) we have

$$\cosh^{-1} \frac{x}{a} = \log \left(\frac{x}{a} + \sqrt{\frac{x^2}{a^2} - 1} \right) = \log \left(\frac{x}{a} + \frac{y}{b} \right);$$

hence the hyperbolic sector (4) is also given by

$$(5) \qquad S = \frac{1}{2} ab \log \left(\frac{x}{a} + \frac{y}{b} \right).$$

127. Functions of Bounded Variation.   A function $f(x)$ is said to be of *bounded variation* in an interval $a \leq x \leq b$ if the sums

$$(1) \qquad S = \sum_1^n |f(x_i) - f(x_{i-1})|$$

are bounded for all possible nets,

$$a = x_0 < x_i < x_2 < \cdots < x_n = b \quad \text{over } (a, b).$$

If $f(x)$ is monotone in (a, b), it is certainly of bounded variation; for all the sums (1) equal $|f(b) - f(a)|$.   The same is true if $f(x)$ is merely *piecewise monotone* in (a, b), that is, if (a, b) can be subdivided into a finite number of intervals in each of which $f(x)$ is monotone.   But a continuous function need not be of bounded variation.   Thus the function

$$f(x) = x \sin \frac{1}{x} \quad (x \neq 0), \qquad f(0) = 0,$$

is not of bounded variation in $0 \leq x \leq 1/\pi$; for, if n is an odd integer, the net

$$0, \quad \frac{2}{n\pi}, \quad \frac{2}{(n-1)\pi}, \quad \cdots \quad \frac{2}{3\pi}, \quad \frac{2}{2\pi}$$

gives

$$S = \frac{4}{\pi}\left(\frac{1}{n} + \frac{1}{n-2} + \cdots + \frac{1}{3}\right),$$

which becomes infinite as $n \to \infty$.

If $f(x)$ has a bounded derivative in (a, b), the mean-value theorem (57.1) shows that $f(x)$ is of bounded variation; for, if $|f'(x)| < M$,

$$S = \sum_1^n |f'(\xi_i)| \, (x_i - x_{i-1}) < M(b - a).$$

Important properties of functions of bounded variation are stated in the following theorems

THEOREM 1.   *A function of bounded variation in a closed interval is also bounded.*

Proof.   For the net a, x, b we have

$$|f(x) - f(a)| + |f(b) - f(x)| < M;$$

hence,

$$|f(x) - f(a)| < M \quad \text{and} \quad |f(x)| < M + |f(a)|.$$

THEOREM 2. *A function of bounded variation in a closed interval can always be expressed as the difference of two increasing functions in this interval.*

Proof. Let $F(x)$ denote the least upper bound of all the sums S over the interval (a, x). Defining $F(0) = 0$, $F(x)$ is then a bounded, monotone increasing function in $a \leq x \leq b$. Now express

$$f(x) = \frac{F(x) + f(x)}{2} - \frac{F(x) - f(x)}{2} = p(x) - q(x).$$

Then $p(x)$ and $q(x)$ increase monotonely in (a, b); for, if $x_2 > x_1$, the definition of $F(x)$ shows that

$$F(x_2) - F(x_1) \geq |f(x_2) - f(x_1)|,$$

and hence

$$p(x_2) - p(x_1) = \tfrac{1}{2}[F(x_2) - F(x_1) + f(x_2) - f(x_1)] \geq 0,$$
$$q(x_2) - q(x_1) = \tfrac{1}{2}[F(x_2) - F(x_1) - f(x_2) + f(x_1)] \geq 0.$$

As a consequence of these theorems and Theorem 115.3 we have

THEOREM 3. *A function of bounded variation in an interval is integrable there.*

128. Length of a Curve. A curve, given by the parametric equations,

$$(1) \qquad x = x(t), \qquad y = y(t), \qquad \alpha \leq t \leq \beta,$$

is said to be *continuous* when the functions $x(t)$, $y(t)$ are continuous in (α, β). If the derivatives $x'(t)$, $y'(t)$ exist and are continuous in (α, β) and never vanish together, the curve is said to be *smooth*.

To define the length of a continuous curve (1), take any net over (α, β),

$$(2) \qquad \alpha = t_0 < t_1 < t_2 < \cdots < t_n = \beta,$$

with n subintervals $\delta_i = t_i - t_{i-1}$. The $n + 1$ points of the curve corresponding to these parameter values form the vertices of a broken line inscribed in the curve. The length of this broken line is

$$(3) \qquad \lambda = \sum_{i=1}^{n} \sqrt{[x(t_i) - x(t_{i-1})]^2 + [y(t_i) - y(t_{i-1})]^2}.$$

The positive numbers λ for all possible choices of the net (2) will have a least upper bound. When this l.u.b. is a finite number L, the curve is said to be *rectifiable* and to have the length L. Then for any $\varepsilon > 0$ we can always find a net t_i for which

$$(4) \qquad L - \varepsilon < \lambda \leq L.$$

THEOREM 1. *The curve* (1) *is rectifiable when and only when the functions* $x(t)$ *and* $y(t)$ *are of bounded variation in* (α, β).

Proof. If we write

$$X = \sum_{1}^{n} |x(t_i) - x(t_{i-1})|, \qquad Y = \sum_{1}^{n} |y(t_i) - y(t_{i-1})|,$$

we have from (3)

$$X \text{ or } Y \leq \lambda \leq X + Y;$$

for the radical in (3) lies between either one of the quantities

$$|x(t_i) - x(t_{i-1})|, \qquad |y(t_i) - y(t_{i-1})|$$

and their sum.†   Hence, when X and Y are bounded, λ is bounded; and, if λ is bounded, both X and Y are bounded.

When the net (2) is refined by adding intermediate t-values, the new vertices added to the broken line will cause λ to increase if it changes at all. This is due to the *triangle inequality*; any two sides of a triangle must exceed the third.‡   We can now prove the important

THEOREM 2. *If the continuous curve* (1) *is rectifiable, its length*

$$(5) \qquad\qquad L = \lim_{\delta \to 0} \lambda,$$

where λ *is the length of the broken line* (3) *corresponding to a net over* (α, β) *of mesh* δ.

Proof. The argument is essentially the same as in Theorem 114.1, with s and I replaced by λ and L. In § 114 a refinement of the net increased s; now it increases λ. We proceed to show that $|L - \lambda| < \varepsilon$ for *all* nets of sufficiently small mesh.

Let N be a net of mesh δ for which λ is given by (3). Next choose a net N_1 for which $\lambda_1 > L - \frac{1}{2}\varepsilon$. Let N_1 have k points between α and β; henceforth we hold N_1 fast so that k is constant. Since the continuous functions $x(t)$, $y(t)$ are also *uniformly* continuous in (α, β), we can find a number Δ so that

$$\sqrt{[x(t) - x(t')]^2 + [y(t) - y(t')]^2} < \frac{\varepsilon}{4k} \quad \text{when} \quad |t - t'| < \Delta.$$

---

† The hypothenuse of a right triangle is greater than either side but less than their sum.

‡ In the triangle formed by the vectors **a**, **b**, **a** + **b**,

$$|\mathbf{a} + \mathbf{b}|^2 = (\mathbf{a} + \mathbf{b}) \cdot (\mathbf{a} + \mathbf{b}) = |\mathbf{a}|^2 + 2\mathbf{a} \cdot \mathbf{b} + |\mathbf{b}|^2$$

$$\leq |\mathbf{a}|^2 + 2|\mathbf{a}||\mathbf{b}| + |\mathbf{b}|^2 = (|\mathbf{a}| + |\mathbf{b}|)^2;$$

hence,

$$|\mathbf{a} + \mathbf{b}| \leq |\mathbf{a}| + |\mathbf{b}|.$$

Denote by N_2 the net formed by the points of both N and N_1; for this net the length of the corresponding broken line,

$$(6) \qquad \lambda_2 \geq \lambda_1 > L - \tfrac{1}{2}\varepsilon.$$

Now N has two types of subintervals:

(i) those with *no* points of N_1 inside,
(ii) those having points of N_1 inside.

The subintervals of type (i) contribute an amount to λ_2 which certainly is less than λ. The subintervals of type (ii) number at most k and contain at most $2k$ subintervals of N_2; and, when $\delta < \Delta$, their total contribution to λ_2 is less than $2k(\varepsilon/4k) = \tfrac{1}{2}\varepsilon$. Hence,

$$\lambda + \tfrac{1}{2}\varepsilon > \lambda_2 \quad \text{or} \quad \lambda > L - \tfrac{1}{2}\varepsilon - \tfrac{1}{2}\varepsilon$$

in view of (6). Thus,

$$L \geq \lambda > L - \varepsilon \quad \text{when} \quad \delta < \Delta,$$

and, consequently, $\lambda \to L$ as $\delta \to 0$.

129. Smooth Curves. Let the equations

$$(1) \qquad x = x(t), \qquad y = y(t), \qquad \alpha \leq t \leq \beta$$

represent a *smooth* curve. It will have no *double points* if each t in the interval $\alpha \leq t < \beta$ gives a different point (x, y). The curve is *closed* when $x(\alpha) = x(\beta)$, $y(\alpha) = y(\beta)$. If a curve without double points consists of a finite number of smooth arcs, the curve is called *regular*. A regular curve is *piecewise smooth*; its tangent turns continuously except at certain points where two arcs meet. We shall show that a regular curve is rectifiable.

THEOREM. *An arc of a smooth curve is rectifiable, and its length from $t = \alpha$ to $t = \beta$ is given by*

$$(2) \qquad L = \int_\alpha^\beta \sqrt{x'(t)^2 + y'(t)^2}\, dt.$$

Proof. Since the curve is smooth, $x'(t)$, $y'(t)$ are continuous in (α, β), and the mean-value theorem (57.1) gives

$$x(t_i) - x(t_{i-1}) = (t_i - t_{i-1})\, x'(\xi_i),$$

$$y(t_i) - y(t_{i-1}) = (t_i - t_{i-1})\, y'(\eta_i),$$

where $t_{i-1} < \xi_i, \eta_i < t_i$. Hence the broken line corresponding to any net t_i has the length

$$\lambda = \sum_{i=1}^n \sqrt{x'(\xi_i)^2 + y'(\eta_i)^2}\, \delta_i,$$

where $\delta_i = t_i - t_{i-1}$. Since

$$\varphi(t, u) = \sqrt{x'(t)^2 + y'(u)^2}$$

is a continuous function of t and u, by Duhamel's theorem (125.1).

$$\lim_{\delta \to 0} \lambda = \int_\alpha^\beta \varphi(t, t)\, dt = \int_\alpha^\beta \sqrt{x'(t)^2 + y'(t)^2}\, dt,$$

where $\delta = \max \delta_i$.

Corollary. *A smooth curve in space,*

$$(3) \qquad x = x(t), \qquad y = y(t), \qquad z = z(t) \qquad (\alpha \le t \le \beta),$$

has the arc length,

$$(4) \qquad L = \int_\alpha^\beta \sqrt{x'(t)^2 + y'(t)^2 + z'(t)^2}\, dt.$$

Proof. The broken line for the net t_i now has the length

$$\lambda = \sum_{i=1}^n \sqrt{x'(\xi_i)^2 + y'(\eta_i)^2 + z'(\zeta_i)^2}\, \delta_i,$$

where $t_{i-1} < \xi_i, \eta_i, \zeta_i < t_i$. In passing to the limit $\delta \to 0$, we use the form (125.3) of Duhamel's theorem.

In vector notation both formulas (2) and (4) are included in

$$(5) \qquad L = \int_\alpha^\beta \left| \frac{d\mathbf{r}}{dt} \right| dt,$$

where $\mathbf{r} = [x, y, z]$ is the position vector to the curve and $d\mathbf{r}/dt$ a tangent vector (101.4). Note that the tangent vector $d\mathbf{r}/dt$ of a smooth curve exists since $x'(t)$, $y'(t)$, $z'(t)$ exist and never vanish together; moreover, the continuity of the derivatives makes the tangent turn continuously as the curve is traversed.

The arc length s from t_0 to a variable point t is

$$(6, 7) \qquad s = \int_{t_0}^t \left| \frac{d\mathbf{r}}{dt} \right| dt \quad \text{and} \quad \frac{ds}{dt} = \left| \frac{d\mathbf{r}}{dt} \right| > 0.$$

Thus s is positive or negative according as $t > t_0$ or $t < t_0$; consequently, the parameter determines a sense on the curve. Moreover, $ds/dt > 0$ shows that s increases with t.

If, in particular, we use the arc s as parameter, we have, from (7),

$$(8) \qquad \left| \frac{d\mathbf{r}}{ds} \right| = \frac{ds}{ds} = 1;$$

hence, $d\mathbf{r}/ds$ is a *unit tangent vector* (102.1).   Moreover, since

$$(9) \qquad \lim_{\Delta s \to 0} \frac{|\Delta \mathbf{r}|}{|\Delta s|} = \left|\frac{d\mathbf{r}}{ds}\right| = 1,$$

we see that:

The ratio of a chord of a smooth curve to the arc it subtends approaches unity as the arc approaches zero.

On squaring (7) and multiplying through by dt^2, we have

$$(10) \qquad ds^2 = |d\mathbf{r}|^2 = dx^2 + dy^2 + dz^2$$

The independent variable implied in this equation is an *arbitrary* parameter t; and dx, dy, dz, ds mean the derivatives $dx/dt, dy/dt, dz/dt, ds/dt$ multiplied by $dt = \Delta t$ where Δt can be chosen at pleasure.

A curve having the Cartesian equation $y = f(x)$ may be put in the parametric form $x = t$, $y = f(t)$. Hence, if $f'(x)$ is continuous in $a \leqq x \leqq b$, the arc in this interval has the length

$$(11) \qquad L = \int_a^b \sqrt{1 + f'(x)^2}\, dx.$$

A curve having the polar equation $r = f(\theta)$ has the parametric form

$$x = r \cos \theta, \qquad y = r \sin \theta \qquad (\alpha \leqq \theta \leqq \beta).$$

If we take $t = \theta$, the polar angle,

$$x' = r' \cos \theta - r \sin \theta, \qquad y' = r' \sin \theta + r \cos \theta,$$

$$x'^2 + y'^2 = r'^2 + r^2,$$

we have from (2)

$$(12) \qquad L = \int_\alpha^\beta \sqrt{r'^2 + r^2}\, d\theta.$$

Example 1. The *cycloid* (Ex. 126.2) has the equations

$$x = a(t - \sin t), \qquad y = a(1 - \cos t),$$

in which a is the radius of the rolling circle and t the angle through which it has rolled. Since

$$x' = a(1 - \cos t)\, dt, \qquad y' = a \sin t\, dt.$$

$$\sqrt{x'^2 + y'^2} = a\sqrt{(1 - \cos t)^2 + \sin^2 t} = a\sqrt{2 - 2 \cos t}$$

$$= a\sqrt{4 \sin^2 \frac{t}{2}} = 2a \sin \frac{t}{2},$$

the length of the cycloidal arc when the circle has rolled through an angle t,

$$s = 4a \int_0^t \sin \frac{t}{2} \frac{dt}{2} = -4a \cos \frac{t}{2}\Big|_0^t = 4a \left(1 - \cos \frac{t}{2}\right).$$

For a complete arch $t = 2\pi$ and $s = 8a$.

Example 2. The *parabola* $2ay = x^2$ has its vertex at the origin. Since

$$2ay' = 2x, \qquad y' = \frac{x}{a}, \qquad \sqrt{1 + y'^2} = \frac{\sqrt{a^2 + x^2}}{a},$$

the arc length from $x = 0$ to $x > 0$ is

$$s = \frac{1}{a} \int_0^x \sqrt{a^2 + x^2} \, dx = a \int_0^t \cosh^2 t \, dt,$$

on putting $x = a \sinh t$. An integration by parts ($u = \cosh t$) now gives

$$s = \frac{a}{2} (\sinh t \cosh t + t) = \frac{x}{2a} \sqrt{x^2 + a^2} + \frac{a}{2} \sinh^{-1} \frac{x}{a}.$$

We may express $\sinh^{-1}(x/a)$ in terms of logarithms by using (53.4),

Example 3. The *catenary* $y = a \cosh \dfrac{x}{a}$ has its vertex at $(0, a)$. Since

$$y' = \sinh \frac{x}{a} \qquad \sqrt{1 + y'^2} = \cosh \frac{x}{a},$$

the arc length from $x = 0$ to $x > 0$ is

$$s = \int_0^x \cosh \frac{x}{a} \, dx = a \sinh \frac{x}{a}.$$

Example 4. An *equiangular spiral* which cuts all polar radii at the constant angle α has the polar equation,

$$r = ae^{k\theta} \quad \text{where} \quad k = \cot \alpha.$$

Since $r' = ake^{k\theta}$,

$$\sqrt{r^2 + r'^2} = a\sqrt{1 + k^2} \, e^{k\theta} = \frac{a}{\sin \alpha} e^{k\theta};$$

and the arc length from (θ_0, r_0) to (θ, r) is

$$s = \frac{a}{\sin \alpha} \int_{\theta_0}^\theta e^{k\theta} \, d\theta = \frac{a}{k \sin \alpha} (e^{k\theta} - e^{k\theta_0}) = \frac{r - r_0}{\cos \alpha}.$$

Thus the arc of an equiangular spiral is proportional to increment in the polar radii. Since $r \to 0$ as $\theta \to -\infty$, the spiral coils infinitely often about the pole without ever reaching it. However as the point (θ_0, r_0) approaches the pole asymptotically, the length remains finite; for

$$\lim_{r_0 \to 0} s = r/\cos \alpha.$$

Example 5. The *ellipse* (Ex. 126.3) has the equations

$$x = a \cos t, \qquad y = b \sin t.$$

The length of the arc AP (Fig. 126b), measured from the end of the major axis, to the point $t = \varphi$, is

$$s(\varphi) = \int_0^\varphi \sqrt{a^2 \sin^2 t + b^2 \cos^2 t} \, dt = a \int_0^\varphi \sqrt{1 - \varepsilon^2 \cos^2 t} \, dt,$$

where $\varepsilon = (a^2 - b^2)^{1/2}/a$ is the eccentricity. To reduce this to an *elliptic integral of the second kind*,

(13) $$E(k, \varphi) = \int_0^\varphi \sqrt{1 - k^2 \sin^2 u}\, du,$$

we put $u = t + \tfrac{1}{2}\pi$; then

(14) $$s(\varphi) = \int_{\frac{1}{2}\pi}^{\varphi + \frac{1}{2}\pi} \sqrt{1 - \varepsilon^2 \sin^2 u}\, du = aE(\varepsilon, \varphi + \tfrac{1}{2}\pi) - aE(\varepsilon, \tfrac{1}{2}\pi).$$

Thus one fourth of the elliptic perimeter is

(15) $$s(\tfrac{1}{2}\pi) = aE(\varepsilon, \pi) - aE(\varepsilon, \tfrac{1}{2}\pi) = aE(\varepsilon, \tfrac{1}{2}\pi).$$

If we add this to $s(\varphi)$, we obtain the

(16) $$\text{arc } BP = aE(\varepsilon, \varphi + \tfrac{1}{2}\pi), \qquad -\frac{\pi}{2} \leq \varphi \leq \frac{3\pi}{2},$$

measured from the lower end B of the minor axis.

Example 6. The *circular helix* of radius a,

$$x = a \cos t, \qquad y = a \sin t, \qquad z = bt,$$

winds about the z-axis, rising $2\pi b$ in each revolution. From (4), the arc length measured from $t = 0$ is

$$s = \int_0^t \sqrt{a^2 + b^2}\, dt = t\sqrt{a^2 + b^2}.$$

PROBLEMS

1. Find the length of the curve $x = t^2$, $y = t^3$ from $t = 0$ to $t = 5$.

2. Find the length of the *involute of a circle*
$$x = a(\cos t + t \sin t), \qquad y = a(\sin t - t \cos t)$$
from $t = 0$ to $t = \pi$.

3. Find the entire length of the *hypocycloid of four cusps*: $x = a \cos^3 t$, $y = a \sin^3 t$.

4. Find the entire length of the curve $r = a \sin^3 \tfrac{1}{3}\theta$.

5. Find the entire length of the *cardioid* $r = a(1 - \cos \theta)$.

6. Find the length of the curve $x = 2t$, $y = t^2$, $z = \log t$ from $t = 1$ to any point t.

7. Find the length of the curve $x = 2t$, $y = 3t^2$, $z = 3t^3$ from $t = 0$ to $t = 1$.

8. Find the length of the curve $x = \cosh t$, $y = \sinh t$, $z = t$ from $t = 0$ to any point t.

9. Find the length of the curves

(a) $y = a \log \dfrac{x}{b} - \dfrac{x^2}{8a}$ from $x = b$ to $x = c$;

(b) $y = \dfrac{x^3}{3a^2} + \dfrac{a^2}{4x}$ from $x = a$ to $x = 2a$.

Do your answers check dimensionally?

130. Approximate Integration. In many cases, and for comparatively simple functions such as

$$f(x) = \sqrt{1 + x^3}, \quad 1/\log x, \quad \sin x^2, \quad \sqrt{\sin x}, \quad e^{-x^2}, \quad e^x/x,$$

the corresponding indefinite integral $F(x)$ cannot be expressed by means of a *finite* number of operations on the elementary functions.†

The fundamental theorem

$$\int_a^b f(x)\, dx = F(b) - F(a)$$

is then of no service in computing the definite integral. In such cases we may resort to approximate integration. One method is to replace $f(x)$ over the interval of integration by a polynomial $P(x)$ such that $P(x) = f(x)$ when $x = a, b$ and at certain intermediate points. The simplest polynomial is $Ax + B$; but a closer approximation is obtained in using the quadratic

$$P(x) = Ax^2 + Bx + C.$$

We shall choose A, B, C so that $f(x) = P(x)$ at three equidistant points. If we take the origin at the central point, the abscissas may be written $x = -h$, $x = 0$, $x = h$. Thus we have the equations

$$f(-h) = Ah^2 - Bh + C, \quad f(0) = C, \quad f(h) = Ah^2 + Bh + C$$

to determine A, B, C. Now

$$\int_{-h}^{h} (Ax^2 + Bx + C)\, dx = \frac{1}{3} Ax^3 + \frac{1}{2} Bx^2 + Cx \Big|_{-h}^{h} = \frac{2}{3} Ah^3 + 2Ch$$

and, since

$$C = f(0), \qquad 2Ah^2 = f(-h) + f(h) - 2f(0),$$

$$(2) \qquad \int_{-h}^{h} P(x)\, dx = \frac{h}{3} [f(-h) + 4f(0) + f(h)].$$

This formula, known as *Simpson's rule*, gives the exact area under the parabola $y = P(x)$ and an approximation to the area

$$(3) \qquad \int_{-h}^{h} f(x)\, dx = F(h) - F(-h),$$

under the curve $y = f(x)$. We may express this rule as follows:

Simpson's Rule. Let a curve lying above the x-axis have the ordinates y_1, y_2, y_3 at equal distances h. Then the area between the curve, the x-axis,

---

† The reader should consult Ritt, *Integration in Finite Terms*, Columbia University Press, 1948, for a systematic treatment of this question, due principally to the researches of Liouville in the period 1833–1841.

and the extreme ordinates is approximately equal to its width multiplied by the mean of the first, last and four times the middle ordinate:

$$(4) \qquad S_1 = 2h \frac{y_1 + 4y_2 + y_3}{6} = \frac{1}{3} h(y_1 + 4y_2 + y_3).$$

We shall estimate the error committed in the next article.

In general, the accuracy is improved by taking the ordinates close together. Thus, in computing $\int_a^b f(x)\, dx$, we may divide $b - a$ into an even number $2n$ of equal subintervals h, apply (3) to successive pairs, and add the results. If the ordinates are written $y_1, y_2, \cdots, y_{2n+1}$, we obtain in this way

$$S_n = \frac{2h}{6}(y_1 + 4y_2 + y_3 + y_3 + 4y_4 + y_5 + \cdots + y_{2n-1} + 4y_n + y_{2n+1})$$

or

$$(5) \qquad S_n = \frac{h}{3}(y_1 + 4y_2 + 2y_3 + \cdots + 2y_{2n-1} + 4y_{2n} + y_{2n+1}).$$

When Simpson's rule is applied to find a volume V regarded as a limiting sum of thin slices of area $f(x)$,

$$(6) \qquad V = \int_a^b f(x)\, dx \sim \frac{b-a}{6}\left[f(a) + f\left(\frac{a+b}{2}\right) + f(b)\right]$$

is known as the *prismoid formula*. It is easy to verify that (6) gives the volume of a cylinder, a cone, a sphere, and an ellipsoid exactly.

Example 1. When $y = x^3 + 2x^2 + 3$,

$$\int_0^2 y\, dx = \frac{1}{4}x^4 + \frac{2}{3}x^3 + 3x \Big|_0^2 = 15\tfrac{1}{3}.$$

In this case Simpson's rule with two intervals is exact:

$$S_1 = \tfrac{1}{3}(3 + 24 + 19) = 15\tfrac{1}{3}.$$

Example 2. One arch of the sine curve has the area $\int_0^\pi \sin x\, dx = 2$. Simpson's rule with 2, 4, and 6 intervals gives

$$S_1 = \frac{\pi}{6}(0 + 4 + 0) = \frac{2}{3}\pi = 2.094,$$

$$S_2 = \frac{\pi}{12}(0 + 2\sqrt{2} + 2 + 2\sqrt{2} + 0) = 2.004,$$

$$S_3 = \frac{\pi}{18}(0 + 2 + \sqrt{3} + 4 + \sqrt{3} + 2 + 0) = 2.0009.$$

Example 3. We next compute an integral that arises in the study of *Gibbs' phenomenon* in Fourier series (§ 229), namely,

$$I = \int_0^\pi \frac{\sin x}{x} \, dx.$$

Since a primitive of $\sin x/x$ cannot be expressed in terms of the elementary functions, we must resort to an approximate method to find I. Note that the integral exists; for, as $x \to 0$, $(\sin x)/x \to 1$ (Theorem 115.2, Cor.).

Simpson's rule with 2, 4, and 6 intervals gives

$$S_1 = \frac{\pi}{6}\left(1 + 4\frac{2}{\pi} + 0\right) = \frac{\pi}{6} + \frac{4}{3} = 1.857;$$

$$S_2 = \frac{\pi}{12}\left(1 + \frac{8\sqrt{2}}{\pi} + \frac{4}{\pi} + \frac{8\sqrt{2}}{3\pi} + 0\right) = \frac{\pi}{12} + \frac{8\sqrt{2}}{9} + \frac{1}{3} = 1.852;$$

$$S_3 = \frac{\pi}{18}\left(1 + \frac{12}{\pi} + \frac{3\sqrt{3}}{\pi} + \frac{8}{\pi} + \frac{3\sqrt{3}}{2\pi} + 0\right) = \frac{\pi}{18} + \frac{\sqrt{3}}{4} + \frac{56}{45} = 1.85198.$$

131. Error in Simpson's Rule. From (130.2–3) we see that the error in Simpson's rule is

$$(1) \qquad E(h) = F(h) - F(-h) - \frac{h}{3}[f(-h) + 4f(0) + f(h)].$$

We now assume that $f(x)$ has continuous derivatives up to the fourth order. Since $F'(x) = f(x)$, $E(h)$ has the same property; and, on differentiating (1) three times, we get

$$E'(h) = \frac{2}{3}[f(h) + f(-h) - 2f(0)] - \frac{h}{3}[f'(h) - f'(-h)],$$

$$E''(h) = \frac{1}{3}[f'(h) - f'(-h)] - \frac{h}{3}[f''(h) + f''(-h)],$$

$$(2) \qquad E'''(h) = -\frac{h}{3}[f'''(h) - f'''(-h)].$$

Thus, when $h = 0$,

$$E(0) = E'(0) = E''(0) = E'''(0) = 0.$$

Consequently, the function

$$(3) \qquad \varphi(t) = E(t) - \frac{E(h)}{h^5} t^5$$

has the property $\varphi(h) = 0$ as well as

$$\varphi(0) = \varphi'(0) = \varphi''(0) = \varphi'''(0) = 0.$$

Thus, from Rolle's theorem we obtain successively

$$\varphi'(\xi_1) = 0, \quad \varphi''(\xi_2) = 0, \quad \varphi'''(\xi_3) = 0, \qquad 0 < \xi_1, \xi_2, \xi_3 < h.$$

Now from (2) and (3)

$$\varphi'''(t) = -\frac{t}{3}[f'''(t) - f'''(-t)] - 60\frac{E(h)}{h^5}t^2,$$

and, on putting $t = \xi_3$, we have

$$E(h) = -\frac{h^5}{180}\frac{f'''(\xi_3) - f'''(-\xi_3)}{\xi_3}.$$

The mean-value theorem (57.1) now gives

$$f'''(\xi_3) - f'''(-\xi_3) = 2\xi_3 f^{iv}(\xi), \qquad -h < \xi < h;$$

hence, the error in Simpson's rule is given by the simple expression,

$$(4) \qquad E(h) = -\frac{h^5}{90}f^{iv}(\xi), \qquad -h < \xi < h.$$

Thus, if the central abscissa is x (instead of 0), we may write

$$(5) \qquad \int_{x-h}^{x+h} f(x)\,dx = \frac{h}{3}[f(x-h) + 4f(x) + f(x+h)] - \frac{h^5}{90}f^{iv}(\xi),$$

where ξ lies between $x - h$ and $x + h$. If $f^{iv}(x)$ does not change sign in $(x - h, x + h)$, the *sign* of the error is known; and, if $f^{iv}(x)$ is bounded, say $|f^{iv}(x)| \leq M$, the error cannot exceed $Mh^5/90$. In practice we divide the range of integration into n parts of width $2h$ and apply (5) to each part.

When $f(x)$ is a polynomial of degree less than four, $f^{iv}(x) = 0$; hence Simpson's rule is *exact* when $f(x) = Ax^3 + Bx^2 + Cx + D$.

Example 1. In Ex. 130.2, $f(x) = \sin x$, $h = \pi/6$ with 6 intervals, and $f^{iv}(x) = \sin x$. Hence, the error for each pair of intervals is

$$E\left(\frac{\pi}{6}\right) = -\frac{1}{90}\left(\frac{\pi}{6}\right)^5 \sin \xi,$$

where ξ lies in the appropriate range. We may estimate the total error as $3E(\pi/6)$ when $\sin \xi$ is given its average value for the interval $(0, \pi)$, namely,

$$\frac{1}{\pi}\int_0^\pi \sin x\,dx = \frac{2}{\pi}.$$

This gives a correction of

$$-3\frac{1}{90}\left(\frac{\pi}{6}\right)^5\frac{2}{\pi} = -\frac{1}{90}\left(\frac{\pi}{6}\right)^4 = -0.00083,$$

so that the corrected area is $2.0009 - 0.0008 = 2.0001$.

Example 2. If we compute $\log 2 = \int_1^2 \dfrac{dx}{x}$ by using Simpson's rule with 10 intervals $(h = 0.1)$,

$$S_5 = \frac{1}{30}\left[1 + \frac{4}{1.1} + \frac{2}{1.2} + \frac{4}{1.3} + \frac{2}{1.4} + \frac{4}{1.5} + \frac{2}{1.6} + \frac{4}{1.7} + \frac{2}{1.8} + \frac{4}{1.9} + \frac{1}{2} \right]$$

$$= 0.693150.$$

Since $f(x) = 1/x$, $f^{\mathrm{iv}}(x) = 24/x^5$, the error for each pair of intervals is

$$E(0.1) = -\frac{(0.1)^5}{90} \frac{24}{\xi^5} = -\frac{0.00004}{15\xi^5},$$

where ξ lies in the appropriate range. We may estimate the total error as $5E(0.1)$ when $1/\xi^5$ is given its average value for the interval $(1, 2)$, namely,

$$\int_1^2 \frac{dx}{x^5} = -\frac{1}{4x^4}\Big|_1^2 = \frac{15}{64}.$$

This gives a correction of

$$-5\frac{0.0004}{64} = -0.000003,$$

which applied to S_5 gives $\log 2 = 0.693147$, correct to the last place.

PROBLEMS

1. As a first approximation to the area $\int_0^h f(x)\, dx$ we may take the area of the inscribed trapezoid of width h and having $f(0)$, $f(h)$ as parallel sides: $\frac{1}{2}h[f(0) + f(h)]$. Show that the error in this *trapezoid* formula,

$$E(h) = \int_0^h f(x)\, dx - \frac{h}{2}[f(0) + f(h)] = -\frac{h^3}{12} f''(\xi), \qquad 0 < \xi < h.$$

$[E''(h) = -\frac{1}{2}h f'''(h);$ consider the function

$$\varphi(t) = E(t) - E(h)t^3/h^3.]$$

2. Let $y_i, y_2, \cdots, y_{n+1}$ be ordinates of a curve at equal distances h. Show that the trapezoid formula gives the approximate area under the curve between y_1 and y_{n+1} as

$$T_n = \frac{h}{2} [y_1 + 2y_2 + 2y_3 + \cdots + 2y_n + y_{n+1}].$$

3. From $\dfrac{\pi}{4} = \displaystyle\int_0^1 \dfrac{dx}{1 + x^2}$ compute π by the trapezoid formula and by Simpson's rule, taking $h = 0.1$. Estimate the error in each case.

4. Compute $\log 2 = \displaystyle\int_1^2 \dfrac{dx}{x}$ by the trapezoid formula taking $h = 0.1$. Estimate the error and compare with Ex. 2.

5. Compute $\int_0^1 \sqrt{1 + x^4}\, dx$ by Simpson's rule using three ordinates; five ordinates.

132. The Integral as a Function of a Parameter. The definite integral with constant limits,

$$(1) \qquad\qquad \varphi(x) = \int_a^b f(x, t)\, dt,$$

is a function of the "parameter" x that occurs in the integrand.

THEOREM. *If $f(x, t)$ is a continuous function of x and t in the rectangle R:* $\alpha \leq x \leq \beta$, $a \leq t \leq b$, $\varphi(x)$ *is continuous in the interval* (α, β).

Proof. We have from (1)

$$\varphi(x + h) - \varphi(x) = \int_a^b [f(x + h, t) - f(x, t)]\, dt.$$

Since $f(x, t)$ is uniformly continuous in R (Theorem 75.2), for every $\varepsilon > 0$ we can find a δ such that

$$\left| f(x + h, t) - f(x, t) \right| < \frac{\varepsilon}{b - a} \quad \text{when} \quad \left| h \right| < \delta$$

and for all values x, t in R. Hence when $\left| h \right| < \delta$,

$$\left| \varphi(x + h) - \varphi(x) \right| \leq \int_a^b \left| f(x + h, t) - f(x, t) \right| dt < \varepsilon;$$

that is, $\varphi(x)$ is continuous in (α, β).

When $\varphi(x)$ is continuous at x_1,

$$\lim_{x \to x_1} \varphi(x) = \varphi(x_1) = \int_a^b f(x_1, t)\, dt. \qquad (45.1).$$

Hence, when $f(x, t)$ is continuous,

$$(2) \qquad \lim_{x \to x_1} \int_a^b f(x, t)\, dt = \int_a^b \lim_{x \to x_1} f(x, t)\, dt,$$

so that the limit operations $x \to x_1$ and integration are commutative.

133. Differentiation of Integrals. When an indefinite integral,

$$F(x, t) = \int f(x, t)\, dt,$$

is known,

$$(1) \qquad \varphi(x) = \int_a^b f(x, t)\, dt = F(x, b) - F(x, a),$$

and

$$\varphi'(x) = F_x(x, b) - F_x(x, a).$$

In many cases $\varphi'(x)$ may be computed without knowledge of $F(x, t)$ by *differentiating under the integral sign.*

THEOREM. *If $f(x, t)$ and its partial derivative $f_x(x, t)$ are continuous in the rectangle R: $\alpha \leq x \leq \beta$, $a \leq t \leq b$,*

$$(2) \qquad \frac{d}{dx} \int_a^b f(x, t)\, dt = \int_a^b f_x(x, t)\, dt$$

and is a continuous function of x in (α, β).

Proof. To compute $\varphi'(x)$ we form

$$\frac{\varphi(x + h) - \varphi(x)}{h} = \int_a^b \frac{f(x + h, t) - f(x, t)}{h}\, dt = \int_a^b f_x(\xi, t)\, dt$$

by the mean-value theorem (57.1); and $|\xi - x| < |h|$. Now let $h \to 0$ on the left; then $\xi \to x$ on the right, and from (132.2) it seems plausible that the limit on the right is $\int_a^b f_x(x, t)\, dt$. But, when x is fixed, ξ is a function of unknown character of both h and t. The above limit is thus open to question. To settle this point, form

$$\left| \int_a^k f_x(\xi, t)\, dt - \int_a^b f_x(x, t)\, dt \right| \leq \int_a^b |f_x(\xi, t) - f_x(x, t)|\, dt.$$

Since $f_x(x, t)$ is *uniformly* continuous in the rectangle R (Theorem 75.2), we can choose δ so that

$$|f_x(\xi, t) - f_x(x, t)| < \frac{\varepsilon}{b - a} \quad \text{when} \quad |\xi - x| < \delta;$$

and, since $|\xi - x| < |h|$,

$$\left| \int_a^b [f_x(\xi, t) - f_x(x, t)]\, dt \right| < \varepsilon \quad \text{when} \quad |h| < \delta.$$

Consequently,

$$\varphi'(x) = \lim_{h \to 0} \int_a^b f_x(\xi, t)\, dt = \int_a^b f_x(x, t)\, dt,$$

and (2) is established. The continuity of $\varphi'(x)$ follows from Theorem 132.

An integral of $f(x, t)$ between variable limits a, b is a function of a and b as well as x:

$$(3) \qquad \varphi(x, a, b) = \int_a^b f(x, t)\, dt.$$

If $f(x, t)$ and $f_x(x, t)$ are continuous in R, the partial derivatives of $\varphi(x, a, b)$ are

(4)
$$\frac{\partial \varphi}{\partial x} = \int_a^b f_x(x, t)\, dt,$$

(5)
$$\frac{\partial \varphi}{\partial b} = f(x, b) \tag{119.3},$$

(6)
$$\frac{\partial \varphi}{\partial a} = -f(x, a) \tag{119.4}.$$

These partial derivatives are all continuous in (α, β); hence $\varphi(x, a, b)$ is a differentiable function (Theorem 78) and, of course, also a continuous function of x, a, b.

When a and b are differentiable functions of x, $\varphi(x, a, b)$ becomes a continuous and differentiable function of x alone whose derivative is given by the chain rule (80.6):

$$\frac{d\varphi}{dx} = \frac{\partial \varphi}{\partial x} + \frac{\partial \varphi}{\partial b}\frac{db}{dx} + \frac{\partial \varphi}{\partial a}\frac{da}{dx}.$$

On substitution from (4), (5), and (6), this gives the important formula,

(7)
$$\frac{d}{dx}\int_a^b f(x, t)\, dt = \int_a^b f_x(x, t)\, dt + f(x, b)\frac{db}{dx} - f(x, a)\frac{da}{dx}.$$

Example 1. By direct computation

$$\varphi(x) = \int_{-x}^{\sin x} \frac{dt}{x + t + 1} = \log(x + t + 1)\Big|_{-x}^{\sin x} = \log(x + \sin x + 1),$$

$$\varphi'(x) = \frac{1 + \cos x}{x + \sin x + 1}.$$

By use of (7):

$$\varphi'(x) = -\int_{-x}^{\sin x} \frac{dt}{(x + t + 1)^2} + \frac{1}{x + \sin x + 1}\cos x - 1\cdot(-1),$$

$$= \frac{1}{(x + t + 1)}\Big|_{-x}^{\sin x} + \frac{\cos x}{x + \sin x + 1} + 1,$$

which gives the previous value.

Example 2. The function

$$\varphi(x) = \int_x^{x^2} \frac{\sin xt}{t}\, dt$$

cannot be expressed in terms of the elementary functions. From (7)

$$\varphi'(x) = \int_x^{x^2} \cos xt\, dt + \frac{\sin x^3}{x^2}\cdot 2x - \frac{\sin x^2}{x}\cdot 1$$

$$= \frac{\sin xt}{x}\Big|_x^{x^2} + \frac{2\sin x^3 - \sin x^2}{x}$$

$$= \frac{3\sin x^3 - 2\sin x^2}{x}.$$

Example 3. A definite integral that depends on a parameter may sometimes be computed by first finding its derivative. Consider, for example, the integral

$$\varphi(y) = \int_0^\pi \log (1 + y \cos x) \, dx;$$

$$\varphi'(y) = \int_0^\pi \frac{\cos x}{1 + y \cos x} \, dx$$

$$= \frac{1}{y} \int_0^\pi \left(1 - \frac{1}{1 + y \cos x} \right) dx$$

$$= \frac{\pi}{y} - \frac{1}{y} \int_0^\pi \frac{dx}{1 + y \cos x} \, .$$

In the last integral put $t = \tan \frac{1}{2}x$; then from (123.5)

$$\int \frac{dx}{1 + y \cos x} = \int \frac{\dfrac{2dt}{1 + t^2}}{1 + y \dfrac{1 - t^2}{1 + t^2}}$$

$$= \int \frac{2dt}{(1 + y) + (1 - y)t^2}$$

$$= \frac{2}{1 - y} \int \frac{dt}{\dfrac{1 + y}{1 - y} + t^2}$$

$$= \frac{2}{\sqrt{1 - y^2}} \tan^{-1} \left(\sqrt{\frac{1 - y}{1 + y}} \tan \frac{x}{2} \right) + C;$$

hence, on inserting the limits,

$$\varphi'(y) = \frac{\pi}{y} - \frac{2}{y\sqrt{1 - y^2}} \frac{\pi}{2} = \pi \left(\frac{1}{y} - \frac{1}{y\sqrt{1 - y^2}} \right).$$

On integration, we have

$$\varphi(y) + C = \pi \left(\log y + \log \frac{1 + \sqrt{1 - y^2}}{y} \right) = \pi \log (1 + \sqrt{1 - y^2}).$$

When $y = 0$, $\varphi(0) = 0$, and $C = \pi \log 2$; hence, the given integral,

$$\varphi(y) = \pi \log \tfrac{1}{2}(1 + \sqrt{1 - y^2}),$$

has been evaluated without benefit of a primitive.

PROBLEMS

1. Find $\varphi'(x)$ in two ways when $\varphi(x)$ is

(a) $\displaystyle \int_0^x \sin (t - x) \, dt;$ (b) $\displaystyle \int_0^{x^2} \tan^{-1} \frac{t}{x^2} \, dt;$ (c) $\displaystyle \int_x^{x^2} e^{xt} \, dt;$

(d) $\displaystyle \int_x^{x^2} (2t + x^2) \, dt$ (e) $\displaystyle \int_0^{\log x} e^{-t} \, dt;$ (f) $\displaystyle \int_{-x}^{x^2} \frac{dt}{x + t + 1}.$

2. Evaluate $\varphi(\alpha) = \displaystyle\int_0^1 \frac{x^\alpha - 1}{\log x}\, dx$ ($\alpha > 0$) by finding $\varphi'(\alpha)$.

3. Find $\partial F/\partial x$ when $F(x, t)$ is

(a) $\displaystyle\int_0^{xt} \sin(x + t)\, dt$;   (b) $\displaystyle\int_0^t \sin^{-1}\frac{t}{x}\, dt$.

4. Show that

$$J_0(x) = \frac{1}{\pi}\int_0^\pi \cos(x \cos\theta)\, d\theta \quad \text{satisfies} \quad J_0'' + \frac{1}{x}J_0' + J_0 = 0.$$

(Find $J_0'(x)$ and integrate by parts.)

5. Show that

$$\int_0^1 \log f(x + t)\, dt = \int_0^x \log\frac{f(x + 1)}{f(x)}\, dx + \int_0^1 \log f(x)\, dx.$$

6. From $\displaystyle\int_0^1 x^\alpha\, dx = \frac{1}{\alpha + 1}$ show that

$$\int_0^1 (\log x)^n\, x^\alpha\, dx = (-1)^n\, n!/(\alpha + 1)^{n+1}.$$

7. Find $\partial\varphi/\partial a$ and $\partial\varphi/\partial b$ when

$$\varphi(a, b) = \int_0^{\pi/2} \frac{dx}{a^2 + b^2 \tan^2 x}.$$

8. If $0 < x < 1$ and

$$\varphi(x) = \int_0^1 f(x, t)\, dt, \qquad f(x, t) = \begin{cases} t & (t \leq x) \\ 1 - t & (t > x) \end{cases},$$

find $\varphi'(x)$ without integration.

9. Let $u(x)$ and $v(x)$ be solutions of the linear differential equation,

$$u'' + P(x)u' + Q(x)u = 0,$$

whose Wronskian

$$W(x) = \begin{vmatrix} u(x) & v(x) \\ u'(x) & v'(x) \end{vmatrix} \neq 0, \qquad a \leq x \leq b.$$

Then, if $a < c < b$, prove that

(8) $$Y(x) = \int_c^x \frac{f(t)}{W(t)} \begin{vmatrix} u(t) & v(t) \\ u(x) & v(x) \end{vmatrix} dt$$

satisfies the equation

(9) $$y'' + P(x)y' + Q(x)y = f(x)$$

and the conditions $Y(c) = Y'(c) = 0$.

10. Using formula (8), obtain the solutions

(i) $$Y(x) = \int_c^x (x - t)f(t)\, dt \quad \text{for} \quad y'' = f(x);$$

(ii) $$Y(x) = \frac{1}{n} \int_c^x f(t) \sin n(x - t)\, dt \quad \text{for} \quad y'' + n^2 y = f(x).$$

Verify each solution by differentiation.

[For (i) take $u = 1$, $v = x$; for (ii) take $u = \cos nx$, $v = \sin nx$.]

134. Application to Differential Equations. The rule (133.7) for differentiating an integral leads to the useful

THEOREM 1. *If $f(x)$ is continuous in a neighborhood of $x = 0$, the differential equation,*

(1) $$y^{(n+1)}(x) = f(x),$$

with the initial conditions,

(2) $$y(0) = 0, \quad y'(0) = 0, \quad \cdots, \quad y^{(n)}(0) = 0,$$

has the unique solution,

(3) $$Y(x) = \frac{1}{n!} \int_0^x (x - t)^n f(t)\, dt,$$

in this neighborhood.

Proof. On differentiating (3) $n + 1$ times, we find

$$Y^{(k)}(x) = \frac{1}{(n - k)!} \int_0^x (x - t)^{n-k} f(t)\, dt, \qquad k = 1, 2, \cdots, n - 1;$$

for the factor $x - t$ in the integrand makes the term from the upper limit vanish. Now

$$Y^{(n)}(x) = \int_0^x f(t)\, dt, \qquad Y^{(n+1)}(x) = f(x);$$

and $Y(x)$ obviously fulfils the conditions (2).

Let $Y_1(x)$ be a second solution of (1) and (2). Then (§ 57, Cor. 2)

$$Y_1^{(n+1)}(x) = Y^{(n+1)}(x), \qquad Y_1^{(n)}(x) = Y^{(n)}(x) + C,$$

and $C = 0$ from (2). Thus step by step we can show that

$$Y_1^{(k)}(x) = Y^{(k)}(x), \qquad k = n, n - 1, \cdots, 2, 1,$$

and finally $Y_1(x) = Y(x)$.

Corollary. *If the initial conditions are*

(4) $$y(0) = a_0, \quad y'(0) = a_1, \quad \cdots, \quad y^{(n)}(0) = a_n,$$

the unique solution of the system (1)–(4) is given by

(5) $$y(x) = a_0 + \frac{a_1}{1!} x + \frac{a_2}{2!} x^2 + \cdots + \frac{a_n}{n!} x^n + \frac{1}{n!} \int_0^x (x - t)^n f(t)\, dt.$$

Proof. If we denote the polynomial in (5) by $P_n(x)$, $y(x) - P_n(x)$ satisfies the system (1)–(2) and must be its unique solution $Y(x)$.

THEOREM 2. *If $f^{(n+1)}(x)$ is continuous in a neighborhood of $x=0$, the Maclaurin expansion of $f(x)$ is given by*

$$(6)\quad f(x)=f(0)+\frac{f'(0)}{1!}x+\cdots+\frac{f^{(n)}(0)}{n!}x^n+\frac{1}{n!}\int_0^x(x-t)^nf^{(n+1)}(t)dt.$$

Proof. The corollary above shows that the function on the right is the unique solution of the system

$$y^{(n+1)}(x)=f^{(n+1)}(x),$$

$$y(0)=f(0),\quad y'(0)=f'(0),\quad\cdots,\quad y^{(n)}(0)=f^{(n)}(0).$$

But $f(x)$ obviously satisfies this system.

Corollary. *The remainder in the Maclaurin expansion of $f(x)$ is given by*

$$(7)\qquad R_n(x)=\frac{1}{n!}\int_0^x(x-t)^nf^{(n+1)}(t)\,dt.$$

The mean-value theorem (121.2), with $g(x)=(x-t)^n$, now gives

$$R_n(x)=\frac{1}{n!}f^{(n+1)}(\xi)\int_0^x(x-t)^n\,dt$$

or

$$(8)\qquad R_n(x)=f^{(n+1)}(\xi)\frac{x^{n+1}}{(n+1)!},\quad 0<\xi<x.$$

This is the *Lagrange remainder*, in agreement with (65.3). This proof, however, requires the *continuity* of $f^{n+1}(x)$ rather than its mere *existence* (cf. § 65).

With $g(x)=1$, the mean-value theorem gives

$$R_n(x)=\frac{1}{n!}f^{(n+1)}(\xi)(x-\xi)^n\int_0^x dx$$

or

$$(9)\qquad R_n(x)=f^{(n+1)}(\xi)\frac{(x-\xi)^n}{n!}x.$$

This is the *Cauchy remainder*.

135. Repeated Integrals.

THEOREM. *If $f(x,y)$ is a continuous function in the rectangle $a\leqq x\leqq b$, $c\leqq y\leqq d$,*

$$(1)\qquad \int_c^d dy\int_a^b f(x,y)\,dx=\int_a^b dx\int_c^d f(x,y)\,dy.$$

Proof. If we set

(2) $$\varphi(u) = \int_c^d dy \int_a^u f(x, y)\, dx, \qquad a \leq u \leq b,$$

$$\varphi'(u) = \int_c^d f(u, y)\, dy,$$

from Theorem 133. If we now integrate $\varphi'(u)$ from $u = a$ to $u = b$ we have

$$\varphi(b) - \varphi(a) = \int_a^b du \int_c^d f(u, y)\, dy,$$

or, since $\varphi(a) = 0$,

$$\varphi(b) = \int_a^b dx \int_c^d f(x, y)\, dy,$$

which is the right-hand member of (1). But, on putting $u = b$ in (2), we get the left-hand member of (1).

136. Integrals. If $f(x)$ is bounded in $a \leq x \leq b$, we define

(1) $$\int_a^b f(x)\, dx = \lim_{\delta \to 0} \sum_{i=1}^n f(\xi_i)\delta_i;$$

the net x_i over (a, b) forms n subintervals δ_i, ξ_i is any point of δ_i, and $\delta = \max \delta_i$ is the *mesh* of the net. If the limit exists, $f(x)$ is integrable over (a, b).

If m_i, M_i are the closest bounds of $f(x)$ in δ_i,

$$s = \sum m_i \delta_i \quad \text{and} \quad S = \sum M_i \delta_i$$

are the *lower* and *upper sums* for the net x_i. As $\delta \to 0$, the sums s and S approach limits I and J; and $f(x)$ is integrable over (a, b) when and only when $I = J$. Moreover, $I = J$ when there is a net for any $\varepsilon > 0$ such that $S - s < \varepsilon$. This test shows that functions of the following types are integrable in (a, b):

(i)   continuous functions;

(ii) bounded functions that have a finite number of discontinuities;

(iii) bounded monotonic functions;

(iv) sums and products of integrable functions;

(v) the reciprocal and square root of positive integrable functions;

(vi) functions of bounded variation.

An interchange of limits reverses the sign of an integral. Moreover, if the integrals exist,

(2)
$$\int_a^b f(x)\, dx + \int_b^c f(x)\, dx = \int_a^c f(x)\, dx.$$

If $f(x)$ is integrable in (a, b),

(3) $\varphi(x) = \int_a^x f(t)\, dt$ is continuous in (a, b);   and

(4) $\varphi'(x) = f(x)$

at every point where $f(x)$ is continuous.   Moreover, if $F(x)$ is any function having $f(x)$ as derivative,

(5)
$$\int_a^b f(x)\, dx = F(b) - F(a).$$

$F(x)$ is called an indefinite integral or *primitive* of $f(x)$ and denoted by $\int f(x)\, dx$.   The function $\varphi(x)$ in (3) is that particular primitive which vanishes at a.

If $f(x)$ is continuous in (a, b), the mean-value theorem,

$$\varphi(b) - \varphi(a) = \varphi'(\xi)(b - a), \qquad a < \xi < b,$$

gives the mean-value theorem for integrals:

(6)
$$\int_a^b f(x)\, dx = f(\xi)(b - a), \qquad a < \xi < b.$$

If $u(x)$, $v(x)$ have integrable derivatives, we may *integrate by parts*:

(7)
$$\int_a^b u\, dv = uv \Big|_a^b - \int_a^b v\, du.$$

The substitution $x = \varphi(t)$ gives

(8)
$$\int_a^b f(x)\, dx = \int_\alpha^\beta f[\varphi(t)]\, \varphi'(t)\, dt,$$

provided $a = \varphi(\alpha)$, $b = \varphi(\beta)$ and the functions $\varphi'(t)$ and $f[\varphi(t)]$ are continuous in (α, β).

The area between the curve $y = f(x)$, the x-axis, and the lines $x = a$, $x = b$ is

(9)
$$A = \int_a^b y\, dx = \int_a^b f(x)\, dx.$$

If the curve has the parametric equations,

(10) $x = x(t), \qquad y = y(t), \qquad \alpha \leq t \leq \beta,$

and $a = x(\alpha)$, $b = x(\beta)$, then

$$(11) \qquad\qquad A = \int_{\alpha}^{\beta} y(t)\, x'(t)\, dt.$$

The curve (10) is *continuous* when $x(t)$ and $y(t)$ are continuous in (α, β); it is *smooth* if $x'(t)$ and $y'(t)$ are continuous in (α, β) and never vanish together. The curve is *rectifiable* if all possible nets over (α, β) produce inscribed broken lines whose lengths λ have a finite least upper bound L; then L is the *length* of the curve. A curve is rectifiable when and only when $x(t)$, $y(t)$ are functions of *bounded variation*; and then $\lambda \to L$ as the mesh of the net approaches zero.

A *smooth* curve is rectifiable, and its length from $t = \alpha$ to $t = \beta$,

$$(12) \qquad\qquad \int_{\alpha}^{\beta} \sqrt{x'^2 + y'^2}\, dt = \int_{\alpha}^{\beta} \left| \frac{d\mathbf{r}}{dt} \right| dt,$$

where $r(t)$ is the position vector to the curve. The arc s, measured from $t = \alpha$, is given by

$$(13) \qquad\qquad s = \int_{\alpha}^{t} \left| \frac{d\mathbf{r}}{dt} \right| dt \quad \text{and} \quad \frac{ds}{dt} = \left| \frac{d\mathbf{r}}{dt} \right|,$$

for plane or space curves; and $ds^2 = dx^2 + dy^2 + dz^2$.

Simpson's rule approximates

$$(14) \qquad \int_{a}^{b} f(x)\, dx \qquad \text{by} \qquad (b - a)\, \frac{f(a) + 4f\left(\dfrac{a+b}{2}\right) + f(b)}{6}.$$

If $f(x, t)$ is continuous in a rectangle R,

$$\varphi(x) = \int_{a}^{b} f(x, t)\, dt$$

is continuous over the range in x. If $f_x(x, t)$ is continuous in R, $\varphi(x)$ has the continuous derivative,

$$(15) \qquad\qquad \varphi'(x) = \int_{a}^{b} f_x(x, t)\, dt.$$

If a and b are also functions of x, add

$$f(x, b)\, b'(x) - f(x, a)\, a'(x)$$

to the right-hand side of (15).

CHAPTER 7

Improper Integrals

137. Types of Improper Integrals. The definition of a definite integral $\int_a^b f(x)\,dx$ presupposes (i) that the limits are finite, (ii) that the integrand is bounded in $a \leq x \leq b$. Hence, when a limit is infinite or the integrand becomes infinite in $a \leq x \leq b$, we need new definitions if such *improper integrals* are to have a meaning.

Improper integrals are of two main types:

I. A limit a or b is infinite;

II. The integrand becomes infinite at a or b.

Under type I we have

$$\int_a^\infty f(x)\,dx, \qquad f(x) \text{ integrable } a \leq x < \infty,$$

$$\int_{-\infty}^b f(x)\,dx, \qquad f(x) \text{ integrable } -\infty < x \leq b;$$

and under type II

$$\int_{a+}^b f(x)\,dx, \qquad f(a+) = \pm\infty, \qquad f(x) \text{ integrable } a < x \leq b,$$

$$\int_a^{b-} f(x)\,dx, \qquad f(b-) = \pm\infty, \qquad f(x) \text{ integrable } a \leq x < b.$$

Any improper integral whose integrand has but a finite number of infinities can be expressed as a sum of simple integrals of types I and II For example,

$$\int_{-\infty}^\infty \frac{dx}{(x+2)(x-1)} = \int_{-\infty}^{-3} + \int_{-3}^{-2} + \int_{-2}^0 + \int_0^1 + \int_1^2 + \int_2^\infty \frac{dx}{(x+2)(x-1)};$$

here the first and last integrals are of type I, the others of type II.

These improper integrals are defined as follows:

(1)
$$\int_a^\infty f(x)\,dx = \lim_{b\to\infty} \int_a^b f(x)\,dx;$$

(2)
$$\int_{-\infty}^b f(x)\,dx = \lim_{a\to-\infty} \int_a^b f(x)\,dx;$$

(3)
$$\int_a^b f(x)\,dx = \lim_{\varepsilon\to 0} \int_{a+\varepsilon}^b f(x)\,dx;$$

(4)
$$\int_a^b f(x)\,dx = \lim_{\varepsilon\to 0} \int_a^{b-\varepsilon} f(x)\,dx.$$

When a limit exists, the corresponding integral is said to *converge* to this value.   When a limit does not exist, the integral is said to *diverge*.

Example 1. If $b > a > 0$,

$$\int_a^b \frac{dx}{x^2} = \frac{1}{a} - \frac{1}{b}, \qquad \int_a^\infty \frac{dx}{x^2} = \frac{1}{a}.$$

Example 2.

$$\int_a^b e^x\,dx = e^b - e^a, \qquad \int_{-\infty}^b e^x\,dx = e^b.$$

Example 3.

$$\int_{a+\varepsilon}^b \frac{dx}{\sqrt{x-a}} = 2\sqrt{b-a} - 2\sqrt{\varepsilon}, \qquad \int_a^b \frac{dx}{\sqrt{x-a}} = 2\sqrt{b-a}.$$

Example 4.

$$\int_a^{b-\varepsilon} \frac{dx}{\sqrt{b-x}} = 2\sqrt{b-a} - 2\sqrt{\varepsilon}, \qquad \int_a^b \frac{dx}{\sqrt{b-x}} = 2\sqrt{b-a}.$$

Example 5. If $a, b > 0$.

$$\int_a^b \frac{dx}{x} = \log b - \log a, \qquad \int_0^\infty \frac{dx}{x} \quad \text{diverges to } \infty.$$

Example 6. Since

$$\int_a^b \sin x\,dx = \cos a - \cos b,$$

$$\int_a^\infty \sin x\,dx \quad \text{and} \quad \int_{-\infty}^b \sin x\,dx \quad \text{diverge by finite oscillation.}$$

Example 7. If $b > a > 0$,

$$\int_{a+\varepsilon}^b \frac{dx}{x-a} = \log(b-a) - \log \varepsilon, \qquad \int_a^b \frac{dx}{x-a} \quad \text{diverges to } \infty.$$

138. Evaluation of Improper Integrals. When a primitive function is known, a *convergent* improper integral may be evaluated by a natural extension of the fundamental theorem,

$$\int_a^b f(x)\, dx = F(b) - F(a). \tag{120.1}$$

For integrals of type I we write

$$\lim_{b \to \infty} F(b) = F(\infty), \qquad \lim_{a \to -\infty} F(a) = F(-\infty);$$

then from (137.1) and (137.2) we have

$$(1) \qquad\qquad \int_a^\infty f(x)\, dx = F(\infty) - F(a),$$

$$(2) \qquad\qquad \int_{-\infty}^b f(x)\, dx = F(b) - F(-\infty),$$

when $F(\infty)$ of $F(-\infty)$ exist.

If both limits become infinite, we write

$$\int_a^b f(x)\, dx = \int_a^c f(x)\, dx + \int_c^b f(x)\, dx,$$

and let $a \to -\infty$, $b \to \infty$ in the right-hand integrals. If both limits exist independently of each other, we have

$$(3) \quad \int_{-\infty}^\infty f(x)\, dx = F(c) - F(-\infty) + F(\infty) - F(c) = F(\infty) - F(-\infty).$$

The choice of c is obviously immaterial.

Example 1.

$$\int_{-\infty}^\infty \frac{dx}{1 + x^2} = \tan^{-1} \infty - \tan^{-1}(-\infty) = \frac{\pi}{2} + \frac{\pi}{2} = \pi.$$

Example 2. As $a \to -\infty$, $b \to \infty$ in

$$\int_a^b \sin x\, dx = \cos a - \cos b,$$

neither limit on the right exists. Hence,

$$\int_{-\infty}^\infty \sin x\, dx \quad \text{diverges by oscillation.}$$

Note that in view of definition (3) we cannot conclude from

$$\int_{-a}^a \sin x\, dx = 0 \quad \text{that} \quad \int_{-\infty}^\infty \sin x\, dx = 0.$$

For integrals of type II we write

$$\lim_{\varepsilon \to 0} F(a + \varepsilon) = F(a+), \qquad \lim_{\varepsilon \to 0} F(b - \varepsilon) = F(b-);$$

then from (135.3) and (135.4) we have

(3) $$\int_{a+}^{b} f(x)\, dx = F(b) - F(a+),$$

(4) $$\int_{a}^{b-} f(x)\, dx = F(b-) - F(a),$$

when $F(a+)$ and $F(b-)$ exist.

If $f(x)$ becomes infinite at both limits, we write

$$\int_{a+\varepsilon}^{b-\varepsilon} f(x)\, dx = \int_{a+\varepsilon}^{c} f(x)\, dx + \int_{c}^{b-\varepsilon} f(x)\, dx,$$

and let $\varepsilon \to 0$ in the right-hand integrals.   If both limits exist independently of each other, we have

$$\int_{a+}^{b-} f(x)\, dx = F(b-) - F(a+).$$

Again the choice of c is immaterial.

Example 3.

$$\int_{-1}^{1} \frac{dx}{\sqrt{1 - x^2}} = \sin^{-1} 1 - \sin^{-1}(-1) = \frac{\pi}{2} + \frac{\pi}{2} = \pi.$$

Example 4. If $0 < c < 1$,

$$\int_{\varepsilon}^{c} \frac{dx}{x \log x} = \log |\log x|\, \Big|_{\varepsilon}^{c} \to -\infty \quad \text{as} \quad \varepsilon \to 0;$$

$$\int_{c}^{1-\varepsilon} \frac{dx}{x \log x} = \log |\log x|\, \Big|_{c}^{1-\varepsilon} \to -\infty \quad \text{as} \quad \varepsilon \to 0.$$

Thus both $\displaystyle\int_{0}^{c} \frac{dx}{x \log x}$ and $\displaystyle\int_{c}^{1} \frac{dx}{x \log x}$ diverge.

PROBLEMS

Which of the following integrals converge; and to what value?

1. $\displaystyle\int_{0}^{\pi/2} \cot x\, dx.$

2. $\displaystyle\int_{0}^{1} \log x\, dx.$

3. $\displaystyle\int_{1}^{2} \frac{dx}{x^2 - 1}.$

4. $\displaystyle\int_{0}^{\infty} \frac{dx}{(x^2 + 1)^2}.$

5. $\displaystyle\int_{0}^{\infty} \frac{3\, dx}{(x^2 + 1)(x^2 + 4)}.$

6. $\displaystyle\int_{-\infty}^{\infty} \frac{x^4\, dx}{(x^2 + 1)^2 (x^2 + 4)}.$

7. $\displaystyle\int_0^\infty e^{-ax}\sin bx\,dx \qquad (a>0).$ **8.** $\displaystyle\int_0^\infty e^{-ax}\cos bx\,dx \qquad (a>0).$

9. $\displaystyle\int_1^2 \frac{2x^2\,dx}{x^4-1}.$ **10.** $\displaystyle\int_2^\infty \frac{2x^2\,dx}{x^4-1}.$

11. $\displaystyle\int_0^1 \frac{1-x^n}{1-x}\,dx.$ **12.** $\displaystyle\int_0^\infty e^{-ax^2}x\,dx.$

13. $\displaystyle\int_0^1 x^n\log x\,dx.$ **14.** $\displaystyle\int_{-\infty}^\infty \frac{e^{x/3}}{1+e^x}\,dx.$

15. Prove that $\displaystyle\int_0^\infty \frac{dx}{(r^2+2rx\cos\theta+x^2)^{3/2}} = \frac{1}{r^2(1+\cos\theta)}.$

[Put $t=(x+r\cos\theta)/r\sin\theta$.]

16. Find the centroid of the volume formed by revolving the entire area under the curve $y=e^{-\frac12 x^2}$ about the y-axis.

139. Analogy with Series. Since

$$\int_1^n f(x)\,dx = \int_1^2 f(x)\,dx + \int_2^3 f(x)\,dx + \cdots + \int_{n-1}^n f(x)\,dx,$$

it is clear that improper integrals of type I are analogous to infinite series. We may therefore expect a certain analogy in their properties. There is, however, an important difference. In an improper integral a *continuous* variable (one of its limits) becomes infinite; whereas in an infinite series an *integral* variable (the sum index) becomes infinite. This brings about some differences in behavior. Thus, when the series

$$\sum_1^\infty f(n) \quad \text{converges,} \quad \lim_{n\to\infty} f(n)=0, \tag{21.5};$$

but, when the integral

$$\int_1^\infty f(x)\,dx \quad \text{converges,} \quad \lim_{n\to\infty} f(n)=\lambda$$

is not necessarily zero, even when $f(x)$ is continuous. In fact the integral may converge when $f(x)$ is not bounded, as we see in the following

Example 1. Let $f(x)=0$ everywhere except in the neighborhood of $x=2,3,4,\cdots$ where $f(x)$ has the triangular graph shown in Fig. 139. Since the area of the triangle about $x=n$ is $1/n^2$,

$$\int_1^\infty f(x)\,dx = \sum_{n=2}^\infty \frac{1}{n^2} = \zeta(2)-1. \tag{26.9}$$

Thus the integral converges, but $f(x)=1$ at $x=2,3,4,\cdots$.

If we alter the triangle so that its height and base are $n^{1/2}$ and $2n^{-5/2}$, we obtain a function which is not bounded but whose integral has the same value as before.

The integral $\int_1^\infty f(x)\,dx$ is usually easier to compute than the series $\sum_1^\infty f(n)$ and may often give information about the series as in the useful *integral test* for convergence.

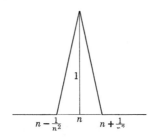

Fig. 139.   Graph of $f(x)$ near $x = n$.

THEOREM 1 (Maclaurin). *If $f(x)$ is a positive decreasing function of x when $x \geq 1$, then the series and the integral*

$$S = \sum_{i=1}^\infty f(i), \qquad I = \int_1^\infty f(x)\,dx$$

both converge or both diverge; and, in the case of convergence,

(1) $$I < S < I + f(1).$$

Proof. When $i \geq 1$,

$$f(i+1) < f(x) < f(i), \qquad i < x < i+1,$$

and hence

$$f(i+1) < \int_i^{i+1} f(x)\,dx < f(i) \qquad\qquad (117.6).$$

If we add these inequalities for $i = 1, 2, \cdots, n-1$, we have

(2) $$f(2) + \cdots + f(n) < \int_1^n f(x)\,dx < f(1) + \cdots + f(n-1).$$

Now let $n \to \infty$; if the integral converges to I, the series on the left will converge (Theorem 23.1) and hence also the series S on the right; and

(3) $$S - f(1) < I < S.$$

Conversely if the series S converges, the integral $\int_k^x f(x)\, dx$ is a positive increasing function of x which is bounded above by S; hence the integral converges. Therefore the series and integral either converge together or diverge together. Moreover, (1) is a consequence of (3).

THEOREM 2. *If $f(x)$ is a positive decreasing function when $x \geq 1$, the sequence*

$$(4) \qquad u_n = f(1) + f(2) + \cdots + f(n) - \int_1^n f(x)\, dx$$

converges to a limit between 0 and $f(1)$.

Proof. From (2) we have

$$0 < f(n) < u_n < f(1);$$

moreover,

$$u_{n+1} - u_n = f(n+1) - \int_n^{n+1} f(x)\, dx < 0,$$

since $f(x) > f(n+1)$ in the interval $n \leq x < n+1$. Thus u_n, a positive decreasing sequence, must converge to a limit between 0 and $f(1)$.

Example 2. When $f(x) = 1/x$, the sequence

$$(5) \qquad u_n = 1 + \frac{1}{2} + \frac{1}{3} + \cdots + \frac{1}{n} - \log n \to \gamma,$$

where $\gamma = 0.57721\,56649 \cdots$, known as *Euler's constant*, has been computed to 263 decimal places. The irrationality of γ has not as yet been proved.

Example 3. When $p \neq 1$,

$$\int_1^x \frac{dx}{x^p} = \frac{x^{1-p}}{1-p}\Big|_1^x = \frac{1}{1-p}\left(\frac{1}{x^{p-1}} - 1\right), \qquad \int_1^x \frac{dx}{x} = \log x;$$

hence, the integral

$$(6) \qquad \int_1^\infty \frac{dx}{x^p} = \begin{cases} \dfrac{1}{p-1} & \text{when} \quad p > 1, \\[2mm] \infty & \text{when} \quad p \leq 1. \end{cases}$$

Therefore, the series

$$(7) \qquad \sum_{n=1}^\infty \frac{1}{n^p} = \begin{cases} \zeta(p)\dagger & \text{when} \quad p > 1, \\[2mm] \infty & \text{when} \quad 0 < p \leq 1. \end{cases}$$

Moreover, from (1)

$$\frac{1}{p-1} < \zeta(p) < \frac{1}{p-1} + 1.$$

---

† $\zeta(p)$ is *Riemann's zeta function* (26.9). When $p \leq 0$, the integral (6) and series (7) obviously diverge.

PROBLEMS

1. If $f(x)$ is positive, decreases steadily to zero, and $f'(x)$ is continuous when $x \geq a$, prove that

$$\int_a^\infty |f'(x)| \, dx = f(a).$$

2. If $f(x)$ is positive and decreases steadily to zero, the convergence of $\int_a^\infty f(x) \, dx$ implies that $\lim_{x \to \infty} x f(x) = 0$.

[Let m be an integer $\geq a$, and apply Abel's theorem (§ 26) to the series

$$\sum_{n=m}^\infty \int_n^{n+1} f(x) \, dx.]$$

140. Comparison Tests: Type I. In analogy with Theorem 23.1 on positive series we have the

THEOREM 1. *If $f(x)$ is a nonnegative integrable function when $x \geq a$ and $\int_a^x f(t) \, dt$ is bounded above, the integral will converge as $x \to \infty$; otherwise it will diverge to ∞.*

Proof. A positive increasing function approaches a limit if bounded above; but, if unbounded, it will approach ∞.

In analogy with the comparison tests on positive series in § 25, we have

THEOREM 2. *Let $f(x)$ and $g(x)$ be integrable functions when $x \geq a$ such that*

$$0 \leq f(x) \leq g(x).$$

Then

(i) $\qquad \int_a^\infty f(x) \, dx$ *converges if* $\int_a^\infty g(x) \, dx$ *converges;*

(ii) $\qquad \int_a^\infty g(x) \, dx$ *diverges if* $\int_a^\infty f(x) \, dx$ *diverges.*

Proof. In case (i),

$$\int_a^x f(x) \, dx \leq \int_a^x g(x) \, dx < \int_a^\infty g(x) \, dx;$$

In case (ii),

$$\int_a^x g(x) \, dx \geq \int_a^x f(x) \, dx \quad \text{which} \quad \to \infty.$$

The conclusions now follow from Theorem 1.

Example. If $\varphi(x)$ is bounded when $x \geq a > 0$ and $p > 1$,

$$\int_a^\infty \frac{|\varphi(x)|}{x^p} \, dx \quad \text{converges.}$$

For, when $x \geq a$, $|\varphi(x)| < M$ and

$$\frac{|\varphi(x)|}{x^p} < \frac{M}{x^p} \quad \text{and} \quad M\int_a^\infty \frac{dx}{x^p} \quad \text{converges} \tag{139.6}.$$

We shall see in the next article that

$$\int_a^\infty \frac{\varphi(x)}{x^p}\, dx \quad \text{also converges.}$$

141. Absolute Convergence: Type I. The integral $\int_a^\infty f(x)\, dx$ is said to *converge absolutely* when $\int_a^\infty |f(x)|\, dx$ converges. The theorem in § 32 on series has its analogue in the

THEOREM. *The absolute convergence of $\int_a^\infty f(x)\, dx$ implies its convergence.*

Proof. Since
$$0 \leq |f(x)| - f(x) \leq 2\,|f(x)|;$$

Theorem 140.2 shows that

$$\int_a^\infty [\,|f(x)| - f(x)]\, dx \quad \text{converges};$$

hence,

$$\int_a^\infty f(x)\, dx = \int_a^\infty [f(x) - |f(x)|]\, dx + \int_a^\infty |f(x)|\, dx \quad \text{converges.}$$

142. Limit Tests: Type I. The comparison tests of § 140 may be put in the convenient limit form.

THEOREM 1. *Let $f(x)$ and $g(x)$ be integrable functions when $x \geq a$ and $g(x)$ be positive. Then, if*

$$(1) \qquad\qquad \lim_{x\to\infty} \frac{f(x)}{g(x)} = \lambda \neq 0,$$

the integrals

$$F = \int_a^\infty f(x)\, dx \quad \text{and} \quad G = \int_a^\infty g(x)\, dx$$

both converge absolutely or both diverge.

If $f/g \to 0$ and G converges, then F converges absolutely.

If $f/g \to \pm\infty$ and G diverges, then F diverges.

Proof. Suppose (1) holds and $\lambda > 0$. Choose ε so that $0 < \varepsilon < \lambda$; then we can always find a number $b > a$ such that

$$\lambda - \varepsilon < \frac{f(x)}{g(x)} < \lambda + \varepsilon \quad \text{when} \quad x \geq b,$$

or, since $g(x) > 0$,

$$(\lambda - \varepsilon)g(x) < f(x) < (\lambda + \varepsilon)g(x), \qquad x \geqq b.$$

Since $f(x) > 0$ when $x \geqq b$, we can apply the comparison tests of Theorem 140.2; according as G converges or diverges, we use

$$f(x) < (\lambda + \varepsilon)g(x) \quad \text{or} \quad (\lambda - \varepsilon)g(x) < f(x)$$

to prove the absolute convergence or divergence of $\int_b^\infty f(x)\, dx$.

If $\lambda < 0$, this argument applies to $-F$; but $-F$ and F converge at the same time.

When $f(x)/g(x) \to 0$, we can find b so that

$$\frac{|f(x)|}{g(x)} < \varepsilon, \qquad |f(x)| < \varepsilon\, g(x), \qquad x \geqq b;$$

hence, if G converges, F converges absolutely.

When $f/g \to +\infty$,

$$\frac{f(x)}{g(x)} > M, \qquad f(x) > M\, g(x), \qquad x \geqq b;$$

hence, if G diverges, F also diverges. If $f/g \to -\infty$, this argument shows that $-F$ diverges; hence F also diverges.

When $g(x) = 1/x^p$ $(x > 0)$, this theorem gives the following useful test.

THEOREM 2. *Let $f(x)$ be an integrable function when $x \geqq a$. Then*

$$F = \int_a^\infty f(x)\, dx \quad \text{converges absolutely if}$$

(2) $$\lim_{x \to \infty} x^p f(x) = \lambda, \qquad p > 1,$$

and F diverges if

(3) $$\lim_{x \to \infty} x^p f(x) = \lambda (\neq 0) \quad \text{or} \quad \pm\infty, \qquad p \leqq 1.$$

Proof. From (139.6),

$$G = \int_a^\infty \frac{dx}{x^p} \quad \begin{matrix} \text{converges} & p > 1, \\ \text{diverges} & p \leqq 1. \end{matrix}$$

When the limit is ∞ in (2) or 0 in (3), we can draw no conclusion. Consider, for example, the integrals

$$\int_1^\infty \frac{dx}{x^2} = -\frac{1}{x}\Big|_1^\infty = 1, \qquad \int_2^\infty \frac{dx}{x \log x} = \log \log x\Big|_2^\infty = \infty.$$

With $p = 3$, $x^3 f(x) \to \infty$ in both; and, with $p = 1$, $x f(x) \to 0$ in both.

Integrals of type I in which the lower limit is $-\infty$ may be transformed so that the upper limit is $+\infty$ by the change of variable $x = -t$; for

(4) $$\int_{-\infty}^{b} f(x)\, dx = \int_{-b}^{\infty} f(-t)\, dt.$$

Since

$$t^p f(-t) = (-x)^p f(x),$$

we see that the integral (4) converges absolutely if

(2)′ $$\lim_{x \to -\infty} (-x)^p f(x) = \lambda, \qquad p > 1;$$

and diverges if

(3)′ $$\lim_{x \to -\infty} (-x)^p f(x) = \lambda(\neq 0) \quad \text{or} \quad \infty, \qquad p \leqq 1.$$

Example 1. $\displaystyle\int_{1}^{\infty} \frac{dx}{x\sqrt{x^2 + 1}}$ converges; for the integrand behaves like x^{-2} at ∞,

and, when $x \to \infty$,

$$x^2 f(x) = \frac{1}{\sqrt{1 + x^{-2}}} \to 1.$$

Example 2. $\displaystyle\int_{0}^{\infty} \frac{x^2\, dx}{\sqrt{x^5 + 1}}$ diverges; for the integrand behaves like $x^{-1/2}$ at ∞,

and, when $x \to \infty$,

$$x^{1/2} f(x) = \frac{1}{\sqrt{1 + x^{-5}}} \to 1.$$

Example 3. $\displaystyle\int_{0}^{\infty} e^{-x^2}\, dx$ converges; for, as $x \to \infty$,

$$x^2 e^{-x^2} \to 0. \tag{61.4}.$$

Example 4. $\displaystyle\int_{1}^{\infty} \frac{\log x}{x^2}\, dx$ converges; for, as $x \to \infty$,

$$x^{3/2} f(x) = (\log x)/x^{1/2} \to 0 \tag{61.5}.$$

Example 5. $\displaystyle\int_{1}^{\infty} x^n e^{-x}\, dx$ converges for all values of n; for, as $x \to \infty$,

$$x^2 f(x) = x^{n+2}/e^x \to 0 \tag{61.4}.$$

143. Comparison Tests: Type II. Improper integrals of type II—those in which the integrand becomes infinite—may be reduced to integrals of type I by a change of variable. Thus, if $f(a+) = \infty$, the substitution $x = a + t^{-1}$ gives

$$\int_{a+\varepsilon}^{b} f(x)\, dx = \int_{(b-a)^{-1}}^{\varepsilon^{-1}} \frac{1}{t^2}\, f\!\left(a + \frac{1}{t}\right) dt;$$

and, as $\varepsilon \to 0$,

(1) $$\int_{a+}^{b} f(x)\, dx = \int_{(b-a)^{-1}}^{\infty} \frac{1}{t^2}\, f\!\left(a + \frac{1}{t}\right) dt.$$

when the integral exists.

Similarly, if $f(b-) = \infty$, the substitution $x = b - t^{-1}$ gives

(2)
$$\int_a^{b-} f(x)\, dx = \int_{(b-a)^{-1}}^{\infty} \frac{1}{t^2}\, f\left(b - \frac{1}{t}\right) dt.$$

By making use of (1) and (2), the entire theory of integrals of type I can be carried over to integrals of type II. Thus the comparison tests of § 140 take the following form.

THEOREM 1. *Let $f(x)$ and $g(x)$ be integrable functions in $(a+, b)$ or $(a, b-)$ such that*
$$0 \leq f(x) \leq g(x).$$
Then

(i)
$$\int_a^b f(x)\, dx \quad \text{converges if} \quad \int_a^b g(x)\, dx \quad \text{converges};$$

(ii)
$$\int_a^b g(x)\, dx \quad \text{diverges if} \quad \int_a^b f(x)\, dx \quad \text{diverges}.$$

Proof. If $f(a+) = \infty$, transform the integral as in (1). By hypothesis
$$0 \leq f\left(a + \frac{1}{t}\right) \leq g\left(a + \frac{1}{t}\right), \qquad a < a + \frac{1}{t} \leq b;$$
hence,
$$0 \leq \frac{1}{t^2}\, f\left(a + \frac{1}{t}\right) \leq \frac{1}{t^2} g\left(a + \frac{1}{t}\right), \qquad \frac{1}{b-a} \leq t < \infty.$$
The theorem now follows from Theorem 140.2. If $f(b-) = \infty$, transform the integral as in (2).

The integrals (1) or (2) are said to be absolutely convergent when $\int_a^b |f(x)|\, dx$ converges. When transformations (1) or (2) are used, Theorem 141 gives the

THEOREM 2. *The absolute convergence of $\int_a^b f(x)\, dx$ implies its convergence.*

144. Limit Tests: Type II. Theorems 142.1 and 142.2 carry over to the following.

THEOREM 1. *Let $f(x)$ and $g(x)$ be integrable functions when $a < x \leq b$ and $g(x)$ be positive. Then, if*

(1)
$$\lim_{x \to a+} \frac{f(x)}{g(x)} = \lambda \neq 0,$$

the integrals
$$F = \int_a^b f(x)\, dx \quad \text{and} \quad G = \int_a^b g(x)\, dx$$
both converge absolutely or both diverge.

If $f/g \to 0$ and G converges, then F converges absolutely.

If $f/g \to \pm\infty$ and G diverges, then F diverges.

THEOREM 2. *Let $f(x)$ be an integrable function when $a < x \leqq b$. Then*

$$F = \int_a^b f(x) \quad \text{converges absolutely if}$$

(2) $$\lim_{x \to a+} (x-a)^q f(x) = \lambda, \qquad 0 < q < 1;$$

and F diverges if

(3) $$\lim_{x \to a+} (x-a)^q f(x) = \lambda (\neq 0) \quad \text{or} \quad \pm\infty, \qquad q \geqq 1.$$

Proof of Theorem 2. This is clearest when the integral F is compared with the following integral of type II. When $q \neq 1$, we have

$$\int_{a+\varepsilon}^b \frac{dx}{(x-a)^q} = \frac{(x-a)^{1-q}}{1-q} \Big|_{a+\varepsilon}^b = \frac{1}{1-q} [(b-a)^{1-q} - \varepsilon^{1-q}],$$

and, when $q = 1$,

$$\int_{a+\varepsilon}^b \frac{dx}{x-a} = \log(b-a) - \log \varepsilon$$

On letting $\varepsilon \to 0$, we see that

(4) $$\int_a^b \frac{dx}{(x-a)^q} = \begin{cases} \dfrac{(b-a)^{1-q}}{1-q} & \text{when} \quad 0 < q < 1, \\ \infty & \text{when} \quad q \geqq 1. \end{cases}$$

When $q \leqq 0$, the integral is proper. If we now take $g(x) = (x-a)^{-q}$ in Theorem 1, we obtain Theorem 2.

Of course, the proof can also be effected by the change of variable $x = a + 1/t$ as in (143.1). We then note that

$$t^{p-2} f\left(a + \frac{1}{t}\right) = (x-a)^q f(x), \qquad q = 2 - p;$$

and $p > 1, p \leqq 1$ correspond to $q < 1, q \geqq 1$. When $q \leqq 0$, the relation (2) shows that $f(x)$ remains finite as $x \to a+$, and hence the integral is proper; hence convergence of the *improper* integral F corresponds to $0 < q < 1$.

Integrals of type II in which the integrand becomes infinite at the upper limit b may be transformed by the change of variable $x = a + b - t$;

(5) $$\int_a^{b-\varepsilon} f(x) \, dx = \int_{a+\varepsilon}^b f(a+b-t) \, dt.$$

Since

$$(t - a)^q f(a + b - t) = (b - x)^q f(x),$$

we see that the integral (5) converges absolutely if

(2)' $\lim\limits_{x \to b-} (b - x)^q f(x) = \lambda,$ $0 < q < 1,$

and diverges if

(3)' $\lim\limits_{x \to b-} (b - x)^q f(x) = \lambda(\neq 0)$ or $\pm\infty,$ $q \geq 1.$

Example 1. If $0 < m, n < 1,$

$$\int_a^b (x - a)^{-m}(b - x)^{-n} \, dx \quad \text{converges absolutely; for}$$

$$\lim\limits_{x \to a+} (x - a)^m f(x) = (b - a)^{-n}, \qquad \lim\limits_{x \to b-} (b - x)^n f(x) = (b - a)^{-m}.$$

If $m, n \geq 1$, the integral diverges

Example 2. $\int_0^1 \dfrac{\log x}{\sqrt{x}} \, dx$ converges; for

$$\lim\limits_{x \to 0+} x^{3/4} \cdot \frac{\log x}{x^{1/2}} = \lim\limits_{x=0+} \frac{\log x}{x^{-1/4}} = \lim\limits_{x=0} (-4x^{1/4}) = 0. \qquad (\S\ 61).$$

Example 3. $\int_1^2 \dfrac{\sqrt{x}}{\log x} \, dx$ diverges; for

$$\lim\limits_{x \to 1+} (x - 1) \cdot \frac{\sqrt{x}}{\log x} = \lim\limits_{x \to 1+} \frac{\frac{3}{2}x^{1/2} - \frac{1}{2}x^{-1/2}}{1/x} = 1. \qquad (\S\ 60).$$

PROBLEMS

Test the following integrals for convergence or divergence.

1. $\int_1^\infty \sin \dfrac{1}{x^2} \, dx.$

2. $\int_1^\infty \dfrac{\sin x}{x^2} \, dx.$

3. $\int_0^1 \dfrac{\sin x}{x^{3/2}} \, dx.$

4. $\int_1^\infty e^{-x} \log x \, dx.$

5. $\int_2^3 \dfrac{x^2 + 1}{x^2 - 4} \, dx.$

6. $\int_3^\infty \dfrac{x^2 + 1}{x^2 - 4} \, dx.$

Find the area between the curve $y = f(x)$ and the x-axis from 1 to ∞ and also the volume obtained by revolving the curve about the x-axis:

7. $y = e^{-x}.$ 8. $y = 1/x.$ 9. $y = 1/x^2.$

10. Show that $\int_0^\infty \dfrac{\sin^2 x}{x^2} \, dx$ converges and is equal to $\int_0^\infty \dfrac{\sin x}{x} \, dx.$ [Integrate by parts.]

11. Prove that $I = \int_0^{\pi/2} \log \sin x \, dx = -\dfrac{\pi}{2} \log 2.$

[Use the identity $\sin 2x = 2 \sin x \cos x$ to get

$$\int_0^{\pi/2} \log \sin 2x \, dx = \frac{\pi}{2} \log 2 + 2I,$$

and put $2x = t$ in the integral.]

12. Prove that $I = \int_0^\pi x \log \sin x \, dx = -\frac{\pi^2}{2} \log 2.$

[Put $x = \pi - t$, and show that $2I = \pi \int_0^\pi \log \sin x \, dx.$]

13. Prove that the ratio $\int_2^x \frac{dt}{\log t} : \frac{x}{\log x} \to 1$ as $x \to \infty.$

14. If $P_m(x)$ and $Q_n(x)$ are polynominals of degree m and n, and $Q_n(x) \neq 0$ $(x \geq a)$,

show that $\int_a^\infty \frac{P_m(x)}{Q_n(x)} \, dx$ converges when and only when $n > m + 1.$

145. Conditional Convergence: Type I. An integral is said to converge conditionally when it converges, but not absolutely. We begin with a theorem that simply establishes convergence.

THEOREM 1. *If $g(x)$ has an indefinite integral $G(x)$ that remains bounded when $x \geq a > 0$, the integral*

(1) $$\int_a^\infty \frac{g(x)}{x^r} \quad \text{converges when} \quad r > 0,$$

Proof. An integration by parts gives

$$\int_a^x \frac{g(x)}{x^r} \, dx = \frac{G(x)}{x^r} \Big|_a^x + r \int_a^x \frac{G(x)}{x^{r+1}} \, dx;$$

and, on passing to the limit $x \to \infty$,

$$\int_a^\infty \frac{g(x)}{x^r} \, dx = -\frac{G(a)}{a^r} + r \int_a^\infty \frac{G(x)}{x^{r+1}} \, dx$$

The last integral converges by (142.2); for choose p so that $1 < p < r + 1$ then

$$\lim_{x \to \infty} x^p \frac{G(x)}{x^{r+1}} = 0$$

Example 1. Since the indefinite integrals of $\sin x$ and $\cos x$ ($-\cos x$ and $\sin x$) are bounded, the integrals

$$\int_0^\infty \frac{\sin x}{\sqrt{x}} \, dx, \qquad \int_a^\infty \frac{\cos x}{\sqrt{x}} \, dx, \qquad a > 0$$

converge. In the first the integrand approaches zero as $x \to 0$.

We next deal with integrals having oscillating integrands. These are analogous to alternating series.

THEOREM 2. *In the interval $a \leq x < \infty$, let $g(x)$ be a continuous function that steadily decreases to zero as x passes from a to ∞; then*

$$(2) \qquad\qquad I = \int_a^\infty g(x) \sin x \, dx$$

converges; and, if the series

$$(3) \qquad\qquad \sum_{k=m}^\infty g(k\pi) \quad \text{diverges} \quad (m\pi \geq a),$$

the integral I converges conditionally.

Proof. By hypothesis $g(x) > 0$ when $x \geq a$. Now select two integers m and n such that

$$a \leq m\pi < n\pi < b < (n+1)\pi,$$

and write

$$I = \int_a^{m\pi} + \int_{m\pi}^{n\pi} + \int_{n\pi}^{b} g(x) \sin x \, dx.$$

As b (and n) $\to \infty$, the last integral approaches zero; for its absolute value is less than

$$g(n\pi) \int_{n\pi}^{(n+1)\pi} \sin x \, dx = 2g(n\pi),$$

for the area under one arch of the sine curve is 2 (Ex. 126.1). Since m is held constant, we need only consider the limit of the series,

$$(4) \qquad\qquad \sum_{k=m}^{k=n-1} \int_{k\pi}^{(k+1)\pi} g(x) \sin x \, dx,$$

as $n \to \infty$. On making the change of variable $x = t + k\pi$, we have

$$\int_{k\pi}^{(k+1)\pi} g(x) \sin x \, dx = (-1)^k \int_0^\pi g(t + k\pi) \sin t \, dt$$

$$= (-1)^k 2g(k\pi + \xi_k), \qquad 0 < \xi_k < \pi,$$

on using the mean-value theorem (121.2). Thus, as $n \to \infty$, (4) becomes the alternating series,

$$(5) \qquad\qquad 2 \sum_{k=m}^\infty (-1)^k g(k\pi + \xi_k),$$

which converges by Leibnitz's test (§ 34). Thus the first part of the theorem is proved.

To prove the second part we must show that, when the series (3) diverges, the integral

$$\int_a^\infty g(x) \, |\sin x| \, dx$$

diverges; that is, the series of absolute values

(6) $$2 \sum_{k=m}^{\infty} g(k\pi + \xi_k)$$

obtained from (5) diverges. But, since

$$g(k\pi + \xi_k) > g[(k+1)\pi],$$

the series (6) diverges by the comparison test of § 25.

Example 2. $\int_0^\infty \dfrac{\sin x}{x}\, dx$ converges conditionally. For $\int_0^\pi \dfrac{\sin x}{x}\, dx$ exists, and

$\int_\pi^\infty \dfrac{\sin x}{x}$ converges by Theorem 1 or 2, but only conditionally since the series $\dfrac{1}{\pi} \sum_{k=1}^{\infty} \dfrac{1}{k}$

diverges (Ex. 26.1).

Example 3. $\int_0^\infty \sin x^2\, dx$ converges conditionally.

Replace the lower limit by $\sqrt{\pi}$ and set $x = \sqrt{t}$; then

$$\int_{\sqrt{\pi}}^\infty \sin x^2\, dx = \frac{1}{2} \int_\pi^\infty \frac{\sin t}{\sqrt{t}}\, dt$$

converges, but only conditionally since the series $\sum_{k=1}^{\infty} 1/\sqrt{k\pi}$ diverges (Ex. 26.1).

Obviously the given integral (with lower limit 0) also converges conditionally.

This integral also shows that $\int_a^\infty f(dx)\, dx$ may converge when $\lim_{x \to \infty} f(x) \neq 0$ (Ex. 139.1). Indeed the next example shows that the integral may converge even when $f(x)$ is not bounded.

Example 4. $\int_e^\infty \dfrac{\log x}{x} \sin x\, dx$ converges by Theorem 2 since $(\log x)/x \downarrow 0$. But the

change of variable $x = e^t$ converts the integral into $\int_1^\infty t \sin e^t\, dt$ in which $t \sin e^t$ is not bounded as $t \to \infty$.

146. Conditional Convergence: Type II. In order to test improper integrals of type II for conditional convergence reduce them to type I by changing the variable as in (143.1) or (143.2), and apply the theorems of § 145.

Consider, for example, the integral

$$I = \int_0^1 \frac{\sin 1/x}{x^r}\, dx, \qquad r > 0.$$

The change of variable $x = 1/t$ gives

$$\int_\varepsilon^1 \frac{\sin 1/x}{x^r}\, dx = \int_1^{1/\varepsilon} \frac{\sin t}{t^{2-r}}\, dt.$$

As $\varepsilon \to 0$, the integral on the right converges when $r < 2$ (Theorem 145.1) and diverges when $r \geq 2$ (why?). Hence I converges when $0 < r < 2$, diverges when $r \geq 2$.

We now apply Theorem 145.2 to test I for conditional convergence. Since the series

$$\sum_{k=1}^\infty \frac{1}{k^{2-r}} \quad \text{diverges when} \quad 2 - r \leq 1 \quad \text{or} \quad r \geq 1,$$

we see that I converges conditionally when $1 \leq r < 2$. Moreover I converges absolutely when $0 < r < 1$. For choose p so that $r < p < 1$; then, as $x \to 0$,

$$x^p \frac{\sin 1/x}{x^r} = x^{p-r} \sin \frac{1}{x} \to 0 \tag{144.2}$$

We thus have the following results; when $r > 0$:

$$\text{(1)} \qquad \int_0^1 \frac{\sin 1/x}{x^r}\, dx \quad \begin{cases} \text{conv. abs.} & 0 < r < 1, \\ \text{conv. cond.} & 1 \leq r < 2, \\ \text{diverges} & r \geq 2. \end{cases}$$

PROBLEMS

Show that the following integrals converge conditionally:

1. $\displaystyle\int_1^\infty \frac{\cos x}{\log x}\, dx.$

2. $\displaystyle\int_0^1 \frac{\cos x^{-1}}{x}\, dx.$

3. $\displaystyle\int_2^\infty \frac{\log \log x}{\log x} \sin x\, dx$

4. $\displaystyle\int_0^1 \frac{\sin x^{-1}}{x^2 \log (1 + x^{-1})}\, dx.$

5. When $g(x)$ satisfies the hypotheses of Theorem 145.2 prove that

$$\int_a^\infty g(x) \sin (ax + b)\, dx \quad \text{and} \quad \int_a^\infty g(x) \cos (ax + b)\, dx$$

converge.

6. When $g(x)$ satisfies the hypotheses of Theorem 145.2 and $n\pi \geq a$, prove that

$$\left| \int_{n\pi}^\infty g(x) \sin x\, dx \right| < 2 g(n\pi).$$

147. Combinations of Types I and II. We have seen in § 137 that an improper integral in which the integrand has but a finite number of discontinuities may be expressed as a sum of simple integrals of types I or

II. These should be dealt with one at a time by the tests that are applicable. The behavior of the given integral may then be determined.

Example 1. $F(x) = \int_0^\infty \frac{t^{x-1}}{1+t}\,dt = \int_0^1 + \int_1^\infty \frac{t^{x-1}}{1+t}\,dt.$

The part I between the limits $t = 0$ and $t = 1$ is proper when $x \geq 1$. When $0 < x < 1$, I converges; for, as $t \to 0$,

$$t^{1-x} f(x, t) = \frac{1}{1+t} \to 1, \quad 0 < 1 - x < 1, \qquad (144.2).$$

When $x \leq 0$, I diverges; for, as $t \to 0$,

$$t^{1-x} f(x, t) = \frac{1}{1+t} \to 1, \quad 1 - x \geq 1, \qquad (144.3).$$

The part J between $t = 1$ and $t = \infty$ diverges when $x \geq 1$; for, as $t \to \infty$,

$$t f(x, t) = \frac{t^x}{1+t} \to \begin{cases} 1 & (x = 1), \\ \infty & (x > 1). \end{cases} \qquad (142.3).$$

When $x < 1$, J converges; for, as $t \to \infty$,

$$t^{2-x} f(x, t) = \frac{t}{1+t} \to 1, \quad 2 - x > 1, \qquad (142.2).$$

The behavior of I, J and $F = I + J$ is therefore as follows:

	$x \leq 0$	$0 < x < 1$	$x \geq 1$
I	div	conv	proper
J	conv	conv	div
F	div	conv	div

Example 2. $F(x) = \int_0^1 \left(\log\frac{1}{t}\right)^x dt = \int_0^{1/2} + \int_{1/2}^1 \left(\log\frac{1}{t}\right)^x dt.$

The part I between the limits $t = 0$ and $t = {}^1/_2$ is proper when $x \leq 0$; for the integrand approaches a finite limit as $t \to 0+$ (1 when $x = 0$, 0 when $x < 0$). When $x > 0$, I converges; for, if we put $t = e^{-u}$,

$$\lim_{t \to 0+} t^{1/2} \left(\log\frac{1}{t}\right)^x = \lim_{u \to \infty} e^{-u/2} u^x = 0.$$

The part J between the limits $t = {}^1/_2$ and $t = 1$ is proper when $x \geq 0$. When $x < 0$, we have

$$\lim_{t \to 1-} (1 - t)^{-x} \left(\log\frac{1}{t}\right)^x = \lim_{t \to 1-} \left(\frac{\log t}{t - 1}\right) \to 1.$$

The tests (144.2)' and (144.3)' now show that J converges when $-1 < x < 0$, diverges when $x \leq -1$.

The behavior of I, J and $F = I + J$ is therefore as follows:

	$x \leqq -1$	$-1 < x < 0$	$x = 0$	$x \geqq 0$
I	proper	proper	proper	conv
J	div	conv	proper	proper
F	div	conv	proper	conv

148. Laplace Transform.   The *Laplace transform* of a function $f(x)$ is defined as

$$(1) \qquad \mathscr{L}\{f(x)\} = \int_0^\infty e^{-xt} f(t)\, dt$$

for all values of x for which the integral converges.   We shall assume that $f(t)$ is *piecewise continuous* when $t > 0$; that is, $f(x)$ is continuous except for a finite number of finite jumps.   If $f(t) \to \infty$ as $t \to 0+$, the integral is an improper integral of type II as well as type I.

Let us write $\mathscr{L}(f)$ as the sum of three integrals:

$$I_2 + I + I_1 = \int_0^\varepsilon + \int_\varepsilon^b + \int_b^\infty e^{-xt} f(t)\, dt;$$

then I_2 is of type II, I is proper, and I_1 is of type I.   Then I_2 will converge absolutely for all $x > 0$ if

$$(2) \qquad f(t) = O(t^{-q})\ddagger, \qquad 0 < q < 1, \qquad 0 < t \leqq \varepsilon;$$

for

$$e^{-xt} < 1, \qquad |f(t)| < M t^{-q}, \qquad\qquad \text{and}$$

$$\int_0^\varepsilon e^{-xt} |f(t)|\, dt < M \int_0^\varepsilon t^{-q}\, dt = M\, \frac{\varepsilon^{1-q}}{1-q}.$$

I_1 will converge absolutely when $x > x_0$ if

$$(3) \qquad\qquad f(t) = O(e^{x_0 t}), \qquad t \geqq b;$$

for

$$\int_b^\infty e^{-xt} |f(t)|\, dt < M \int_b^\infty e^{-(x-x_0)t}\, dt = M\, \frac{e^{-(x-x_0)b}}{x - x_0}.$$

The tests (2) and (3) are usually adequate to cope with the Laplace integrals that occur in practice, but more refined tests are sometimes needed.

---

‡ Cf. 64.10 for the O-notation.

If we evaluate the following integrals with c as upper limit and then let $c \to \infty$, we obtain the following transforms:

$$\text{(4)} \qquad \mathscr{L}(1) = \int_0^\infty e^{-xt}\, dt = \frac{1}{x}, \qquad\qquad x > 0;$$

$$\text{(5)} \qquad \mathscr{L}(x^n) = \int_0^\infty e^{-xt}\, t^n\, dt = \frac{n!}{x^{n+1}}, \quad n = 1, 2, \cdots, x > 0;$$

$$\text{(6)} \qquad \mathscr{L}(e^{ax}) = \int_0^\infty e^{(a-x)t}\, dt = \frac{1}{x - a}, \qquad\qquad x > a;$$

$$\text{(7)} \qquad \mathscr{L}(\sin bx) = \int_0^\infty e^{-xt} \sin bt\, dt = \frac{b}{x^2 + b^2}, \qquad x > 0;$$

$$\text{(8)} \qquad \mathscr{L}(\cos bx) = \int_0^\infty e^{-xt} \cos bt\, dt = \frac{x}{x^2 + b^2}, \qquad x > 0.$$

We may regard $\mathscr{L}$ as an operator that transforms a function $f(x)$ into another $\mathscr{L}\{f(x)\}$. Equation (1) shows that $\mathscr{L}$ is a *linear operator*:

$$\text{(9)} \qquad \mathscr{L}\{a f(x) + b\, g(x)\} = a\,\mathscr{L}\{f(x)\} + b\,\mathscr{L}\{g(x)\}.$$

Thus from (6) we find the transforms

$$\text{(10)} \qquad \mathscr{L}(\sinh bx) = \frac{1}{2}\mathscr{L}(e^{bx}) - \frac{1}{2}\mathscr{L}(e^{-bx}) = \frac{b}{x^2 - b^2}, \qquad x > b;$$

$$\text{(11)} \qquad \mathscr{L}(\cosh bx) = \frac{1}{2}\mathscr{L}(e^{bx}) + \frac{1}{2}\mathscr{L}(e^{-bx}) = \frac{x}{x^2 - b^2}, \qquad x > b.$$

The transform of $f(ax)$ is given by

$$\int_0^\infty e^{-xt} f(at)\, dt = \frac{1}{a} \int_0^\infty e^{-(x/a)u} f(u)\, du$$

on making the substitution $u = at$; hence,

$$\text{(12)} \qquad \mathscr{L}\{f(ax)\} = \frac{1}{a}\, \mathscr{L}\{f(u)\}_{u=x/a}.$$

The following theorems show how the transforms of $f'(x)$ and $\int_0^x f(t)\, dt$ may be computed from $\mathscr{L}\{f(x)\}$.

THEOREM 1. *If $\mathscr{L}\{f(x)\}$ converges when $x > x_0$, and*

(i) *$f(t)$ is continuous and $f'(t)$ is piecewise continuous when $t \geqq 0$,*

(ii) *$\lim\limits_{t \to \infty} e^{-xt} f(t) = 0$; then*

$$\text{(13)} \qquad \mathscr{L}\{f'(x)\} = x\mathscr{L}\{f(x)\} - f(0), \qquad\qquad x > x_0.$$

Proof. An integration by parts gives

$$\int_0^b e^{-xt} f'(t)\, dt = e^{-xt} f(t) \Big|_0^b + x \int_0^b e^{-xt} f(t)\, dt;$$

and, on letting $b \to \infty$, we obtain (13).

THEOREM 2. *If $\mathscr{L}\{f(x)\}$ converges when $x > x_0$, and*

$$\lim_{t \to \infty} e^{-xt} F(t) = 0, \quad \text{where} \quad F(t) = \int_0^t f(u)\, du; \quad \text{then}$$

(14) $$\mathscr{L}\{F(x)\} = \frac{1}{x} \mathscr{L}\{f(x)\}, \qquad\qquad x > x_0.$$

Proof. Replace $f(x)$ by $F(x)$ in (13); then $F'(x) = f(x)$ is piecewise continuous and $F(0) = 0$.

Finally let us differentiate the Laplace integral (1) with respect to x under the integral sign (the validity of which is proved in Ex. 190.2):

$$\frac{d}{dx} \mathscr{L}\{f(x)\} = - \int_0^\infty e^{-xt} t f(t)\, dt.$$

This can be written

(15) $$\mathscr{L}\{x f(x)\} = -D\, \mathscr{L}\{f(x)\},$$

where $D = d/dx$. Hence, on differentiating n times, we get

(16) $$\mathscr{L}\{x^n f(x)\} = (-1)^n\, D^n\, \mathscr{L}\{f(x)\}.$$

PROBLEMS

1. Verify formulas (4)–(8).
2. Find $\mathscr{L}(ax^2 + bx + c)$ and $\mathscr{L}(a - x)(b - x)$.
3. Prove that
    (i) $\mathscr{L}(e^{-x^2})$ converges absolutely for all x;
    (ii) $\mathscr{L}(e^{x^2})$ diverges for all x.
4. Verify (13) on formulas (7) and (8).
5. Show that under appropriate conditions

$$\mathscr{L}\{f''(x)\} = x^2 \mathscr{L}\{f(x)\} - xf(0) - f'(0).$$

6. If the function $h(x) = 1/h$ when $0 < x < h$ and is zero elsewhere, show that

$$\mathscr{L}\{h(x)\} = \frac{1 - e^{-hx}}{hx}.$$

As $h \to 0$, $h(x)$ approaches a "function" $j(x)$ which is infinite when $x = 0$ and zero elsewhere: the *unit impulse function.* If we *define*

$$\mathscr{L}\{j(x)\} = \lim_{h \to 0} \mathscr{L}\{h(x)\}, \text{ show that } \mathscr{L}\{j(x)\} = 1.$$

7. Use formula (14) to obtain (5) from (4).

8. Prove that

$$\mathscr{L}\{e^{ax}f(x)\} = \mathscr{L}\{f(t)\}_{\,t=x-a}.$$

9. Prove the formulas:

$$\mathscr{L}(e^{ax}x^n) = \frac{n!}{(x-a)^{n+1}}, \qquad x > a;$$

$$\mathscr{L}(e^{ax}\sin bx) = \frac{b}{(x-a)^2 + b^2}, \qquad x > a;$$

$$\mathscr{L}(e^{ax}\cos bx) = \frac{x-a}{(x-a)^2 + b^2}, \qquad x > a.$$

10. If $f(x) = \begin{cases} 0 & t < a \\ \varphi(x-a) & t \geq a \end{cases}$, prove that

$$f(x) = e^{-ax}\mathscr{L}\{\varphi(x)\}.$$

11. By applying (15) to (7) and (8) show that

$$\mathscr{L}(x\sin bx) = \frac{2bx}{(x^2+b^2)^2}, \qquad x > 0,$$

$$\mathscr{L}(x\cos bx) = \frac{x^2-b^2}{(x^2+b^2)^2}, \qquad x > 0.$$

12. Apply (16) to (4) to obtain (5).

13. If $\varphi(x) = \mathscr{L}\{f(x)\}$ is given, there is only one *continuous* function $f(x)$ which satisfies this *integral equation*; and we write $f(x) = \mathscr{L}^{-1}\{\varphi(x)\}$. From the Laplace transforms already given, show that

$$\mathscr{L}^{-1}\{(x-a)^{-n}\} = \frac{e^{ax}x^{n-1}}{(n-1)!};$$

$$\mathscr{L}^{-1}\left\{\frac{cx+d}{(x-a)^2+b^2}\right\} = e^{ax}\left[c\cos bx + \frac{ac+d}{b}\sin bx\right];$$

$$\mathscr{L}^{-1}\left\{\frac{cx+d}{(x-a)^2-b^2}\right\} = e^{ax}\left[c\cosh bx + \frac{ac+d}{b}\sinh bx\right].$$

14. If $y(x)$ satisfies the differential equation

$$y'' + b^2 y = 0, \quad \text{and} \quad y(0) = k, \quad y'(0) = 0,$$

find $\mathscr{L}(y)$ and y.

15. If $y(x)$ satisfies the differential equation

$$y'' + 2ay' + b^2 y = 0, \quad \text{and} \quad y(0) = k, \quad y'(0) = 0,$$

find $\mathscr{L}(y)$ and y.

149. Convergence of Improper Integrals. When the integrand $f(x)$ has a known primitive $F(x)$, a convergent improper integral may be evaluated by a natural extension of the fundamental theorem of the integral calculus.

The *absolute convergence* of an improper integral implies its convergence.

An integral that converges, but not absolutely, is said to *converge conditionally*.

Table of Tests

$$\int_a^\infty f(x)\,dx \qquad$$

conv abs ($p > 1$) $\lim_{x \to \infty} x^p f(x) = \lambda$ (finite);

div ($0 < p \le 1$) $\lim_{x \to \infty} x^p f(x) = \begin{cases} \lambda \ne 0, \\ \pm\infty. \end{cases}$

$$\int_{-\infty}^b f(x)\,dx: \qquad \text{Replace } x^p \text{ by } (-x)^p.$$

$$\int_{a+}^b f(x)\,dx \qquad$$

conv abs ($0 < q < 1$) $\lim_{x \to a+} (x - a)^q f(x) = \lambda$ (finite);

div ($q \ge 1$) $\lim_{x \to a+} (x - a)^q f(x) = \begin{cases} \lambda \ne 0, \\ \pm\infty. \end{cases}$

$$\int_a^{b-} f(x)\,dx: \qquad \text{Replace } x - a \text{ by } b - x, \text{ and let } x \to b-.$$

$$\int_a^\infty \frac{g(x)}{x^r}\,dx \qquad \text{conv if } r > 0 \text{ and } \int_a^x g(x)\,dx \text{ is bounded.}$$

$$\int_a^\infty g(x) \sin x\,dx \qquad \text{conv if } g(x) \text{ is continuous } (x \ge a) \text{ and steadily decreases to } 0;$$

$$\text{conv cond if } \sum_{k=m}^\infty g(k\pi) \text{ diverges.}$$

If the Laplace transform is only of type I,

$$\mathscr{L}\{f(x)\} = \int_0^\infty e^{-xt} f(t)\,dt \qquad \text{conv abs when} \qquad x > x_0,$$

$$\text{if } f(t) = O(e^{x_0 t}), \qquad t \ge b.$$

If $f(x)$ is *positive* and *decreasing* when $x \ge 1$, the series and the integral,

$$\sum_{k=1}^\infty f(k) \quad \text{and} \quad \int_1^\infty f(x)\,dx, \qquad \text{both converge or diverge;}$$

and
$$\sum_{k=1}^n f(k) - \int_1^n f(x)\,dx \to \lambda \quad \text{where} \quad 0 < \lambda < f(1).$$

When $f(x) = 1/x$, λ is *Euler's constant*.

CHAPTER 8

Line Integrals

150. Line Integrals. Let the functions $P(x, y)$, $Q(x, y)$ be defined at every point of a curve C given by the parametric equations

(1) $$x = x(t), \qquad y = y(t), \qquad \alpha \leq t \leq \beta.$$

Form a net

$$\alpha = t_0 < t_1 < t_2 < \cdots < t_n = \beta$$

of mesh δ over (α, β), and let x_i, y_i, P_i, Q_i denote the values of x, y, P, Q, when $t = t_i$. Then the equation

(2) $$\int_C P\,dx + Q\,dy = \lim_{\delta \to 0} \sum_{i=1}^{n} [P_i(x_i - x_{i-1}) + Q_i(y_i - y_{i-1})]\dagger$$

defines the *line integral* over C in the left member whenever the limit exists. We shall show that the line integral reduces to a Riemann integral.

THEOREM. *Let $P(x, y)$ and $Q(x, y)$ reduce to continuous functions $P(t)$ $Q(t)$ on the curve C. Then, if C is a regular curve,*

(3) $$\int_C P\,dx + Q\,dy = \int_\alpha^\beta [P(t)\,x'(t) + Q(t)\,y'(t)]\,dt.$$

Proof. Suppose first that the entire curve C is smooth; then $x'(t)$, $y'(t)$ are continuous in (α, β). From the mean-value theorem (57.1)

$$x_i - x_{i-1} = x(t_i) - x(t_{i-1}) = x'(\tau_i)\,(t_i - t_{i-1}) = x'(\tau_i)\,\delta_i$$

where $t_{i-1} < \tau_i < t_i$. Hence, from (2)

$$\int_C P\,dx = \lim_{\delta \to 0} \sum_{i=1}^{n} P(t_i)x'(\tau_i)\delta_i,$$

---

† This definition may be generalized by replacing $P(t_i)$, $Q(t_i)$ by $P(\bar{t}_i)$, $Q(\bar{t}_i)$, where $\bar{t}_i$ is any point of the interval (t_{i-1}, t_i).

and, by Duhamel's theorem (125.1),

(4) $$\int_C P \, dx = \int_\alpha^\beta P(t) \, x'(t) \, dt.$$

In the same way we find that

(5) $$\int_C Q \, dy = \int_\alpha^\beta Q(t) \, y'(t) \, dt.$$

On adding (4) and (5), we obtain (3).

If the curve is piecewise smooth, (3) holds over each smooth arc and hence, by addition, over the entire curve. The finite discontinuities of $x'(t)$, $y'(t)$ at corners of the curve do not affect the existence of the integral (Theorem 115.2).

The curve C may be any broken line, for example a *step line* consisting of two adjacent sides of a rectangle. Moreover, C may be a closed curve or polygon; in this case

$$\oint_C P \, dx + Q \, dy$$

denotes the circuit integral about C taken *counterclockwise*.

If P, Q, R are functions of three variables x, y, z, we define the line integral,

(6) $$\int_C P \, dx + Q \, dy + R \, dz$$

$$= \lim_{\delta \to 0} \sum_{i=1}^n [P_i(x_i - x_{i-1}) + Q_i(y_i - y_{i-1}) + R_i(z_i - z_{i-1})],$$

by an obvious extension of (2). When P, Q, R are continuous on a regular curve C, the line integral (6) reduces to the Riemann integral,

(7) $$\int_\alpha^\beta [P(t) \, x'(t) + Q(t) \, y'(t) + R(t) \, z'(t)] \, dt.$$

In terms of the vectors

$$\mathbf{f} = [P, Q, R], \qquad \mathbf{r} = [x, y, z],$$

(7) may be written in the form

(8) $$\int_C \mathbf{f} \cdot d\mathbf{r} = \int_\alpha^\beta \mathbf{f}(t) \cdot \mathbf{r}'(t) \, dt,$$

which applies equally well to two or three dimensions.

Example 1. Compute $\int x\,dy - y\,dx$ from (0, 0) to (1, 1) over the

(i)   Straight line $y = x$;
(ii) the step path $y = 0$, $x = 1$ (⌐);
(iii) the step path $x = 0$, $y = 1$ (⌐);
(iv) the parabola $y = x^2$;

(v)   the circle $x = \cos t$, $y = 1 + \sin t$ $\left(-\dfrac{\pi}{2} \le t \le 0\right)$.

(i) $I = \displaystyle\int_0^1 (x - x)\,dx = 0$; note that the vector $[x, -y]$ is always perpendicular to

the path.

(ii) $I = 0 + \displaystyle\int_0^1 dy = 1$;

(iii) $I = 0 - \displaystyle\int_0^1 dx = -1$;

(iv) $I = \displaystyle\int_0^1 (2x^2 - x^2)\,dx = \dfrac{1}{3}$;

(v) $I = \displaystyle\int_{-\pi/2}^0 [\cos^2 t + (1 + \sin t)\sin t]\,dt = \dfrac{\pi}{2} - 1$.

Example 2. Compute $\int x\,dy + y\,dx$ over the paths of Ex. 1.

(i) $\displaystyle\int_0^1 2x\,dx = 1$; (ii) $\displaystyle\int_0^1 dy = 1$; (iii) $\displaystyle\int_0^1 dx = 1$;

(iv) $\displaystyle\int_0^1 3x^2\,dx = 1$; (v) $\displaystyle\int_{-\pi/2}^0 [\cos^2 t - (1 + \sin t)\sin t]\,dt = 1$.

The results are all equal because this line integral is independent of the path; thus, for any regular curve C between (0, 0) and (1, 1),

$$\int_C x\,dy + y\,dx = \int_{(0,0)}^{(1,1)} d(xy) = xy\,\Big|_{(0,0)}^{(1,1)} = 1.$$

Example 3. Compute $\oint (y^2\,dx - x^2\,dy)$ over the circle $x = \cos t$, $y = 1 + \sin t$. The circuit integral taken counterclockwise equals

$$-\int_0^{2\pi} (1 + \sin t)^2 \sin t\,dt - \int_0^{2\pi} \cos^3 t\,dt =$$

$$-\int_0^{2\pi} (\sin t + 2\sin^2 t + \sin^3 t + \cos^3 t)\,dt.$$

The integrals of $\sin t$, $\sin^3 t$, $\cos^3 t$, namely,

$$-\cos t, \quad -\cos t + \tfrac{1}{3}\cos^3 t, \quad \sin t - \tfrac{1}{3}\sin^3 t,$$

have the same value at both limits and contribute nothing to the result. The integral therefore equals

$$-2 \int_0^{2\pi} \sin^2 t \, dt = \int_0^{2\pi} (\cos 2t - 1) \, dt = \left[\frac{\sin 2t}{2} - t \right]_0^{2\pi} = -2\pi.$$

Example 4. Compute $\oint (y^2 \, dx - x^2 \, dy)$ about the triangle whose vertices are (1, 0), (0, 1), (−1, 0).

Beginning the circuit at (1, 0), the sides of the triangle are segments of the lines

$$x + y = 1, \qquad y - x = 1, \qquad y = 0;$$

hence, the circuit integral equals

$$\int_1^0 [(1 - x)^2 + x^2] \, dx + \int_0^{-1} [(1 + x)^2 - x^2] \, dx$$

$$= \left[\frac{x^3}{3} - \frac{(1 - x)^3}{3} \right]_1^0 + \left[\frac{(1 + x)^3}{3} - \frac{x^3}{3} \right]_0^{-1} = -\frac{2}{3}$$

Example 5. The line integral,

$$\int y \, dx + z \, dy + x \, dz,$$

over the twisted cubic,

$$x = t, \qquad y = t^2, \qquad z = t^3,$$

from $t = 0$ to $t = 1$ equals

$$\int_0^1 (t^2 + 2t^4 + 3t^3) \, dt = \frac{1}{3} + \frac{2}{5} + \frac{3}{4} = \frac{89}{60}.$$

151. Line Integrals Independent of Path.

THEOREM 1. *If $P \, dx + Q \, dy$ is the total differential dF of a single-valued†
differentiable function $F(x, y)$ defined in a region R, the line integral has
the same value over all regular curves in R having the same endpoints (x_1, y_1),
(x_2, y_2); and*

(1) $$\int_{(x_1, y_1)}^{(x_2, y_2)} P \, dx + Q \, dy = F(x_2, y_2) - F(x_1, y_1).$$

Proof. Over the curve

$$x = x(t), \qquad y = y(t), \qquad t_1 \leq t \leq t_2,$$

the integral (1) is evaluated as a Riemann integral:

$$\int_{t_1}^{t_2} \left[P(x, y) \frac{dx}{dt} + Q(x, y) \frac{dy}{dt} \right] dt = \int_{t_1}^{t_2} \frac{dF}{dt} \, dt = F(x, y) \Big|_{t_1}^{t_2}$$

in view of the chain rule (79.3); and, since $x(t_1) = x_1$, $y(t_1) = y_1$, etc., we obtain (1).

---

† We use 'single-valued' for emphasis only; the definition we have adopted for 'function' implies one-to-one correspondence.

Corollary. *Under the conditions of the theorem*

$$(2) \qquad\qquad \oint P \, dx + Q \, dy = 0$$

over any regular closed curve in R.

Proof. If the closed curve is $ABCDA$,

$$\oint = \int_{ABC} + \int_{CDA} = \int_{ABC} - \int_{ADC} = 0.$$

THEOREM 2. *If the line integral $\int P \, dx + Q \, dy$ is independent of the path for all regular curves in an open region R, in which P, Q are continuous, then*

$$(3) \qquad\qquad \int_{(x_1,y_1)}^{(x,y)} P \, dx + Q \, dy = F(x, y)$$

is a differentiable function of x and y such that $dF = P \, dx + Q \, dy$, and hence

$$(4) \qquad\qquad P = F_x, \qquad Q = F_y.$$

Proof. With a fixed lower limit (x_1, y_1), the integral (3) is a function $F(x, y)$ of the upper limit (x, y); for the integral is independent of the path. Now, if we compute $F(x + h, y)$ over a curve from (x_1, y_1) to (x, y) and a horizontal line segment from (x, y) to $(x + h, y)$, we have

$$F(x + h, y) - F(x, y) = \int_{(x,y)}^{(x+h,y)} P \, dx + Q \, dy = \int_{x}^{x+h} P(t, y) \, dt.$$

But, by the mean-value theorem (121.2),

$$\int_{x}^{x+h} P(t, y) \, dt = h \, P(\xi, y), \qquad x < \xi < x + h;$$

hence,

$$F_x = \lim_{h \to 0} \frac{F(x + h, y) - F(x, y)}{h} = P(x, y).$$

Similarly, we find

$$F_y = \lim_{k \to 0} \frac{F(x, y + k) - F(x, y)}{k} = Q(x, y).$$

The three-dimensional forms of these theorems are obvious. We shall restate them in a vector form which applies to two or three dimensions.

THEOREM 3. *If F is a differentiable single-valued function in a region R, the integral*

$$(5) \qquad\qquad \int_{\mathbf{r}_1}^{\mathbf{r}_2} \nabla F \cdot d\mathbf{r} = F(\mathbf{r}_2) - F(\mathbf{r}_1)$$

is the same over all regular curves from P_1 to P_2.

Conversely, if the line integral $\int \mathbf{f} \cdot d\mathbf{r}$ is the same over all regular curves in R with the same end points,

(6) $$\mathbf{f} = \nabla F \quad \text{where} \quad F = \int_{\mathbf{r_1}}^{\mathbf{r}} \mathbf{f} \cdot d\mathbf{r}$$

Proof: Since $\nabla F = [F_x, F_y, F_z]$ by definition,

$$\nabla F \cdot d\mathbf{r} = F_x \, dx + F_y \, dy + F_z \, dz = dF,$$

and (5) follows from Theorem 1. Moreover, (6) follows from Theorem 2 which states that the components of $\mathbf{f}$ are the partial derivatives of F:

$$\mathbf{f} = [F_x, F_y, F_z] = \nabla F.$$

PROBLEMS

1. Compute the line integral,

$$\int_{(0,0)}^{(1,1)} (x^2 + y) \, dx + (2x + y) \, dy,$$

over the five paths given in Ex. 150.1.

2. Compute the circuit integral in Ex. 150.3 over
(a) the circle $x = x_0 + r \cos t$, $y = y_0 + r \sin t$;
(b) the ellipse $x = a \cos t$, $y = b \sin t$;
(c) the square included by the lines $x = 0$, $y = 0$, $x = 1$, $y = 1$.

3. Compute $\oint \dfrac{x \, dy - y \, dx}{x^2 + y^2}$ over the circles

(a) $x = 1 + \cos t$, $y = 2 + \sin t$;
(b) $x = 1 + \cos t$, $y = \sin t$;
(c) $x = 1 + 2 \cos t$, $y = 2 \sin t$.

4. Compute $\int x \, dy + y \, dx + z \, dz$ over the curve

$$x = \cos t, \qquad y = \sin t, \qquad z = t(2\pi - t)$$

from $t = 0$ to $t = \pi$, and from $t = 0$ to $t = 2\pi$.

5. Show that the line integral,

$$\int_{(0,0)}^{(x,y)} (x + y^2) \, dx + 2xy \, dy,$$

is independent of the path, and express it as a function of x and y.

6. Compute the line integral in Prob. 5 from $(0, 0)$ to $(1, 2)$
(a) along the line $y = 2x$;
(b) along the parabola $y^2 = 4x$.
Check answers by the result of Prob. 5.

7. Compute $\oint \left(\dfrac{dx}{y} + \dfrac{dy}{x} \right)$ over the triangle included by the lines $y = 1$, $x = 4$, $y = x$.

8. Compute the line integral

$$\int_{(1,0)}^{(2,1)} \left(\frac{1+y^2}{x^3} \, dx - \frac{1+x^2}{x^2} \, y \, dy \right)$$

(a) over the step path $(1, 0) - (2, 0) - (2, 1)$;
(b) over the step path $(1, 0) - (1, 1) - (2, 1)$;
(c) over the line $y = x - 1$.

9. Compute the line integral

$$\int_{(1,2)}^{(3,4)} x^2 y \, dx + y^3 \, dy$$

(a) over the step path $(1, 2) - (3, 2) - (3, 4)$;
(b) over the step path $(1, 2) - (1, 4) - (3, 4)$;
(c) over the line $x - y + 1 = 0$.

10. Compute $I = \int_{r_1}^{r_2} \mathbf{f} \cdot d\mathbf{r}$ over any path (on which $\mathbf{f}$ remains finite) when

     (a) $\mathbf{f} = [a, b, c]$;   (b) $\mathbf{f} = \mathbf{r}$;   (c) $\mathbf{f} = -\mathbf{r}/r^3$.

In each case express $\mathbf{f}$ as ∇F.

[Since $\mathbf{r} \cdot \mathbf{r} = r^2$, $\mathbf{r} \cdot d\mathbf{r} = r \, dr$.]

152. Field of Force. When a vector $\mathbf{f}(\mathbf{r})$ is associated with each point of a region R, we have a *vector field* in the region. In particular, when $\mathbf{f}$ is a force, we speak of a *field of force* in R. If $\mathbf{f}$ is the force exerted on a particle of mass m, the *work* W done by the field on the particle as it moves from P_0 to P over a curve $\mathbf{r} = \mathbf{r}(t)$ is defined as the line integral,

(1) $$W = \int_C \mathbf{f} \cdot d\mathbf{r} = \int_{t_0}^t \mathbf{f} \cdot \frac{d\mathbf{r}}{dt} \, dt.$$

In general, the work depends upon the path C; but, if $\mathbf{f}$ is the gradient of a scalar, the work is independent of the path. Thus, if we write

(2) $$\mathbf{f} = -\nabla \varphi,$$

(the reason for the minus sign will appear later) we have from (151.5)

(3) $$W = \int_{r_0}^r \mathbf{f} \cdot d\mathbf{r} = - \int_{r_0}^r d\varphi = \varphi(\mathbf{r}_0) - \varphi(\mathbf{r}).$$

If a particle of mass m has the position vector $\mathbf{r}$, its *velocity* and *acceleration* are the time derivatives;

$$\mathbf{v} = \frac{d\mathbf{r}}{dt}, \qquad \mathbf{a} = \frac{d\mathbf{v}}{dt} \qquad (103.10-11).$$

From the fundamental law of dynamics

(4)
$$\mathbf{f} = m\mathbf{a} = m\frac{d\mathbf{v}}{dt},$$

and hence

(5)
$$\mathbf{f} \cdot \frac{d\mathbf{r}}{dt} = m\mathbf{v} \cdot \frac{d\mathbf{v}}{dt} = \tfrac{1}{2}\, m\frac{d}{dt}\,(\mathbf{v} \cdot \mathbf{v}) = \frac{d}{dt}(\tfrac{1}{2}\, mv^2).$$

Now the work of the field as the particle moves from P_0 to P over any path is

$$W = \int_{\mathbf{r}_0}^{\mathbf{r}} d(\tfrac{1}{2}\, mv^2) = \tfrac{1}{2}\, mv^2 - \tfrac{1}{2}\, mv_0{}^2.$$

Hence, from (3),

$$\tfrac{1}{2}mv^2 + \varphi(\mathbf{r}) = \tfrac{1}{2}mv_0{}^2 + \varphi(\mathbf{r}_0);$$

that is,

(6)
$$\tfrac{1}{2}mv^2 + \varphi(\mathbf{r}) = c,$$

a constant, at all points of the path.

The quantities $\tfrac{1}{2}mv^2$ and $\varphi(\mathbf{r})$ are defined the *kinetic energy* and *potential energy* of the particle; the former is the energy due to its *motion*, the latter to its *position*. The minus sign in (2) was introduced so that the energies *add* in the left member of (6). This equation is a special case of the

LAW OF CONSERVATION OF ENERGY. *When a field of force has a potential function, the sum of the kinetic and potential energies of a particle moving in the field is constant.*

A field of force having a potential function is said to be *conservative* because the *total* energy, kinetic plus potential, is conserved. If the function $\varphi(\mathbf{r})$ satisfies (2), all functions $\varphi(\mathbf{r}) +$ const do likewise. The additive constant is usually determined, once for all, by prescribing the value of $\varphi(\mathbf{r})$ at some point $\mathbf{r}_0$, say where $\varphi(\mathbf{r}_0) = 0$.

Example 1. *The Localized Field of the Earth.* In the neighborhood of a given locality, the earth attracts a particle of mass m with the force

(7)
$$\mathbf{f} = m\mathbf{g} = -mg\mathbf{e}_3 = -mg\,\nabla z,$$

where $\mathbf{g}$ is the local acceleration of gravity and $\mathbf{e}_3 = \nabla z$ is the unit vertical vector directed upward (106.9). For a particle at a distance z above the earth we have

(8)
$$\mathbf{f} = -\nabla(mgz), \qquad \varphi = mgz.$$

Thus the field of the earth is conservative and has the potential $\varphi = mgz$. A particle moving in the earth's local field has constant total energy

(9)
$$\tfrac{1}{2}mv^2 + mgz = c.$$

Thus, if a particle is projected with an initial speed v_0 at the height $z = h$, it will strike the ground with a speed V given by

(10)
$$\tfrac{1}{2}mV^2 + 0 = \tfrac{1}{2}mv_0{}^2 + mgh, \qquad V^2 = v_0{}^2 + 2gh.$$

Example 2. The Field of the Sun. The sun of mass M attracts a particle of mass m with the force

(11)
$$\mathbf{f} = -\gamma \frac{mM}{r^2} \frac{\mathbf{r}}{r} \, ;$$

r is the radial distance of m from the sun's center, $\mathbf{r}/r$ is a unit radial vector, and γ the constant of gravitation. Since $\mathbf{r}/r = \nabla r$ (106.13),

$$\mathbf{f} = -\frac{\gamma mM}{r^2} \nabla r = \nabla \left(\frac{\gamma mM}{r} \right).$$

Therefore, the field of the sun is conservative and has the potential

(12)
$$\varphi = -\frac{\gamma mM}{r} \, .$$

A particle moving in the sun's field has the constant total energy

(13)
$$\tfrac{1}{2} mv^2 - \frac{\gamma mM}{r} = c.$$

Thus, if a meteor of mass m is at rest when $r = \infty$, $c = 0$; and it will strike the sun's surface $r = R$ with speed V given by

(14)
$$\tfrac{1}{2} mV^2 - \frac{\gamma mM}{R} = 0, \qquad V = \sqrt{\frac{2\gamma M}{R}} \, .$$

If M and R denote the mass and radius of the earth, (14) gives the speed V at which a meteor enters the earth's atmosphere. Near the earth's surface $r = R$ we have from (7) and (11)

$$g = \frac{\gamma M}{R^2} \, ; \qquad \text{hence} \quad V = \sqrt{2gR}.$$

Conversely, in order that a particle may leave the earth it must be projected with at least the speed V (neglecting air resistance). Thus V is often called the "speed of escape."

153. Irrotational Vectors. A vector function $\mathbf{f}(\mathbf{r})$ is said to be *irrotational* in a region if rot $\mathbf{f} = \mathbf{0}$ there. We have seen in § 109 that, if the scalar function $\varphi(\mathbf{r})$ has continuous second partial derivatives in a region, its gradient $\nabla \varphi$ is irrotational:

(1)
$$\text{rot } \nabla \varphi = \mathbf{0} \tag{109.12}.$$

This raises the question: Can every vector $\mathbf{f}$ that is irrotational in a region be expressed as a gradient vector? The answer is *yes* provided the region is *simply connected* in the sense of the following

Definition. A closed curve is called *reducible* in a region if it can be shrunk continuously to a point within the region. A region in which all regular closed curves are reducible is called *simply connected*.

Thus a plane region bounded by a single regular closed curve is simply connected; but the region between two concentric circles is not. In three dimensions, the region inside a sphere or rectangular prism is simply connected. The same is true of the region between two concentric spheres.

But the region within a torus is not simply connected; for all closed curves that encircle its axis are irreducible.

If $\mathbf{f} = [P, Q, R]$, we have from (109.7)

(2)
$$\operatorname{rot} \mathbf{f} = \begin{vmatrix} \mathbf{e}_1 & \mathbf{e}_2 & \mathbf{e}_3 \\ \partial/\partial x & \partial/\partial y & \partial/\partial z \\ P & Q & R \end{vmatrix} ;$$

hence, rot $\mathbf{f} = \mathbf{0}$ implies

(3) $$R_y = Q_z, \qquad P_z = R_x, \qquad Q_x = P_y.$$

When $R = 0$, $\mathbf{f}$ is a plane vector and rot $\mathbf{f} = \mathbf{0}$ implies $Q_x = P_y$.

We consider the plane case first and for a simple rectangular region.

THEOREM 1. *If the plane vector* $\mathbf{f} = [P, Q, 0]$ *has continuous first partial derivatives in a rectangle and*

$$\operatorname{rot} \mathbf{f} = [0, 0, Q_x - P_y] = \mathbf{0},$$

then $\mathbf{f} = \nabla F$, *a gradient vector.*

Proof. Define $F(x, y)$ by the line integral,

$$\int \mathbf{f} \cdot d\mathbf{r} = \int P \, dx + Q \, dy,$$

taken over the step path $(x_0, y_0) - (x, y_0) - (x, y)$:

(4) $$F(x, y) = \int_{x_0}^{x} P(t, y_0) \, dt + \int_{y_0}^{y} Q(x, t) \, dt.$$

Now, from (119.3) and (133.2)

$$F_x = P(x, y_0) + \int_{y_0}^{y} Q_x(x, t) \, dt, \qquad F_y = Q(x, y).$$

By hypothesis $Q_x(x, t) = P_t(x, t)$; hence,

$$F_x = P(x, y_0) + \int_{y_0}^{y} P_t(x, t) \, dt = P(x, y).$$

We thus have

$$\mathbf{f} = [P, Q] = [F_x, F_y] = \nabla F.$$

Theorem 151.3 now shows that $\int_{\mathbf{r}_0}^{\mathbf{r}} \mathbf{f} \cdot d\mathbf{r}$ is independent of the path.

Therefore $F(x, y)$ is given by this line integral taken over any path in the rectangle from (x_0, y_0) to (x, y). For example, the step path $(x_0, y_0) - (x_0, y) - (x, y)$ yields

(5) $$F(x, y) = \int_{x_0}^{x} P(x, y) \, dx + \int_{y_0}^{y} Q(x_0, y) \, dy,$$

where the integrals are written in reverse order.

In three dimensions we have the corresponding

THEOREM 2. *If the vector* $\mathbf{f} = [P, Q, R]$ *has continuous first partial derivatives in a rectangular prism and*

$$\operatorname{rot} \mathbf{f} = [R_y - Q_z, P_z - R_x, Q_x - P_y] = \mathbf{0},$$

then $\mathbf{f} = \nabla F$, *a gradient vector.*

Proof. Define $F(x, y, z)$ by the line integral,

$$\int \mathbf{f} \cdot d\mathbf{r} = \int P\, dx + Q\, dy + R\, dz,$$

taken over the step path $(x_0, y_0, z_0) - (x, y_0, z_0) - (x, y, z_0) - (x, y, z)$. Then

$$(6) \quad F(x, y, z) = \int_{x_0}^{x} P(t, y_0, z_0)\, dt + \int_{y_0}^{y} Q(x, t, z_0)\, dt + \int_{z_0}^{z} R(x, y, t)\, dt,$$

and we find $F_x = P$, $F_y = Q$, $F_z = R$ as in Theorem 1.

With the step path $(x_0, y_0, z_0) - (x_0, y_0, z) - (x_0, y, z) - (x, y, z)$ we obtain

$$(7) \quad F(x, y, z) = \int_{x_0}^{x} P(x, y, z)\, dx + \int_{y_0}^{y} Q(x_0, y, z)\, dy + \int_{z_0}^{z} R(x_0, y_0, z)\, dz,$$

where the integrals are written in reverse order. Formulas (5) and (7) are readily remembered; they give a function F which vanishes at $\mathbf{r}_0$. Changing the initial point $\mathbf{r}_0$ merely alters F by a constant. In practice, $\mathbf{r}_0$ is chosen to simplify the integrals as much as possible. If the region in which $\operatorname{rot} \mathbf{f} = \mathbf{0}$ includes the origin, this point is usually a good choice.

In order to extend these theorems from rectangular to simply connected regions in general, requires a rather detailed argument which we shall only sketch in outline. Consider a variable polygon $\Gamma(t)$ in the region whose vertices are given by position vectors $\mathbf{r}_1(t), \mathbf{r}_2(t), \cdots, \mathbf{r}_k(t)$ which are continuous functions of the time t such that $\Gamma(t)$ shrinks to a point as t varies 0 to 1. If we can show that the circuit integral,

$$(8) \quad I(t) = \oint_{\Gamma(t)} \mathbf{f} \cdot d\mathbf{r} = \text{const},$$

while $\Gamma(t)$ shrinks to a point, the constant must be zero. Since the original polygon given by $\mathbf{r}_1(0), \mathbf{r}_2(0), \cdots, \mathbf{r}_k(0)$ is arbitrary, this means that $\oint \mathbf{f} \cdot d\mathbf{r} = 0$ over any closed polygon in the region. Hence $\int_{\mathbf{r}_0}^{\mathbf{r}} \mathbf{f} \cdot d\mathbf{r}$ is the same for all polygonal paths between $\mathbf{r}_0$ and $\mathbf{r}$ and defines a function $F(\mathbf{r})$. We can now show as in Theorem 151.3 that $\mathbf{f} = \nabla F$ and that the line integral has the same value over all regular curves between the same end points.

Thus the proof hinges on (8). To establish this result consider two polygons $\Gamma(t)$ and $\Gamma(t')$. Choose n points on $\Gamma(t)$, $A_1, A_2, \cdots, A_n$ so that each pair $A_i A_{i+1}$ lies in a rectangle R_i (or rectangular prism) inside our region. Let $A_i \to A_i'$ on $\Gamma(t')$. Now choose $t' - t$ so small that $A_i' A'_{i+1}$ also lies in R_i $(i = 1, 2, \cdots, n)$. When rot $\mathbf{f} = \mathbf{0}$, the circuit integral over the quadrangle $A_i A_{i+1} A'_{i+1} A_i' A_i$ is zero by Theorem 1 or 2; and, if we add all such circuit integrals, we have $I(t) - I(t') = 0$.

Example 1. When

$$\mathbf{f} = [2x^2 + 6xy,\ 3x^2 - y^2,\ 0], \qquad \text{rot } \mathbf{f} = \mathbf{0}.$$

Integrating over the step path $(0, 0) - (x, 0) - (x, y)$, we get

$$F(x, y) = \int_0^x 2x^2\, dx + \int_0^y (3x^2 - y^2)\, dy = \frac{2}{3} x^3 + 3x^2 y - \frac{1}{3} y^3,$$

and, over $(0, 0) - (0, y) - (x, y)$,

$$F(x, y) = \int_0^x (2x^2 + 6xy)\, dx - \int_0^y y^2\, dy = \frac{2}{3} x^3 + 3x^2 y - \frac{1}{3} y^3,$$

where the integrals are written in reverse order.

We can now compute the line integral between any two points:

$$\int_{\mathbf{r}_1}^{\mathbf{r}_2} \mathbf{f} \cdot d\mathbf{r} = F(x_2, y_2) - F(x_1, y_1).$$

Example 2. When

$$\mathbf{f} = [2xz,\ 2yz^2,\ x^2 + 2y^2 z - 1], \qquad \text{rot } \mathbf{f} = \mathbf{0}.$$

Integrating over the step path $(0, 0, 0) - (x, 0, 0) - (x, y, 0) - (x, y, z)$, we get

$$F(x, y, z) = 0 + 0 + \int_0^z (x^2 + 2y^2 z - 1)\, dz = x^2 z + y^2 z^2 - z;$$

and, over the step path $(0, 0, 0) - (0, 0, z) - (0, y, z) - (x, y, z)$,

$$F(x, y, z) = \int_0^x 2xz\, dx + \int_0^y 2yz^2\, dy - \int_0^z dz = x^2 z + y^2 z^2 - z,$$

where the integrals are written in reverse order.

Example 3. When $\mathbf{f} = cr^{n-1} \mathbf{r}$, we can show that rot $\mathbf{f} = \mathbf{0}$ without resorting to rectangular coordinates. Since $\mathbf{r}/r = \nabla r$ (106.13),

$$\mathbf{f} = cr^n\, \nabla r = \begin{cases} \nabla \left(\dfrac{cr^{n+1}}{n + 1} \right), & (n \neq -1), \\ \nabla (c \log r), & (n = -1), \end{cases}$$

we see that $\mathbf{f}$ is a gradient vector.

Example 4. The function $u(x, y)$ and $v(x, y)$ are said to be *conjugate* when they satisfy the *Cauchy-Riemann equations*,

$$(9) \qquad\qquad u_x = v_y, \qquad u_y = -v_x \qquad\qquad (195.6),$$

and have continuous second partial derivatives. Then u and v satisfy Laplace's equation (109.11); for example,

$$\nabla^2 u = u_{xx} + u_{yy} = v_{yx} - v_{xy} = 0 \qquad\qquad \text{(Theorem 82.1)}.$$

When $u(x, y)$ is a solution of $\nabla^2 u = 0$, its conjugate $v(x, y)$ can be found by line integration; thus the integral

$$\int_{(x_0,y_0)}^{(x,y)} v_x \, dx + v_y \, dy = \int_{(x_0,y_0)}^{(x,y)} u_x \, dy - u_y \, dx$$

is independent of the path and gives a function conjugate to u.

For example, if $u = x^3 - 3xy^2$,

$$v(x, y) = \int_{(0,0)}^{(x,y)} 6xy \, dx + (3x^2 - 3y^2) \, dy.$$

If we integrate over the step path $(0, 0) - (x, 0) - (x, y)$, we find

$$v = 0 + \int_0^y 3(x^2 - 3y^2) \, dy = 3x^2y - y^3;$$

whereas, over the step path $(0, 0) - (0, y) - (x, y)$,

$$v = -3 \int_0^y y^2 \, dy + \int_0^x 6xy \, dx = -y^3 + 3x^2y.$$

PROBLEMS

1. Show that the line integral,

$$\int_{(1,0)}^{(x,y)} \left(\frac{1 + y^2}{x^3} \, dx - \frac{1 + x^2}{x^2} \, y \, dy \right),$$

is independent of the path, find its value $F(x, y)$ and use it to solve Prob. 151.8.

2. Show that the line integral,

$$F(x, y) = \int_{(0,1)}^{(x,y)} \left(\frac{xy + 1}{y} \, dx + \frac{2y - x}{y^2} \, dy \right),$$

is independent of the path, and find $F(x, y)$. Compute the integral over a curve from $(1, 2)$, $(2, 4)$.

3. Show that the force

$$\mathbf{f} = [yz^2, xz^2 - 1, 2xyz - 2]$$

is conservative, and find its potential function $\varphi(x, y, z)$ for which $\varphi(0, 0, 0) = 0$.

4. Find the work done by the force $\mathbf{f}$ in Prob. 3 over a path from $(1, 2, 3)$ to $(3, 5, -1)$

5. Show that the vector

$$\mathbf{f} = [2x + 2y, 2x - z^2, -2yz]$$

is irrotational, and find its potential.

6. At a distance r from the center of the sun a body is launched into its field with a speed $v_0 < \sqrt{2\gamma M/r}$. Show that it cannot leave the solar system.

7. If the speed of a planet at perihelion $(r = r_1)$ is v_1, find its speed v_2 at aphelion $(r = r_2)$.

8. In elliptic harmonic motion a particle of mass m is attracted to the origin with a force $\mathbf{f} = -m k \mathbf{r}$ $(k > 0)$. Find the potential, and show thal $v^2 + kr^2 = \text{const.}$

9. A body is projected vertically upward from the earth with a speed v_0. What height would it reach if air resistance were negligible? [Cf. Ex. 152.1]

10. Find the conjugates of the following functions:

$$(a)\ \log \sqrt{x^2 + y^2}; \quad (b)\ e^x \cos y; \quad (c)\ \frac{x}{x^2 + y^2}; \quad (d)\ \cos x \sinh y.$$

11. If $u(x, y)$ and $v(x, y)$ are conjugate functions, show that the integrals

$$U(x, y) = \int_{(x_0,y_0)}^{(x,y)} u\,dx - v\,dy, \qquad V(x, y) = \int_{(x_0,y_0)}^{(x,y)} v\,dx + u\,dy$$

define functions of x, y which are also conjugate.

154. Area of a Sector. Let a curve have the polar equation,

$$(1) \qquad\qquad r = f(\varphi),$$

where $f(\varphi)$ is continuous in the interval $\alpha \leq \varphi \leq \beta$. We shall express the area of the sector S, bounded by the curve and two radial lines $\varphi = \alpha$, $\varphi = \beta$, as an integral. We first note that the area of circular sector of central angle φ is

$$\pi r^2 \cdot \frac{\varphi}{2\pi} = \tfrac{1}{2} r^2 \varphi.$$

Form the net

$$\alpha = \varphi_0 < \varphi_1 < \varphi_2 < \cdots < \varphi_n = \beta$$

over (α, β), and draw the radial lines $\varphi = \varphi_i$. These divide S into n elementary sectors of central angle $\delta_i = \varphi_i - \varphi_{i-1}$; and, if m_i and M_i are the least and greatest values of $f(\varphi)$ in the interval $\varphi_{i-1} \leq \varphi \leq \varphi_i$, the ith sector of the curve is included between the circular sectors $\tfrac{1}{2} m_i^2 \delta_i$ and $\tfrac{1}{2} M_i^2 \delta_i$. As the mesh δ of the net approaches zero, these sums have the common limit,

$$(2) \qquad\qquad S = \tfrac{1}{2} \int_\alpha^\beta r^2\,d\varphi,$$

which is defined as the area of the sector.

If the curve has the parametric equations,

$$(3) \qquad\qquad x = x(t), \quad y = y(t), \qquad t_1 \leq t \leq t_2,$$

and we make the change of variable,

$$\tan \varphi = \frac{y(t)}{x(t)}, \qquad \frac{r^2\,d\varphi}{x^2\,dt} = \frac{xy' - yx'}{x^2},$$

(for $\sec^2 \varphi = r^2/x^2$), the integral (2) becomes

$$(4) \qquad S = \tfrac{1}{2} \int_{t_1}^{t_2} r^2 \frac{d\varphi}{dt}\,dt = \tfrac{1}{2} \int_{t_1}^{t_2} (xy' - yx')\,dt. \qquad (123.1).$$

Thus S is expressed as a line integral,

$$(5) \qquad\qquad S = \tfrac{1}{2} \int x\,dy - y\,dx,$$

along the curve. If the curve is closed the area within the curve is

(6) $$S = \tfrac{1}{2} \oint x \, dy - y \, dx.$$

The area defined by (4) is in harmony with our previous definition in § 126. For, if we integrate $\int xy' \, dt$ by parts, we have from (4)

(7) $$S = \tfrac{1}{2} xy \Big|_{t_1}^{t_2} - \int_{t_1}^{t_2} yx' \, dt = \int_{t_2}^{t_1} yx' \, dt + \tfrac{1}{2} x_2 y_2 - \tfrac{1}{2} x_1 y_1.$$

From (126.2), $\displaystyle\int_{t_2}^{t_1} yx' \, dt$ is the area $P_2 Q_2 Q_1 P_1$ between the curve and the x-axis, and $\tfrac{1}{2} x_2 y_2$, $\tfrac{1}{2} x_1 y_1$ are the areas of the triangles $OP_2 Q_2$, $OP_1 Q_1$. A glance at Fig. 154a now shows that (7) is correct.

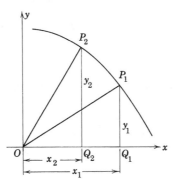

FIG. 154a.   Area of a sector

Since

$$\mathbf{r} \times \frac{d\mathbf{r}}{dt} = [x, y, 0] \times [x', y', 0] = (xy' - yx')\mathbf{e}_3,$$

the vector form of (4) is

(8) $$S\mathbf{e}_3 = \tfrac{1}{2} \int_{t_1}^{t_2} \mathbf{r} \times \frac{d\mathbf{r}}{dt} \, dt.$$

The vector $S\mathbf{e}_3$, normal to the plane of the curve, is called the *vector area* of the sector S.

Example 1. The sector of the *ellipse* $x = a \cos t$, $y = b \sin t$ from $t = 0$ to $t = \phi$ is

$$S = \frac{1}{2} \int_0^\varphi ab \, (\cos^2 t + \sin^2 t) \, dt = \frac{1}{2} ab\varphi \qquad (126.3)$$

Example 2. The sector of the *hyperbola* $x = a \cosh t$, $y = b \sinh t$ from $t = 0$ to $t = \varphi$ is

$$S = \frac{1}{2} \int_0^\varphi ab (\cosh^2 t - \sinh^2 t) \, dt = \frac{1}{2} ab\varphi \qquad (126.4).$$

Example 3. The *folium of Descartes* (Fig. 83*b*),

$$x^3 + y^3 - 3axy = 0,$$

admits the parametric equations,

$$x = \frac{3at}{1 + t^3}, \qquad y = \frac{3at^2}{1 + t^3},$$

found by putting $y = tx$. With this parameter

$$x \, dy - y \, dx = x(t \, dx + x \, dt) - tx \, dx = x^2 \, dt.$$

The folium has a double point at the origin which corresponds to both $t = 0$ and $t = \infty$. The loop of the folium therefore has the area

$$A = \frac{1}{2} \int_0^\infty x^2 \, dt = \frac{3}{2} a^2 \int_0^\infty \frac{3t^2 \, dt}{(1 + t^3)^2} = - \frac{\frac{3}{2} a^2}{1 + t^3} \Big|_0^\infty = \frac{3}{2} a^2.$$

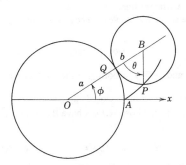

FIG. 154*b*.   Epicycloid

Example 4. An *epicycloid* is the locus of a point on a circle of radius b as it rolls without slipping over a circle of radius a (Fig. 154*b*). While the point of contact Q advances through an angle φ on the fixed circle, the moving circle turns through an angle θ given by $a\varphi = b\theta$. From the figure we have

$$\mathbf{r} = \overrightarrow{OP} = \overrightarrow{OB} + \overrightarrow{BP} = (a + b)\mathbf{R}(\varphi) + b\mathbf{R}(\varphi + \pi + \theta)$$

where $\mathbf{R}(\varphi)$ and $\mathbf{R}(\varphi + \pi + \theta) = -\mathbf{R}(\varphi + \theta)$ are unit radial vectors along $\overrightarrow{OB}$ and $\overrightarrow{BP}$. If we put $b = ka$, $\varphi = k\theta$, the vector equation of the curve becomes

$$\mathbf{r} = a(1 + k)\mathbf{R}(k\theta) - ak\mathbf{R}(1 + k)\theta.$$

The parametric equations of the epicycloid are therefore

$$x = a(1 + k) \cos k\theta - ak \cos (1 + k)\theta,$$

$$y = a(1 + k) \sin k\theta - ak \sin (1 + k)\theta.$$

A simple calculation now shows that

$$x \, dy - y \, dx = a^2 k(1 + k)(1 + 2k)(1 - \cos \theta) \, d\theta.$$

Hence the sector subtended by one loop of the epicycloid has the area

$$S = \frac{a^2}{2} k(1 + k)(1 + 2k) \int_0^{2\pi} (1 - \cos \theta) \, d\theta = \pi a^2 k(1 + k)(1 + 2k),$$

(9) $$S = \frac{\pi b}{a}(a + b)(a + 2b);$$

and, if we subtract the area of the included circular sector

$$\frac{1}{2} a^2 \frac{2\pi b}{a} = \pi ab,$$

we get the area of an epicycloidal arch exterior to the fixed circle:

(10) $$A = 3\pi b^2 + \frac{2\pi b^3}{a}.$$

If we let $a \to \infty$, this gives the area $3\pi b^2$ of one arch of a *cycloid* generated by a circle of radius b rolling over a straight line.

When $b = a$ the curve is called a *cardioid*. The area within the cardioid is given by (9), namely $6\pi a^2$.

PROBLEMS

1. Prove that $\oint x \, dy$ and $- \oint y \, dx$ both give the area within a closed regular curve.

2. From (8), the vector area within a closed curve is $A\mathbf{e}_3 = \frac{1}{2} \oint \mathbf{r} \times d\mathbf{r}$. Hence show that a polygon in the xy-plane, whose vertices have the position vectors $\mathbf{r}_1, \mathbf{r}_2, \cdots, \mathbf{r}_n$, has the vector area

$$A\mathbf{e}_3 = \tfrac{1}{2}(\mathbf{r}_1 \times \mathbf{r}_2 + \mathbf{r}_2 \times \mathbf{r}_3 + \ldots + \mathbf{r}_n \times \mathbf{r}_1).$$

3. Find the area of the loop in the following curves which have a double point at the origin:

    (a) $a(x^2 - y^2) = x^3$;   (b) $b^2 y^2 = x^2(a^2 - x^2)$;   (c) $y(x^2 + y^2) = a(x^2 - y^2)$.

In (c) compute the area in two ways: from the polar equation and from the parametric equations ($t = y/x$).

4. A parabola has its focus at the origin and directrix $x = 2p$. Show that its polar equation is $r = p \sec^2 \dfrac{\varphi}{2}$; and that the area of a parabolic sector included between the axis and the ray $\varphi = \alpha$ is

$$S = p^2 \tan \frac{\alpha}{2} \left(1 + \frac{1}{3} \tan^2 \frac{\alpha}{2}\right).$$

5. If the force acting on a particle is $\mathbf{f} = \mathbf{e}_3 \times \mathbf{r}$, show that $\mathbf{f}$ does the work $2A$ on a particle as it makes a counterclockwise circuit of a simple closed curve in the xy-plane which encloses an area A.

CHAPTER 9

Multiple Integrals

155. Double Integral over a Rectangle. Let $f(x, y)$ be a bounded function in the rectangle

$$R: \quad a \leq x \leq b, \qquad c \leq y \leq d.$$

Then $f(x, y)$ will have a greatest lower bound m and a least upper bound M in R. These bounds are the closest numbers for which

$$m \leq f(x, y) \leq M \quad \text{in } R;$$

and $M - m$ is called the *oscillation* of $f(x, y)$ in R. For any net

$$a = x_0 < x_1 < \cdots < x_m = b,$$
$$c = y_0 < y_1 < \cdots < y_n = d,$$

the lines $x = x_i$, $y = y_j$ divide R into mn rectangles of area

$$r_k = (x_i - x_{i-1})(y_j - y_{j-1}), \qquad k = 1, 2, \cdots, mn.†$$

Let (ξ_k, η_k) be any point in r_k, and consider the sum

$$(1) \qquad \sum_{k=1}^{mn} f(\xi_k, \eta_k) r_k,$$

which depends on the net and the choice of the points (ξ_k, η_k). The *mesh* δ of the net is defined as the greatest dimension of any one of the rectangles r_k. If the sum (1) approaches a limit as $\delta \to 0$ (and consequently as m, $n \to \infty$) which is independent of the way in which the net and the points (ξ_k, η_k) are chosen, we say that $f(x)$ is *integrable* over R, and we write

$$(2) \qquad \iint_R f(x, y)\, dx\, dy = \lim_{\delta \to 0} \sum f(\xi_k, \eta_k) r_k.$$

When this limit exists, it is called the *double integral* of $f(x, y)$ over the

---

† We may take $k = (j-1)m + i$ where $i = 1, 2, \ldots, m, j = 1, 2, \ldots, n$.

rectangle R. Then to any $\varepsilon > 0$ there corresponds a positive number Δ such that

(3) $$\left| \iint_R f(x, y)\, dx\, dy - \sum f(\xi_k, \eta_k) r_k \right| < \varepsilon \quad \text{when} \quad \delta < \Delta.$$

This inequality must hold for all nets over R whose mesh $\delta < \Delta$, irrespective of how the points (ξ_k, η_k) are chosen.

156. Condition for Integrability. The discussion for double integrals parallels that for simple integrals in § 114. We define the *lower sum s* and the *upper sum S* by

(1) $$s = \sum m_k\, r_k, \qquad S = \sum M_k\, r_k,$$

where m_k and M_k are the g.l.b. and l.u.b. of $f(x, y)$ in r_k. These sums depend only on the net x_i, y_j and not on the points ξ_k, η_k; and, since $m_k \leq f(\xi_k, \eta_k) \leq M_k$,

(2) $$s \leq \sum f(\xi_k, \eta_k) r_k \leq S.$$

Hence, if both s and S approach the same limit L as $\delta \to 0$, then $\sum f(\xi_k, \eta_k) r_k$ also approaches L.

When $f(x, y) = C$, a constant,

$$s = S = \sum C r_k = CA,$$

where A is the area of R; hence,

(3) $$\iint_R C\, dx\, dy = CA.$$

In the following we exclude this simple case; then always $s < S$.

We now prove just as in § 114 the two lemmas.

Lemma 1. If a net over R having the sums s', S' is refined by the addition of new points of division, the new sums s, S satisfy

(4) $$s' \leq s < S \leq S'.$$

Lemma 2. If we consider all possible nets over a rectangle, a lower sum is less than any upper sum.

Since s and S lie between fixed bounds,

(5) $$mA \leq s < S \leq MA,$$

the set of lower sums s has a least upper bound I and the set of upper sums S has a greatest lower bound J; and from Lemma 2

(6) $$s \leq I \leq J \leq S.$$

We now have three theorems entirely analogous to Theorems 114.1–2–3.

THEOREM 1. *If we consider s and S for all possible nets over a rectangle R,*

$$(7) \qquad \lim_{\delta \to 0} s = I, \qquad \lim_{\delta \to 0} S = J.$$

Proof. The argument parallels *exactly* that given for Theorem 114.1. We first consider a function $f(x)$ in R for which $M > m > 0$. Let N be any net of mesh δ giving a lower sum s. Choose $\varepsilon > 0$ arbitrarily small; then, since I is the l.u.b. of s, we can find a net N_1 whose lower sum $s_1 > I - \frac{1}{2}\varepsilon$. Let N_1 have k points inside the rectangle R; henceforth we hold N_1 fast so that k is a constant. We must now show that $s > I - \varepsilon$ for *all* nets N whose mesh δ is less than some fixed number.

Denote by N_2 the net formed by the points of both N and N_1; then, by Lemma 1, its lower sum,

$$(8) \qquad s_2 \geqq s_1 > I - \tfrac{1}{2}\varepsilon.$$

Now N has two types of subrectangles
- (i) those with *no* points of N_1 inside,
- (ii) those having points of N_1 inside.

The rectangles of type (i) contribute to s_2 an amount that certainly is less than s. The rectangles of type (ii) number at most k and contain at most $4k$ rectangles of N_2; and, since the mesh of N is δ, their total contribution to s_2 cannot exceed $4kM\delta^2$; hence,

$$s + 4kM\delta^2 > s_2,$$

or, in view of (8),

$$s > I - \tfrac{1}{2}\varepsilon - 4kM\delta^2.$$

If we now choose $\delta^2 < \varepsilon/8kM$, we have

$$I \geqq s > I - \varepsilon \quad \text{when} \quad \delta^2 < \frac{\varepsilon}{8kM}.$$

Consequently, $s \to I$ as $\delta \to 0$.

The rest of the proof is exactly like that of Theorem 114.1.

I is called the *lower integral*, J the *upper integral* of $f(x, y)$ over R.

THEOREM 2. *A bounded function is integrable over a finite rectangle when and only when its lower and upper integrals are equal; and*

$$(9) \qquad \iint_R f(x, y)\, dx\, dy = I = J.$$

Finally we have the criterion for integrability in

THEOREM 3. *In order that a bounded function be integrable, it is necessary and sufficient that, for any $\varepsilon > 0$, there is a net for which*

$$(10) \qquad S - s < \varepsilon.$$

With this criterion we can easily prove

THEOREM 4. *A continuous function $f(x, y)$ is integrable in a rectangle.*

Proof. Since $f(x, y)$ is uniformly continuous in the closed rectangle (Theorem 75.2), we can find Δ so that

$$\left|f(x, y) - f(x', y')\right| < \varepsilon/A \quad \text{when} \quad \left|x - x'\right|, \left|y - y'\right| < \Delta.$$

Hence, for any net over R of mesh $\delta < \Delta$,

$$M_k - m_k < \varepsilon/A \quad \text{in each } r_k;$$

for $f(x, y)$ actually assumes the values m_k and M_k in each r_k (Theorem 75.4). Consequently,

(11) $$S - s = \sum(M_k - m_k)r_k < \varepsilon$$

for all nets whose mesh $\delta < \Delta$.

157. Continuity of an Integral. Let $f(x, y)$ be bounded in the rectangle

$$R: \quad a \leqq x \leqq b, \qquad c \leqq y \leqq d.$$

We have seen in § 132 that, when $f(x, y)$ is continuous in R,

(1) $$F(y) = \int_a^b f(x, y)\, dx$$

is continuous in (c, d). We now proceed to generalize this theorem.

Lemma (Vallée Poussin). *If $f(x, y)$ is continuous along any line AB, to every number $\varepsilon > 0$ there corresponds a δ such that the oscillation of $f(x, y) < \varepsilon$ in any circle of radius $< \delta$ and having its center on the line.*

Proof. Suppose that the theorem is false for some given ε. Divide AB in two halves; the theorem must be false in at least one of them, and, in case of a choice, let A_1B_1 be the half nearest A. Divide A_1B_1 in two halves; again the theorem must be false in one of them; in case of a choice, let A_2B_2 be the half nearest A_1. We can continue this process of bisection indefinitely and obtain a nest of intervals AB, A_1B_1, A_2B_2, $\cdots$ whose lengths form a null sequence and define a unique point (α, β) of the line (Theorem 20). Now (α, β) is a point of every one of the nested intervals, and *every* circle about it contains an infinite number of segments A_iB_i in which the theorem is false. Thus in every circle about (α, β), no matter how small, the oscillation of $f(x, y) \geqq \varepsilon$. Consequently $f(x, y)$ is discontinuous at (α, β), contradicting the hypothesis.

THEOREM 1. *If $f(x, y)$ is bounded in the rectangle R and has but a finite number of discontinuities for each value of y, the integral $F(y)$ is continuous in (c, d).*

Proof. The integral $F(y)$ exists under the stated conditions (Theorem 115.2). Suppose that, on the line $y = y_1, f(x, y_1)$ has just *one* discontinuity at x_1. We shall prove that $F(y)$ is continuous at y_1.

From (1) we have

$$\left| F(y) - F(y_1) \right| \leq \int_a^{x_1-\varepsilon} + \int_{x_1-\varepsilon}^{x_1+\varepsilon} + \int_{x_1+\varepsilon}^b \left| f(x, y) - f(x, y_1) \right| dx.$$

If $\left| f(x, y) \right| < M$ in R, equation (117.4) shows that the second integral $< 2M \cdot 2\varepsilon = 4M\varepsilon$. As to the first and third, since $f(x, y)$ is continuous on the segments

$$y = y_1, \quad a \leq x \leq x_1 - \varepsilon; \qquad y = y_1, \quad x_1 + \varepsilon \leq x \leq b;$$

the oscillation of $f(x, y) < \varepsilon$ in any circle about (x, y_1) whose radius $< \delta$. Hence, in these integrals

$$\left| f(x, y) - f(x, y_1) \right| < \varepsilon \quad \text{when} \quad \left| y - y_1 \right| < \delta,$$

and, consequently, their sum $< (b - a)\varepsilon$. Thus,

$$\left| F(y) - F(y_1) \right| < (4M + b - a)\varepsilon, \quad \text{when} \quad \left| y - y_1 \right| < \delta,$$

and $F(y)$ is continuous at y_1. Obviously this is still true when $f(x, y_1)$ has a limited number of discontinuities in the interval (a, b).

THEOREM 2. *If $a(y)$, $b(y)$ are continuous functions of y in the interval $c \leq y \leq d$,*

$$(2) \qquad\qquad F(y) = \int_{a(y)}^{b(y)} f(x, y) \, dx$$

is continuous in (c, d) provided $f(x, y)$ is bounded in the region R_1 between the curves

$$x = a(y), \qquad x = b(y), \qquad c \leq y \leq d$$

and has but a finite number of discontinuities for each value of y.

Proof. Since the functions $a(y)$, $b(y)$ are continuous in (c, d), we can find two constants A, B such that

$$A < a(y), b(y) < B, \qquad c \leq y \leq d.$$

Now in the rectangle $A \leq x \leq B$, $c \leq y \leq d$ define the function

$$(3) \qquad\qquad f_1(x, y) = \begin{cases} f(x, y) & \text{in } R_1, \\ 0 & \text{outside of } R_1. \end{cases}$$

On any line $y = y_1$, $f_1(x, y)$ has the same discontinuities as $f(x, y)$ with possibly two others at $x = a(y_1)$ and $b(y_1)$ where the line cuts the curve boundaries of R_1. Hence by Theorem 1,

$$\int_A^B f_1(x, y) \, dx = \int_{a(y)}^{b(y)} f(x, y) \, dx = F(y)$$

is continuous in (c, d).

158. Double Integral within a Curve. Let the function $f(x, y)$ be continuous in the closed region R_1 bounded by a simple closed rectifiable curve C of length L. Enclose C in a rectangle R of area A; and define

(1) $$f_1(x, y) = \begin{cases} f(x, y) & \text{in } R_1, \\ 0 & \text{outside of } R_1. \end{cases}$$

Then the discontinuities of $f_1(x, y)$ are confined to the curve C. We shall show that $f_1(x, y)$ is integrable in R.

As in § 155 we form a net of mesh δ over R which subdivides it into mn rectangles r_k. These subrectangles are of two types:

(i) $r_{k'}$, which contain points of C;

(ii) $r_{k''}$, which do not contain points of C.

Since $f(x, y)$ is continuous in the latter, their contribution to $S - s$ can be made $< \varepsilon/2$ by choosing δ so small, say $\delta < \Delta$, that $M_{k''} - m_{k''} < \varepsilon/2A$ (see proof of Theorem 156.4).

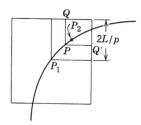

FIG. 158.   Arc $P_1P_2 = L/p$.

Consider now the rectangles $r_{k'}$ of type (i). Divide C into p parts of equal length L/p, and at each point of division draw a square whose sides are parallel to the axes and of length $4L/p$. If P_1, P_2 are consecutive points of division, every point P of the arc P_1P_2 is at least a distance L/p away from the perimeter of the square about P_1 (Fig. 158); for

$$\overline{P_1P} + \overline{PQ} \geq 2L/p, \qquad \overline{P_1P} \leq L/p; \qquad \therefore \overline{PQ} \geq L/p.$$

Similarly $\overline{PQ'} \geq L/p$. Thus, when $\delta < L/p$, all rectangles $r_{k'}$ that contain points of the arc P_1P_2 will lie in the square about P_1. Consequently, *all* the rectangles $r_{k'}$ will be in the set of p squares whose total area is $p(4L/p)^2 = 16L^2/p$; and their contribution to $S - s$ is at most $(M - m)16L^2/p$. If p is chosen sufficiently large, this also can be made $< \varepsilon/2$.

Thus, when the mesh $\delta < \Delta$ *and* L/p, $S - s < \varepsilon$ and $f_1(x, y)$ is integrable over the rectangle R. Hence, from the definition of $f_1(x, y)$ in (1), we have the

THEOREM. *If* $f(x, y)$ *is continuous within and on any simple closed rectifiable curve, it is integrable in this region.*

159. Double and Repeated Integrals. Let $f(x, y)$ be continuous within and on a simple closed rectifiable curve C which is met by any line parallel to an axis in at most two points (Fig. 159a). Enclose C in the smallest possible rectangle,

$$R: \quad a \leq x \leq b, \quad c \leq y \leq d,$$

with sides parallel to the axes. Define $f_1(x, y)$ as in (158.1); then from Theorem 157.1

$$F(y) = \int_a^b f_1(x, y) \, dx$$

is continuous in (c, d). If we integrate $F(y)$ between c and d, we get the

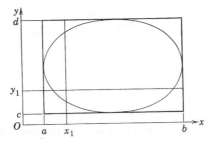

FIG. 159a.   Region bounded by a simple closed curve

repeated integral,

$$(1) \qquad\qquad P = \int_c^d dy \int_a^b f_1(x, y) \, dx.$$

Two properties of P are given in the following lemmas.

Lemma 1. If $m \leq f(x, y) \leq M$, and A is the area of R,

$$mA \leq P \leq MA.$$

Proof. From equation (117.4), applied first to $F(y)$ and then to P, we have

$$m(b - a) \leq F(y) \leq M(b - a);$$

$$m(b - a)(d - c) \leq \ P \ \leq M(b - a)(d - c).$$

Lemma 2. If the rectangle R is divided into subrectangles by vertical or horizontal lines, P equals the sum of the repeated integrals over these rectangles:

Proof. For a vertical line $x = x_1$,

$$P = \int_c^d dy \left[\int_a^{x_1} f_1(x, y) \, dx + \int_{x_1}^b f_1(x, y) \, dx \right];$$

and, for a horizontal line $y = y_1$,

$$P = \int_c^{y_1} + \int_{y_1}^d F(y)\, dy = \int_c^{y_1} dy \int_a^b f_1(x, y)\, dx + \int_y^d dy \int_a^b f_1(x, y)\, dx.$$

The general case is reducible to these two.

We can now prove

THEOREM 1. *The double integral over a rectangle R*,

$$(2) \qquad \int\int_R f_1(x, y)\, dx\, dy = \int_c^d dy \int_a^b f_1(x, y)\, dx.$$

Proof. Divide R into subrectangles r_k by a net $x_i,\, y_j$. Then the repeated

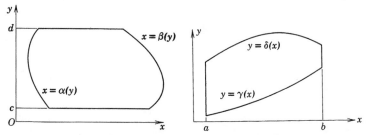

FIGS. 159b and 159c.   Horizontal or vertical segments form part of boundary

integral P is equal to the sum of the repeated integrals P_k over r_k. Now, by Lemma 1,

$$m_k r_k \leqq P_k \leqq M_k r_k;$$

and, on adding these inequalities, we get by Lemma 2

$$\sum m_k r_k \leqq P \leqq \sum M_k r_k.$$

As the mesh of the net approaches zero, both sums approach the double integral; hence P, which always lies between the sums, must equal the double integral.

If the left and right portions of C have the equations

$$x = \alpha(y), \qquad x = \beta(y),$$

equation (2) is equivalent to

$$(2)' \qquad \int\int_R f(x, y)\, dx\, dy = \int_c^d dy \int_{\alpha(y)}^{\beta(y)} f(x, y)\, dx.$$

This equation is still valid when the top and bottom portions of C are segments parallel to the x-axis (Fig. 159b).

If we interchange the order of integrations in (1), we get the repeated integral,

$$(3) \qquad Q = \int_a^b dx \int_c^d f_1(x, y)\, dy,$$

which has the same properties expressed by the two lemmas. Hence, if we replace P by Q in the proof of Theorem 1, we get

THEOREM 2. *The double integral,*

(4)
$$\iint_R f_1(x, y)\, dx\, dy = \int_a^b dx \int_c^d f_1(x, y)\, dy.$$

If the lower and upper portions of C have the equations

$$y = \gamma(x), \qquad y = \delta(x),$$

equation (4) is equivalent to

(4)′
$$\iint_R f(x, y)\, dx\, dy = \int_a^b dx \int_{\gamma(x)}^{\delta(x)} f(x, y)\, dy.$$

This equation is still valid when the left and right portions of C are segments parallel to the y-axis (Fig. 159c).

Incidentally we have proved the important

THEOREM 3. *A repeated integral of a continuous function over a region bounded by a simple closed rectifiable curve is independent of the order in which the integrations are performed.*

Finally we have the mean-value theorem for double integrals:

THEOREM 4. *If $f(x, y)$ and $g(x, y)$ are integrable in a closed region R bounded by a simple rectifiable curve C, and $g(x, y) > 0$, then*

(5)
$$\iint_R f(x, y)\, g(x, y)\, dx\, dy = \mu \iint_R g(x, y)\, dx\, dy, \qquad m < \mu < M,$$

where m, M are the closest bounds of $f(x, y)$ in the region. If $f(x, y)$ is continuous in the region,

(6)
$$\iint_R f(x, y)\, g(x, y)\, dx\, dy = f(\xi, \eta) \iint_R g(x, y)\, dx\, dy, \qquad (\xi, \eta) \quad \text{in } R.$$

The proof is essentially the same as the corresponding Theorem 121 for functions of one variable.

In particular, when $g(x, y) = 1$, we have from (5)

(7)
$$\iint_R f(x, y)\, dx\, dy = \mu\, A,$$

where A is the area enclosed by C.

Example 1. The double integral $\iint_R f(x, y)\, dx\, dy$ taken over a region R with given boundaries may be interpreted as the mass of a plane lamina R of density $\rho = f(x, y)$. When $f(x, y) = 1$, the integral

(8)
$$\iint_R dx\, dy = A, \quad \text{the enclosed area.}$$

When $f(x, y) = x^2$ or y^2, we obtain the *moments of inertia* of the region about the y- and x-axes, respectively:

$$(9) \qquad I_y = \iint_R x^2 \, dx \, dy, \qquad I_x = \iint_R y^2 \, dx \, dy;$$

and their sum gives the moment of inertia about the z-axis:

$$(10) \qquad I_z = I_x + I_y = \iint_R (x^2 + y^2) \, dx \, dy.$$

The *mean value* of $f(x, y)$ over the region R is defined as

$$(11) \qquad \text{Mean } f(x, y) = \frac{1}{A} \iint_R f(x, y) \, dx \, dy.$$

When $f(x, y) = x$ or y, (11) gives the coordinates of the *centroid* of R:

$$(12) \qquad x^* = \frac{1}{A} \iint_R x \, dx \, dy, \qquad y^* = \frac{1}{A} \iint_R y \, dx \, dy.$$

When $f(x, y) = y^2$ or x^2 in (11), we obtain the squares of the *radii of gyration* about the axes:

$$(13) \qquad k_x^2 = \frac{I_x}{A}, \qquad k_y^2 = \frac{I_y}{A}.$$

Example 2. Let $f(x, y)$ be integrable in the right isosceles triangle T bounded by the lines $y = 0, x = a, x - y = 0$. Then the double integral of $f(x, y)$ over T is given by the repeated integrals,

$$\int_0^a \int_y^a f(x, y) \, dx \, dy = \int_0^a \int_0^x f(x, y) \, dy \, dx.$$

When $f(x, y)$ has the following values, the integral I is given below:

$$f(x, y): \quad 1 \quad\quad x \quad\quad y \quad\quad x^2 \quad\quad y^2,$$

$$I \ : \quad \tfrac{1}{2}a^2 \quad \tfrac{1}{3}a^3 \quad \tfrac{1}{6}a^3 \quad \tfrac{1}{4}a^4 \quad \tfrac{1}{12}a^4.$$

Hence for the triangle T we have (Ex. 1)

$$A = \tfrac{1}{2}a^2, \quad x^* = \tfrac{2}{3}a, \quad y^* = \tfrac{1}{3}a, \quad I_y = \tfrac{1}{4}a^4, \quad I_x = \tfrac{1}{12}a^4.$$

PROBLEMS

1. Describe and compute the area $\displaystyle\int_{-1}^{1} \int_{x^2}^{2-x^2} dy \, dx$. Find I_x and I_y for this area.

2. Find the area of the segment of the parabola $ay^2 = b^2x$ in the first quadrant bounded by the lines $x = a, y = 0$. What are the coordinates of its centroid? Compute I_x and I_y.

3. Find I_x for the triangle bounded by the lines $x = 0, y = 0, bx + ay = ab$. Perform the repeated integration in two ways.

4. Find I_x, I_y, and I_z for the ellipse $b^2x^2 + a^2y^2 = a^2b^2$. What is the mean value of $r^2 = x^2 + y^2$ over the ellipse?

5. Find A, x^*, y^*, I_y, I_x for the area included between the parabola $y^2 = 4x$ and the line $x + y = 3$.

6. If r is the distance of a point from the center of a square of side a, find the mean value of r^2 over the square.

7. If $f(x, y)$ is continuous within the rectangle $a \leqq x \leqq A$, $b \leqq y \leqq B$, and

$$\varphi(x, y) = \int_b^y \int_a^x f(u, v) \, du \, dv,$$

show that $\varphi_{xy}(x, y) = f(x, y)$. Conversely, if $\varphi = F(x, y)$ is any solution of the last equation, show that

$$\int_b^y \int_a^x f(u, v) \, du \, dv = F(x, y) - F(a, y) - F(x, b) + F(a, b).$$

8. If x, y are numbers $\geqq 0$ such that $x + y \leqq 3$, find the mean value of $x + y$. Verify the answer.

9. Verify that $\displaystyle\int_0^1 dx \int_x^1 e^{x/v} \, dy = \frac{1}{2}(e - 1); \quad \int_0^1 dy \int_y^1 e^{-x^2} \, dx = \frac{1}{2}\left(1 - \frac{1}{e}\right).$

160. Green's Theorem in the Plane. Let R be a closed region that is bounded by one or more simple closed curves. Then the *positive sense* on the boundary is the one that leaves R to the left. Thus for the region within a closed curve C (Fig. 160a) the positive sense over C is counter-

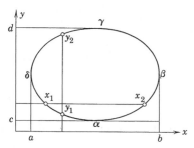

FIG. 160a. Region bounded by a simple closed curve

clockwise; but, for the region interior to C_1, exterior to C_2 and C_3 (Fig. 160c), the positive sense is counterclockwise for the outer boundary, clockwise for the inner boundaries.

GREEN'S THEOREM. *Let R be a closed region bounded by one or more closed regular† curves and capable of dissection into subregions whose*

---

† A *regular curve* is piecewise smooth and without double points (§ 129).

boundaries are cut in at most two points by parallels to the axes. Then, if
$P(x, y)$, $Q(x, y)$, P_y, Q_x *are continuous in* R,

(1)
$$\iint_R \left(\frac{\partial Q}{\partial x} - \frac{\partial P}{\partial y}\right) dx \, dy = \oint P \, dx + Q \, dy,$$

where the circuit integral is taken over the boundary of R *in the positive sense.*

Proof. Suppose first that R is bounded by single closed curve C which is cut in at most two points by any line parallel to the axes (Fig. 160a). Then from (159.2)′

$$\iint \frac{\partial Q}{\partial x} \, dx \, dy = \int_c^d dy \int_{x_1}^{x_2} \frac{\partial Q}{\partial x} \, dx$$

$$= \int_c^d \{Q(x_2, y) - Q(x_1, y)\} \, dy$$

$$= \int_c^d Q(x_2, y) \, dy + \int_d^c Q(x_1, y) \, dy,$$

where the integrals are taken over the arcs $\alpha\beta\gamma$ and $\gamma\delta\alpha$, respectively; hence,

(2)
$$\iint_R \frac{\partial Q}{\partial x} \, dx \, dy = \oint_C Q(x, y) \, dy.$$

In similar fashion

$$\iint_R \frac{\partial P}{\partial y} \, dx \, dy = \int_a^b dx \int_{y_1}^{y_2} \frac{\partial P}{\partial y} \, dy$$

$$= \int_a^b \{P(x, y_2) - P(x, y_1)\} \, dx$$

$$= -\int_a^b P(x, y_1) \, dx - \int_b^a P(x, y_2) \, dx,$$

where the integrals are taken over the arcs $\delta\alpha\beta$ and $\beta\gamma\delta$, respectively; hence,

(3)
$$-\iint_R \frac{\partial P}{\partial y} \, dx \, dy = \oint_C P(x, y) \, dx.$$

On adding (2) and (3), we get (1).

We may now extend equation (3) to regions that may be dissected into a finite number of subregions having the property stated in the theorem. For each subregion equation (3) is valid; and, when these equations are added, the double integrals combine into an integral over the entire region; but the line integrals over the internal boundaries cancel, since each is traversed twice but in opposite directions, leaving only the circuit integral over the external boundary (Fig. 160*b*).

FIG. 160*b*.   Dissected region

A region R bounded by a single closed curve is *simply connected* (§ 153); for any closed curve in R can be continuously contracted into a point within the region. But, if the boundary of R consists of two or more closed curves (Fig. 160*c*) it is *multiply connected*; for a closed curve

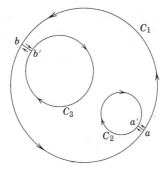

FIG. 160*c*.   Multiply connected region made simply connected by cross cuts

about C_2 or C_3 cannot be contracted to a point without crossing C_2 or C_3. But the region R of Fig. 160*c* can be made simply connected by two cross cuts aa', bb' joining C_1 to C_2 and C_3; and, if we pass around the new boundary as shown by the arrows (each cross cut being traversed twice), the entire circuit is made in the positive sense.

We can now apply Green's theorem (1) to region of Fig. 160*c* made simply connected with cross cuts. In the left member the double integral covers the region interior to C_1 and exterior to C_2 and C_3. In the right member the line integral comprises circuits in the positive sense over C_1, C_2, and C_3; for the cross-cut integrals cancel as they are traversed twice and in opposite senses.

An internal boundary of a region may consist of a single point P_0 which must be excluded from R to meet the continuity requirements of

Green's theorem. Such a *punctured* region is multiply connected; for a curve about P_0 cannot be contracted to a point within the region without crossing P_0. If the region consists of the entire plane punctured at P_0, a cross cut from P_0 to infinity will make it simply connected; for all closed curves not crossing the cut may be contracted to a point.

THEOREM 2. *If $P_y = Q_x$ throughout a simply connected region R, the circuit integral,*

$$\oint P \, dx + Q \, dy = 0,$$

over any closed, piecewise smooth curve in R. The curve may have double points.

Proof. If the curve does not cross itself it is *regular* and (1) shows that the circuit integral is zero. In order to apply Green's theorem we assume that a simply connected region may be dissected into subregions of the type postulated in the statement of the theorem. The proof of this fact is omitted.

Even if the curve has double points, the circuit integral over each loop is zero, and hence the integral over the entire curve vanishes. The sense of integration is immaterial.

Corollary. *If $P_y = Q_x$ throughout a simply connected region R the line integral,*

(4)
$$\int_{(x_0, y_0)}^{(x, y)} P \, dx + Q \, dy = F(x, y),$$

is independent of the path in R joining (x_0, y_0) to (x, y) and defines a single-valued function $F(x, y)$.

Proof. Let C_1 and C_2 denote any two curves in R joining P_0 to P; then the integral over the circuit $P_0 C_1 P C_2 P_0$ is zero and

$$\int_{P_0 C_1 P} P \, dx + Q \, dy - \int_{P_0 C_2 P} P \, dx + Q \, dy = 0.$$

Hence the line integral (4), being independent of the path, defines a single-valued function $F(x, y)$.

By means of this corollary we can extend Theorem 153.1 from rectangles to simply connected regions in general.

THEOREM 3. *If the vector $\mathbf{f} = [P, Q, 0]$ has continuous partial derivatives in a simply connected region and $P_y = Q_x$, then $\mathbf{f} = \nabla F$, a gradient vector.*

Proof. The function $F(x, y)$ defined in (4) has the derivatives $F_x = P$, $F_y = Q$ (Theorem 151.2); hence

$$\mathbf{f} = [F_x, F_y, 0] = \nabla F.$$

Although step paths usually afford the simplest method of computing the integral (4), a straight segment joining the end points is sometimes preferable. If $\mathbf{f}(\mathbf{r})$ is continuous when $\mathbf{r} = \mathbf{0}$, we can integrate along the radial line from $\mathbf{0}$ to $\mathbf{r}$. The points of this line have position vectors $t\mathbf{r}(0 \leq t \leq 1)$, and hence

$$(5) \qquad F(\mathbf{r}) = \int_0^1 \mathbf{f}(t\mathbf{r}) \cdot d(t\mathbf{r}) = \mathbf{r} \cdot \int_0^1 \mathbf{f}(t\mathbf{r}) \, dt.$$

In particular when $\mathbf{f} = [P, Q]$ is homogeneous of degree n, $\mathbf{f}(t\mathbf{r}) = t^n \mathbf{f}(\mathbf{r})$ and (5) becomes

$$(6) \qquad F(\mathbf{r}) = \frac{\mathbf{r} \cdot \mathbf{f}(\mathbf{r})}{n+1} = \frac{xP + yQ}{n+1}, \qquad n > -1.$$

THEOREM 4. *If P, Q, P_y, Q_x are continuous throughout a simply connected region and $\oint P \, dx + Q \, dy = 0$ over all regular closed paths in the region, then $P_y = Q_x$.*

Proof. If the continuous function $Q_x - P_y \neq 0$ at any interior point (a, b) of the region, we can always find a circle about (a, b) in which $Q_x - P_y$ keeps the same sign (§ 74). On integrating over its boundary we have

$$\oint P \, dx + Q \, dy = \iint (Q_x - P_y) \, dx \, dy \neq 0,$$

contrary to hypothesis. Thus $Q_x - P_y = 0$ at all interior points and, since $Q_x - P_y$ is continuous, also on the boundary.

Green's theorem admits of two vector interpretations.

1. Let us put

$$\mathbf{f} = [P, Q, 0], \qquad \text{rot } \mathbf{f} = [0, 0, Q_x - P_y].$$

Since the unit tangent vector to the boundary of R is

$$\mathbf{T} = [dx/ds, dy/ds, 0], \qquad (102.1)$$

equation (1) becomes

$$(7) \qquad \iint_R \mathbf{e_3} \cdot \text{rot } \mathbf{f} \, dA = \oint \mathbf{f} \cdot \mathbf{T} \, ds,$$

where $dA = dx \, dy$ is the element of area. We shall see that (7) is a special case of Stokes' theorem (166.1).

2. Let us put

$$\mathbf{f} = [Q, -P, 0], \qquad \text{div } \mathbf{f} = Q_x - P_y.$$

Since the unit exterior normal vector to the boundary is

$$\mathbf{n} = \mathbf{\tau} \times \mathbf{e}_3 = [dy/ds, -dx/ds, 0],$$

equation (1) becomes

(8)
$$\iint_R \operatorname{div} \mathbf{f} \, dA = \oint \mathbf{f} \cdot \mathbf{n} \, ds.$$

We shall see that (8) is the plane version of the divergence theorem (171.1).

Thus, if we apply (7) and (8) to the position vector $\mathbf{r} = [x, y, 0]$ in the xy-plane, we obtain

$$0 = \oint \mathbf{r} \cdot \mathbf{\tau} \, ds, \qquad 2A = \oint \mathbf{r} \cdot \mathbf{n} \, ds.$$

Example 1. When $Q = x/2$, $P = -y/2$ the double integral gives the area of R:

(9)
$$A = \iint_R dx \, dy = \frac{1}{2} \oint (x \, dy - y \, dx)$$

in agreement with (154.6). On putting $Q = x$, $P = 0$ or $Q = 0$, $P = -y$, we have also

(10)
$$A = \oint x \, dy = -\oint y \, dx.$$

Example 2. Let us consider the examples 150.3–4 in the light of Green's theorem. With $P = y^2$, $Q = -x^2$,

$$\oint (y^2 \, dx - x^2 \, dy) = -2 \iint (x + y) \, dx \, dy.$$

Now

$$\iint x \, dx \, dy = x^* A, \qquad \iint y \, dx \, dy = y^* A,$$

where (x^*, y^*) is the centroid of the area bounded by the curve.

For the circle of Ex. 150.3, $A = \pi$, $x^* = 0$, $y^* = 1$; hence the circuit integral equals -2π.

For the triangle of Ex. 150.4, $A = 1$, $x^* = 0$, $y^* = \frac{1}{3}$; and the circuit integral equals $-\frac{2}{3}$.

Example 3. If $f(x, y), f_x, f_y, f_{xx}, f_{yy}$ are continuous in a region R, we have, on putting $Q = f_x$, $P = -f_y$ in (3),

(11)
$$\iint_R (f_{xx} + f_{yy}) \, dx \, dy = \oint_C (f_x \, dy - f_y \, dx).$$

If we use the arc s as parameter along C, the circuit integral becomes

$$\oint (f_x \sin \theta - f_y \cos \theta) \, ds = -\oint \left[f_x \cos \left(\theta + \frac{\pi}{2} \right) + f_y \sin \left(\theta + \frac{\pi}{2} \right) \right] ds,$$

where $\theta = $ angle $(\mathbf{e}_1, \boldsymbol{\tau})$ between the x-axis and the tangent to C in the positive direction (102.2). Now $f_{xx} + f_{yy} = \nabla^2 f$ (109.10), and

$$f_x \cos\left(\theta + \frac{\pi}{2}\right) + f_y \sin\left(\theta + \frac{\pi}{2}\right) = \frac{df}{dn} \tag{106.2}'$$

where df/dn is the directional derivative of f in the direction of the *internal* normal to the curve C. Hence (11) becomes

(12)
$$\iint_R \nabla^2 f \, dx \, dy = -\oint_C \frac{df}{dn} \, ds.$$

Example 4. $P_y = Q_x$ for the plane vector,

$$\mathbf{f} = \left[\frac{-y}{x^2 + y^2}, \frac{x}{x^2 + y^2}, 0\right], \qquad x, y \neq 0,$$

in the xy-plane punctured at the origin. But, since the punctured plane is not simply connected we cannot conclude that the line integral $\int \mathbf{f} \cdot d\mathbf{r}$ is independent of the path. If we cut the plane along the negative x-axis, we obtain a simply connected region in which $\int \mathbf{f} \cdot d\mathbf{r}$ *is* independent of the path. In polar coordinates we have

$$\mathbf{f} = \left[\frac{-\sin\varphi}{r}, \frac{\cos\varphi}{r}, 0\right] = \frac{\mathbf{P}}{r} = \nabla\varphi \tag{106.14}$$

and hence

$$\int_{\mathbf{r}_1}^{\mathbf{r}_2} \mathbf{f} \cdot d\mathbf{r} = \varphi_2 - \varphi_1 \qquad (-\pi < \varphi < \pi)$$

along any path from $\mathbf{r}_1$ to $\mathbf{r}_2$ which does not cross the cut. Note that $\varphi \to \pi$ or $-\pi$, according as the cut is approached from above or below.

PROBLEMS

Compute the circuit integral over the given boundary, and verify the result by Green's theorem.

1. $\oint (2x - y) \, dx + (x + 3y) \, dy$;   ellipse: $x = 2 \cos t, y = \sin t$.

2. $\oint (x^2 + 2y^2) \, dx$;   triangle: $y = 0, 2x - y = 2, x = 0$.

3. $\oint (xy - x^2) \, dx + x^2 y \, dy$;   triangle: $y = 0, x = 1, y = x$.

4. $\oint x^3 \, dy - y^3 \, dx$;   rectangle: $x = \pm a, y = \pm b$.

5. $\oint (x^2 + xy) \, dx + (x^2 + y^2) \, dy$;   rectangle: $x = \pm a, y = \pm b$.

Also compute the integral over any closed curve having the y-axis as axis of symmetry.

6. Compute $\iint x^2 \, dx \, dy$ over the area inside the square $(0, 0) - (4, 0) - (4, 4) - (0, 4) - (0, 0)$ and outside of the rectangle $(1, 1) - (2, 1) - (2, 3) - (1, 3) - (1, 1)$. Verify by transforming the integral into a line integral over the entire boundary.

7. Prove that

$$\tfrac{1}{3} \oint x^3 \, dy - y^3 \, dx = I_z,$$

the moment of inertia about the z-axis of the enclosed area.

161. Element of Area. The equations

$$(1) \qquad\qquad x = F(u, v), \qquad y = G(u, v)$$

for a change of variables $(x, y) \to (u, v)$ may be regarded as setting up a correspondence between the points of the xy-plane and uv-plane. We assume that F, G and their first partial derivatives are continuous in a region R' of the uv-plane, and that the Jacobian

$$J = \frac{\partial(x, y)}{\partial(u, v)} = \begin{vmatrix} F_u & F_v \\ G_u & G_v \end{vmatrix} \neq 0 \quad \text{in } R'.$$

Since J is also continuous, it maintains the same sign throughout R'.

We shall suppose that equations (1) map R', bounded by a curve C', on a region R of the xy-plane bounded by a curve C; and that the points (u, v) and (x, y) correspond one-to-one. This means that the inverse transformation,

$$(2) \qquad\qquad u = f(x, y), \qquad v = g(x, y),$$

obtained by solving equations (1) is unique (cf. § 87). The Jacobian

$$\frac{\partial(u, v)}{\partial(x, y)} = \begin{vmatrix} f_x & f_y \\ g_x & g_y \end{vmatrix} = \frac{1}{J} \qquad\qquad (87.4);$$

and equations (1) or (2) map the regions R, R' and their boundaries C, C' one-to-one on each other.

As C is described in the positive sense, C' may be described in the positive or negative sense. The correspondence $C \leftrightarrow C'$ is said to be *direct* or *inverse* in the respective cases.

The area A of the region R is given by

$$A = \iint_R dx\, dy = \oint_C x\, dy, \qquad\qquad (160.10).$$

If we regard u and v as functions of t along the curve C' such that (u, v) completes a circuit of C' as t varies from α to β, then x and y are also functions of t through (1), and (x, y) completes a circuit of C as t varies from α to β. Thus A is given by the ordinary integral,

$$A = \int_\alpha^\beta x\, \frac{dy}{dt}\, dt = \int_\alpha^\beta x \left(\frac{\partial y}{\partial u}\frac{du}{dt} + \frac{\partial y}{\partial v}\frac{dv}{dt} \right) dt.$$

The latter, in which x and y are given by (1), is equivalent to the circuit integral over C':

$$A = \pm \oint_{C'} x\, \frac{\partial y}{\partial u}\, du + x\, \frac{\partial y}{\partial v}\, dv,$$

where the sign is $+$ or $-$, according as the correspondence between C and C' is direct or inverse. By use of Green's theorem (160.3) this becomes the double integral,

$$A = \pm \int\int_{R'} \left[\frac{\partial}{\partial u}\left(x\,\frac{\partial y}{\partial v} \right) - \frac{\partial}{\partial v}\left(x\,\frac{\partial y}{\partial u} \right) \right] du\, dv,$$

or, if $y_{uv} = y_{vu}$ (§ 82),

(3) $$A = \pm \int\int_{R'} J\, du\, dv.$$

Now $A > 0$, and J keeps the same sign in R'; hence the sign to be chosen must agree with the sign of J: Consequently,

The correspondence $C \leftrightarrow C'$ is direct or inverse according as $J > 0$ or $J < 0$; and we may put (3) in the unambiguous form:

(3)′ $$A = \int\int_{R'} |J|\, du\, dv = \int\int_{R'} \left| \frac{\partial(x, y)}{\partial(u, v)} \right| du\, dv.$$

We speak of $|J|\, du\, dv$ as the *element of area* for the curvilinear coordinates u, v.

Example 1. Let $x = u + v$, $y = u - v$; then $J = -2$, and the correspondence $C \leftrightarrow C'$ is inverse. Thus the triangle T, $(0, 0) - (1, 0) - (1, 1)$, in the xy-plane is mapped on the triangle T', $(0, 0)$, $(1/2, 1/2)$, $(1, 0)$, in the uv-plane. The area

$$T = 2\int\int du\, dv = 2T' = \frac{1}{2}.$$

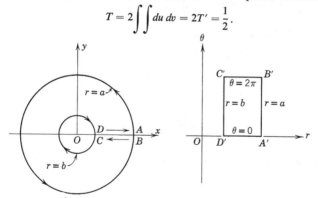

FIG. 161. A cut annulus mapped on a rectangle

Example 2. Polar Coordinates r, θ are connected with rectangular by the equations

$$x = r\cos\theta, \qquad y = r\sin\theta;$$

$$J = \begin{vmatrix} x_r & x_\theta \\ y_r & y_\theta \end{vmatrix} = \begin{vmatrix} \cos\theta & -r\sin\theta \\ \sin\theta & r\cos\theta \end{vmatrix} = r,$$

and hence the element of area is $r\, dr\, d\theta$.

$J > 0$ in any region that does not include the origin. The circle $x^2 + y^2 = a^2$ apparently corresponds to the vertical line $r = a$ in the $r\theta$-plane. This paradox is due to the fact that $J = 0$ at the center of the circle. The closed curve $ABCDA$ formed by the circles $x^2 + y^2 = a^2$, $x^2 + y^2 = b^2$ together with the segment ("canal") $b \leq x \leq a$ traversed twice, maps directly on the rectangle $A'B'C'D'A'$ whose sides are $r = a$, $\theta = 2\pi$, $r = b$, $\theta = 0$ (Fig. 161).

162. Change of Variables in a Double Integral. We propose to change the variables in the double integral,

$$\iint_R f(x, y) \, dx \, dy,$$

taken over a region R of the xy-plane from x, y to u, v by means of the transformation

(1) $$x = F(u, v), \qquad y = G(u, v),$$

whose Jacobian $J \neq 0$. If we define

$$\varphi(x, y) = \int_a^x f(t, y) \, dt, \qquad \frac{\partial \varphi}{\partial x} = f(x, y),$$

and we may apply Green's theorem:

$$\iint_R f(x, y) \, dx \, dy = \iint_R \frac{\partial \varphi}{\partial x} \, dx \, dy = \oint_C \varphi(x, y) \, dy.$$

As in § 161, let $R \leftrightarrow R'$, $C \leftrightarrow C'$; and, as before, the circuit integral over C equals the ordinary integral,

$$\int_\alpha^\beta \varphi \frac{dy}{dt} \, dt = \int_\alpha^\beta \varphi \left(\frac{\partial y}{\partial u} \frac{du}{dt} + \frac{\partial y}{\partial v} \frac{dv}{dt} \right) dt$$

$$= \pm \oint_{C'} \varphi \frac{\partial y}{\partial u} \, du + \varphi \frac{\partial y}{\partial v} \, dv,$$

where the sign chosen is that of J. We now apply Green's theorem to the last integral; then, since

$$\frac{\partial}{\partial u} (\varphi y_v) - \frac{\partial}{\partial v} (\varphi y_u) = \varphi_u y_v - \varphi_v y_u = \varphi_x (x_u y_v - x_v y_u) = f(x, y)J,$$

we finally obtain the basic equation for changing variables in a double integral:

(2) $$\iint_R f(x, y) \, dx \, dy = \iint_{R'} f(x, y) \left| \frac{\partial(x, y)}{\partial(u, v)} \right| du \, dv.$$

On the right x and y must be expressed in terms of u and v.

Example. To prove that

(3)
$$I = \int_0^\infty e^{-x^2}\, dx = \frac{\sqrt{\pi}}{2},$$

consider the double integral of $e^{-x^2-y^2}$ in rectangular and polar coordinates. Since

$$x = r\cos\theta, \qquad y = r\sin\theta, \qquad \frac{\partial(x, y)}{\partial(r, \theta)} = \begin{vmatrix} \cos\theta & -r\sin\theta \\ \sin\theta & r\cos\theta \end{vmatrix} = r,$$

we have over any region

$$\iint e^{-x^2-y^2} dx\, dy = \iint e^{-r^2} r\, dr\, d\theta.$$

The left-hand integral over the square S $(0 \le x, y \le a)$ equals

$$\int_0^a e^{-x^2} dx \int_0^a e^{-y^2} dy = \left[\int_0^a e^{-x^2} dx \right]^2 = I_a^2.$$

The right-hand integral over a circular quadrant of radius b is

$$\int_0^{\pi/2} d\theta \int_0^b re^{-r^2} dr = \frac{\pi}{4}\left(1 - e^{-b^2}\right).$$

Since the square S lies between quadrants of radii a and $\sqrt{2}\,a$ (Fig. 162),

$$\frac{\pi}{4}\left(1 - e^{-a^2}\right) < I_a^2 < \frac{\pi}{4}\left(1 - e^{-2a^2}\right),$$

and, as $a \to \infty$, $I_a^2 \to I^2 = \pi/4$.

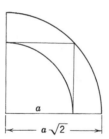

FIG. 162.   Square between two quadrants

PROBLEMS

1. Transform the double integral $\iint f(x, y)\, dx\, dy$ over the circle $x^2 + y^2 \le 1$ by means of the following changes of variable. What is the region of integration in the uv-plane?

(a) $x = u + a, y = v + b$;

(b) $x = au, y = bv$;

(c) $x = u\cos\alpha - v\sin\alpha, y = u\sin\alpha + v\cos\alpha$;

(d) $x = \alpha u + \beta v, y = \gamma u + \delta v$

(e) $x = \dfrac{u}{u^2 + v^2}, y = \dfrac{v}{u^2 + v^2}.$

2. Compute $\iint e^{-(x^2+y^2)}\, dx\, dy$ over the circle $x^2 + y^2 \leq a^2$ by transforming to polar coordinates.

3. Show that
$$\iint_R (f_x^2 + f_y^2)\, dx\, dy = \iint_{R'} (f_u^2 + f_v^2)\, du\, dv$$
under an inversion in the unit circle about the origin (transformation (e) of Prob. 1). Verify when $f = x$ and R is the circle $x^2 + y^2 \leq 1$.

4. Solve Prob. 3 by first transforming to polar coordinates r, θ and then making the inversion in the form $r = 1/\rho$, $\theta = \varphi$

5. Compute $\iint e^{-(x-y)/(x+y)}\, dx\, dy$ taken over the triangle bounded by the lines $x = 0$, $y = 0$, $x + y = 1$. [Put $u = x + y$, $v = x - y$.]

6. Compute
$$\iint \frac{dx\, dy}{(1 + x^2 + y^2)^2}$$

(a) over the triangle in Prob. 5;

(b) over the circle $x^2 + y^2 \leq 1$.

[Use polar coordinates.]

7. Show that

$$\iint f(x, y)\, dx\, dy \qquad \text{over the ellipse} \quad \frac{x^2}{a^2} + \frac{y^2}{b^2} \leq 1$$

$$= ab \iint f(ax, by)\, dx\, dy \quad \text{over the circle} \quad x^2 + y^2 \leq 1.$$

[Put $x = au$, $y = bv$.]

8. Show that $\displaystyle\iint_R (x^2 - y^2)^2\, dx\, dy = \frac{4}{45}$ when R is the triangle bounded by the lines $y = 0$, $x = 1$, $x - y = 0$:

(a) by computing a repeated integral;

(b) by the change of variable $u = x + y$, $v = x - y$.

163. Curves on a Surface.

The equations of a surface are (§ 105),

$$(1) \qquad x = x(u, v), \qquad y = y(u, v), \qquad z = z(u, v),$$

where the functions are defined over a certain region D of the uv-plane. If $\mathbf{r} = [x, y, z]$, the vector equation of the surface is

$$(2) \qquad\qquad\qquad \mathbf{r} = \mathbf{r}(u, v).$$

The *parametric curves* on the surface,

$$(3) \qquad\qquad\qquad v = b, \qquad u = a,$$

have the vector equations,

$$\mathbf{r} = \mathbf{r}(u, b), \qquad \mathbf{r} = \mathbf{r}(a, v);$$

their tangent vectors are $\mathbf{r}_u$ and $\mathbf{r}_v$. At any *regular point* of the surface $\mathbf{r}_u \times \mathbf{r}_v \neq 0$ and the vector $\mathbf{r}_u \times \mathbf{r}_v$ is normal to surface. Since

$$\mathbf{r}_u = [x_u, y_u, z_u], \qquad \mathbf{r}_v = [x_v, y_v, z_v],$$

(4) $$\mathbf{r}_u \times \mathbf{r}_v = [A, B, C],$$

where

(5) $$A = \frac{\partial(y, z)}{\partial(u, v)}, \qquad B = \frac{\partial(z, x)}{\partial(u, v)}, \qquad C = \frac{\partial(x, y)}{\partial(u, v)}.$$

At a regular point at least one of these Jacobians is not zero.

A curve on the surface is given by the equations

(6) $$u = f(t), \qquad v = g(t),$$

or by the vector equation $\mathbf{r} = \mathbf{r}(t)$ derived from (2) by substituting these values of u and v. Its tangent vector,

$$\frac{d\mathbf{r}}{dt} = \mathbf{r}_u \frac{du}{dt} + \mathbf{r}_v \frac{dv}{dt},$$

and, denoting t-derivatives by primes,

(7) $$\mathbf{r}' \cdot \mathbf{r}' = \mathbf{r}_u \cdot \mathbf{r}_u \, u'^2 + 2\mathbf{r}_u \cdot \mathbf{r}_v \, u'v' + \mathbf{r}_v \cdot \mathbf{r}_v \, v'^2.$$

The arc along the curve is given by (129.6):

$$s = \int_{t_0}^{t} |\mathbf{r}'| \, dt,$$

and $ds/dt = |\mathbf{r}'|$; hence,

(8) $$ds^2 = E \, du^2 + 2F \, du \, dv + G \, dv^2,$$

where the notation

(9) $$E = \mathbf{r}_u \cdot \mathbf{r}_u, \qquad F = \mathbf{r}_u \cdot \mathbf{r}_v, \qquad G = \mathbf{r}_v \cdot \mathbf{r}_v$$

is standard in surface geometry. Equation (8) is the *fundamental quadratic form* for the surface, and E, F, G are called the *fundamental quantities of the first order*. They depend on the parameters u, v but are not altered by a change of rectangular coordinates x, y, z; for $ds^2 = dx^2 + dy^2 + dz^2$ is invariant to such a change.

From (99.5) we have

$$(\mathbf{r}_u \times \mathbf{r}_v) \cdot (\mathbf{r}_u \times \mathbf{r}_v) = (\mathbf{r}_u \cdot \mathbf{r}_u)(\mathbf{r}_v \cdot \mathbf{r}_v) - (\mathbf{r}_u \cdot \mathbf{r}_v)^2,$$

or, in view of (9),

(10) $$|\mathbf{r}_u \times \mathbf{r}_v|^2 = EG - F^2.$$

This shows that the vector $\mathbf{r}_u \times \mathbf{r}_v$, normal to the surface, is also invariant to a change of rectangular axes. Moreover from (4) we have the identity

(11) $A^2 + B^2 + C^2 = EG - F^2.$

Example 1. *Curves on a Sphere.* The equation of a sphere of radius a is

(12) $\mathbf{r} = a[\sin\theta\cos\varphi,\ \sin\theta\sin\varphi,\ \cos\theta]$

where the colatitude $u = \theta$ and longitude $v = \varphi$ are parameters. From (9) we have

(13) $E = a^2, \qquad F = 0, \qquad G = a^2\sin^2\theta;$

and from (8)

(14) $ds^2 = a^2\,(d\theta^2 + \sin^2\theta\,d\varphi^2)$

for any curve on the sphere.

Example 2. *Isogonal Trajectories.* Find the differential equation of the curves on the surface $r = r(u, v)$ that cut the parametric curves $v = b$ at a constant angle α.

Assume that the parametric curves (3) cut at right angles; then $F = \mathbf{r}_u \cdot \mathbf{r}_v = 0$. A surface curve $v = f(u)$ has the vector equation $\mathbf{r} = \mathbf{r}(u, f(u))$, and hence the tangent vector $\mathbf{r}_u + \mathbf{r}_v f'(u)$. Since $v = b$ has the tangent vector $\mathbf{r}_u$,

$$\cos\alpha = \frac{\mathbf{r}_u\cdot(\mathbf{r}_u + \mathbf{r}_v f')}{|\mathbf{r}_u|\,|\mathbf{r}_u + \mathbf{r}_v f'|} = \frac{E}{E^{\frac{1}{2}}(E + Gf'^2)^{\frac{1}{2}}},$$

and hence

$$\frac{E + Gf'^2}{E} = \sec^2\alpha, \qquad \frac{G}{E}f'^2 = \tan^2\alpha.$$

Thus,

(15) $$\frac{dv}{du} = \sqrt{\frac{E}{G}}\tan\alpha$$

is the differential equation of the required curves.

From (8) we find their element of arc,

(16) $$ds = \frac{\sqrt{E}}{\cos\alpha}\,du = \frac{\sqrt{G}}{\sin\alpha}\,dv.$$

Example 3. The *rhumb lines* on a sphere cut its meridians $\varphi = $ const at a fixed angle α. From (15) we have their differential equation (cf. Ex. 1),

$$\frac{d\varphi}{d\theta} = \frac{\tan\alpha}{\sin\theta}; \quad \text{whence} \quad \log\tan\tfrac{1}{2}\theta = \varphi\cot\alpha + C$$

on integration.

The length of a rhumb line can be found from (16):

$$ds = \frac{a}{\cos\alpha}\,d\theta = \frac{a\sin\theta}{\sin\alpha}\,d\varphi.$$

Thus, if $\alpha \neq \pi/2$, $s = a\sec\alpha\,|\theta_2 - \theta_1|.$

PROBLEMS

1. On the sphere (12) find the length

(a) of the "small" circle $\theta = \beta$:

(b) of the rhumb line when $\alpha = \pi/4$ between the parallels $\theta = \pi/4$ and $\theta = \pi/2$.

(c) of the curve $\theta = \varphi$ from $\theta = 0$ to $\theta = \tfrac{1}{2}\pi$.

164. Area of a Surface. Let the surface $\sum$ be given by the vector equation,

(1) $$\mathbf{r} = \mathbf{r}(u, v), \qquad (u, v) \text{ in } D.$$

As (u, v) ranges over the region D of the uv-plane, the position vector $\mathbf{r} = [x, y, z]$ describes the surface (the portion corresponding to D). We assume that the tangent vectors $\mathbf{r}_u, \mathbf{r}_v$ to the parametric curves are continuous on the surface; then the normal vector $\mathbf{r}_u \times \mathbf{r}_v$ is also continuous.

We choose as element of area on the surface the area of a parallelogram having the tangent vectors $\mathbf{r}_u\, du, \mathbf{r}_v\, dv$ as sides, namely $| \mathbf{r}_u \times \mathbf{r}_v |\, du\, dv$ (98.3). The area of the surface is then *defined* as the double integral,

(2) $$S = \iint_D | \mathbf{r}_u \times \mathbf{r}_v |\, du\, dv = \iint_D \sqrt{EG - F^2}\, du\, dv.$$

The area thus defined is independent of the choice of rectangular axes; for we have seen that $\mathbf{r}_u \times \mathbf{r}_v$ has this property (§ 163). Moreover, S is independent of the choice of parameters u, v. For, if we change parameters from u, v to p, q by the equations

$$u = u(p, q), \qquad v = v(p, q),$$

we have

$$\mathbf{r}_p = \mathbf{r}_u u_p + \mathbf{r}_v v_p, \qquad \mathbf{r}_q = \mathbf{r}_u u_q + \mathbf{r}_v v_q,$$

(3) $$\mathbf{r}_p \times \mathbf{r}_q = \mathbf{r}_u \times \mathbf{r}_v \frac{\partial(u, v)}{\partial(p, q)} ;$$

hence, if the region D of the uv-plane corresponds to D' in the pq-plane, we have from (162.2)

$$S = \iint_{D'} | \mathbf{r}_u \times \mathbf{r}_v | \frac{\partial(u, v)}{\partial(p, q)}\, dp\, dq = \iint_{D'} | \mathbf{r}_p \times \mathbf{r}_q |\, dp\, dq.$$

Let the direction cosines of

$$\mathbf{r}_u \times \mathbf{r}_v = [A, B, C]$$

be $\cos\alpha, \cos\beta, \cos\gamma$. If the part of surface considered has no vertical tangent planes, $\cos\gamma \neq 0$; then, by virtue of the continuity of $\mathbf{r}_u \times \mathbf{r}_v$, $\cos\gamma$ will always have the same sign. Then the points of the surface correspond one-to-one with their vertical projections on the xy-plane. If these occupy a region Z of the xy-plane and we adopt x, y as surface parameters, we have

$$S = \iint_Z | \mathbf{r}_u \times \mathbf{r}_v | \left| \frac{\partial(u, v)}{\partial(x, y)} \right| dx\, dy,$$

or, since

$$| \mathbf{r}_u \times \mathbf{r}_v | \cos\gamma = C = \frac{\partial(x, y)}{\partial(u, v)},$$

(4)
$$S = \int\int_Z \frac{dx\,dy}{|\cos \gamma|}.$$

Since the points of the surface are in one-to-one correspondence with the points of Z, the equation of the surface may be written

(5) $z = f(x, y),$ (x, y) in Z.

The surface normal is parallel to

$$\nabla(z - f(x, y)) = [-f_x, -f_y, 1] = [-p, -q, 1],$$

where $p = \partial z/\partial x, q = \partial z/\partial y$; hence,

$$|\cos \gamma| = \frac{1}{\sqrt{p^2 + q^2 + 1}},$$

and (4) becomes

(4)′ $$S = \int\int_Z \sqrt{p^2 + q^2 + 1}\; dx\,dy.$$

When y, z or z, x are chosen as parameters, we obtain obvious analogues of (4).

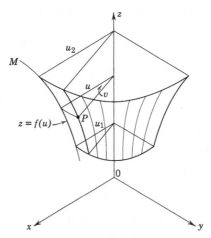

Fig. 164a.   Surface of revolution

Example 1. A surface of revolution about the z-axis has the parametric equations,

$$x = u \cos v,   y = u \sin v,   z = f(u),$$

where $u = \rho, v = \varphi$ are plane-polar coordinates (Fig. 164a). Now

$$\mathbf{r}_u = [\cos v,\, \sin v,\, f'(u)],$$
$$\mathbf{r}_v = u[-\sin v,\, \cos v,\, 0],$$
$$\mathbf{r}_u \times \mathbf{r}_v = u[-f'(u)\cos v,\, -f'(u)\sin v,\, 1].$$
$$|\mathbf{r}_u \times \mathbf{r}_v| = u\sqrt{\left(\frac{dz}{du}\right)^2 + 1}.$$

Hence if D is the rectangle $u_1 \leq u \leq u_2,\ 0 \leq v \leq 2\pi$,

$$S = \int_0^{2\pi} dv \int_{u_1}^{u_2} u \sqrt{\left(\frac{dz}{du}\right)^2 + 1}\ du = 2\pi \int_{u_1}^{u_2} u\ ds,$$

where ds is the element of arc along a meridian M: $z = f(u)$.

Thus, on a sphere of radius a, $u = a \sin \theta$, $ds = a\ d\theta$, where $\theta = $ angle (z, OP); hence,

$$S = 2\pi \int_0^\pi a^2 \sin \theta\ d\theta = 2\pi a^2(-\cos \theta)\Big|_0^\pi = 4\pi a^2.$$

Example 2. An *ellipsoid* has the parametric equations

$$x = a \sin u \cos v, \qquad y = b \sin u \sin v, \qquad z = c \cos u$$

where $u = \theta$, $v = \varphi$ are spherical coordinates (Ex. 111.2). We find that

$$\mathbf{r}_u \times \mathbf{r}_v = \sin u\ [bc \sin u \cos v,\ ac \sin u \sin v,\ ab \cos u];$$

and since D is the rectangle $0 \leq u \leq \pi,\ 0 \leq v \leq 2\pi$,

$$S = \int_0^\pi \sin u\ du \int_0^{2\pi} \{c^2 \sin^2 u\,(a^2 \sin^2 v + b^2 \cos^2 v) + a^2 b^2 \cos^2 u\}^{1/2}\ dv,$$

which can be expressed in terms of elliptic integrals. When $a = b$, we have an ellipsoid of revolution about the z-axis whose area is

$$S = 2\pi a \int_0^\pi \sin u \sqrt{c^2 \sin^2 u + a^2 \cos^2 u}\ du.$$

If $a > c$, we have an *oblate ellipsoid* of eccentricity, $\varepsilon^2 = (a^2 - c^2)/a^2$, and

$$S = 2\pi a \int_0^\pi \sin u \sqrt{c^2 + a^2 \varepsilon^2 \cos^2 u}\ du$$

$$= 2\pi a \int_{-1}^1 \sqrt{c^2 + a^2 \varepsilon^2 t^2}\ dt \qquad (t = \cos u)$$

$$= 2\pi \left(a^2 + \frac{c^2}{2\varepsilon} \log \frac{1 + \varepsilon}{1 - \varepsilon}\right).$$

If $a < c$, we have a *prolate ellipsoid* of eccentricity, $\varepsilon^2 = (c^2 - a^2)/c^2$, and

$$S = 2\pi a \int_0^\pi \sin u \sqrt{c^2 - c^2 \varepsilon^2 \cos^2 u}\ du$$

$$= 2\pi ac \int_{-1}^1 \sqrt{1 - \varepsilon^2 t^2}\ dt \qquad (t = \cos u)$$

$$= 2\pi ac \left(\sqrt{1 - \varepsilon^2} + \frac{\sin^{-1} \varepsilon}{\varepsilon}\right).$$

As $\varepsilon \to 0$, $c \to a$, and in both cases $S \to 4\pi a^2$, the surface of a sphere.

Example 3. Find the area of the surface of the sphere $x^2 + y^2 + z^2 = a^2$ included within the cylinder $x^2 + y^2 = ay$ (Fig. 164b).

Since the radius is normal to the sphere, $\cos \gamma = z/a$; hence, from (4)

$$\frac{1}{4} S = a \int \int_Z \frac{dx\, dy}{z},$$

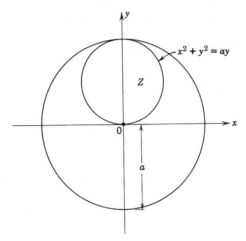

FIG. 164*b*. Projection of sphere and cylinder on xy plane

where Z is the semicircle $x^2 + y^2 = ay$ to the right of the y-axis. In cylindrical co-ordinates (Ex. 111.1) the equations of sphere and cylinder become

$$\rho^2 + z^2 = a^2, \qquad \rho = a \sin \varphi;$$

and

$$\frac{1}{4} S = a \int \int_Z \frac{\rho \, d\rho \, d\varphi}{\sqrt{a^2 - \rho^2}} = a \int_0^{\pi/2} d\varphi \int_0^{a \sin \varphi} \frac{\rho \, d\rho}{\sqrt{a^2 - \rho^2}}$$

$$= a \int_0^{\pi/2} d\varphi \left[-\sqrt{a^2 - \rho^2} \right]_0^{a \sin \varphi} = a^2 \int_0^{\pi/2} (1 - \cos \varphi) \, d\varphi$$

$$= a^2 \left[\varphi - \sin \varphi \right]_0^{\pi/2} = a^2 \left(\frac{\pi}{2} - 1 \right).$$

Hence $S = (2\pi - 4)a^2$; since $2\pi a^2$ is the area of a hemisphere, the hemisphere less the hole cut by the cylinder has a residual area of $4a^2$. This is the first known instance of a curved surface that admits an "exact" quadrature, that is, without involving irrationals such as π.

PROBLEMS

1. Find the area of the cylinder $x^2 + y^2 = a^2$ in the first octant included between the planes $z = 0$ and $z = mx$.

2. Find the area of the surface cut from the cylinder $x^2 + y^2 = a^2$ by the cylinder $y^2 + z^2 = a^2$.

3. Find the area of the surface of the cone $x^2 + y^2 - z^2 = 0$ included within the cylinder $x^2 + y^2 - 2ax = 0$.

4. Find the area of the surface of a sphere $x^2 + y^2 + z^2 = 2cz$ lying within the paraboloid $az = x^2 + y^2$ ($0 < a < 2c$).

5. Find the surface of the cylinder $y^2 = 4ax$ in the first octant cut off by the cylinder $z^2 = 4ax$ and the plane $x = 3a$.

165. Surface Integral. A surface is said to be *bilateral* if it is possible to distinguish one of its sides from the other. Not all surfaces are bilateral. The simplest unilateral surface is the *Möbius strip*; this may be materialized by taking a strip of paper, giving it one twist, and pasting its ends together. This surface has but *one side*; thus, if we attempt to paint one side of the surface without crossing its edges, we automatically paint the entire strip. This is not the case with a closed surface such as a sphere or cube; this has an *inside* and an *outside*. Any portion of surface bounded by a simple closed *plane* curve C is also bilateral; for we cannot pass from one side to the other without crossing C.

Let S be a bilateral surface whose vector equation is

$$\mathbf{r} = \mathbf{r}(u, v), \qquad (u, v) \text{ in } D;$$

then we shall call

$$(1) \qquad\qquad d\mathbf{S} = \mathbf{r}_u \times \mathbf{r}_v \, du \, dv$$

its *vector element of area*. The direction of $d\mathbf{S}$ or $\mathbf{r}_u \times \mathbf{r}_v$ is normal to the surface, and its magnitude is the element of area defined in § 164. The side of S toward which $d\mathbf{S}$ points is called its *positive side*.

If $\mathbf{f}(\mathbf{r})$ is any vector function defined over the surface S, the integral

$$(2) \qquad\qquad \iint_S \mathbf{f} \cdot d\mathbf{S} = \iint_D \mathbf{f} \cdot \mathbf{r}_u \times \mathbf{r}_v \, du \, dv$$

is called the *surface integral over the positive side of S*. If $\mathbf{f} = [P, Q, R]$, we see from (163.4) that

$$(3) \qquad\qquad \iint_S \mathbf{f} \cdot d\mathbf{S} = \iint_D (PA + QB + RC) \, du \, dv,$$

an ordinary double integral in the uv-plane.

The three terms of this integral may be expressed as double integrals over the areas X, Y, Z in the yz-, zx-, xy-planes into which the surface projects. We note first that, if $\cos \alpha$, $\cos \beta$, $\cos \gamma$ are the direction cosines of the normal vector $\mathbf{r}_u \times \mathbf{r}_v = [A, B, C]$, the Jacobians A, B, C have the signs of $\cos \alpha$, $\cos \beta$, $\cos \gamma$. Now, if $\cos \gamma > 0$ over S, $C > 0$ and

$$(4) \qquad \iint_D RC \, du \, dv = \iint_D R \left| \frac{\partial(x, y)}{\partial(u, v)} \right| du \, dv = \iint_Z R \, dx \, dy.$$

But, if $\cos \gamma > 0$ over a part of S which projects into Z^+ and $\cos \gamma < 0$ over a part which projects into Z^-, then

$$(5) \qquad \iint_D RC \, du \, dv = \iint_{Z^+} R^+ \, dx \, dy - \iint_{Z^-} R^- \, dx \, dy.$$

For a *closed* surface S, Z^+ and Z^- are the same; and we write simply

$$(5)' \qquad \iint_D RC \, du \, dv = \iint_{Z^\pm} R \, dx \, dy.$$

We can name the parameters u, v so that $\mathbf{r}_u \times \mathbf{r}_v$ points *outside the* surface; then $\cos \gamma > 0 \; (C > 0)$ on the upper side, $\cos \gamma < 0 \; (C < 0)$ on the lower side; the integral (5)' is then an *external* surface integral.

With this same notation we have

$$(3)' \qquad \iint_D (PA + QB + RC) \, du \, dv = \iint_{X^\pm} P \, dy \, dz$$
$$+ \iint_{Y^\pm} Q \, dz \, dx + \iint_{Z^\pm} R \, dx \, dy.$$

If $\mathbf{f}$ denotes the *velocity* of a fluid flowing through the surface S,

$$\mathbf{f} \cdot d\mathbf{S} = \mathbf{f} \cdot \mathbf{n} \, dS = \text{(normal comp. of } \mathbf{f}\text{) (area } dS\text{)}$$

gives the rate at which fluid volume is passing through the element dS in the direction $\mathbf{n}$. Hence we often speak of the surface integral (1) as the *flux of the vector* $\mathbf{f}$ *through the surface.* When $\mathbf{f}$ is an actual fluid velocity, the integral (1) gives the instantaneous rate of flow through S (in units of volume per unit of time: as cubic centimeters per second).

Example 1.   If S is the sphere $x^2 + y^2 + z^2 = a^2$, Z the circle $x^2 + y^2 = a^2$, we have from (5):

$$\iint_S \frac{dx \, dy}{z} = \iint_Z \frac{dx \, dy}{\sqrt{a^2 - x^2 - y^2}} - \iint_Z \frac{dx \, dy}{-\sqrt{a^2 - x^2 - y^2}}$$

$$= 2 \int_{-a}^{a} dx \int_{-\sqrt{a^2 - x^2}}^{\sqrt{a^2 - x^2}} \frac{dy}{\sqrt{a^2 - x^2 - y^2}}$$

$$= 2 \int_{-a}^{a} dx \left[\sin^{-1} \frac{y}{\sqrt{a^2 - x^2}} \right]_{-\sqrt{a^2 - x^2}}^{\sqrt{a^2 - x^2}}$$

$$= 2\pi \, 2a = 4\pi a.$$

Complications in sign are avoided by using the parametric equations of S:

$$(6) \qquad x = a \sin \theta \cos \varphi, \qquad y = a \sin \theta \sin \varphi, \qquad z = a \cos \theta,$$

where θ, φ are the colatitude and longitude (Ex. 163).

Since

$$C = \frac{\partial(x, y)}{\partial(\theta, \varphi)} = a^2 \sin \theta \cos \theta,$$

the given integral equals

$$\int\int_D \frac{C}{z} \, d\theta \, d\varphi = \int_0^{2\pi} d\varphi \int_0^\pi a \sin \theta \, d\theta = 4\pi a.$$

from (4); D is the rectangle $0 \leqq \theta \leqq \pi$, $0 \leqq \varphi \leqq 2\pi$. Since $C > 0$ on the upper side of the sphere, $C < 0$ on the lower side, the surface integral is external.

Example 2. From (6) we find that

(7) $\mathbf{r}_\theta \times \mathbf{r}_\varphi = a^2 \sin \theta \, [\sin \theta \cos \varphi, \sin \theta \sin \varphi, \cos \theta].$

The outward flux of the vector $\mathbf{f}(\theta, \varphi)$ through the sphere may now be computed directly as

$$\int_0^{2\pi} d\varphi \int_0^\pi \mathbf{f} \cdot \mathbf{r}_\theta \times \mathbf{r}_\varphi \, d\theta.$$

166. Stokes' Theorem. Consider a bilateral surface that is *piecewise smooth*. Such a surface consists of portions S having an equation $\mathbf{r} = \mathbf{r}(u, v)$ and over which the normal vector $\mathbf{r}_u \times \mathbf{r}_v$ is continuous; furthermore, we assume that $\mathbf{r}_{uv} = \mathbf{r}_{vu}$.‡ The *positive side* of S is the side toward which $\mathbf{r}_u \times \mathbf{r}_v$ points; and a person, erect on the positive side, will have S to his left when he traverses its boundary in the *positive sense*.

The vector form of Green's theorem,

$$\int\int \mathbf{e}_3 \cdot \operatorname{rot} \mathbf{f} \, dA = \oint \mathbf{f} \cdot d\mathbf{r}, \tag{160.7}$$

which converts the surface integral over a plane region into a circuit integral over its boundary, may be generalized to apply to curved surfaces; we need only replace the vector element of area,

$$\mathbf{e}_3 \, dA = \mathbf{r}_x \times \mathbf{r}_y \, dx \, dy \quad \text{by} \quad \mathbf{n} \, dS = \mathbf{r}_u \times \mathbf{r}_v \, du \, dv,$$

where $\mathbf{n}$ is a unit normal in the direction of $\mathbf{r}_u \times \mathbf{r}_v$.

STOKES' THEOREM. *Let S be a surface consisting of portions that are piecewise smooth and bounded by regular closed curves. Then, if the vector function $\mathbf{f}(\mathbf{r})$ has a continuous gradient $\nabla\mathbf{f}$ over S, the surface integral of* rot $\mathbf{f}$ *over S is equal to the circuit integral of $\mathbf{f}$ over its outer boundary C in the positive sense:*

(1) $$\int\int_S \mathbf{n} \cdot \operatorname{rot} \mathbf{f} \, dS = \oint_C \mathbf{f} \cdot d\mathbf{r}.$$

---

‡ This is certainly the case when the functions in (163.1) have continuous second partial derivatives.

Proof. It will suffice to prove (1) for a smooth portion of the surface over which $\mathbf{r}_{uv} = \mathbf{r}_{vu}$; for, if such equations for all the smooth portions are added, we obtain (1). Note that in adding the circuit integrals all internal boundaries are traversed twice and in opposite directions so that the corresponding line integrals cancel.

We may therefore assume that S is smooth and bounded by a regular closed curve C. Let D be the region of the uv-plane which contains all parameter values u, v corresponding to points of S (Fig. 166a). Moreover,

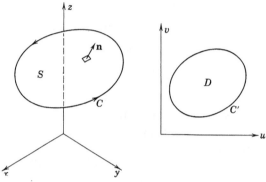

FIG. 166a.   Surface S mapped on plane region D

we suppose that S is so restricted that the correspondence $(u, v) \leftrightarrow (x, y, z)$ is one-to-one. If we put

$$\mathbf{n} \, dS = d\mathbf{S} = \mathbf{r}_u \times \mathbf{r}_v \, du \, dv,$$

the surface integral in (1) becomes

(2) $$\iint_D (\mathbf{r}_u \times \mathbf{r}_v) \cdot \operatorname{rot} \mathbf{f} \, du \, dv \qquad (165.2).$$

taken over the region D of the uv-plane. The curve C forming the boundary of S maps into a curve C', the boundary of D; and the circuit integral in (1) may be computed as a circuit integral over C':

(3) $$\oint_{C'} \mathbf{f} \cdot (\mathbf{r}_u \, du + \mathbf{r}_v \, dv).$$

We now apply Green's theorem in the uv-plane to show that the integrals (2) and (3) are equal.

On making use of formula (99.5),

$$(\mathbf{a} \times \mathbf{b}) \cdot (\mathbf{c} \times \mathbf{d}) = (\mathbf{a} \cdot \mathbf{c})(\mathbf{b} \cdot \mathbf{d}) - (\mathbf{a} \cdot \mathbf{d})(\mathbf{b} \cdot \mathbf{c}),$$

we find that

$$(\mathbf{r}_u \times \mathbf{r}_v) \cdot \text{rot } \mathbf{f} = (\mathbf{r}_u \times \mathbf{r}_v) \cdot (\mathbf{e}_1 \times \mathbf{f}_x + \mathbf{e}_2 \times \mathbf{f}_y + \mathbf{e}_3 \times \mathbf{f}_z)$$
$$= x_u \, \mathbf{r}_v \cdot \mathbf{f}_x - x_v \, \mathbf{r}_u \cdot \mathbf{f}_x +$$
$$y_u \, \mathbf{r}_v \cdot \mathbf{f}_y - y_v \, \mathbf{r}_u \cdot \mathbf{f}_y +$$
$$z_u \, \mathbf{r}_v \cdot \mathbf{f}_z - z_v \, \mathbf{r}_u \cdot \mathbf{f}_z,$$

and hence from the chain rule

(4) $$(\mathbf{r}_u \times \mathbf{r}_v) \cdot \text{rot } \mathbf{f} = \mathbf{r}_v \cdot \mathbf{f}_u - \mathbf{r}_u \cdot \mathbf{f}_v = (\mathbf{r}_v \cdot \mathbf{f})_u - (\mathbf{r}_u \cdot \mathbf{f})_v,$$

since $\mathbf{r}_{vu} = \mathbf{r}_{uv}$. If we substitute this value in (2), we may transform the integral by Green's theorem letting $u, v, \mathbf{r}_u \cdot \mathbf{f}, \mathbf{r}_v \cdot \mathbf{f}$ correspond to x, y, P, Q in (160.1); thus we find

$$\iint_D \{(\mathbf{r}_v \cdot \mathbf{f})_u - (\mathbf{r}_u \cdot \mathbf{f})_v\} \, du \, dv = \oint_{C'} (\mathbf{r}_u \cdot \mathbf{f}) \, du + (\mathbf{r}_v \cdot \mathbf{f}) \, dv,$$

which is precisely the integral (3). Thus the proof is complete.

With the same conditions on the function $\mathbf{f}$ and the surface, we can prove the more general transformation,

(5) $$\iint_S d\mathbf{S} \times \nabla \mathbf{f} = \oint_C d\mathbf{r} \, \mathbf{f}.$$

between the integrals of *dyadics*. Such integrals require no new definitions; for any dyadic can be reduced to the form $\mathbf{e}_1 \mathbf{f}_1 + \mathbf{e}_2 \mathbf{f}_2 + \mathbf{e}_3 \mathbf{f}_3$ where $\mathbf{f}_1, \mathbf{f}_2, \mathbf{f}_3$ are vector functions.† Thus, on making use of the formula (98.7),

$$(\mathbf{a} \times \mathbf{b}) \times \mathbf{c} = (\mathbf{a} \cdot \mathbf{c}) \, \mathbf{b} - (\mathbf{b} \cdot \mathbf{c}) \, \mathbf{a},$$

we find that

(6) $$\begin{aligned}(\mathbf{r}_u \times \mathbf{r}_v) \times \nabla \mathbf{f} &= (\mathbf{r}_u \times \mathbf{r}_v) \times (\mathbf{e}_1 \mathbf{f}_x + \mathbf{e}_2 \mathbf{f}_y + \mathbf{e}_3 \mathbf{f}_z) \\ &= (x_u \mathbf{r}_v - x_v \mathbf{r}_u) \, \mathbf{f}_x + (y_u \mathbf{r}_v - y_v \mathbf{r}_u) \, \mathbf{f}_y + (z_u \mathbf{r}_v - z_v \mathbf{r}_u) \, \mathbf{f}_z \\ &= \mathbf{r}_v (x_u \mathbf{f}_x + y_u \mathbf{f}_y + z_u \mathbf{f}_z) - \mathbf{r}_u (x_v \mathbf{f}_x + y_v \mathbf{f}_y + z_v \mathbf{f}_z) \\ &= \mathbf{r}_v \mathbf{f}_u - \mathbf{r}_u \mathbf{f}_v \\ &= (\mathbf{r}_v \mathbf{f})_u - (\mathbf{r}_u \mathbf{f})_v. \end{aligned}$$

Hence, on applying Green's theorem, which still holds when P and Q are dyadics, we have

$$\begin{aligned}\iint_S d\mathbf{S} \times \nabla \mathbf{f} &= \iint_D [(\mathbf{r}_v \mathbf{f})_u - (\mathbf{r}_u \mathbf{f})_v] \, du \, dv \\ &= \oint_{C'} \mathbf{r}_u \mathbf{f} \, du + \mathbf{r}_v \mathbf{f} \, dv \\ &= \oint_{C'} (\mathbf{r}_u \, du + \mathbf{r}_v \, dv) \mathbf{f} = \oint_C d\mathbf{r} \, \mathbf{f}.\end{aligned}$$

---

† Cf. Brand, L. *Vector and Tensor Analysis*, Wiley, 1948, p. 138.

The dyadic equation (6) gives two others on taking scalar and vector invariants (§ 108). A comparison of (6) and (4) shows that the former is Stokes' theorem (1). The latter is new (see Prob. 5):

(7)
$$\iint_S \{\nabla\mathbf{f}\cdot\mathbf{n} - \mathbf{n}\,\mathrm{div}\,\mathbf{f}\}\,dS = \oint_C d\mathbf{r}\times\mathbf{f}.$$

Example. Let us verify Stokes' theorem for the vector $\mathbf{f} = [x + y,\ 2x - z,\ y + z]$ taken over the triangle ABC cut from the plane $3x + 2y + z = 6$ by the coordinate planes (Fig. 166b).

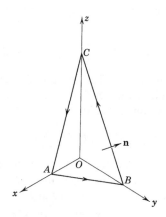

FIG. 166b.   Circuit integral computed by Stokes' theorem

If the parametric equations of the plane are taken as

$$x = u, \qquad y = v, \qquad z = 6 - 3u - 2v,$$

its positive normal n has the direction of

$$\mathbf{r}_u \times \mathbf{r}_v = [1, 0, -3] \times [0, 1, -2] = [3, 2, 1],$$

and the vector element of area is

$$\mathbf{n}\,dS = \mathbf{r}_u \times \mathbf{r}_v\,du\,dv = [3, 2, 1]\,dx\,dy.$$

Since rot $\mathbf{f} = [2, 0, 1]$,

$$\iint_{ABC} \mathbf{n}\cdot\mathrm{rot}\,\mathbf{f}\,dS = 7\iint_{AOB} dx\,dy = 7\cdot 3 = 21;$$

for the area of the triangle AOB (the projection of ABC on the xy-plane) is 3.

Corresponding to the direction of **n**, the positive sense of circuit is ABC. The circuit integral,

$$\oint_{ABC} (x + y)\,dx + (2x - z)\,dy + (y + z)\,dz,$$

is the sum of the line integrals over AB, BC, CA which prove to be 1, 36, -16, respectively. Thus the theorem is verified ($21 = 1 + 36 - 16$).

PROBLEMS

1. Verify Stokes' theorem for $\mathbf{f} = [2y + z, x - z, y - x]$ taken over the triangle ABC cut from the plane $x + y + z = 1$ by the coordinate planes.

[If $A(1, 0, 0)$, $B(0, 1, 0)$, $C(0, 0, 1)$, the line integrals over AB, BC, CA, are $-1/2, 1, 1$; hence $\oint \mathbf{f} \cdot d\mathbf{r} = \frac{3}{2}$ over ABC.]

2. Verify Stokes' theorem for the vector $\mathbf{f} = [z, x, y]$ taken over the half of the sphere $x^2 + y^2 + z^2 = a^2$ lying above the xy-plane.

[$\oint \mathbf{f} \cdot d\mathbf{r}$ around the circle $x^2 + y^2 = a^2$ reduces to $\oint \dot{x}\, dy = \pi a^2$.]

3. Compute $\oint \mathbf{f} \cdot d\mathbf{r}$ around the circle $(x - 1)^2 + y^2 = 1$, $z = 3$ when $\mathbf{f} = [-y, x, 2]$.

4. If $u(x, y, z)$, $v(x, y, z)$ have continuous gradients, show that

$$(8) \qquad \int\int_S \mathbf{n} \cdot \nabla u \times \nabla v\, dS = \oint_C u\, dv,$$

where S is a portion of a surface bounded by a closed curve C. [rot $(u\, \nabla v) = \nabla u \times \nabla v$.]

5. Prove formula (7) by showing that the vector invariant of the dyadic in (6) is $\nabla \mathbf{f} \cdot (\mathbf{r}_u \times \mathbf{r}_v) - (\text{div}\, \mathbf{f})\, \mathbf{r}_u \times \mathbf{r}_v$

6. When $u = \rho^2/2$, $v = \varphi$ in (8), show that

$$\int\int_S \mathbf{n} \cdot \mathbf{e}_3\, dS = \frac{1}{2} \oint_C \rho^2\, d\varphi.$$

[See Ex. 111.1 and Prob. 111.6.]

167. Line Integrals in Space. Consider the line integral,

$$(1) \qquad \int_{\mathbf{r}_0}^{\mathbf{r}} \mathbf{f} \cdot d\mathbf{r} = \int_{t_0}^{t} \mathbf{f} \cdot \frac{d\mathbf{r}}{dt}\, dt,$$

taken over the curve $\mathbf{r} = \mathbf{r}(t)$ from t_0 to t. Theorem 153.2 states that, when rot $\mathbf{f} = \mathbf{0}$ in a rectangular prism R, $\mathbf{f} = \nabla F$, and hence

$$(2) \qquad \int_{\mathbf{r}_0}^{\mathbf{r}} \mathbf{f} \cdot d\mathbf{r} = F(\mathbf{r}) - F(\mathbf{r}_0)$$

is independent of the path. We now extend this result to simply connected regions in which rot $\mathbf{f} = \mathbf{0}$.

THEOREM. *If $\mathbf{f}$ has continuous first partial derivatives and rot $\mathbf{f} = \mathbf{0}$ in a simply connected region, the circuit integral,*

$$\oint \mathbf{f} \cdot d\mathbf{r} = 0,$$

over any regular closed curve in the region.

Proof. In a simply connected region all regular closed curves are *reducible* (§ 153), that is, may be continuously contracted to a point of the region. Thus we can span a surface S over the closed curve; for in shrinking the curve to a point such a surface is generated. We now assume that this surface fulfils the requirements of Stokes' theorem; then

$$\oint \mathbf{f} \cdot d\mathbf{r} = \int_S \mathbf{n} \cdot \text{rot } \mathbf{f} \, dS = 0.$$

Corollary. If $\mathbf{f}(\mathbf{r})$ *has continuous first partial derivatives and* rot $\mathbf{f} = 0$ *in a simply connected region, the line integral,*

$$(3) \qquad\qquad \int_{\mathbf{r}_0}^{\mathbf{r}} \mathbf{f} \cdot d\mathbf{r} = F(\mathbf{r}),$$

is independent of path in this region and defines single-valued function $F(\mathbf{r})$.

The proof is the same as for the corollary of Theorem 160.2. The constant $-F(\mathbf{r}_0)$ in (2) is assimilated into the function $F(\mathbf{r})$ in (3).

With $\mathbf{f} = [P, Q, R]$, we proved in § 153 that, when the line integral (3) was taken over a step path, $F_x = P, F_y = Q, F_z = R$, and hence $\bigtriangledown F = \mathbf{f}$. We can readily deduce this result when the integral in (3) is taken over any path from $\mathbf{r}_0$ to $\mathbf{r}$. Since the integral is independent of the path,

$$F(x+h, y, z) - F(x, y, z) = \int_{(x,y,z)}^{(x+h,y,z)} \mathbf{f} \cdot d\mathbf{r} = \int_x^{x+h} P(t, y, z) \, dt$$

if we integrate along a straight segment from (x, y, z) to $(x+h, y, z)$. Since $P(x, y, z)$ is continuous, the mean-value theorem (117.5) gives

$$\int_x^{x+h} P(t, y, z) \, dt = h \, P(\xi, y, z), \qquad x < \xi < x + h;$$

hence,

$$\frac{\partial F}{\partial x} = \lim_{h \to 0} \frac{F(x+h, y, z) - F(x, y, z)}{h} = P(x, y, z).$$

Similarly, we find $\partial F/\partial y = Q, \partial F/\partial z = R$; and hence $\bigtriangledown F = [P, Q, R] = \mathbf{f}$.

Since the path of integration is immaterial in (3), we may integrate along a straight line from $\mathbf{r}_0$ to $\mathbf{r}$, provided it lies in the simply connected region. The calculation is especially simple when $\mathbf{r}_0 = \mathbf{0}$. We then find, exactly as in § 160, that

$$(4) \qquad\qquad F(\mathbf{r}) = \mathbf{r} \cdot \int_0^1 \mathbf{f}(t\mathbf{r}) \, dt;$$

and, when $\mathbf{f}(\mathbf{r})$ is homogeneous of degree n,

$$(5) \qquad\qquad F(\mathbf{r}) = \frac{\mathbf{r} \cdot \mathbf{f}(\mathbf{r})}{n+1} = \frac{xP + yQ + zR}{n+1}, \qquad n \neq -1.$$

168. Triple Integral over a Rectangular Prism. Let $f(x, y, z)$ be a bounded function in the rectangular prism,

$$R: \quad a \leq x \leq b, \qquad c \leq y \leq d, \qquad e \leq z \leq f.$$

For any net

$$a = x_0 < x_1 < \cdots < x_l = b,$$
$$c = y_0 < y_1 < \cdots < y_m = d,$$
$$e = z_0 < z_1 < \cdots < z_n = e,$$

the planes $x = x_i$, $y = y_j$, $z = z_k$ divide R into lmn prisms of volume

$$r_h = (x_i - x_{i-1})(y_j - y_{j-1})(z_k - z_{k-1}), \qquad h = 1, 2, \cdots, lmn.$$

Let (ξ_h, η_h, ζ_h) be any point in r_h, and consider the sum

$$(1) \qquad \sum_{h=1}^{lmn} f(\xi_h, \eta_h, \zeta_h) r_h,$$

which depends upon the net and the choice of the points (ξ_h, η_h, ζ_h). The *mesh* δ of the net is defined as the greatest dimension of any one of the prisms r_h. If the sum (1) approaches a limit as $\delta \to 0$ (and consequently as $l, m, n \to \infty$) which is independent of the way in which the net and the points (ξ_h, η_h, ζ_h) are chosen, we say that $f(x, y, z)$ is *integrable* over R, and we write

$$(2) \qquad \iiint_R f(x, y, z) \, dx \, dy \, dz = \lim_{\delta \to 0} \sum f(\xi_h, \eta_h, \zeta_h) r_h.$$

When this limit exists, it is called the triple integral of $f(x, y, z)$ over the prism R. Then to any $\varepsilon > 0$ there corresponds a positive number Δ such that

$$(3) \qquad \left| \iiint f(x, y, z) \, dx \, dy \, dz - \sum f(\xi_h, \eta_h, \zeta_h) r_h \right| < \varepsilon \quad \text{when} \quad \delta < \Delta.$$

If m_h, M_h are the g.l.b. and l.u.b. of $f(x, y, z)$ in r_h, we define the *lower sum* s and the upper sum S by

$$s = \sum m_h r_h \qquad S = \sum M_h r_h.$$

These sums depend only upon the net x_i, y_j, z_k and not upon the points (ξ_h, η_h, ζ_h); and

$$s \leq \sum f(\xi_h, \eta_h, \zeta_h) r_h \leq S.$$

Hence, if both s and S approach the same limit L as $\delta \to 0$, then the sum $\sum f(\xi_h, \eta_h, \zeta_h) r_h$ also approaches L.

When $f(x, y, z) = C$, a constant,

$$s = S = \sum C r_h = CV$$

where V is the volume of R; hence,

(4) $$\iiint_R C\, dx\, dy\, dz = CV.$$

If we exclude this simple case, then always

$$mV \leq s < S \leq MV,$$

where m and M are the closest bounds of $f(x, y, z)$ in R.

Now the sums s have a l.u.b. I and the sums S a g.l.b. J; and we can show as in § 156 that

$$\lim_{\delta \to 0} s = I, \qquad \lim_{\delta \to 0} S = J;$$

Moreover $f(x, y, z)$ is integrable in R when, and only when, $I = J$; and then

(5) $$\iiint_R f(x, y, z)\, dx\, dy\, dz = I = J.$$

In order that $I = J$ it is necessary and sufficient that for any $\varepsilon > 0$ there is a net for which $S - s < \varepsilon$. This is always the case when $f(x, y, z)$ is continuous in R; hence any continuous function is integrable. Moreover, $f(x, y, z)$ is integrable if its discontinuities lie on any closed surface S within R, provided that all sections of S by planes parallel to the sides of R are regular, closed curves. Consequently, if $f(x, y, z)$ is continuous on and within such a closed surface, it is integrable in the region R_1 so defined; for, if we define

(6) $$f_1(x, y, z) = \begin{cases} f(x, y, z) & \text{in } R_1, \\ 0 & \text{outside of } R_1, \end{cases}$$

$$\iiint_{R_1} f(x, y, z)\, dx\, dy\, dz = \iiint_R f_1(x, y, z)\, dx\, dy\, dz.$$

The mean-value theorem 159.4 is readily extended to triple integrals. The special case (159.7) now becomes

(7) $$\iiint_{R_1} f(x, y, z)\, dx\, dy\, dz = \mu V, \qquad m < \mu < M,$$

where V is the volume enclosed by the surface S.

169. Element of Volume. Let us change from rectangular to curvilinear coordinates through the equations (§ 111)

(1) $x = F(u, v, w), \qquad y = G(u, v, w), \qquad z = H(u, v, w)$

with a positive Jacobian

(2) $$J = \frac{\partial(x, y, z)}{\partial(u, v, w)} = \mathbf{r}_u \times \mathbf{r}_v \cdot \mathbf{r}_w > 0.$$

Then the tangent vectors $\mathbf{r}_u$, $\mathbf{r}_v$, $\mathbf{r}_w$ to the coordinate curves form a *dextral* set (§ 99). The element of volume is taken as the volume of a prism formed by the differentials $\mathbf{r}_u\, du$, $\mathbf{r}_v\, dv$, $\mathbf{r}_w\, dw$, namely,

(3) $$dV = \mathbf{r}_u \times \mathbf{r}_v \cdot \mathbf{r}_w \, du\, dv\, dw = J\, du\, dv\, dw.$$

When $u = x$, $v = y$, $w = z$, this gives $dV = dx\, dy\, dz$; for

$$\mathbf{r}_x \times \mathbf{r}_y \cdot \mathbf{r}_z = \mathbf{e}_1 \times \mathbf{e}_2 \cdot \mathbf{e}_3 = 1.$$

The curvilinear coordinates are said to be *orthogonal* if the coordinate curves always cut at right angles. Then their tangent vectors $\mathbf{r}_u$, $\mathbf{r}_v$, $\mathbf{r}_w$ are mutually orthogonal:

$$\mathbf{r}_u \cdot \mathbf{r}_v = \mathbf{r}_v \cdot \mathbf{r}_w = \mathbf{r}_w \cdot \mathbf{r}_u = 0.$$

The prism formed by $\mathbf{r}_u$, $\mathbf{r}_v$, $\mathbf{r}_w$ is then rectangular and has the volume $|\mathbf{r}_u|\,|\mathbf{r}_v|\,|\mathbf{r}_w|$ and

(4) $$dV = |\mathbf{r}_u|\,|\mathbf{r}_v|\,|\mathbf{r}_w|\, du\, dv\, dw.$$

If the region R' of uvw-space corresponds to the region R in xyz-space by means of equations (1), the triple integral,

(5) $$\iiint_R f(x, y, z)\, dx\, dy\, dz = \iiint_{R'} f(x, y, z) \left| \frac{\partial(x, y, z)}{\partial(u, v, w)} \right| du\, dv\, dw.\dagger$$

On the right x, y, z must be expressed in terms of u, v, w. This is the analogue of the formula (162.2) for changing variables in a double integral.

Example 1. *Cylindrical Coordinates* (Ex. 111.1) ρ, φ, z are related to x, y, z by the equations

(5) $$x = \rho \cos \varphi, \qquad y = \rho \sin \varphi, \qquad z = z.$$

Since

(6) $$J = \frac{\partial(x, y, z)}{\partial(\rho, \varphi, z)} = \begin{vmatrix} \cos \varphi & -\rho \sin \varphi & 0 \\ \sin \varphi & \rho \cos \varphi & 0 \\ 0 & 0 & 1 \end{vmatrix} = \rho,$$

the element of volume is

(7) $$dV = \rho\, d\rho\, d\varphi\, dz.$$

Cylindrical coordinates are orthogonal.

Example 2. Spherical coordinates (Ex. 111.2) r, θ, φ are related to x, y, z by the equations

(8) $$x = r \sin \theta \cos \varphi, \qquad y = r \sin \theta \sin \varphi, \qquad z = r \cos \theta.$$

Since

$$\frac{\partial(x, y, z)}{\partial(r, \theta, \varphi)} = \begin{vmatrix} \sin \theta \cos \varphi & r \cos \theta \cos \varphi & -r \sin \theta \sin \varphi \\ \sin \theta \sin \varphi & r \cos \theta \sin \varphi & r \sin \theta \cos \varphi \\ \cos \theta & -r \sin \theta & 0 \end{vmatrix},$$

(9) $$J = \frac{\partial(x, y, z)}{\partial(r, \theta, \varphi)} = r^2 \sin \theta,$$

---

† This formula applies in general, irrespective of the sign of J.

the element of volume is

(10) $$dV = r^2 \sin \theta \, dr \, d\theta \, d\varphi.$$

Spherical coordinates are orthogonal. The components of $\mathbf{r}_r, \mathbf{r}_\theta, \mathbf{r}_\varphi$ form the columns of the Jacobian, and

$$|\mathbf{r}_r| = 1, \qquad |\mathbf{r}_\theta| = r, \qquad |\mathbf{r}_\varphi| = r \sin \theta$$

thus verifying (10).

PROBLEMS

1. Compute $\iiint xyz \, dx \, dy \, dz$ over the tetrahedron bounded by the coordinate planes

and $$\frac{x}{a} + \frac{y}{b} + \frac{z}{c} = 1 \ (a, b, c > 0).$$

2. Show that $\iiint f(x, y, z) \, dx \, dy \, dz$ over the ellipsoid $\dfrac{x^2}{a^2} + \dfrac{y^2}{b^2} + \dfrac{z^2}{c^2} \leqq 1$

$= abc \iiint f(ax, by, cz) \, dx \, dy \, dz$ over the sphere $x^2 + y^2 + z^2 \leqq 1$.

3. The volume of a unit sphere is $4\pi/3$, and its moment of inertia about a diameter is $8\pi/15$. Use Prob. 2 to find the volume of an ellipsoid and its moments of inertia I_x, I_y, I_z.

4. Compute $\iiint xyz \, dx \, dy \, dz$ over the ellipsoid of Prob. 2.

170. Triple and Repeated Integrals. In much the same way that we converted a double integral into a repeated integral (in two ways), a triple integral can be converted into a repeated integral (in 3! or six ways). Consider the integral of a continuous function $f(x, y, z)$ over the region R_1 bounded by a closed surface S. Enclose S in the smallest possible rectangular prism

$$R: \quad a \leqq x \leqq b, \qquad c \leqq y \leqq d, \qquad e \leqq z \leqq f$$

with sides parallel to the coordinate planes. Define $f_1(x, y, z)$ as in 168.6; then

(1) $$I = \iiint_{R_1} f(x, y, z) \, dx \, dy \, dz = \iiint_R f_1(x, y, z) \, dx \, dy \, dz$$

$$= \int_a^b dx \int_c^d dy \int_e^f f_1(x, y, z) \, dz$$

to give but one of six possible repeated integrals.

We must now adjust the limits to accord with the definition of f_1. Let the surface S project into the area Z of the xy-plane bounded by the closed curve C. If (x, y) is a point of Z, let the line parallel to the z-axis through (x, y) meet S in two points $z_1(x, y)$, $z_2(x, y)$. Since $f_1 = 0$ outside of S, the last integral of (1) becomes

$$\iint_Z dx \, dy \int_{z_1(x,y)}^{z_2(x,y)} f(x, y, z) \, dz.$$

Next let the line parallel to the y-axis through (x, y) meet the curve C in two points $y_1(x)$ and $y_2(x)$; we thus obtain

$$(2) \qquad I = \int_a^b dx \int_{y_1(x)}^{y_2(x)} dy \int_{z_1(x,y)}^{z_2(x,y)} f(x, y, z).$$

The first integration is with respect to z with x and y held constant; after the limits are inserted, we next integrate the resulting function with respect to y, holding x constant; and, after the limits of y are inserted, we finally integrate with respect to x between constant limits. We may think of the first integration as a summation of elements forming a prismatic *bar* parallel to the z-axis; the next integration sums these elementary bars into a *plate* parallel to the yz-plane; the last integration sums these elementary plates into the desired volume integral.

Example 1. The integral I of $f(x, y, z)$ over the triangular pyramid V bounded by the planes $x = 0$, $y = 0$, $z = 0$ and

$$x/a + y/b + z/c = 1, \qquad a, b, c > 0$$

may be computed as

$$I = \int_0^a dx \int_0^{b(1-x/a)} dy \int_0^{c(1-x/a-y/b)} f(x, y, z)\, dz$$

If $f = 1$, the integral gives the volume

$$V = c \int_0^a dx \int_0^{b(1-x/a)} \left(1 - \frac{x}{a} - \frac{y}{b}\right) dy$$

$$= c \int_0^a \left[-\frac{b}{2} \left(1 - \frac{x}{a} - \frac{y}{b}\right)^2 \right]_0^{b(1-x/a)} dx$$

$$= \frac{bc}{2} \int_0^a \left(1 - \frac{x}{a}\right)^2 dx$$

$$= -\frac{abc}{6} \left(1 - \frac{x}{a}\right)^3 \Big|_0^a = \frac{1}{6} abc.$$

If $f = x$, the integral gives Vx^*, where x^* refers to the centroid of V. The first two integrations are the same as before; hence,

$$Vx^* = \frac{bc}{2} \int_0^a x \left(1 - \frac{x}{a}\right)^2 dx = \frac{1}{24} a^2bc, \qquad x^* = \frac{1}{4} a.$$

Example 2. Find the volume common to two circular cylinders of radius a whose axes are the x- and y-axes (Fig. 170).

Consider the part in the first octant. The plane $x = y$ divides it in half; hence its volume is

$$2 \int_0^a dx \int_0^x dy \int_0^{\sqrt{a^2-x^2}} dz = 2 \int_0^a x \sqrt{a^2 - x^2}\, dx = \frac{2}{3} a^3.$$

The total volume is therefore $16a^3/3$.

Example 3. *Problem of Viviani.* Find the volume of the cylinder $x^2 + y^2 = ay$ included within the sphere $x^2 + y^2 + z^2 = a^2$.   (Cf. Ex. 164.3.)

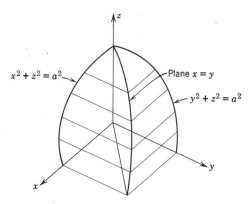

FIG. 170.   Intersecting cylinders

In cylindrical coordinates the equations of the cylinder and sphere are

$$\rho = a \sin \varphi, \qquad \rho^2 + z^2 = a^2,$$

and $dV = \rho \, d\rho \, d\varphi \, dz$.   Hence,

$$\frac{1}{4} V = \int_0^{\pi/2} d\varphi \int_0^{a \sin \varphi} \rho \, d\rho \int_0^{\sqrt{a^2 - \rho^2}} dz = \int_0^{\pi/2} d\varphi \int_0^{a \sin \varphi} \sqrt{a^2 - \rho^2} \, \rho \, d\rho$$

$$= \int_0^{\pi/2} d\varphi \left[-\frac{1}{3} (a^2 - \rho^2)^{3/2} \right]_0^{a \sin \varphi} = \frac{1}{3} a^3 \int_0^{\pi/2} (1 - \cos^3 \varphi) \, d\varphi$$

$$= \frac{1}{3} a^3 \int_0^{\pi/2} (1 - \cos \varphi + \sin^2 \varphi \cos \varphi) \, d\varphi$$

$$= \frac{1}{3} a^3 \left[\varphi - \sin \varphi + \frac{1}{3} \sin^3 \varphi \right]_0^{\pi/2},$$

$$V = \frac{4}{3} a^3 \left(\frac{\pi}{2} - \frac{2}{3} \right) = \left(\frac{2}{3} \pi - \frac{8}{9} \right) a^3.$$

Thus the hemisphere with the hole cut out by the cylinder has a residual volume of $\frac{8}{9} a^3$, an example of a solid with curved boundaries whose volume does not involve irrationalities.

PROBLEMS

1. Find the volume between the cylinder $x^2 + y^2 - 2ax = 0$, the paraboloid $x^2 + y^2 - az = 0$, and plane $z = 0$.

2. Find the volume in the first octant bounded by the cylinder $x^2 + y^2 = a^2$ and the planes $z = 0$, $x = 0$, $y - z = 0$.

3. A circular hole of radius b is drilled through a sphere of radius a along a diameter as axis. Show that the volume remaining is $\frac{4}{3}\pi (a^2 - b^2)^{3/2}$.

4. A plane cuts off a cap of height h from a sphere of radius r. Show that its volume is $\frac{1}{3}\pi h^2(3r - h)$.

5. If k is the radius of gyration of a solid of revolution about its axis, show that $k^2 = 3r^2/10, 4r^2/10, 5r^2/10$ for a cone, sphere, and cylinder of radius r.

6. Find the volume included within both sphere $x^2 + y^2 + z^2 = a^2$ and cylinder $x^2 + y^2 - ax = 0$.

$$\left[\frac{1}{4} V = \int_0^{1/2\,\pi} \int_a^{a\cos\theta} \sqrt{a^2 - r^2}\, r\, dr\, d\theta \right].$$

7. Solve Ex. 170.1 by making the change of variable $x = au, y = bv, z = cw$.

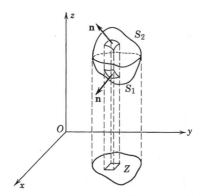

FIG. 171. Region bounded by a closed surface

171. Divergence Theorem. The basic transformation of volume to surface integrals is given by the

DIVERGENCE THEOREM (Gauss). *If the vector function* $\mathbf{f(r)}$ *has continuous first partial derivatives in the region* V *bounded by closed surface* S *over which the unit external normal* $\mathbf{n}$ *is sectionally continuous, then the volume integral of* div $\mathbf{f}$ *within* S *is equal to the external surface integral of* $\mathbf{f}$ *over* S:

(1) $$\iiint_V \operatorname{div} \mathbf{f}\, dV = \iint_S \mathbf{n} \cdot \mathbf{f}\, dS.$$

Proof. If $\mathbf{f} = [P, Q, R]$, from (109.3)

(2) $$\operatorname{div} \mathbf{f} = \frac{\partial P}{\partial x} + \frac{\partial Q}{\partial y} + \frac{\partial R}{\partial z}.$$

Consider, first, a surface S which is cut in at most two points by a line parallel to the z-axis (Fig. 171); denote them by (x, y, z_1) and (x, y, z_2) where $z_1 < z_2$. Then S has a lower portion S_1 consisting of the points

(x, y, z_1) and an upper portion S_2 consisting of the points (x, y, z_2). The points for which $z_1 = z_2$ form a closed curve separating S_1 from S_2. Now, if S_1 and S_2 project into the area Z of the xy-plane,

$$\iiint_V \frac{\partial R}{\partial z} dx\, dy\, dz = \iint_Z dx\, dy \int_{z_1}^{z_2} \frac{\partial R}{\partial z} dz$$

$$= \iint_Z R(x, y, z_2)\, dx\, dy - \iint_Z R(x, y, z_1)\, dx\, dy$$

$$= \iint_{Z\pm} R(x, y, z)\, dx\, dy;$$

hence, from (165.5)′,

$$(3) \qquad\qquad \iiint_V \frac{\partial R}{\partial z} dx\, dy\, dz = \iint_D RC\, du\, dv.$$

Since $\cos\gamma > 0$ $(C > 0)$ over S_2, $\cos\gamma < 0$ $(C < 0)$ over S_1, the parameters u, v must be named so that $\mathbf{r}_u \times \mathbf{r}_v$ is an *external* normal to the surface.

This formula is also valid when S is bounded laterally by a part of a cylinder parallel to the z-axis and separating S_1 from S_2, for $\cos\gamma = 0$ and hence $C = 0$ over the cylinder, so that it contributes nothing to the integral over S_1.

We may now remove the condition that S is cut in only two points by a parallel to the z-axis. For, if we divide V into parts bounded by surfaces which satisfy this condition and apply formula (3) to each part and add the results, the volume integrals will combine to the left member of (3); the surface integrals over the boundaries between the parts cancel (for each appears twice but with opposed values of $\mathbf{n}$), whereas the remaining surface integrals combine to the right member of (1).

Finally we may extend (1) to regions bounded by two or more closed surfaces (regions with cavities), by introducing additional surfaces so that all parts of V are bounded by a single closed surface; the surface integrals over these will cancel in pairs as before.

We may transform the other terms of the volume integral in the same way and obtain the formulas

$$(4) \qquad\qquad \iiint_V \frac{\partial P}{\partial x} dx\, dy\, dz = \iint_D PA\, du\, dv,$$

$$(5) \qquad\qquad \iiint_V \frac{\partial Q}{\partial y} dx\, dy\, dz = \iint_D QB\, du\, dv.$$

On adding (3), (4), (5), we get

$$\iiint_V \operatorname{div} \mathbf{f} \, dx \, dy \, dz = \iint_D (PA + QB + RC) \, du \, dv,$$

which is the same as (1) in view of (165.3).

Volume and surface integrals are sometimes denoted by a single integral sign when the differential dV or dS indicates their character. Moreover $\oint \cdots dS$ is used to denote integration over a closed surface. In this notation the divergence theorem becomes

(1)′
$$\int_V \operatorname{div} \mathbf{f} \, dV = \oint_S \mathbf{n} \cdot \mathbf{f} \, dS.$$

THEOREM 2. *If the scalar $f(x, y, z)$ has continuous first partial derivatives in a region V,*

(6)
$$\int_V \nabla f \, dV = \oint_S \mathbf{n} f \, dS.$$

Proof. The vector integrals forming the two members of (6) have their x-, y-, and z-components the same by virtue of (4), (5), and (3); for

$$\nabla f = [f_x, f_y, f_z], \qquad \mathbf{n} \, dS = [A, B, C] \, du \, dv.$$

THEOREM 3. *If the vector $\mathbf{f}(x, y, z)$ has continuous first partial derivatives in a region V,*

(7)
$$\int_V \nabla \mathbf{f} \, dV = \oint_S \mathbf{n} \, \mathbf{f} \, dS.$$

Proof. Let $\mathbf{f} = [P, Q, R]$; then from (6)

$$\int \nabla P \, dV = \oint \mathbf{n} P \, dS,$$

$$\int \nabla Q \, dV = \oint \mathbf{n} Q \, dS,$$

$$\int \nabla R \, dV = \oint \mathbf{n} R \, dS,$$

and hence

$$\int (\nabla P \, \mathbf{e}_1 + \nabla Q \, \mathbf{e}_2 + \nabla R \, \mathbf{e}_3) \, dV = \oint \mathbf{n} \, (P\mathbf{e}_1 + Q\mathbf{e}_2 + R\mathbf{e}_3) \, dS.$$

Now $\mathbf{f} = P\mathbf{e}_1 + Q\mathbf{e}_2 + R\mathbf{e}_3$, and from (107.4)

$$\nabla \mathbf{f} = \mathbf{e}_1(P_x\mathbf{e}_1 + Q_x\mathbf{e}_2 + R_x\mathbf{e}_3) + \cdots$$
$$= (\mathbf{e}_1 P_x + \mathbf{e}_2 P_y + \mathbf{e}_3 P_z)\mathbf{e}_1 + \cdots$$
$$= \nabla P \, \mathbf{e}_1 + \nabla Q \, \mathbf{e}_2 + \nabla R \, \mathbf{e}_3.$$

Thus (7) is proved. The dyad $\mathbf{n} f$ on the right must not be reversed.

The dyadic equation (7) gives the divergence theorem (1)' on equating the scalar invariants of both members. On equating the vector invariants, we get the new theorem:

(8) $$\int_V \operatorname{rot} \mathbf{f} \, dV = \oint_S \mathbf{n} \times \mathbf{f} \, dS.$$

Since both dot and cross multiplication are distributive with respect to addition, the process of taking either invariant commutes with addition and also with integration.

Example 1. When $\mathbf{f} = \mathbf{r}$, div $\mathbf{r} = 3$; then from (1)

(9) $$V = \frac{1}{3} \oint_S \mathbf{n} \cdot \mathbf{r} \, dS = \frac{1}{3} \oint_S p \, dS,$$

where p is the perpendicular from O on the tangent plane to S at the end point of $\mathbf{r}$. Thus the volume within S is expressed as a surface integral over S.

For a *cone* of base B, altitude h, take O at the vertex. Then $p = 0$ over the lateral surface, $p = h$ over the base, and $V = \frac{1}{3}Bh$.

For a *sphere* of radius a, take O at the center. Then $p = a$ and $V = \frac{1}{3}aS = \frac{4}{3}\pi a^3$.

Example 2. Put $f = 1$ in (7); then

(10) $$\oint_S \mathbf{n} \, dS = 0.$$

Put $\mathbf{f} = \mathbf{r}$ in (8); since rot $\mathbf{r} = 0$,

(11) $$\oint_S \mathbf{r} \times \mathbf{n} \, dS = 0.$$

From (10) and (11) we conclude that the closed surface is in equilibrium under any system of normal pressures $-\mathbf{n}p$ of constant magnitude p; for their vector sum and vector moment about O are zero.

Example 3. Put $f = \frac{1}{2}r^2$ in (7); then $\nabla f = \mathbf{r}$, and the left member,

$$\int \mathbf{r} \, dV = V\mathbf{r}^*,$$

where $\mathbf{r}^*$ locates the centroid of V; hence,

(12) $$V\mathbf{r}^* = \frac{1}{2} \oint_S r^2 \mathbf{n} \, dS.$$

PROBLEMS

1. Verify the divergence theorem for the vector $\mathbf{f} = [x^2, y^2, z^2]$ taken over the cube $0 \le x, y, z \le 1$.

2. Find $\int$ div $\mathbf{f} \, dV$ for the vector $\mathbf{f} = [x, -y, 2z]$ within the sphere

$$x^2 + y^2 + (z - 1)^2 = 1.$$

3. If $\mathbf{f} = [u(x, y), v(x, y), 0]$, prove that

$$\oint_C \mathbf{f} \times d\mathbf{r} = \mathbf{e}_3 \int\int_A \operatorname{div} \mathbf{f} \, dx \, dy.$$

where C is a closed curve in the xy-plane enclosing the region A.

[Apply the divergence theorem to $\mathbf{f}$ over a cylinder of base A and height $z = 1$.]

4. Find a vector $\mathbf{f} = g(r)\,\mathbf{r}$ such that $\operatorname{div}\mathbf{f} = r^m$ $(m \neq -3)$. Prove that

$$\int r^m \, dV = \frac{1}{m+3} \oint r^m \, \mathbf{r} \cdot \mathbf{n} \, dS.$$

5. If φ and ψ are scalar point functions having continuous derivatives of the first and second orders, prove *Green's identities*:

(13)
$$\int \nabla\varphi \cdot \nabla\psi \, dV + \int \varphi \, \nabla^2\psi \, dV = \oint \varphi \frac{d\psi}{dn} \, dS;$$

(14)
$$\int (\varphi\nabla^2\psi - \psi\nabla^2\varphi) \, dV = \oint \left(\varphi\frac{d\psi}{dn} - \psi\frac{d\varphi}{dn} \right) dS.$$

The normal derivatives $d\varphi/dn$, $d\psi/dn$ are in the direction of the external normal to the bounding surface.

[Apply the divergence theorem to $\mathbf{f} = \varphi\nabla\psi$].

6. Prove that, if $m \neq -1$,

$$\int r^{m-1} \mathbf{r} \, dV = \frac{1}{m+1} \oint r^{m+1} \mathbf{n} \, dS.$$

172. Solenoidal Vectors. A vector function $\mathbf{f}(\mathbf{r})$ is said to be *solenoidal* in a region R if $\operatorname{div}\mathbf{f} = 0$ in R.

THEOREM 1. *The vector* $\operatorname{rot}\mathbf{g}$ *is solenoidal in any region in which* $\mathbf{g}$ *has continuous second partial derivatives.*

Proof. Let $\mathbf{g} = [p, q, r]$; then

(1)
$$\operatorname{rot}\mathbf{g} = [r_y - q_z, p_z - r_x, q_x - p_y],$$
$$\operatorname{div}\operatorname{rot}\mathbf{g} = (r_y - q_z)_x + (p_z - r_x)_y + (q_x - p_y)_z = 0,$$

for the mixed derivatives cancel in pairs

Corollary. The vector $\nabla u \times \nabla v$ is solenoidal; for

(2)
$$\nabla u \times \nabla v = \operatorname{rot}(u \, \nabla v). \qquad \text{(Prob. 109.1)}$$

THEOREM 2. *Every solenoidal vector can be put in the form*

(3)
$$\mathbf{f} = \operatorname{rot}\mathbf{g}.$$

Proof. If $\mathbf{f} = [P, Q, R]$,

(4)
$$\operatorname{div}\mathbf{f} = P_x + Q_y + R_z = 0.$$

We shall find a vector $\mathbf{g}_0 = [p, q, 0]$ such that $\mathbf{f} = \operatorname{rot}\mathbf{g}_0$; that is,

(5)
$$P(x, y, z) = -\frac{\partial q}{\partial z}, \quad Q(x, y, z) = \frac{\partial p}{\partial z}, \quad R(x, y, z) = \frac{\partial q}{\partial x} - \frac{\partial p}{\partial y}.$$

The first two equations of (5) are satisfied when

(6) $$p = \int_{z_0}^{z} Q(x, y, t)\, dt, \qquad q = -\int_{z_0}^{z} P(x, y, t)\, dt + \alpha(x, y),$$

where $\alpha(x, y)$ is an arbitrary function of x, y. In order that these values of p and q satisfy the third equation of (5),

$$-\int_{z_0}^{z} \{P_x(x, y, t) + Q_y(x, y, t)\}\, dt + \frac{\partial \alpha}{\partial x} = R$$

or, in view of (4),

$$\int_{z_0}^{z} R_t(x, y, t)\, dt + \frac{\partial \alpha}{\partial x} = R(x, y, z).$$

When the integration is performed, this reduces to

$$\frac{\partial \alpha}{\partial x} = R(x, y, z_0),$$

so that we may take

$$\alpha(x, y) = \int R(x, y, z_0)\, dx.$$

Thus the vector

(7) $$\mathbf{g}_0 = \left[\int_{z_0}^{z} Q(x, y, t)\, dt, \quad \int R(x, y, z_0)\, dx - \int_{z_0}^{z} P(x, y, t)\, dt, \quad 0 \right]$$

is a particular solution of the equation $\mathbf{f} = \operatorname{rot} \mathbf{g}$.

If $\mathbf{g}$ is any other solution, $\operatorname{rot}(\mathbf{g} - \mathbf{g}_0) = 0$, and hence $\mathbf{g} - \mathbf{g}_0 = \nabla\varphi$; thus the general solution is

(8) $$\mathbf{g} = \mathbf{g}_0 + \nabla\varphi$$

where φ is an arbitrary scalar function having second partial derivatives.

In mathematical physics the solenoidal vector $\mathbf{f} = \operatorname{rot} \mathbf{g}$ is said to be derived from the *vector potential* $\mathbf{g}$.

We shall also sketch a vector solution of the problem.† If $\operatorname{div} \mathbf{f} = 0$, the divergence theorem shows that

(9) $$\oint_S \mathbf{n} \cdot \mathbf{f}\, dS = 0$$

over any closed surface S. Let S_1 denote the lateral surface of a cone with vertex at O and having the closed curve C as directrix and boundary

---

† See Brand, L., The Vector Potential of a Solenoidal Vector, *Am. Math. Monthly*, vol. 57, 1950, pp. 161–167.

(Fig. 172*a*). We now apply (9) to the *closed* surface formed by S_1 and the "base" S_2 of the cone (a portion of a surface bounded by C). By Stokes' theorem, S_2 contributes

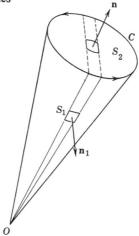

Fig. 172*a*.   Closed conical region

(10) $$\int_{S_2} \mathbf{n} \cdot \mathbf{f} \, dS = \int_{S_2} \mathbf{n} \cdot \mathrm{rot}\, \mathbf{g} \, dS = \oint_C \mathbf{g} \cdot \boldsymbol{\tau} \, ds$$ (166.1)

to the integral (9). If $\mathbf{r} = \mathbf{r}(s)$ is the equation of C, S_1 has the vector equation

$$\mathbf{r}_1 = t\,\mathbf{r}(s), \qquad 0 < t \leq 1;$$

and its vector element of area in the direction of the external normal $\mathbf{n} = \mathbf{n}_1$ is

$$\mathbf{n}\, dS = \frac{\partial \mathbf{r}_1}{\partial s} \times \frac{\partial \mathbf{r}_1}{\partial t} \, ds\, dt = t\,\boldsymbol{\tau} \times \mathbf{r}(s) \, ds\, dt.$$

Hence the contribution of S_1 to (9) is

(11) $$\int_{S_1} \mathbf{n} \cdot \mathbf{f} \, dS = \int\int_{S_1} \mathbf{f}(t\mathbf{r}) \cdot \boldsymbol{\tau} \times \mathbf{r}\, t\, dt\, ds = \oint_C \left[\int_0^1 \mathbf{r} \times \mathbf{f}(t\mathbf{r}) t \, dt \right] \cdot \boldsymbol{\tau} \, ds$$

Since the sum of the integrals (10) and (11) is zero, we have

$$\oint_C \left[\mathbf{g} + \int_0^1 \mathbf{r} \times \mathbf{f}(t\mathbf{r})\, t \, dt \right] \cdot \boldsymbol{\tau} \, ds = 0.$$

As this holds for an arbitrary closed curve C, the vector in brackets is zero; hence,

(12) $$\mathbf{g} = -\,\mathbf{r} \times \int_0^1 \mathbf{f}(t\mathbf{r})\, t \, dt$$

whenever the integral exists.

When $\mathbf{f}(\mathbf{r})$ is homogeneous of degree n, $\mathbf{f}(t\mathbf{r}) = t^n\,\mathbf{f}(\mathbf{r})$, and

(13) $$\mathbf{g} = -\mathbf{r} \times \mathbf{f}(\mathbf{r}) \int_0^1 t^{n+1}\,dt = \frac{\mathbf{f} \times \mathbf{r}}{n+2}, \qquad n > -2.$$

Example 1. The vector

$$\mathbf{f} = \mathbf{a} \times \mathbf{r} = [a_2 z - a_3 y,\ a_3 x - a_1 z,\ a_1 y - a_2 x]$$

is solenoidal. Since it is homogeneous of degree 1, its vector potential is, from (13),

$$\mathbf{g} = \tfrac{1}{3}(\mathbf{a} \times \mathbf{r}) \times \mathbf{r}.$$

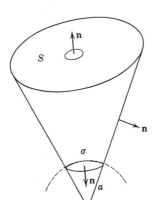

Fig. 172b.   Solid angle

Example 2. When $\mathbf{f} = [y,\ z,\ x]$, div $\mathbf{f} = 0$. The solution (7) gives

$$\mathbf{g}_0 = [\tfrac{1}{2}z^2,\ \tfrac{1}{2}x^2 - yz,\ 0].$$

while from (13)

$$\mathbf{g} = \frac{\mathbf{f} \times \mathbf{r}}{3} = \frac{1}{3}\,[z^2 - xy,\ x^2 - yz,\ y^2 - zx].$$

The difference

$$\mathbf{g}_0 - \mathbf{g} = \tfrac{1}{3}[\tfrac{1}{2}z^2 + xy,\ \tfrac{1}{2}x^2 - 2yz,\ -y^2 + zx]$$

is irrotational, namely,

$$\tfrac{1}{3}\nabla\,(\tfrac{1}{2}xz^2 + \tfrac{1}{2}x^2 y - y^2 z).$$

Example 3. *Solid Angle.* The rays from a point O through the points of a closed curve generate a cone; and the surface of a unit sphere about O intercepted by this cone is called the *solid* angle Ω of the cone.

Let us apply the divergence theorem to the vector

$$\mathbf{f} = -\nabla\,\frac{1}{r} = \frac{\mathbf{r}}{r^3}$$

in the region interior to a cone of solid angle Ω which is limited externally by a surface S, internally by a portion σ of a small sphere of radius a about O (Fig. 172b). Within this region $\mathbf{f}$ has continuous first partial derivatives, and from Ex. 111.2

$$\text{div } \mathbf{f} = -\nabla^2\,\frac{1}{r} = 0 \qquad \text{and hence} \qquad \oint \mathbf{n} \cdot \mathbf{f}\,dS = 0.$$

Over the small sphere the external normal $\mathbf{n} = -\mathbf{r}/a$; while over the conical surface $\mathbf{n} \cdot \mathbf{r} = 0$; hence,

$$\oint_S \mathbf{n} \cdot \mathbf{f} \, dS = \int_S \frac{\mathbf{n} \cdot \mathbf{r}}{r^3} \, dS - \int_\sigma \frac{\mathbf{r}}{a} \cdot \frac{\mathbf{r}}{a^3} \, dS = \int_S \frac{\mathbf{n} \cdot \mathbf{r}}{r^3} \, dS - \frac{\sigma}{a^2} = 0.$$

But, since $\sigma/a^2 = \Omega/1^2$, we have

(14) $$\Omega = \int_S \frac{\mathbf{n} \cdot \mathbf{r}}{r^3} \, dS.$$

If S is closed surface, we have

(15) $$\oint \frac{\mathbf{n} \cdot \mathbf{r}}{r^3} \, dS = \begin{cases} 4\pi, & O \text{ inside of } S, \\ 0, & O \text{ outside of } S. \end{cases}$$

When O is outside of S, $\mathbf{f}$ has continuous partial derivatives throughout its interior, and (15) follows at once from (9). In this case the elements of solid angle $d\Omega = \mathbf{n} \cdot \mathbf{r} \, dS/r^3$ corresponding to the same ray cancel in pairs.

Example 4. *Electric Point Charges.* The force $\mathbf{E}$ exerted by an electric charge e at O upon a unit change at P is called the *electric intensity* at P due to the charge e. By Coulomb's law

(16) $$\mathbf{E} = \frac{e}{r^2} \mathbf{R} = -\nabla \frac{e}{r}, \ddagger$$

so that $\mathbf{E}$ has the potential function $\varphi = e/r$ (§ 152). The vector $\mathbf{E}$ is both irrotational and solenoidal; for

$$\text{rot } \mathbf{E} = 0, \qquad \text{div } \mathbf{E} = -e\nabla^2 \frac{1}{r} = 0.$$

If S is a closed surface, we have from (14),

(17) $$\oint \mathbf{n} \cdot \mathbf{E} \, dS = \begin{cases} 4\pi e, & e \text{ inside of } S, \\ 0, & e \text{ outside of } S. \end{cases}$$

The electric intensity $\mathbf{E}$ due to a system of point charges $e_1, e_1, \cdots, e_n$ is the vector sum of their separate intensities; and, if the charges are within a closed surface, we have from (17)

(18) $$\oint \mathbf{n} \cdot \mathbf{E} \, dS = 4\pi \, \Sigma \, e_i.$$

The normal flux of electric intensity through a closed surface is equal to 4π times the sum of the enclosed charges.

PROBLEMS

1. Show that a constant vector $\mathbf{c}$ has the scalar potential $\varphi = -\mathbf{c} \cdot \mathbf{r}$ and the vector potential $\mathbf{g} = \frac{1}{2}(\mathbf{c} \times \mathbf{r})$.

2. Find the vector potential of $\mathbf{f} = (\mathbf{r} - \mathbf{a}) \times (\mathbf{r} - \mathbf{b})$.

---

‡ In a vacuum, if the charges are measured in statcoulombs and the distance in centimeters, the force is given in dynes.

3. Find the vector potential of the given solenoidal vector $\mathbf{f}$ from (7) and also from (13), and verify that $\mathbf{g} - \mathbf{g}_0$ is irrotational.

(a) $\mathbf{f} = [x, x, -z]$;   (b) $\mathbf{f} = [z - y, x - z, y - x]$;

(c) $\mathbf{f} = [xy, -y^2, yz]$.   (d) $\mathbf{f} = [z, x, y]$.

4. Find the vector potential of $\mathbf{f}$ from (13) when

(a) $\mathbf{f} = r^3 \mathbf{c} \times \mathbf{r}$;   (b) $\mathbf{f} = [x, 1, -z]$.

173. Line, Surface, and Volume Integrals. A line integral over a curve C, $\mathbf{r} = \mathbf{r}(t)$ ($t_0 \leq t \leq t_1$), may be computed as an ordinary integral:

$$(1) \qquad \int_C \mathbf{f}(\mathbf{r}) \cdot d\mathbf{r} = \int_{t_0}^{t_1} \mathbf{f}(\mathbf{r}) \cdot \frac{d\mathbf{r}}{dt} \, dt.$$

If $\mathbf{f}(\mathbf{r})$ has continuous first partial derivatives in a simply connected region, a necessary and sufficient condition that the line integral be independent of the path is rot $\mathbf{f} = 0$. Then the *irrotational* vector $\mathbf{f} = \nabla F$, and

$$(2) \qquad \int_{\mathbf{r}_0}^{\mathbf{r}_1} \mathbf{f}(\mathbf{r}) \cdot d\mathbf{r} = F(\mathbf{r}_1) - F(\mathbf{r}_0).$$

The scalar F may be conveniently found by integrating $\mathbf{f}(\mathbf{r})$ along a step path from P_0 to P, or along the straight line $P_0 P$. The latter method gives

$$(3, 4) \qquad F = \mathbf{r} \cdot \int_0^1 \mathbf{f}(t\mathbf{r}) \, dt; \quad \text{and} \quad F = \frac{\mathbf{r} \cdot \mathbf{f}(\mathbf{r})}{n + 1}$$

when $\mathbf{f}(\mathbf{r})$ is homogeneous of degree $n(\neq -1)$. The function $-F$ is called the *scalar potential* of $\mathbf{f}$.

If $f(x, y)$ is bounded in a closed rectangle $R: a \leq x \leq b, c \leq y \leq d$, we define

$$(5) \qquad \iint_R f(x, y) \, dx \, dy = \lim_{\delta \to 0} \sum_{k=1}^{mn} f(\xi_k, \eta_k) r_k;$$

the net x_i, y_j over R forms mn subrectangles r_k; ξ_k, η_k is any point of r_k, and δ, the greatest dimension of r_k, is the *mesh* of the net. If this limit exists, $f(x, y)$ is *integrable* over R; for this it is necessary and sufficient that for any $\varepsilon > 0$ there is a net for which $S - s < \varepsilon$.

A continuous function is integrable in a rectangle. Moreover, if $f(x, y)$ is continuous within and on any simple, closed rectifiable curve, it is integrable in this region. In either case the double integral may be computed in two ways as a repeated integral.

To change variables in a double integral, $x = F(u, v)$, $y = G(u, v)$, use

$$(6) \qquad \iint_R f(x, y) \, dx \, dy = \iint_{R'} f(x, y) \left| \frac{\partial(x, y)}{\partial(u, v)} \right| du \, dv.$$

The area S of a surface $\mathbf{r} = \mathbf{r}(u, v)$ corresponding to the region D in the uv-plane is defined as

$$(7) \qquad S = \int\int_D |\mathbf{r}_u \times \mathbf{r}_v|\, du\, dv.$$

The surface integral of $\mathbf{f}(\mathbf{r})$ over the positive side of surface (indicated by $\mathbf{r}_u \times \mathbf{r}_v$) is

$$(8) \qquad \int\int_S \mathbf{f} \cdot d\mathbf{S} = \int\int_D \mathbf{f} \cdot \mathbf{r}_u \times \mathbf{r}_v\, du\, dv$$

When $\mathbf{f}(\mathbf{r})$ has continuous first partial derivatives,

$$(9) \qquad \int\int_S \operatorname{rot}\mathbf{f} \cdot d\mathbf{S} = \oint_C \mathbf{f} \cdot d\mathbf{r} \qquad \text{(Stokes' theorem),}$$

where the circuit integral over C, the boundary of S, is taken in the positive sense. In the xy-plane, with $f = [P, Q, 0]$, this becomes Green's theorem,

$$(10) \qquad \int\int_R (Q_x - P_y)\, dx\, dy = \oint_C P\, dx + Q\, dy.$$

The definition of a triple integral is an obvious extension of (5). The criterion for integrability is the same; and a continuous function $f(x, y, z)$ is integrable in a rectangular prism or in a suitably restricted closed surface. In either case the triple integral may be computed in six ways as a repeated integral.

In curvilinear coordinates u, v, w, a volume is given by

$$(11) \qquad V = \int\int\int \mathbf{r}_u \times \mathbf{r}_v \cdot \mathbf{r}_w\, du\, dv\, dw,$$

where the Jacobian $\mathbf{r}_u \times \mathbf{r}_v \cdot \mathbf{r}_w > 0$. A triple integral in x, y, z is changed to one in u, v, w by the formula

$$(12) \qquad \int\int\int_R f(x, y, z)\, dx\, dy\, dz = \int\int\int_{R'} f(x, y, z)\left|\frac{\partial(x, y, z)}{\partial(u, v, w)}\right| du\, dv\, dw,$$

which is an obvious generalization of (6).

When the vector function $\mathbf{f}(\mathbf{r})$ has continuous first partial derivatives,

$$(13) \qquad \int_V \operatorname{div}\mathbf{f}\, dV = \int_S \mathbf{n} \cdot \mathbf{f}\, dS \qquad \text{(Divergence theorem).}$$

More generally,

$$(14) \qquad \int \nabla \mathbf{f}\, dV = \oint \mathbf{n}\mathbf{f}\, dS,$$

when $\mathbf{f}$ is a scalar or a vector function.

If div $\mathbf{f} = 0$, the vector $\mathbf{f(r)}$ is called *solenoidal*. Both rot $\mathbf{g}$ and $\nabla u \times \nabla v$ are solenoidal vectors. When $\mathbf{f(r)}$ is solenoidal, we can always find $\mathbf{g}$, its *vector potential*, so that $\mathbf{f} = $ rot $\mathbf{g}$. A vector $\mathbf{g_0}$ can be found by quadratures if we set any one of its components zero; then $\mathbf{g} = \mathbf{g_0} + \nabla \varphi$ where φ is an arbitrary scalar function. Another particular solution is

$$(15) \qquad \mathbf{g} = -\mathbf{r} \times \int_0^1 \mathbf{f}(t\mathbf{r})\, t \, dt; \quad \text{and} \qquad \mathbf{g} = \frac{\mathbf{f} \times \mathbf{r}}{n+2},$$

when $\mathbf{f(r)}$ is homogeneous of degree $n \, (\neq -2)$.

CHAPTER 10

Uniform Convergence

174. Reversal of Order in Limiting Processses. Let the sequence of functions $f_n(x)$ have $f(x)$ as a limit:

$$f(x) = \lim_{n \to \infty} f_n(x).$$

As $x \to \xi$, is it true that

$$(1) \qquad \lim_{x \to \xi} f(x) = \lim_{x \to \xi} \lim_{n \to \infty} f_n(x) = \lim_{n \to \infty} \lim_{x \to \xi} f_n(x)?$$

Here the question is sharp: Can the order of taking the two limits be reversed? But differentiation and integration are also limiting processes; and again we inquire if the equations

$$(2) \qquad \frac{d}{dx} f(x) = \frac{d}{dx} \lim_{n \to \infty} f_n(x) = \lim_{n \to \infty} \frac{d}{dx} f_n(x),$$

$$(3) \qquad \int_{x_0}^{x} f(x) \, dx = \int_{x_0}^{x} \lim_{n \to \infty} f_n(x) \, dx = \lim_{n \to \infty} \int_{x_0}^{x} f_n(x) \, dx$$

are true. Here again the question is essentially the same: Can we reverse the order in two limiting processes? We cannot expect the answer to be an unqualified *yes*; for we have already seen in § 73, Ex. 1, that

$$\lim_{y \to 0} \lim_{x \to 0} \frac{x + y}{x - y} = -1, \qquad \lim_{x \to 0} \lim_{y \to 0} \frac{x + y}{x - y} = 1.$$

Such questions continually arise in analysis, and a complete answer is often very difficult. However, in order to give even sufficient conditions for the validity of reversing certain limits, it is necessary to define a new concept: *uniform convergence*. The student should spare no pains to understand this concept thoroughly for future developments hinge largely upon it.

175. Uniform Convergence of a Sequence. If the sequence of functions $f_n(x)$ converges to $f(x)$ over the interval $a \leq x \leq b$, for every x in the interval and for any choice of $\varepsilon > 0$, there exists a number N such that

(1) $$|f(x) - f_n(x)| < \varepsilon \quad \text{when} \quad n > N.$$

In general, N depends upon both ε and x; thus, when ε is given, N still depends upon the value of x in the interval (a, b). In some cases, however, it is possible to choose an N that is applicable to the entire range of x-values; *then N depends only upon ε*. In this case $f_n(x)$ is said to *converge uniformly* to $f(x)$ in the interval (a, b).

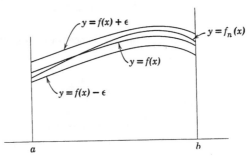

FIG. 175.   Graphical interpretation of uniform convergence

DEFINITION.   *The sequence of functions $\{f_n(x)\}$ converges uniformly to the function $f(x)$ in the interval $a \leq x \leq b$, when for any $\varepsilon > 0$ there exists a number N, independent of x, such that*

(2) $$|f(x) - f_n(x)| < \varepsilon \quad \text{when} \quad n > N(\varepsilon), \qquad a \leq x \leq b.$$

Of course, N is not a *function* of ε due to the multiplicity of possible choices. But, when ε is given, the least of the integers N for which (2) holds is actually a function $N(\varepsilon)$ of ε.

Note that, while the convergence $f_n(x) \to f(x)$ has a meaning for each point x, uniform convergence has no meaning unless an interval (or set of points) is specified.

Uniform convergence has a simple graphical interpretation (Fig. 175). For, if $f_n(x) \to f(x)$ uniformly in the interval (a, b), all the approximation curves,

$$y = f_n(x), \qquad a \leq x \leq b,$$

will lie in the strip between the curves

$$y = f(x) - \varepsilon, \qquad y = f(x) + \varepsilon,$$

when $n > N$. As $\varepsilon \to 0$, the approximation curves of sufficiently high index are confined to progressively thinner strips about the limit curve $y = f(x)$. The convergence of $y = f_n(x)$ to the limit curve is thus effectively *uniform* throughout $a \leq x \leq b$.

When the limit function is known, the following test for uniform convergence is often applicable.

THEOREM. *The sequence $\{f_n(x)\}$ converges uniformly to $f(x)$ in (a, b) if there is a null sequence $\{\varphi(n)\}$, independent of x, such that*

$$(3) \qquad \left| f(x) - f_n(x) \right| \leq \varphi(n) \quad \text{in } (a, b).$$

Proof. Since $\varphi(n) \to 0$, for any $\varepsilon > 0$ we can choose N, independent of x, so that $\varphi(n) < \varepsilon$ when $n > N$.

Example 1. In the interval $0 \leq x \leq 1$,

$$f_n(x) = x - \frac{x^n}{n} \to x \quad \text{uniformly};$$

for

$$\left| x - f_n(x) \right| = \frac{x^n}{n} \leq \frac{1}{n} < \varepsilon \quad \text{when} \quad n > \frac{1}{\varepsilon}.$$

Thus, when $n > 1/\varepsilon$, the approximation curves $y = x - x^n/n$ lie within the strip bounded by the lines $y = x \pm \varepsilon$, $x = 0$, $x = 1$.

Example 2. In the interval $0 \leq x \leq 1$,

$$f_n(x) = n^2 x e^{-nx} \to 0. \qquad (61.4)$$

The convergence is nonuniform since we cannot have

$$\left| 0 - n^2 x e^{-nx} \right| < \varepsilon, \qquad n > N,$$

throughout the interval; for, when $x = 1/n$, we obtain the impossible inequality:

$$n/e < \varepsilon \quad \text{when} \quad n > N.$$

The approximation curves,

$$y = n^2 x e^{-nx},$$

all pass through the origin and have a maximum of n/e at $x = 1/n$. As $n \to \infty$, the maximum point $(1/n, n/e)$ gets higher without limit while moving towards the y-axis, while the curve descends more and more steeply to the origin. At any *fixed* point $(x, 0)$, no matter how close to the origin, the ordinate y_n eventually approaches zero, while the constantly rising peak presses closer to the y-axis. Thus the approach of the approximation curves to the x-axis is fantastically nonuniform. While the *function* $f_n(x) \to 0$ at every point of the interval $(0, 1)$, the *curves* $y = f_n(x)$ approach the segment $0 \leq x \leq 1$ of the x-axis *plus the entire positive y-axis*.

Example 3. In the interval $0 \leq x < \infty$

$$f_n(x) = \frac{nx}{nx + 1} \to f(x) = \begin{cases} 0 & x = 0, \\ 1 & x > 0. \end{cases}$$

We cannot have

$$1 - f_n(x) = \frac{1}{nx + 1} < \varepsilon, \qquad n > N,$$

when $0 < x < \delta$; for when $x = 1/n$, $1/(nx + 1) = \frac{1}{2}$. The convergence is therefore nonuniform in any neighborhood of the origin.

The nth approximation curve is a part of the equilateral hyperbola

$$y = \frac{nx}{nx + 1} \qquad \text{or} \qquad \left(x + \frac{1}{n}\right)(y - 1) = -\frac{1}{n}$$

issuing from the origin and rising toward the asymptote $y = 1$; and, as $n \to \infty$, this part approaches the segment $0 \leq y \leq 1$ of the y-axis plus the positive half-line $y = 1$.

In the interval $0 < a \leq x < \infty$ however, $f_n(x) \to 1$ uniformly; for

$$|1 - f_n(x)| \leq \frac{1}{na + 1} < \varepsilon \quad \text{when} \quad n > \frac{1 - \varepsilon}{a\varepsilon}.$$

Example 4. In any interval $0 \leq x \leq b$

(4) $$\left(1 + \frac{x}{n}\right)^n \to e^x \quad \text{uniformly.}$$

For, if we put $t = n/x$, we have from (19.4)

$$\lim_{n \to \infty} \left(1 + \frac{x}{n}\right)^n = \lim_{t \to \infty} \left(1 + \frac{1}{t}\right)^{tx} = e^x.$$

For a fixed x, we can show as in § 19 that $(1 + x/n)^n$ increases steadily with n; hence

$$r_n(x) = e^x - \left(1 + \frac{x}{n}\right)^n > 0 \quad \text{and also} \quad r_n'(x) = e^x - \left(1 + \frac{x}{n}\right)^{n-1} > 0.$$

Thus $r_n(x)$ is an increasing function of x and

$$e^x - \left(1 + \frac{x}{n}\right)^n < e^b - \left(1 + \frac{b}{n}\right)^n < \varepsilon, \qquad 0 \leq x \leq b$$

when n is taken sufficiently large. The convergence (4) is therefore uniform.

176. Continuity of the Limit Function.

THEOREM. *If the sequence* $f_n(x)$ *converges uniformly to* $f(x)$ *in the interval* $a \leq x \leq b$, $f(x)$ *is continuous at all points where the functions* $f_n(x)$ *are continuous:*

(1) $$\lim_{x \to \xi} f(x) = f(\xi).$$

Proof. If ξ and x are points of (a, b),

$$|f(x) - f(\xi)| = |f(x) - f_n(x) + f_n(x) - f_n(\xi) + f_n(\xi) - f(\xi)|$$
$$\leq |f(x) - f_n(x)| + |f_n(x) - f_n(\xi)| + |f_n(\xi) - f(\xi)|.$$

Since $f_n(x) \to f(x)$ uniformly, choose N so that

$$|f(x) - f_n(x)| < \frac{\varepsilon}{3}, \qquad n > N$$

for all x in (a, b). Then, if $f_n(x)$ is continuous at ξ, we can choose δ so that

$$\left| f_n(x) - f_n(\xi) \right| < \frac{\varepsilon}{3}, \qquad \left| x - \xi \right| < \delta,$$

where n has any specific value $> N$. Hence,

$$\left| f(x) - f(\xi) \right| < \frac{\varepsilon}{3} + \frac{\varepsilon}{3} + \frac{\varepsilon}{3} = \varepsilon, \qquad \left| x - \xi \right| < \delta,$$

and this is precisely the content of (1).

If the functions $f_n(x)$ are continuous throughout $a \leq x \leq b$, $f(x)$ is continuous in this interval. Of course, when $\xi = a$ or b, the limit (1) is one-sided.

We can write (1) in the form

$$(2) \qquad \lim_{x \to \xi} \lim_{n \to \infty} f_n(x) = \lim_{n \to \infty} \lim_{x \to \xi} f_n(x).$$

Consequently, sufficient conditions for this reversal in the order of limits are

(i) the functions $f_n(x)$ are continuous at ξ;

(ii) $f_n(x) \to f(x)$ uniformly in (a, b).

The theorem is often stated in the form: *A uniformly convergent sequence of continuous functions defines a continuous function.*

If $f_n(x) \to f(x)$ uniformly on the infinite interval $a \leq x < \infty$ and we write $\lim_{x \to \infty} f_n(x) = f_n(\infty)$, equation (2), with $\xi = \infty$, gives

$$(3) \qquad \lim_{x \to \infty} f(x) = \lim_{n \to \infty} f_n(\infty).$$

In the proof of the theorem for this case we must replace $\left| x - \xi \right| < \delta$ by $x, \xi > G$, G taken suitably large.

If the continuous functions $f_n(x) \to f(x)$ in $a \leq x \leq b$, $f(x)$ may be continuous in (a, b) even when the convergence is nonuniform (Ex. 2). However, if $f_n(x)$ converges *monotonely* to a continuous function $f(x)$ in $a \leq x \leq b$, the convergence must be uniform. This is the

THEOREM OF DINI. *Let the monotonic sequence $f_n(x) \to f(x)$ in the closed interval $a \leq x \leq b$: then either*

$$f_{n+1}(x) \geq f_n(x) \qquad or \qquad f_{n+1}(x) \leq f_n(x).\dagger$$

Then, if the functions $f_n(x)$ and $f(x)$ are continuous in $a \leq x \leq b$, the convergence is uniform.

---

† Such monotone approach is often denoted by

$$f_n(x) \uparrow f(x) \qquad or \qquad f_n(x) \downarrow f(x) \quad respectively.$$

Proof (indirect). Assume that the convergence $f_n(x) \uparrow f(x)$ is not uniform; then, for any $\varepsilon > 0$, however small,

$$f(x) - f_n(x) < \varepsilon, \qquad n > N$$

cannot be fulfilled for all x in (a, b). Thus, for each choice of N, say $N = N_i$, there is a number $n_i > N_i$ and a value of x_i such that

(4) $f(x_i) - f_{n_i}(x_i) \geqq \varepsilon$

Consider, now, the infinite set $n_1 < n_2 < n_3 < \cdots$ and a corresponding point set $x_1, x_2, x_3, \cdots$ for which (4) holds. This *bounded* set has at least one cluster point ξ by the Bolzano-Weierstrass theorem (§ 13). Choose a fixed value for N; then, if $n_i \geqq N$,

$$f_N(x_i) \leqq f_{n_i}(x_i) \leqq f(x_i) - \varepsilon \qquad \text{from (4).}$$

Now let $x_i \to \xi$ through suitable values†; then, from the continuity of $f_N(x)$ and $f(x)$,

(5) $f_N(\xi) \leqq f(\xi) - \varepsilon.$

But, since N can be chosen arbitrarily large, (5) conflicts with $f_N(\xi) \to f(\xi)$.

Let us now examine the examples of § 175 in the light of these theorems.

Example 1. The continuous functions $x - x^n/n \to x$ uniformly in the interval $0 \leqq x \leqq 1$. The limit function x is continuous as required by the theorem.

Example 2. Although the continuous functions $n^2 xe^{-nx} \to 0$ nonuniformly in the interval $0 \leqq x \leqq 1$, the limit function 0 is continuous. Thus uniformity of convergence is sufficient but not necessary for the continuity of the limit function.

Example 3. When $f_n(x) = nx/(nx + 1)$, the limit function is discontinuous at $x = 0$. The convergence is therefore nonuniform in $0 \leqq x < \infty$ (Ex. 175.3). However, in the interval $0 < a \leqq x < \infty$, the convergence is uniform and the limit function continuous. Moreover, (3) is verified since

$$\lim_{x \to \infty} f(x) = 1, \qquad f_n(\infty) = 1.$$

We readily verify that $f_{n+1}(x) > f_n(x)$; Dini's theorem now shows that $f_n(x) \to 1$ uniformly in $0 < a \leqq x \leqq b$.

Example 4. Since $(1 + x/n)^n \uparrow e^x$, Dini's theorem shows that the convergence is uniform in $(0 \leqq x \leqq b)$.

Example 5. Uniform convergence in an interval $-b \leqq x \leqq b$, however large, is not the same as uniform convergence for all x $(-\infty < x < \infty)$. Thus

$$f_n(x) = \frac{2x + n}{x + n} \to 1 \quad \text{uniformly in} \quad -b \leqq x \leqq b; \quad \text{for}$$

$$|f_n(x) - 1| = \left| \frac{x}{x + n} \right| \leqq \frac{b}{n - b} < \varepsilon \quad \text{when} \quad n > b\left(1 + \frac{1}{\varepsilon}\right).$$

---

† See the theorem of Appendix 1.

But, since $f_n(\infty) = 2$, $f(\infty) = 1$, equation (3) does not hold and the convergence is *nonuniform* in $-\infty < x < \infty$.

Since we have $f_{n+1}(x) < f_n(x)$, Dini's theorem shows that $f_n(x) \to 1$ uniformly in the *closed* interval $-b \leq x \leq b$ but gives no information about the character of the convergence in the *open* interval $-\infty < x < \infty$.

177. Integrals in a Sequence. If the integrable functions $f_n(x) \to f(x)$ in $a \leq x \leq b$, does

$$\int_a^x f_n(x)\, dx \to \int_a^x f(x)\, dx \ ?$$

THEOREM 1. *If the integrable functions $f_n(x)$ converge uniformly to $f(x)$ in the interval $a \leq x \leq b$, the sequence of integrals,*

$$(1) \qquad \int_a^x f_n(x)\, dx \to \int_a^x f(x)\, dx \quad \text{uniformly.}$$

Proof. Choose N so that

$$|f(x) - f_n(x)| < \frac{\varepsilon}{b - a} \quad \text{when} \quad n > N.$$

Then

$$(2) \qquad \left| \int_a^x f(x)\, dx - \int_a^x f_n(x)\, dx \right| \leq \int_a^x |f(x) - f_n(x)|\, dx < \frac{\varepsilon}{b - a}(x - a) \leq \varepsilon.$$

This proves the uniform convergence of the sequence of integrals to the integral of the limit function. Moreover $\int_a^x f(x)\, dx$ is continuous at all points where the functions $f_n(x)$ are continuous (Theorem 176).

THEOREM 2. *If the integrable functions $f_n(x) \to f(x)$ in the interval $a \leq x \leq b$ and the convergence is uniform in all intervals $a + \delta \leq x \leq b$, (1) is valid when $|f_n(x)| < M$ in (a, b) for all n.*

Proof. Choose N as in Theorem 1 with ε replaced by $\frac{1}{3}\varepsilon$; and choose δ so that $M\delta < \frac{1}{3}\varepsilon$. Then the difference in (2) cannot exceed

$$\int_a^{a+\delta} |f(x)|\, dx + \int_a^{a+\delta} |f_n(x)|\, dx + \int_{a+\delta}^x |f(x) - f_n(x)|\, dx < \frac{1}{3}\varepsilon + \frac{1}{3}\varepsilon + \frac{1}{3}\varepsilon = \varepsilon.$$

Consider again the examples of § 175.

Example 1. In the interval $0 \leq x \leq 1$,

$$f_n(x) = x - \frac{x^n}{n} \to x \quad \text{uniformly,}$$

and

$$\int_0^x f_n(x)\, dx = \frac{x^2}{2} - \frac{x^{n+1}}{n(n + 1)} \to \frac{x^2}{2} \quad \text{uniformly,}$$

as required by the theorem.

Example 2. In the interval $0 \leq x \leq 1$,

$$f_n(x) = n^2 x e^{-nx} \to 0 \quad \text{nonuniformly.}$$

But now an integration by parts shows that

$$\int_0^x f_n(x)\,dx = n^2 \int_0^x x e^{-nx}\,dx = n^2 \left[-\frac{x e^{-nx}}{n} - \frac{e^{-nx}}{n^2} \right]_0^x$$

$$= 1 - (nx + 1)e^{-nx} \to 1,$$

which does *not* coincide with $\int_0^x f(x)dx = 0$. Thus when $f_n(x) \to f(x)$ nonuniformly the validity of (1) is in doubt.

Example 3. In the interval $0 \leq x \leq \infty$,

$$f_n(x) = \frac{nx}{nx + 1} \to f(x) = \begin{cases} 0 & (x = 0), \\ 1 & (x > 0). \end{cases}$$

The convergence is nonuniform; nevertheless,

$$\int_0^x f_n(x)\,dx = \int_0^x \left(1 - \frac{1}{nx + 1} \right) dx = x - \frac{1}{n} \log (nx + 1) \to x = \int_0^x f(x)\,dx.$$

This was to be expected from Theorem 2 since $f_n(x) \to 1$ uniformly in $\delta \leq x < \infty$ and $|f_n(x)| < 1$.

Example 4. $(1 + x/n)^n \to e^x$ uniformly; and

$$\int_0^x \left(1 + \frac{x}{n} \right)^n dx = \frac{n}{n + 1} \left[\left(1 + \frac{x}{n} \right)^{n+1} - 1 \right] \to e^x - 1 = \int_0^x e^x\,dx.$$

178. Derivatives of a Sequence. If the differentiable functions $f_n(x) \to f(x)$ in $a \leq x \leq b$, does

$$f_n'(x) \to f'(x) ?$$

THEOREM. *If the functions $f_n(x)$ converge to $f(x)$ in the interval $a \leq x \leq b$ and have continuous derivatives $f_n'(x)$ that converge uniformly in this interval, the sequence derivatives*

(1) $$f_n'(x) \to f'(x).$$

Proof. Since $\{f_n'(x)\}$ converges uniformly, by Theorem 176

$$f_n'(x) \to \varphi(x), \quad \text{a continuous function;}$$

and, by Theorem 177,

$$\int_a^x f_n'(x)\,dx = f_n(x) - f_n(a) \to \int_a^x \varphi(x)\,dx.$$

But, since

$$f_n(x) - f_n(a) \to f(x) - f(a)$$

by hypothesis, we have

$$f(x) = f(a) + \int_a^x \varphi(x)\, dx.$$

Since $\varphi(x)$ is continuous, Theorem 119.2 shows that the right-hand member has the derivative $\varphi(x)$; hence $f(x)$ is differentiable and $f'(x) = \varphi(x)$. Thus $f_n'(x) \to f'(x)$ as required by the theorem.

Let us examine the examples of § 175 with reference to this theorem.

Example 1. In the interval $0 \leqq x \leqq 1$,

$$f_n(x) = x - \frac{x^n}{n} \to x \quad \text{uniformly,}$$

$$f_n'(x) = 1 - x^{n-1} \to \begin{cases} 1 & (x < 1) \\ 0 & (x = 1) \end{cases}$$

nonuniformly (Theorem 176). Thus $f_n'(x)$ does not approach $f'(x)$ when $x = 1$.

In the interval $0 \leqq x \leqq b < 1$,

$$1 - x^{n-1} \to 1 \quad \text{uniformly;}$$

for $x^{n-1} \leqq b^{n-1}$ and $b^{n-1} \to 0$ (Theorem 175); now $f_n'(x) \to f'(x)$ throughout $0 \leqq x \leqq b$.

Example 2. In the interval $0 \leqq x \leqq 1$,

$$f_n(x) = n^2 x e^{-nx} \to 0 \quad \text{nonuniformly,}$$

$$f_n'(x) = n^2(1 - nx)e^{-nx} \to 0 \quad \text{nonuniformly.}$$

Thus equation (1), now $0 = 0$, is valid in spite of the fact that $f_n(x) \to 0$ and $f_n'(x) \to 0$ nonuniformly.

Example 3. In the interval $0 \leqq x < \infty$,

$$f_n(x) = \frac{nx}{nx + 1} \to f(x) = \begin{cases} 0 & (x=0), \\ 1 & (x > 0), \end{cases}$$

and

$$f_n'(x) = \frac{n}{(nx + 1)^2} \to \begin{cases} \infty & (x = 0), \\ 0 & (x > 0). \end{cases}$$

Thus both $f(x)$ and $f'(x)$ are discontinuous at $x = 0$.

In the interval $0 < a \leqq x < \infty$, $f(x) = 1$ and $f'(x) = 0$ so that (1) is valid. This follows from the theorem since $f_n'(x) \to 0$ uniformly in $a \leqq x < \infty$; for

$$\frac{n}{(nx + 1)^2} \leqq \frac{n}{(na + 1)^2} \to 0 \qquad \text{(Theorem 175).}$$

Example 4. If $f_n(x) = (1 + x/n)^n$,

$$f_n(x) \to e^x \quad \text{and} \quad f_n'(x) = (1 + x/n)^{n-1} \to e^x$$

uniformly; thus $f_n'(x) \to f'(x)$.

PROBLEMS

Find the limit function $f(x)$, and discuss the nature of the convergence $f_n(x) \to f(x)$ in the given interval (a, b). Does

$$\int_a^x f_n(x)\, dx \to \int_a^x f(x)\, dx; \qquad f_n'(x) \to f'(x)?$$

1. $f_n(x) = \dfrac{1}{nx + 1}$ $(0, 1);\quad (1, \infty).$

2. $f_n(x) = \dfrac{nx^2 + 1}{nx + 1}$ $(0, 1).$

3. $f_n(x) = 2n^2 x e^{-n^2 x^2}$ $(0, 1).$

4. $f_n(x) = \tanh nx$ $(-1, 1).$

5. $f_n(x) = \dfrac{nx}{1 + n^2 x^4}$ $(0, b).$

6. $f_n(x) = \dfrac{2x + n}{x + n}$ $(0, b).$

7. (a) Prove that $f_n(x) = \dfrac{nx}{1 + n^2 x^2} \to 0$ nonuniformly in $0 \leq x < \infty$, but uniformly in $(\delta \leq x < \infty)$ for any $\delta > 0$. Does $f_n(\infty) \to f(\infty)$?

(b) Show that the approximation curve $y = f_n(x)$ has the maximum point $(1/n, 1/2)$. What is the limiting *curve*?

(c) Prove that, as $n \to \infty$,

$$\int_0^x f_n(x)\, dx \to 0; \qquad f_n'(x) \to 0 \qquad (x \neq 0).$$

8. (a) Prove that $f_n(x) = x e^{-nx^2} \to 0$ uniformly in $-\infty < x < \infty$.

(b) Show that the approximation curve $y = f_n(x)$ has the maximum point $((2n)^{-\frac{1}{2}}, (2ne)^{-\frac{1}{2}})$. What is the limiting curve?

(c) Prove that, as $n \to \infty$,

$$\int_0^{\tilde{x}} f_n(x)\, dx \to 0, \qquad f_n'(x) \to \begin{cases} 1, & x = 0, \\ 0, & x \neq 0. \end{cases}$$

9. (a) Prove that $f_n(x) = nx(1 - x^2)^n \to 0$ nonuniformly in $0 \leq x \leq 1$ by finding the maximum point of the approximation curves.

(b) Show that $\displaystyle\lim_{n \to \infty} \int_0^1 f_n(x)\, dx = \dfrac{1}{2}.$

179. Uniform Convergence of a Series.

The *series* of functions,

$$(1) \qquad \sum_{k=1}^{\infty} u_k(x) = u_1(x) + u_2(x) + \cdots \qquad (a \leq x \leq b),$$

is said to converge to a function $s(x)$, its *sum*, when the sequence

$$s_n(x) = u_1(x) + \cdots + u_n(x)$$

converges to $s(x)$; and we write

$$(2) \qquad \sum_{k=1}^{\infty} u_k(x) = s(x), \qquad a \leq x \leq b.$$

If the sequence $s_n(x)$ converges uniformly to $s(x)$ in $a \leq x \leq b$, we say that *the series converges uniformly* in this interval. We can then choose N, independent of x, so that

$$(3) \qquad \left| s(x) - s_n(x) \right| < \varepsilon, \qquad n > N,$$

for all x in the interval (a, b).

When the series (1) converges absolutely in an interval, the following "M-test" will often establish its uniform convergence.

THEOREM (Weierstrass). *If* $\displaystyle\sum_{k=1}^{\infty} M_k$ *is a convergent series of positive constants such that*

$$\left| u_k(x) \right| \leq M_k, \qquad a \leq x \leq b,$$

the series $\displaystyle\sum_{k=1}^{\infty} u_k(x)$ *converges absolutely and uniformly in* (a, b).

Proof. The series $\sum u_k(x)$ converges absolutely in (a, b) as we see by comparison with $\sum M_k$ (§ 25). Let the sums

$$\sum_{k=1}^{n} u_k(x) = s_n(x), \qquad \sum_{k=1}^{n} M_k = S_n$$

converge to $s(x)$ and S, respectively. Then

$$\left| s_{n+p}(x) - s_n(x) \right| \leq \left| u_{n+1}(x) \right| + \cdots + \left| u_{n+p}(x) \right|$$
$$\leq M_{n+1} + \cdots + M_{n+p}$$
$$= S_{n+p} - S_n;$$

hence, as $p \to \infty$, we have

$$\left| s(x) - s_n(x) \right| \leq S - S_n.$$

Since $S - S_n \to 0$ by hypothesis, we can choose N so that $S - S_n < \varepsilon$ when $n > N$; and N is obviously independent of x.

When the series to be tested is alternating throughout $a \leq x \leq b$, we can often use the test in

THEOREM 2. *Let the functions $u_n(x)$ be positive and increasing (or de-creasing) in $a \leqq x \leqq b$. Then, if the alternating series,*

$$u_1(x) - u_2(x) + u_3(x) - u_4(x) + \cdots,$$

converges by Leibnitz's Test (§ 34) for each value of x in (a, b), it will also converge uniformly there.

Proof. If the series converges to $s(x)$, we have from (34.2)

$$\left| s(x) - s_n(x) \right| < u_{n+1}(x) \leqq u_{n+1}(b);$$

and, since $u_{n+1}(b) \to 0$, we can choose N so that $u_{n+1}(b) < \varepsilon$ when $n > N$.

If the functions $u_n(x)$ are decreasing, we replace $u_{n+1}(b)$ by $u_{n+1}(a)$ in the proof.

Example 1. The exponential series

$$e^x = 1 + x + \frac{x^2}{2!} + \frac{x^3}{3!} + \cdots$$

converges absolutely and uniformly in any interval $-r \leqq x \leqq r$, however large, for we may take $M_k = \dfrac{r^k}{k!}$.

The series for $\sin x$, $\cos x$, $\sinh x$, $\cosh x$ obviously enjoy the same property.

Example 2. The series

$$\sum_{k=1}^{\infty} \frac{\sin kx}{k^2} \quad \text{and} \quad \sum_{k=1}^{\infty} \frac{\cos kx}{k^2}$$

converge uniformly for all x, for we may take $M_k = 1/k^2$.

Example 3. The geometric series

$$\sum_{k=0}^{\infty} x^k = \frac{1}{1 - x} \quad \text{when} \quad |x| < 1 \tag{22.2}.$$

The convergence is uniform in an interval $-b \leqq x \leqq b < 1$, for we may take $M_k = b^k$.

Example 4. The alternating series

$$(4) \qquad\qquad \log (1 + x) = x - \frac{x^2}{2} + \frac{x^3}{3} - \frac{x^4}{4} + \cdots \tag{67.2}.$$

converges when $0 \leqq x \leqq 1$, but not absolutely; for, when $x = 1$, the series of absolute values $1 + {}^1\!/_2 + {}^1\!/_3 + \cdots$ diverges. Therefore the M-test is not applicable. Nevertheless the series converges uniformly in (0, 1) since

$$\left| \log (1 + x) - s_n(x) \right| < \frac{x^{n+1}}{n + 1} \leqq \frac{1}{n + 1} \tag{34.2}.$$

PROBLEMS

Find the sum of the following series in the given interval, and test for uniform convergence.

1. $\displaystyle\sum_{k=0}^{\infty} \frac{1}{(x+k)(x+k+1)}$, $0 \leq x \leq a$.

2. $\displaystyle\sum_{k=0}^{\infty} (1-x)x^k$, $0 \leq x < 1$.

3. Prob. 2 for the interval $0 \leq x \leq 1 - \delta$.

4. $\displaystyle\sum_{k=1}^{\infty} x(1+x)^{-k}$, $0 \leq x \leq a$.

5. Prob. 4 for the interval $\delta \leq x \leq a$.

6. $\displaystyle\sum_{k=0}^{\infty} (1-x^2)x^{3k}$, $-1 < x < 1$.

7. Prob. 6 for the interval $-1 + \delta \leq x \leq 1 - \delta$.

8. $\displaystyle\sum_{k=1}^{\infty} x(1-x)^k$, $0 < x < 2$.

9. Prob. 8 for the interval $\delta \leq x \leq 2 - \delta$.

Test for uniform convergence in the given interval.

10. $\displaystyle\sum_{k=1}^{\infty} \frac{x^k}{\sqrt{k}}$ $(-1 \leq x \leq 0)$.

11. $\displaystyle\sum_{k=0}^{\infty} (-1)^k \frac{x^{2k+1}}{2k+1}$ $(-1 \leq x \leq 1)$.

12. Show that the series $\displaystyle\sum_{k=1}^{\infty} \frac{x}{k(1+kx^2)}$ converges uniformly for all values of x. Find the maximum value of the general term.

13. Show that the series

$$\sum_{k=-1}^{\infty} x^k \to \frac{1}{x} + \frac{1}{1-x} \quad \text{uniformly in} \quad 0 < x \leq \frac{1}{2}.$$

14. Prove that the series $\displaystyle\sum_{k=1}^{\infty} \frac{x}{k(x+k)} \to f(x)$ uniformly in $0 \leq x \leq 1$, and that, when $x = n$, a positive integer, $\displaystyle f(n) = \sum_{k=1}^{n} \frac{1}{k}$.

15. If $\displaystyle\sum_{k=1}^{\infty} u_k(x) \to f(x)$ uniformly and $g(x)$ is bounded in $a \leqq x \leqq b$, show that

$$\sum_{k=1}^{\infty} g(x)\, u_k(x) \to g(x) f(x) \text{ uniformly in } (a, b).$$

16. Prove that the series for the zeta function, $\displaystyle\zeta(x) = \sum_{k=1}^{\infty} k^{-x}$ converges uniformly when $x \geqq a > 1$. Find $\zeta(\infty)$.

180. Continuity of the Sum. When Theorem 176 is applied to the sequence $s_n(x)$, the sum of n terms of a series, we obtain the

THEOREM. *If the series* $\displaystyle\sum_{n=1}^{\infty} u_n(x)$ *converges uniformly to the sum $s(x)$ in* $a \leqq x \leqq b$, *$s(x)$ is continuous at all points where the terms $u_n(x)$ are continuous.*

We can conclude, for example, that

$$s(b) = \lim_{x\to b} \lim_{n\to\infty} s_n(x) = \lim_{n\to\infty} \lim_{x\to b} s_n(x) = u_1(b) + u_2(b) + \cdots.$$

Example. Since the series (179.4) for $\log(1 + x)$ converges uniformly in $0 \leqq x \leqq 1$,

(1) $$\log 2 = 1 - \tfrac{1}{2} + \tfrac{1}{3} - \tfrac{1}{4} + \cdots.$$

181. Integration of Series. When Theorem 177 is applied to the sequence $s_n(x)$, we obtain the

THEOREM. *If the series* $\displaystyle\sum_{n=1}^{\infty} u_n(x)$ *of integrable functions converges uniformly to the sum $s(x)$ in the interval $a \leqq x \leqq b$, then*

(1) $$\int_a^x s(x)\, dx = \sum_{n=1}^{\infty} \int_a^x u_n(x)\, dx, \qquad a \leqq x \leqq b,$$

and the convergence is again uniform in (a, b).

We say that series (1) is obtained from $s(x) = \sum u_n(x)$ by integrating "term by term."

Example 1. The geometric series,

(2) $$\frac{1}{1 + x^2} = 1 - x^2 + x^4 - x^6 + \cdots,$$

is uniformly convergent in an interval $-b \leqq x \leqq b < 1$; for we may take $1 + b^2 + b^4 + b^6 + \cdots$ as the M-series. Hence, when $|x| < 1$, we may integrate term by term between the limits 0 and x:

(3) $$\tan^{-1} x = x - \frac{x^3}{3} + \frac{x^5}{5} - \frac{x^7}{7} + \cdots \qquad |x| < 1.$$

This series of odd powers is alternating throughout $-1 \le x \le 1$ and uniformly (but not absolutely) convergent there; for the remainder after n terms is less than $1/(2n + 1)$ in absolute value. Hence the series represents a continuous function throughout $-1 \le x \le 1$; and, as $x \to 1$, we get

$$\tan^{-1} 1 = \lim_{x \to 1} \lim_{n \to \infty} s_n(x) = \lim_{n \to \infty} \lim_{x \to 1} s_n(x),$$

(4) $\tfrac{1}{4}\pi = 1 - \tfrac{1}{3} + \tfrac{1}{5} - \tfrac{1}{7} + \cdots .$

Note that we cannot prove (4) by integrating (2) between the limits 0 and 1; for (2) is not even convergent when $x = 1$.

Example 2. The geometric series,

$$\frac{1}{1 + x} = 1 - x + x^2 - x^3 + \cdots,$$

$$\frac{1}{1 - x} = 1 + x + x^2 + x^3 + \cdots,$$

are both uniformly convergent when $|x| < 1$. We may therefore integrate these series termwise from 0 to x ($-1 < x < 1$):

(5) $\log(1 + x) = x - \dfrac{x^2}{2} + \dfrac{x^3}{3} - \dfrac{x^4}{4} + \cdots$

(6) $-\log(1 - x) = x + \dfrac{x^2}{2} + \dfrac{x^3}{3} + \dfrac{x^4}{4} + \cdots$

On adding these series we have

(7) $\log \dfrac{1 + x}{1 - x} = 2\left(x + \dfrac{x^3}{3} + \cdots\right), \qquad |x| < 1.$

If r is any positive number and we put

$$\frac{1 + x}{1 - x} = r, \quad \text{then} \quad x = \frac{r - 1}{r + 1} ;$$

since $(r - 1)/(r + 1) < 1$, $\log r$ is given by (7).

Example 3. From (5) we have

$$\frac{\log(1 + x)}{x} = 1 - \frac{x}{2} + \frac{x^2}{3} - \frac{x^3}{4} + \cdots$$

a series that converges in $0 \le x \le 1$. By Theorem 179.2 it also converges uniformly in this interval and hence may be integrated term by term from 0 to 1. Thus

$$\int_0^1 \frac{\log(1 + x)}{x}\, dx = 1 - \frac{1}{2^2} + \frac{1}{3^2} - \frac{1}{4^2} + \cdots,$$

a convergent series whose sum is known to be $\pi^2/12$. [Cf. (223.14).]

182. Differentiation of Series. When Theorem 178 is applied to the sequence $s_n(x)$, we obtain the

THEOREM. *Let the series of differentiable functions* $\displaystyle\sum_{n=1}^{\infty} u_n(x)$ *converge to*

$s(x)$ in the interval $a \leq x \leq b$. Then, if the derivatives $u_n'(x)$ are continuous and the series $\sum u_n'(x)$ converges uniformly in (a, b),

$$(1) \qquad s'(x) = \sum_{n=1}^{\infty} u_n'(x).$$

We say that the series (1) is obtained from $s(x) = \sum_{n=1}^{\infty} u_n(x)$ by differentiating "term by term."

Example 1. By differentiating

$$\frac{1}{1 + x^2} = 1 - x^2 + x^4 - x^6 + \cdots, \qquad -b \leq x \leq b < 1.$$

term by term, we get (on canceling -2)

$$(2) \qquad \frac{x}{(1 + x^2)^2} = x - 2x^3 + 3x^5 - \cdots.$$

To justify (2) we note that this series converges uniformly when $|x| \leq b$; for we may take $b + 2b^3 + 3b^5 + \cdots$ as the M-series.

Example 2. By differentiating

$$(1 + x)^{-1} = \sum_{n=0}^{\infty} (-1)^n x^n, \qquad -b \leq x \leq b < 1$$

term by term, we get

$$-(1 + x)^{-2} = \sum_{n=1}^{\infty} (-1)^n n x^{n-1};$$

or, on replacing n by $n + 1$,

$$(3) \qquad (1 + x)^{-2} = \sum_{n=0}^{\infty} (-1)^n (n + 1) x^n.$$

Another differentiation gives

$$-2(1 + x)^{-3} = \sum_{n=1}^{\infty} (-1)^n n(n + 1) x^{n-1};$$

or, on replacing n by $n + 1$,

$$(4) \qquad (1 + x)^{-3} = \frac{1}{2} \sum_{n=0}^{\infty} (-1)^n (n + 1)(n + 2) x^n.$$

All these series converge uniformly when $|x| \leq b < 1$.

Example 3. The series for the Riemann zeta function (26.9),

$$(5) \qquad \zeta(x) = \sum_{n=1}^{\infty} \frac{1}{n^x}, \qquad x > 1,$$

converges uniformly when $1 < a \leqq x$; for we may take $\Sigma 1/n^a$ as the M-series. Therefore $\zeta(x)$ is continuous when $x > 1$ (Theorem 180). Moreover,

$$(6) \qquad\qquad \zeta'(x) = - \sum_{n=2}^{\infty} \frac{\log n}{n^x}, \qquad x > 1;$$

for we may take $\Sigma \log n/n^a$ as the M-series to establish the uniform convergence of (6).

Example 4. In order to evaluate the series $\displaystyle\sum_{n=1}^{\infty} \frac{n}{(n+1)!}$ we define $f(x)$ by the power series,

$$f(x) = \sum_{n=1}^{\infty} \frac{n}{(n+1)!} x^{n+1},$$

which converges for all x (why?). Then

$$f'(x) = \sum_{n=1}^{\infty} \frac{x^n}{(n-1)!} = x \sum_{n=1}^{\infty} \frac{x^{n-1}}{(n-1)!} = xe^x,$$

for the convergence is uniform in any interval $-r \leqq x \leqq r$ (Ex. 179.1). Since $f(0) = 0$,

$$f(x) = \int_0^x xe^x dx = (x-1)e^x \Big|_0^x = (x-1)e^x + 1,$$

and hence the sum of the given series is $f(1) = 1$.

PROBLEMS

1. Find $f(x) = \displaystyle\sum_{n=0}^{\infty} x(1-x)^n$, $0 \leqq x \leqq 1$. Is the convergence uniform? Does termwise integration give $\displaystyle\int_0^1 f(t)\, dt = 1$?

2. Show that $\displaystyle\sum_{n=1}^{\infty} x/(1 + n^2 x)$ converges uniformly when $x \geqq 0$ to an increasing function $f(x)$. Find $f(0+)$ and $f(\infty)$.

Prove that $f'(x) = \displaystyle\sum_{n=1}^{\infty} (1 + n^2 x)^{-2}$, $\qquad x \geqq a > 0$.

3. Show that $\displaystyle\sum_{n=1}^{\infty} \frac{x}{n(x+n)}$ converges uniformly when $0 \leqq x \leqq a$ to an increasing function $g(x)$. Compute $g(x)$ when $x = 0, 1, 2, \cdots, n$. Prove that

(a) $g'(x) = \displaystyle\sum_{n=1}^{\infty} (x+n)^{-2}$, $\qquad 0 \leqq x \leqq a$;

(b) $g'(0+) = \pi^2/6$ [cf. (223.15)];

(c) $\displaystyle\int_0^1 g(x)\, dx = \gamma$, Euler's constant [cf. (139.5)].

4. Show that $\displaystyle\sum_{n=0}^{\infty} (x + n)^{-1}(x + n + 1)^{-1}$ converges uniformly to $1/x$ when $x \geqq 0$.

5. Show that $\displaystyle\sum_{n=0}^{\infty} e^{-n^2x}$ and $\displaystyle\sum_{n=0}^{\infty} n^2 e^{-n^2x}$ converge uniformly to functions $f(x)$ and $-f'(x)$ when $x \geqq a > 0$. Find $f''(x)$

$$\left[\text{Since } e^{n^2x} = 1 + n^2x + \tfrac{1}{2}n^4x^2 + \cdots, \quad \frac{1}{e^{n^2x}} \leqq \frac{1}{e^{n^2a}} < \frac{1}{n^2a}, \quad \frac{n^2}{e^{n^2x}} \leqq \frac{n^2}{e^{n^2a}} < \frac{1}{\tfrac{1}{2}n^2a^2} \right]$$

183. Power Series. We have seen in §§ 38 and 41 that the real or complex power series,

$$(1) \qquad\qquad \sum_{n=0}^{\infty} a_n z^n = a_0 + a_1 z + a_2 z^2 + \cdots,$$

has the radius of convergence,

$$(2) \qquad\qquad R = 1/\bar\rho \quad \text{where} \quad \bar\rho = \overline{\lim} \left| a_n \right|^{1/n}.$$

This result is entirely general; but in practice we first see if the ratio $|a_{n+1}/a_n|$ approaches a limit τ. If this limit exists, the radius of convergence $R = 1/\tau$;

A real series then converges within the interval $-R < x < R$, diverges outside; but its behavior at the end points $x = \pm R$ must be determined by special tests.

A complex series converges within the circle $\left| z \right| = R$, diverges outside, and its behavior on the circumference must be otherwise determined. In particular, if $\tau = 0$, $R = \infty$ and the series converges throughout the complex plane; and, if $\tau = \infty$, the series converges only at $z = 0$.

If the series

$$\sum a_n z^n = s(z), \qquad \left| z \right| < R,$$

it is said to converge uniformly in the circle $\left| z \right| = r \, (r < R)$ if we can choose N independent of z so that

$$\left| s(z) - s_n(z) \right| < \varepsilon, \qquad n > N,$$

when $\left| z \right| \leqq r$. The M-test for uniform convergence, and the theorems on the continuity of the sum, termwise integration, and differentiation of series still hold for complex series. The proofs are essentially the same.

THEOREM 1. *A power series* $\displaystyle\sum_{n=0}^{\infty} a_n z^n$ *converges uniformly in and on any circle about the origin of radius* $r < R$, *and represents a continuous function in and on this circle.*

Proof. When $|z| \leq r$, we have

$$|a_n||z|^n \leq |a_n|r^n.$$

Since $\sum |a_n| r^n$ is a convergent series of positive constants, take $M_n = |a_n|r^n$ in the M-test. Thus the series converges uniformly in and on this circle; and, since the terms $a_n z^n$ are continuous functions, the sum of the series is continuous throughout the circle, circumference included.

THEOREM 2. *If the power series,*

$$(3) \qquad \sum_{n=0}^{\infty} a_n z^n = f(z), \qquad |z| < R,$$

the derived series will have the same radius of convergence R, and

$$(4) \qquad \sum_{n=1}^{\infty} n a_n z^{n-1} = f'(z), \qquad |z| < R.$$

Proof. For the derived series (4) the radius of convergence $R' = R$; for

$$\frac{1}{R'} = \overline{\lim} \, |na_n|^{1/n} = \overline{\lim} \, |a_n|^{1/n} = \frac{1}{R}$$

since $\lim n^{1/n} = 1$ (15.11). From Theorem 1 the derived series is uniformly convergent when $|z| \leq r \, (r < R)$, and hence (3) may be differentiated term by term to give (4); for the theorem of § 182 also applies to complex series.

THEOREM 3. *If the series* $\displaystyle\sum_{n=0}^{\infty} a_n z^n = f(z)$ *within its circle of convergence,* *then*

$$(5) \qquad a_n = \frac{f^{(n)}(0)}{n!}.$$

Proof. From Theorem 2, the power series may be differentiated repeatedly; for each derived series has the same radius of convergence as its predecessor. On differentiating the given series n times, we have

$$n! \, a_n + \text{terms in } z, z^2, \cdots = f^{(n)}(z).$$

This series converges uniformly inside the circle of convergence and represents the continuous function $f^{(n)}(z)$. Hence, as $z \to 0$, we obtain (5).

Corollary. *The power series that represents $f(z)$ is unique.* For its coefficients are those of the Maclaurin series determined by $f(z)$.

Example 1. When $|z| < 1$, the complex geometric series,

$$\Sigma a z^n = a + az + az^2 + \cdots = \frac{a}{1-z} \qquad (22.3).$$

The series converges within the unit circle $|z| = 1$ but diverges everywhere on its circumference for the nth term, of absolute value $|a|$, does not approach zero.

Example 2. The series $\Sigma z^n/n$ converges within the unit circle, for

$$\tau = \lim \frac{a_{n+1}}{a_n} = \lim \frac{n}{n+1} = 1,$$

but evidently diverges when $z = 1$. On the unit circle $|z| = 1$, $z = \cos\theta + i\sin\theta$, and the series becomes

$$\sum \frac{(\cos\theta + i\sin\theta)^n}{n} = \sum \frac{\cos n\theta}{n} + i\sum \frac{\sin n\theta}{n},$$

by De Moivre's theorem (9.9). In Ex. 2 of the next article we show that both real series on the right converge at all points of the circle except $z = 1$ ($\theta = 0$) where the cosine series diverges.

Example 3. The series $\Sigma z^n/n^2$ again has the circle of convergence $|z| = 1$, but converges absolutely at all points of its circumference since $\Sigma 1/n^2$ converges (Ex. 26.1).

PROBLEMS

1. From the binomial series (§ 39)

$$(1 - x^2)^{-\frac{1}{2}} = 1 + \frac{1}{2}x^2 + \frac{1\cdot 3}{2\cdot 4}x^4 + \frac{1\cdot 3\cdot 5}{2\cdot 4\cdot 6}x^6 + \cdots, \qquad |x| \leq 1,$$

deduce power series for $\sin^{-1}x$ and $(1 - x^2)^{-\frac{3}{2}}$, and give their precise convergence intervals.

2. From the binomial series for $(1 + x^2)^{-\frac{1}{2}}$, deduce power series for $\sinh^{-1}x$ and $(1 + x^2)^{-\frac{3}{2}}$, and give their precise convergence intervals.

3. From the power series for $(1 - x^2)^{-1}$ deduce a power series for $\tanh^{-1}x$ and $(1 - x^2)^{-2}$, and give their precise convergence intervals.

4. What functions do the following power series represent when $|x| < 1$:

(a) $\displaystyle\sum_{k=0}^{\infty} x^k;$ (b) $\displaystyle\sum_{k=0}^{\infty} (k+1)x^k;$ (c) $\displaystyle\sum_{k=0}^{\infty} (k+1)(k+2)x^k;$

(d) $\displaystyle\sum_{k=0}^{\infty} (-1)^k x^k;$ (e) $\displaystyle\sum_{k=1}^{\infty} kx^k;$ (f) $\displaystyle\sum_{k=1}^{\infty} k^2 x^k?$

5. Show that the series $\displaystyle\sum_{k=1}^{\infty} \frac{z^{3k}}{k}$ converges at all points of the circle $|z| = 1$ except where $z^3 = 1$.

6. Show that $\displaystyle\sum_{k=1}^{\infty} \frac{k}{x^k}$ converges uniformly to $x/(x-1)^2$ in the interval $1 < a \leq x < \infty$. Test equation (176.3).

184. Abel's Theorem. *Let the partial sums of the series* $\displaystyle\sum_{n=1}^{\infty} a_n$ *of real or complex terms be bounded:*

(1) $\qquad |s_n| = |a_1 + a_2 + \cdots + a_n| \leq M, \qquad n = 1, 2, 3, \cdots.$

Then, if $\{k_n\}$ is a positive, nonincreasing sequence, the series

(2) $$S = k_1 a_1 + k_2 a_2 + k_3 a_3 + \cdots$$

converges to a sum S whose absolute value $|S| < Mk_1$, provided

   (a) *the series $\sum a_n$ converges to a sum s (then (1) is automatically fulfilled);*

or

   (b) $k_n \to 0$ *as* $n \to \infty$.

Proof. Series (2) has the partial sums,

$$S_n = k_1 a_1 + k_2 a_2 + \cdots + k_n a_n$$
$$= k_1 s_1 + k_2 (s_2 - s_1) + \cdots + k_n (s_n - s_{n-1});$$

hence,

(3) $S_n - k_n s_n = s_1 (k_1 - k_2) + s_2 (k_2 - k_3) + \cdots + s_{n-1}(k_{n-1} - k_n).$

As $n \to \infty$, the right-hand member becomes an infinite series,

(4) $s_1 (k_1 - k_2) + s_2 (k_2 - k_3) + s_3 (k_3 - k_4) + \cdots,$

whose terms in absolute value do not exceed those of the positive telescopic series,

$$M(k_1 - k_2) + M(k_2 - k_3) + M(k_3 - k_4) + \cdots,$$

which converges to $M(k_1 - k)$, where $k = \lim k_n \geq 0$ (§ 18). Since series (4) converges absolutely, it also converges (simply) to some value L such that $|L| < M(k_1 - k)$. As $n \to \infty$ in (3),

$$\lim_{n \to \infty} (S_n - k_n s_n) = L.$$

Hence, in both cases the series (2) converges to a sum

$$S = L + \lim k_n s_n = \begin{cases} L + ks & \text{in case } (a); \\ L & \text{in case } (b). \end{cases}$$

If we let $n \to \infty$ in (1), we have $|s| \leq M$; thus,

$$|S| \leq |L| + k|s| < M(k_1 - k + k) = Mk_1 \quad \text{in case } (a);$$
$$|S| = |L| < M_1 k_1 \qquad\qquad\qquad\qquad\qquad \text{in case } (b).$$

   Corollary. When $\sum a_n$ converges to s and $\{k_n\}$ is a bounded monotone sequence, the series $\sum k_n a_n$ converges.

   Proof. Let $m < k_n < M$; then either $\{M - k_n\}$ or $\{k_n - m\}$ is a positive decreasing sequence. In the first case

$$S' = (M - k_1)a_1 + (M - k_2)a_2 + \cdots$$

converges by Abel's theorem, and

$$\Sigma k_n a_n = M \Sigma a_n - S' = Ms - S'.$$

The other case is dealt with in similar fashion.

Abel's theorem for series has an exact analogue for integrals which is proved in much the same way.

ABEL'S INTEGRAL THEOREM. *Let $f(x)$ be piecewise continuous and the integral $F(x) = \int_a^x f(t)\, dt$ be bounded when $x \geq a$:*

$$(1)' \qquad\qquad |F(x)| = \left| \int_a^x f(t)\, dt \right| \leq M.$$

Then, if $k(x)$ is positive, differentiable, and decreasing as x varies from a to ∞, the integral

$$(2)' \qquad\qquad I = \int_x^\infty k(x) f(x)\, dx$$

converges to a limit I whose absolute value $|I| < Mk(a)$, provided

(a) the integral $\int_a^\infty f(x)\, dx$ converges to F (then $(1)'$ is automatically fulfilled); or

(b) $k(x) \to 0$ as $x \to \infty$.

Proof. $F(x)$ is continuous and $F'(x) = f(x)$ except at isolated points where $f(x)$ has finite jumps. Hence,

$$\int_a^b k(x) f(x)\, dx = \int_a^b k(x) F'(x)\, dx = k(x) F(x) \Big|_a^b - \int_a^b F(x) k'(x)\, dx,$$

on integrating by parts (see end of § 122); and, since $F(a) = 0$,

$$(3)' \qquad \int_a^b k(x) f(x)\, dx - k(b) F(b) = - \int_a^b F(x) k'(x)\, dx.$$

As $b \to \infty$, the last integral converges absolutely; for $-k'(x) > 0$, and from $(1)'$

$$- \int_a^b |F(x)| \, k'(x)\, dx \leq M[k(a) - k(b)] < M[k(a) - k],$$

where $k = k(\infty)$. Hence $- \int_a^\infty F(x) k'(x)\, dx$ converges (simply) to some value L such that $|L| < M[k(a) - k]$. As $b \to \infty$ in $(3)'$,

$$\lim_{b \to \infty} \left[\int_a^b k(x) f(x)\, dx - k(b) F(b) \right] = L.$$

Hence, in both cases the integral $(2)'$ converges to a limit

$$I = L + \lim k(b)\, F(b) = \begin{cases} L + kF & \text{in case } (a), \\ L & \text{in case } (b). \end{cases}$$

If we let $x \to \infty$ in $(1)'$, we have $|F| \leq M$; thus
$|I| \leq |L| + k\,|F| < M[k(a) - k + k] = Mk(a)$ in case (a);
$|I| = |L| < Mk(a)$ in case (b).

Corollary. When $\displaystyle\int_a^\infty f(x)\,dx$ converges to F and $k(x)$ is a bounded monotone function, the integral $\displaystyle\int_a^\infty k(x) f(x)\,dx$ converges.

Example 1. *Leibnitz's theorem* on alternating series (§ 34) is an immediate consequence of Abel's theorem. For the partial sums of the series $\Sigma(-1)^{n-1}$ are 1 or 0; and, if $\{k_n\}$ is any decreasing null sequence, $\Sigma(-1)^{n-1} k_n$ will converge to a sum $< k_1$.

Example 2. *If $\{k_n\}$ is a decreasing null sequence, the series*

$$(5) \qquad \sum_{n=1}^{\infty} k_n \sin nx \quad \text{converges for every} \quad x$$

$$(6) \qquad \sum_{n=1}^{\infty} k_n \cos nx \quad \text{converges for every} \quad x \neq 2m\pi.$$

Proof. When x has a fixed value $\neq 2m\pi$, all partial sums of the $\displaystyle\sum_{n=1}^{\infty} \sin nx$ and $\displaystyle\sum_{n=1}^{\infty} \cos nx$ are *bounded*; for the sum formulas 201.5 show that

$$\left| \sum_{j=1}^{n} \sin jx \right| \text{ and } \left| \sum_{j=1}^{n} \cos jx \right| \leq \frac{1}{|\sin \frac{1}{2}x|} \qquad (201.5)$$

for all values of n. Hence the series (5) and (6) converge by Abel's theorem when $x \neq 2m\pi$. When $x = 2m\pi$, all terms of (5) are zero, and the series obviously converges but the cosine series (6) becomes Σk_n, which may or may not converge.

Example 3. A *Dirichlet series* has the form

$$(7) \qquad \sum_{n=0}^{\infty} a_n e^{-\lambda_n x} \quad \text{where} \quad \{\lambda_n\} \to \infty \quad \text{monotonely.}$$

If series (7) converges when $x = r$, Abel's theorem shows that it converges when $x > r$; for, when we write (7)

$$\sum_{n=0}^{\infty} a_n e^{-\lambda_n r} \cdot e^{-\lambda_n (x-r)},$$

the series $\displaystyle\sum_{n=0}^{\infty} a_n e^{-\lambda_n r}$ converges and consequently has bounded partial sums, and $\{e^{-\lambda_n (x-r)}\}$ is a decreasing null sequence.

Moreover, if (7) diverges when $x = r$, it will diverge when $x < r$; for convergence when $x < r$ would imply convergence when $x = r$. It now follows easily that every Dirichlet series has an *abscissa of convergence* ξ (which may be $-\infty$ or $+\infty$), such that the series converges when $x > \xi$, diverges when $x < \xi$. In fact, ξ is the greatest lower bound of all values of x for which the series converges.

Example 4. The *Laplace transform*,

$$(8) \qquad \mathscr{L}\{f(x)\} = \int_0^\infty e^{-xt} f(t)\, dt \qquad\qquad (\S\ 148),$$

is the integral analogue of a Dirichlet series and also has an abscissa of convergence ξ (which may be $-\infty$ or $+\infty$) such that $\mathscr{L}\{f(x)\}$ converges when $x > \xi$, diverges when $x < \xi$. To prove this we first show that, if $\mathscr{L}\{f(x)\}$ converges when $x = r$, it will converge for all $x > r$. But, if we write

$$\mathscr{L}\{f(x)\} = \int_0^\infty e^{-(x-r)t}\, e^{-rt} f(t)\, dt,$$

this follows at once from Abel's integral theorem on taking $k(t) = e^{-(x-r)t}$, which decreases to 0 as $t \to \infty$. The existence of ξ now follows as in the case of Dirichlet series. The examples (for which $\xi = 0$)

$$(9) \qquad \frac{1}{x} = \int_0^\infty e^{-xt} \cdot 1\, dt, \qquad\qquad x > 0,$$

$$(10) \qquad \tan^{-1}\frac{1}{x} = \int_0^\infty e^{-xt} \frac{\sin t}{t}\, dt, \qquad\qquad x \geqq 0.$$

show that the Laplace integral may diverge or converge when $x = \xi$. We defer the proof of (10) to Ex. 189.2, but Ex. 145.2 shows that the integral converges when $x = 0$.

185. Consequences of Abel's Theorem.

THEOREM 1 (Abel). *If the power series,*

$$s(x) = a_0 + a_1 x + a_2 x^2 + \cdots,$$

converges for $x = R$, it will converge uniformly in the interval $0 \leqq x \leqq R$ and represent a continuous function there.

Proof. Cauchy's convergence criterion (21.4) shows that

$$\left| a_n R^n + a_{n+1} R^{n+1} + \cdots + a_{n+p} R^{n+p} \right| < \varepsilon, \qquad n > N$$

for $p = 1, 2, 3, \cdots$; hence all partial sums of the infinite series $a_n R^n + a_{n+1} R^{n+1} + \cdots$ have ε as an upper bound. Now, if x lies in the interval $0 \leqq x \leqq R$, $(x/R)^n$, $(x/R)^{n+1}$, $\cdots$ is a positive nonincreasing sequence. Hence, by Abel's theorem, the sum of the series,

$$a_n R^n (x/R)^n + \cdots = a_n x^n + a_{n+1} x^{n+1} + \cdots = s(x) - s_n(x),$$

is less than $\varepsilon(x/R)^n$ in absolute value; that is,

$$\left| s(x) - s_n(x) \right| < \varepsilon, \qquad n > N$$

for all points of the interval $(0, R)$. Thus the series converges uniformly and represents a continuous function in the interval $0 \leq x \leq R$.

THEOREM 2. *If the complex power series represents the function*

$$(1) \qquad f(z) = \sum_{n=0}^{\infty} a_n z^n$$

within its circle of convergence $|z| = R$, *and converges at a point* ζ *on its circumference, then*

$$(2) \qquad f(\zeta) = \sum_{n=0}^{\infty} a_n \zeta^n.$$

Proof. If t is a real variable $(0 \leq t < 1)$, the point $z = t\zeta$ lies within the circle of convergence, and

$$(3) \qquad f(t\zeta) = \sum_{n=0}^{\infty} (a_n \zeta^n) t^n.$$

Although (3) has complex coefficients $a_n \zeta^n$, Theorem 1 still applies; hence the power series (3) represents a continuous function in the interval $0 \leq t \leq 1$, and, on letting $t \to 1$, we obtain (2).

We are now able to prove Abel's theorem on the multiplication of series (§ 37).

THEOREM 3 (Abel). *If the real or complex series,* $\sum_{n=1}^{\infty} a_n, \sum_{n=1}^{\infty} b_n$ *and their Cauchy product,*

$$\sum_{n=1}^{\infty} c_n = \sum (a_1 b_n + a_2 b_{n-1} + \cdots + a_n b_1),$$

converge to A, B, and C, then $AB = C$.

Proof. Consider the power series,

$$\sum_{n=1}^{\infty} a_n z^n = a(z), \qquad \sum_{1}^{\infty} b_n z^n = b(z),$$

which converge for $z = 1$. Since their radius of convergence is at least 1, they converge absolutely when $|z| < 1$ (§ 41). Hence, their Cauchy product converges to $a(z) b(z)$ by Theorem 37:

$$(4) \qquad \sum_{n=1}^{\infty} c_n z^{n+1} = a(z) b(z), \qquad |z| < 1.$$

When $z = 1$, the three power series in (4) converge to C, A, and B, respectively; hence from Theorem 2,

$$C = a(1)\, b(1) = AB.$$

THEOREM 4 (Dirichlet). *The series* $\sum_{n=0}^{\infty} k_n u_n(x)$ *converges uniformly in any interval* $a \leq x \leq b$ *if*

(i) *the partial sums of the series* $\sum_{n=1}^{\infty} u_n(x)$ *are uniformly bounded in* (a, b), *and*

(ii) *the constants* k_n *form a decreasing null sequence.*

Proof. Condition (i) means that

$$\left| \sum_{j=1}^{n} u_j(x) \right| \leq M, \qquad n = 1, 2, 3, \cdots, \qquad a \leq x \leq b.$$

Moreover, in (a, b) and for any value of n,

$$\left| \sum_{j=n+1}^{n+p} u_j(x) \right| = \left| \sum_{j=1}^{n+p} u_j(x) - \sum_{j=1}^{n} u_j(x) \right| \leq 2M, \qquad p = 1, 2, 3, \cdots.$$

Hence, from Abel's theorem, case (b),

$$\sum_{j=1}^{\infty} k_j\, u_j(x) \to S(x), \qquad \sum_{j=n+1}^{\infty} k_j\, u_j(x) \to S(x) - S_n(x)$$

throughout (a, b); and

$$\left| S(x) - S_n(x) \right| < 2M k_{n+1}.$$

Since $k_n \to 0$, we can find a number N so that $k_{n+1} < \varepsilon/2M$ when $n > N$; and then

$$\left| S(x) - S_n(x) \right| < \varepsilon, \qquad n > N, \quad a \leq x \leq b.$$

Example 1. Theorem 1 shows at once that the expansions

(5) $$\tan^{-1} x = x - \frac{x^3}{3} + \frac{x^5}{5} - \frac{x^7}{7} + \cdots$$ (181.3)

(6) $$\log(1 + x) = x - \frac{x^2}{2} + \frac{x^3}{3} - \frac{x^4}{4} + \cdots$$ (181.5)

derived when $|x| < 1$ are also valid when $x = 1$; for both series converge when $x = 1$. We thus obtain series for $\pi/4$ and $\log 2$.

Since (6) converges *absolutely* when $|x| < 1$, we may rearrange its terms (Theorem 36.1); for example,

(7) $$\log(1 + x) = x + \frac{x^3}{3} - \frac{x^2}{2} + \frac{x^5}{5} + \frac{x^7}{7} - \frac{x^4}{4} + \cdots \qquad |x| < 1.$$

Moreover, this series also converges when $x = 1$, *but not to* log 2; in fact,

$$\tfrac{3}{2}\log 2 = 1 + \tfrac{1}{3} - \tfrac{1}{2} + \tfrac{1}{5} + \tfrac{1}{7} - \tfrac{1}{4} + \cdots \qquad (36.6).$$

Abel's theorem does not apply to the series (7) for it is not a *power series* (in which the powers of x must steadily increase).

Example 2. In order to evaluate the integral $\int_0^1 x \log (1 + x)\, dx$, we have from (6)

$$f(x) = \int_0^x t \log (1 + t)\, dt = \frac{x^3}{1 \cdot 3} - \frac{x^4}{2 \cdot 4} + \frac{x^5}{3 \cdot 5} - \cdots, \qquad 0 \leq x < 1.$$

When $x = 1$, this series converges to $1/4$ (Prob. 24.5); hence $f(x)$ is continuous in the interval $0 \leq x \leq 1$ and $f(1) = 1/4$.

Example 3. If the constants k_n form a decreasing null sequence, the series

$$(8) \qquad \sum_{n=1}^{\infty} k_n \sin nx, \qquad \sum_{n=1}^{\infty} k_n \cos nx$$

converge uniformly in any interval $0 < \delta \leq x \leq 2\pi - \delta$.

Proof. This follows from Theorem 4; for the partial sums

$$\left| \sum_{j=1}^{n} \sin jx \right| \text{ and } \left| \sum_{j=1}^{n} \cos jx \right| < \frac{1}{\sin \tfrac{1}{2}\delta}, \qquad \delta \leq x \leq 2\pi - \delta,$$

as we see from (201.5).

PROBLEMS

1. If k_n is a decreasing null sequence show that the series $(\pm k_a)$ converges when the signs follow the cycle $(+ \ + \ - \ + \ - \ -)$.

2. Find the abscissa of convergence for the following Dirichlet series (Ex. 184.3):

$$(a) \ \sum_{n=0}^{\infty} e^{-nx}; \qquad (b) \ \sum_{n=0}^{\infty} e^{-n^2 x}; \qquad (c) \ \sum_{n=1}^{\infty} ne^{-nx};$$

$$(d) \ \sum_{n=1}^{\infty} n^{-x}; \qquad (e) \ \sum_{n=1}^{\infty} 2^{-n} n^{-x}; \qquad (f) \ \sum_{n=1}^{\infty} 2^n n^{-x}.$$

3. For what values of x are the following series convergent:

$$(a) \ \sum_{n=2}^{\infty} \frac{\sin nx}{\log n}; \qquad (b) \ \sum_{n=1}^{\infty} \frac{\cos nx}{n}; \qquad (c) \ \sum_{n=1}^{\infty} \sin \frac{x}{n} \cos nx?$$

4. Show that the series $\displaystyle\sum_{n=1}^{\infty} \frac{\log n}{n} \sin nx$ converges uniformly in $0 < \delta \leq x \leq 2\pi - \delta$.

5. Show that $\displaystyle\int_0^1 t \log (1 - t)\, dt = -\frac{3}{4}$. [Cf. Ex. 2.]

6. Expand $\tfrac{1}{2}[\sqrt{1 + x} - \sqrt{1 - x}]$ in a power series. Hence evaluate

$$\frac{1}{2} + \frac{1 \cdot 3}{2 \cdot 4 \cdot 6} + \frac{1 \cdot 3 \cdot 5 \cdot 7}{2 \cdot 4 \cdot 6 \cdot 8 \cdot 10} + \cdots.$$

7. The series $\displaystyle\sum_{n=0}^{\infty} a_n$ is said to be *summable by the method of Abel* to A when

$$\sum_{n=0}^{\infty} a_n x^n \to f(x), \quad |x| < 1 \quad \text{and} \quad \lim_{x \to 1^-} f(x) = A.$$

If Σa_n *converges* to A, its Abel-sum will also be A by Theorem 185.1. However, the method of Abel assigns a "sum" to certain divergent series. Verify this fact by finding the Abel-sum of the following divergent series:

(a) $1 - 1 + 1 - 1 + 1 - 1 + \cdots$;   (b) $1 - 1 + 0 + 1 - 1 + 0 + \cdots$;

(c) $1 + 0 - 1 + 1 + 0 - 1 + \cdots$;   (d) $1 - 2 + 3 - 4 - 5 + 6 + \cdots$.

8. Show that $1 - 1 + 2 - 2 + 3 - 3 + \cdots$ is not summable by the method of Abel.

9. Prove that the integrals

$$\int_1^\infty \frac{\sin x}{x^p}\,dx, \quad \int_1^\infty \frac{\cos x}{x^p}\,dx \quad \text{converge if } p > 0.$$

$$\left[\left|\int_1^x \sin x\,dx\right| = |\cos 1 - \cos x| \le 2\right].$$

10. Prove that the integrals

$$\int_1^\infty \frac{\sin x^m}{x^{p+1-m}}\,dx, \quad \int_1^\infty \frac{\cos x^m}{x^{p+1-m}}\,dx \quad \text{converge if } p > 0.$$

$$\left[\left|\int_1^x x^{m-1}\sin x^m\,dx\right| = \frac{1}{m}|\cos 1 - \cos x^m| \le \frac{2}{m}.\right]$$

11. Apply Prob. 10 to prove the convergence of

(a) $\displaystyle\int_0^\infty \sin x^2\,dx$;   (b) $\displaystyle\int_0^\infty \frac{\sin\sqrt{x}}{x}\,dx$;   (c) $\displaystyle\int_0^\infty \sqrt{x}\cos x^2\,dx$.

186. Uniform Convergence of Improper Integrals. If the integral of type I,

(1) $$F(x) = \int_a^\infty f(x, t)\,dt,$$

converges for $x = r$, we can choose a number B so that

$$\left|\int_a^\infty f(r, t)\,dt - \int_a^b f(r, t)\,dt\right| = \left|\int_b^\infty f(r, t)\,dt\right| < \varepsilon \quad \text{when} \quad b > B.$$

Now, if the integral $F(x)$ converges in an *interval* I of x-values, it is said to *converge uniformly in the interval* I when there is a number B, *independent of x*, such that

(2) $$\left|\int_b^\infty f(x, t)\,dt\right| < \varepsilon, \qquad b > B, \ x \text{ in I.}$$

Similarly, if the integral of type II,

$$(3) \qquad\qquad G(x) = \int_{a+}^{b} g(x, t)\, dt,$$

converges for $x = r$, we can choose a number δ so that

$$\left| \int_{a+}^{b} g(r, t)\, dt - \int_{a+h}^{b} g(r, t)\, dt \right| = \left| \int_{a+}^{a+h} g(r, t)\, dt \right| < \varepsilon \quad \text{when} \quad 0 < h < \delta.$$

Now, if the integral $G(x)$ converges in an *interval* I of x-values, it is said to *converge uniformly in the interval* I when there is a number δ, *independent of x*, such that

$$(4) \qquad\qquad \left| \int_{a+}^{a+h} g(x, t)\, dt \right| < \varepsilon \quad \text{when} \quad 0 < h < \delta, \ x \text{ in I.}$$

If the increasing sequence $\{b_n\} \to \infty$ $(b_1 > a)$, the integral (1) can be written as an infinite series:

$$(1)' \qquad \int_{a}^{\infty} f(x, t)\, dt = \int_{a}^{b_1} + \int_{b_1}^{b_2} + \cdots + \int_{b_{n-1}}^{b_n} f(x, t)\, dt + \cdots.$$

Similarly, if the decreasing sequence $\{a_n\} \to a$ $(a_1 < b)$, the integral (3) can be written as a series:

$$(3)' \qquad \int_{a+}^{b} g(x, t)\, dt = \int_{a_1}^{b} + \int_{a_2}^{a_1} + \cdots + \int_{a_n}^{a_{n-1}} g(x, t)\, dt + \cdots.$$

If, for all monotone sequences $\{b_n\} \to \infty$ and $\{a_n\} \to a$, the series converge uniformly, the integrals on the left converge uniformly; and conversely. Thus the theory and implications of uniform convergence in the case of series have exact analogues in the case of improper integrals. The M-test, the theorems on continuity, integration, and differentiation are, except for notation, precisely parallel. The proofs in the case of series can be imitated step by step to give the proofs for integrals.

An important instance of uniform convergence is given in the

THEOREM. *If the Laplace transform $\mathscr{L}\{f(x)\}$ converges for $x = r$, it will converge uniformly when $x \geq r$. Moreover,*

$$(5) \qquad\qquad \lim_{x \to \infty} \mathscr{L}\{f(x)\} = 0.$$

Proof. Although the theorem is correct as stated, we shall prove it only when $\mathscr{L}\{f(x)\}$ is not of type II. Write

$$\mathscr{L}\{f(x)\} = \int_{0}^{\infty} e^{-(x-r)t}\, e^{-rt} f(t)\, dt;$$

since this integral converges when $x = r$, we can choose B so that

$$\left| \int_{b}^{\infty} e^{-rt} f(t)\, dt \right| < \varepsilon, \quad \text{when} \quad b > B.$$

Now, for any $x > r$, $e^{-(x-r)t}$ decreases steadily to zero as t varies from b to ∞; hence, by Abel's integral theorem,

(6) $\qquad \left| \int_b^\infty e^{-xt} f(t)\, dt \right| < \varepsilon e^{-(x-r)b} \leqq \varepsilon \quad$ when $\quad \begin{aligned} & b > B, \\ & x \geqq r. \end{aligned}$

Thus $\mathscr{L}\{f(x)\}$ converges uniformly when $x \geqq r$.

Since $\mathscr{L}\{f(x)\}$ is not of type II, $|f(t)| < M$ in $(0, b)$. Now with b chosen as in (6),

$$\left| \mathscr{L}\{f(x)\} \right| < \int_0^b e^{-xt} |f(t)|\, dt + \varepsilon < M \int_0^b e^{-xt}\, dt + \varepsilon$$
$$= M \frac{1 - e^{-xb}}{x} + \varepsilon < \frac{M}{x} + \varepsilon < 2\varepsilon \quad \text{when} \quad x > \frac{M}{\varepsilon}.$$

Thus $\mathscr{L}\{f(x)\} \to 0$ as $x \to \infty$.

187. M-Test for Integrals. When the improper integrals are absolutely convergent, the following "M-tests" will often establish uniform convergence.

THEOREM 1 (Vallée Poussin). *Let $f(x, t)$ be continuous in the region*

$$R: \quad \alpha \leqq x \leqq \beta, \qquad a \leqq t < \infty.$$

Then, if a function $M(t)$ has the properties

(1) $\qquad\qquad |f(x, t)| \leqq M(t)$ in R, *and*

(2) $\qquad\qquad \displaystyle\int_a^\infty M(t)\, dt \quad$ converges,

the integral

(3) $\qquad\qquad F(x) = \displaystyle\int_a^\infty f(x, t)\, dt$

converges absolutely and uniformly in the interval $\alpha \leqq x \leqq \beta$.

Proof. From (1) and (2)

$$\left| \int_b^\infty f(x, t)\, dt \right| \leqq \int_b^\infty |f(x, t)|\, dt \leqq \int_b^\infty M(t)\, dt < \varepsilon$$

when $b > B$; B is obviously independent of x.

THEOREM 2 (Vallée Poussin). *Let $f(x, t)$ be continuous in the region*

$$R: \quad \alpha \leqq x \leqq \beta, \qquad a < t \leqq b.$$

Then, if a function $M(t)$ has the properties

(4) $\qquad\qquad |f(x, t)| \leqq M(t)$ *in R, and*

(5) $\qquad\qquad \displaystyle\int_a^b M(t)\, dt \quad$ converges,

the integral

(6) $$F(x) = \int_a^b f(x, t) \, dt$$

converges absolutely and uniformly in the interval $\alpha \leqq x \leqq \beta$.

Proof. From (4) and (5),

$$\int_a^{a+h} f(x, t) \, dt \leqq \int_a^{a+h} \left| f(x, t) \right| dt \leqq \int_a^{a+h} M(t) \, dt < \varepsilon$$

when $h < \delta$ (independent of x).

Remark. The integral (3) converges uniformly in (α, β) if

$$t^p \left| f(x, t) \right| < K, \qquad p > 1;$$

for we may take $M(t) = K/t^p$. Similarly the integral (6) converges uniformly in (α, β) if

$$(t - a)^q \left| f(x, t) \right| < K, \qquad 0 < q < 1;$$

for we may take $M(t) = K/(t - a)^q$.

Example 1. From (148.4)

$$\mathscr{L}(1) = \int_0^\infty e^{-xt} \, dt = \frac{1}{x}, \qquad x > 0,$$

and the convergence is uniform in the interval $a \leqq x < \infty$ $(a > 0)$. Since $e^{-xt} \leqq e^{-at}$, take $M(t) = e^{-at}$; then

$$\int_0^\infty e^{-at} \, dt = \frac{1}{a}.$$

Example 2. From (122.6) we have

$$\int_0^\infty e^{-t} \cos xt \, dt = e^{-t} \left. \frac{x \sin xt - \cos xt}{1 + x^2} \right|_0^\infty = \frac{1}{1 + x^2},$$

and the convergence is uniform for all x. Since $e^{-t} \left| \cos xt \right| \leqq e^{-t}$, take $M(t) = e^{-t}$.

Example 3. The integral

$$\int_0^1 e^{-t} t^{x-1} \, dt, \qquad 0 < x \leqq 1$$

converges uniformly to a function $F(x)$ in the interval $a \leqq x \leqq 1$ $(a > 0)$. Since $t^{x-1} < t^{a-1}$, take $M(t) = e^{-t} t^{a-1}$; then, from (144.2),

$$\int_0^1 e^{-t} t^{a-1} \, dt \quad \text{converges:} \qquad \lim_{t \to 0+} t^{1-a} \cdot e^{-t} t^{a-1} = 1.$$

Example 4. The integral

$$\int_0^1 \log xt \, dt = \log x + \int_0^1 \log t \, dt = \log x - 1$$

and the convergence is uniform in the interval $1/b \leqq x \leqq b \ (b > 1)$. Since
$$| \log xt | \leqq | \log x | + | \log t | \leqq \log b - \log t,$$
we may take $M(t) = \log b - \log t$.

188. Continuity of Improper Integrals. In this and the next two articles we shall limit our discussion to integrals of type I. The corresponding results for integrals of type II will be obvious.

THEOREM. *Let $f(x, t)$ be continuous in the region*
$$R: \quad \alpha \leqq x \leqq \beta, \qquad a \leqq t < \infty.$$
Then the integral

(1) $$F(x) = \int_a^\infty f(x, t) \, dt, \qquad \alpha \leqq x \leqq \beta$$

will be continuous in the interval $\alpha \leqq x \leqq \beta$ if it converges uniformly in this interval.

Proof. If ξ and x are points in (α, β),

(2) $$F(x) - F(\xi) = \int_a^\infty \{f(x, t) - f(\xi, t)\} \, dx.$$

When the integral (1) is uniformly convergent, we can choose B, independent of x, so that
$$\left| \int_b^\infty f(x, t) \, dt \right| < \frac{\varepsilon}{3} \quad \text{when} \quad b > B,$$
and for all x in (α, β). Now from (2)
$$| F(x) - F(\xi) | \leqq \int_a^b |f(x, t) - f(\xi, t)| \, dt + \left| \int_b^\infty f(x, t) \, dt \right| + \left| \int_b^\infty f(\xi, t) \, dt \right|.$$
Since $f(x, t)$ is uniformly continuous in the rectangle $\alpha \leqq x \leqq \beta, a \leqq t \leqq b$ (Theorem 75.2), we can choose δ so that
$$|f(x, t) - f(\xi, t)| < \frac{\frac{1}{3}\varepsilon}{b - a} \quad \text{when} \quad |x - \xi| < \delta.$$
Therefore,
$$| F(x) - F(\xi) | < \frac{\varepsilon}{3} + \frac{\varepsilon}{3} + \frac{\varepsilon}{3} = \varepsilon \quad \text{when} \quad |x - \xi| < \delta;$$
this proves that $F(x)$ is continuous at $x = \xi$. Thus,
$$\lim_{x \to \xi} F(x) = F(\xi),$$
where the limit is naturally one-sided when $\xi = \alpha$ or β.

Under the conditions of the theorem, *we may pass to the limit under the integral sign,*

(3) $$\lim_{x \to \xi} \int_a^\infty f(x, t) \, dt = \int_a^\infty f(\xi, t) \, dt.$$

PROBLEMS

Prove that the following integrals converge uniformly in the given interval ($a > 0$ arbitrarily small, $b > a$ arbitrarily large.)

1. $\displaystyle\int_0^\infty \frac{\cos t}{t^2 + x^2}\, dt \quad (x \geq a);$

2. $\displaystyle\int_0^\infty \frac{\cos xt}{t^2 + 1}\, dt \quad \text{(all } x);$

3. $\displaystyle\int_0^\infty \log \frac{x}{t}\, dt \quad (1 \leq x \leq b);$

4. $\displaystyle\int_1^\infty \frac{\sin xt}{t^2}\, dt \quad \text{(all } x);$

5. $\displaystyle\int_{0+}^1 \frac{\sin (t/x)}{t}\, dt \quad (x \geq a);$

6. $\displaystyle\int_0^\infty \frac{dt}{(t^2 + x^2)^2} \quad (x \geq a);$

7. $\displaystyle\int_1^\infty \frac{\sin t}{t^x}\, dt \quad (x \geq a);$

8. $\displaystyle\int_0^\infty e^{-xt} \frac{\sin t}{t}\, dt \quad (x \geq 0).$

9. Find $F(x) = \displaystyle\int_0^\infty xe^{-xt}\, dt, \quad x \geq 0.$

Prove that the convergence is nonuniform in $0 \leq x \leq b$, but that $F(x) \to 1$ uniformly in $0 < a \leq x \leq b$.

10. Find $F(x) = \displaystyle\int_0^\infty \frac{x\, dt}{t^2 + x^2} \quad$ for all $x.$

Prove that the convergence in nonuniform in any interval containing the origin, but that the integral $\to \frac{1}{2}\pi$ uniformly in $0 < a \leq x \leq b$.

11. Find $F(x) = \displaystyle\int_0^\infty x^2\, t\, e^{-xt}dt, \quad x \geq 0.$

Prove that the convergence is nonuniform in $0 \leq x \leq b$, but that the integral $\to 1$ uniformly in $0 < a \leq x \leq b$.

12. If $\displaystyle\int_0^\infty f(x, t)\, dt \to F(x)$ uniformly when $x \geq a$, and $\lim_{x \to \infty} f(x, t) = \varphi(t)$, prove that

$$\lim_{x \to \infty} F(x) = \int_0^\infty \varphi(t)\, dt.$$

13. If $\displaystyle\int_0^\infty e^{-xt} f(t)\, dt$ converges absolutely when $x = r$, prove that the integral converges uniformly to $\mathscr{L}\{f(x)\}$ when $x \geq r$.

189. Integration of Improper Integrals.

THEOREM. *Let $f(x, t)$ be continuous in the strip*

$$R: \quad \alpha \leq x \leq \beta, \qquad a \leq t < \infty.$$

Then, if the integral

(1) $$F(x) = \int_a^\infty f(x, t)\, dt$$

converges uniformly to $F(x)$ *in* (α, β),

$$(2) \qquad \int_\alpha^\beta F(x)\, dx = \int_a^\infty dt \int_\alpha^\beta f(x, t)\, dx,$$

and this new improper integral is again uniformly convergent in (α, β).

Proof. Since the convergence is uniform, we can choose B, independent of x, so that

$$\left| \int_b^\infty f(x, t)\, dt \right| < \frac{\varepsilon}{\beta - \alpha}, \qquad b > B.$$

Hence,

$$(3) \qquad \left| \int_\alpha^\beta dx \int_b^\infty f(x, t)\, dt \right| < \varepsilon, \qquad b > B,$$

or

$$\left| \int_\alpha^\beta dx \left\{ F(x) - \int_a^b f(x, t)\, dt \right\} \right| < \varepsilon, \qquad b > B.$$

Hence, as $b \to \infty$, the integral on the left approaches zero; that is,

$$\int_\alpha^\beta F(x)\, dx = \lim_{b \to \infty} \int_\alpha^\beta dx \int_a^b f(x, t)\, dt.$$

If we now reverse the order of integrations on the right (Theorem 135) and let $b \to \infty$, we get

$$(4) \qquad \int_\alpha^\beta dx \int_a^\infty f(x, t)\, dt = \int_a^\infty dt \int_\alpha^\beta f(x, t)\, dx.$$

If we reverse the order of integrations in (3), which is permitted by the theorem,

$$\left| \int_b^\infty dt \int_\alpha^\beta f(x, t)\, dt \right| < \varepsilon, \qquad b > B;$$

this shows that the integral on the right of (4) is uniformly convergent.

Under the conditions of the theorem we may *integrate under the integral sign.*

Example 1. From Ex. 187.1

$$\int_0^\infty e^{-xt}\, dt = \frac{1}{x} \quad \text{uniformly in } 0 < a \leq x \leq b.$$

Hence we may integrate with respect to x, from a to b, under the integral sign. Since

$$\int_a^b e^{-xt}\, dx = -\frac{e^{-xt}}{t} \bigg|_a^b = \frac{e^{-at} - e^{-bt}}{t}, \qquad \int_a^b \frac{dx}{x} = \log\frac{b}{a},$$

we have

$$\int_a^b \frac{e^{-at} - e^{-bt}}{t}\, dt = \log\frac{b}{a}.$$

Example 2. From Ex. 187.2

$$\int_0^\infty e^{-t} \cos xt \, dt = \frac{1}{1+x^2} \qquad \text{uniformly when } x \geqq 0.$$

Hence, we may integrate both members with respect to x, from 0 to x, under the integral sign:

(5) $$\int_0^\infty e^{-t} \frac{\sin xt}{t} \, dt = \tan^{-1} x, \qquad x \geqq 0.$$

With a new variable of integration $u = xt$, this becomes

$$\int_0^\infty e^{-\frac{u}{x}} \frac{\sin u}{u} \, du = \tan^{-1} x, \qquad x \geqq 0.$$

When x is replaced by $1/x$, this gives

(6) $$\mathscr{L}\left(\frac{\sin x}{x}\right) = \int_0^\infty e^{-xu} \frac{\sin u}{u} \, du = \tan^{-1} \frac{1}{x}, \qquad x > 0,$$

which proves (184.10).   Since the integral converges when $x = 0$ (Ex. 145.2), theorem 186 shows that it converges uniformly when $x \geqq 0$ and represents a continuous function. Hence, on letting $x \to 0$, we have

(7) $$\int_0^\infty \frac{\sin u}{u} \, du = \frac{\pi}{2}.$$

From (7) we show that

(8) $$\int_0^\infty \frac{\sin xt}{t} \, dt = \begin{cases} \pi/2 & x > 0, \\ 0 & x = 0, \\ -\pi/2 & x < 0. \end{cases}$$

When $x > 0$, put $u = xt$, $du = x \, dt$ in (7) to obtain the upper result.   When $x = 0$, the integrand vanishes.   Finally, when $x < 0$, put $x = -y$; the integral then becomes $-\int_0^\infty \frac{\sin yt}{t} \, dy = -\frac{\pi}{2}$.   Since the integral (8) represents a function discontinuous at $x = 0$, it must converge nonuniformly in any interval including the origin.

PROBLEMS

1. Integrate (189.5) to get

$$\int_0^\infty e^{-t} \frac{1 - \cos xt}{t^2} \, dt = x \tan^{-1} x - \frac{1}{2} \log (1 + x^2), \qquad x \geqq 0.$$

2. From Prob. 1 deduce

$$\mathscr{L}\left(\frac{1 - \cos x}{x^2}\right) = \tan^{-1} \frac{1}{x} - \frac{x}{2} \log \left(1 + \frac{1}{x^2}\right), \qquad x > 0.$$

3. Show that

$$I = \int_0^\infty \frac{1 - \cos u}{u^2} \, du = \int_0^\infty \left(\frac{\sin t}{t}\right)^2 dt$$

and that the last integral converges. Hence prove that $\mathscr{L}\left(\dfrac{1-\cos x}{x^2}\right)$ converges uniformly when $x \geq 0$ and that $I = \pi/2$.

4. Prove that $\displaystyle\int_0^1 \frac{\log x}{1-x}\,dx = -\frac{\pi^2}{6}.$

$$\left[\int_0^1 \frac{\log x}{1-x}\,dx = \int_0^1 \log x \left(1 + x + \cdots + x^{n-1} + \frac{x^n}{1-x}\right)dx,\right.$$

$$= -\left(1 + \frac{1}{2^2} + \cdots + \frac{1}{n^2}\right) + \int_0^1 \frac{x^n}{1-x}\log x\,dx;$$

from Prob. 62.11,

$$\int_0^1 \frac{x^n\,|\log x|}{1-x}\,dx \leq \int_0^1 x^{n-1}\,dx = \frac{1}{n}.$$

Let $n \to \infty$ and use (223.15).$\Big]$

5. Prove that $\displaystyle\int_0^1 \frac{\log x}{1+x}\,dx = -\frac{12}{\pi^2}.$ [Cf. (223.14].)

6. Solve Prob. 4 by the change of variable $1 - x = t$.
   [Apply Abel's theorem to

$$\int_0^u \frac{\log(1-t)}{t}\,dt = -\sum_{n=1}^{\infty} \frac{u^n}{n^2}, \qquad 0 \leq u < 1.]$$

190. Differentiation of Improper Integrals.

THEOREM. *Let $f(x, t)$ and $f_x(x, t)$ be continuous within the strip:* $\alpha \leq x \leq \beta, a \leq t < \infty$. *Then, if the integral*

$$F(x) = \int_a^{\infty} f(x, t)\,dt \quad \text{converges in} \quad \alpha \leq x \leq \beta,$$

its derivative

$$F'(x) = \int_a^{\infty} f_x(x, t)\,dt, \qquad \alpha \leq x \leq \beta,$$

provided the last integral converges uniformly in $\alpha \leq x \leq \beta$.

Proof. The function

$$\varphi(x) = \int_a^{\infty} f_x(x, t)\,dt$$

is continuous in $\alpha \leq x \leq \beta$ (Theorem 188) and may be integrated from $x = \alpha$ to $x \leq \beta$ *within* the integral sign (Theorem 189); hence,

$$\int_\alpha^x \varphi(x)\,dx = \int_a^{\infty} dt \int_\alpha^x f_x(x, t)\,dx$$

$$= \int_a^{\infty} [f(x, t) - f(a, t)]\,dt$$

$$= F(x) - F(a).$$

On differentiating this equation, we get $\varphi(x) = F'(x)$, which was to be proved.

Example 1. If we differentiate (189.8) under the integral sign, we get $\int_0^\infty \cos xt\, dt = 0$. This is false since

$$\lim_{b \to \infty} \int_0^b \cos xt\, dt = \lim_{b \to \infty} \frac{\sin xb}{x} \quad \text{does not exist.}$$

Example 2. If we differentiate

$$(1) \qquad \mathscr{L}\{f(x)\} = \int_0^\infty e^{-xt} f(t)\, dt$$

under the integral sign, we get

$$(2) \qquad \frac{d}{dx} \mathscr{L}\{f(x)\} = -\int_0^\infty e^{-xt}\, t f(t)\, dt = -\mathscr{L}\{xf(x)\}.$$

If $\mathscr{L}\{xf(x)\}$ converges when $x = r$, we shall show that (2) is valid whenever $x \geq r$. Since $\mathscr{L}\{xf(x)\}$ converges uniformly when $x \geq r$ (Theorem 186), we may integrate

$$\varphi(x) = -\int_0^\infty e^{-xt}\, t f(t)\, dt$$

from $x = r$ to x under the integral sign; thus,

$$(3) \qquad \int_r^x \varphi(x)\, dx = \int_0^\infty (e^{-xt} - e^{-rt}) f(t)\, dt = \mathscr{L}\{f(x)\} - \mathscr{L}\{f(x)\}_{x=r}.$$

Hence, on differentiating (3) with respect to x, we get

$$\varphi(x) = \frac{d}{dx} \mathscr{L}\{f(x)\}, \quad \text{which is (2).}$$

Example 3. We shall prove that

$$(4) \qquad F(x) = \int_0^\infty e^{-a^2 t^2} \cos 2xt\, dt = \frac{\sqrt{\pi}}{2a} e^{-x^2/a^2},$$

If we differentiate $F(x)$ under the integral sign,

$$(5) \qquad F'(x) = -2 \int_0^\infty t e^{-a^2 t^2} \sin 2xt\, dt.$$

To justify this step we must show that (4) converges and that (5) converges uniformly for all values of x. In fact, *both* integrals converge uniformly; for we may take $M(t)$ as $e^{-a^2 t^2}$ in (4), $t e^{-a^2 t^2}$ in (5), and $\int_0^\infty M(t)\, dt$ converges in both cases (142.2). Hence (5) is valid; and, on integration by parts with

$$u = \sin 2xt, \qquad dv = -2t e^{-a^2 t^2}\, dt,$$

we find

$$F'(x) = -\frac{2x}{a^2} \int_0^\infty e^{-a^2 t^2} \cos 2xt\, dt = -\frac{2x}{a^2} F(x).$$

This differential equation for F gives on integration

$$F(x) = Ce^{-x^2/a^2} ;$$

and, when $x = 0$,

$$C = F(0) = \int_0^\infty e^{-a^2 t^2} dt = \frac{1}{a} \int_0^\infty e^{-u^2} du = \frac{\sqrt{\pi}}{2a} \qquad (162.3).$$

PROBLEMS

1. Show that $\displaystyle\int_0^\infty e^{-\alpha t^2} dt = \frac{\sqrt{\pi}}{2} \alpha^{-\frac{1}{2}}$ (162.3).

2. By differentiating the equation of Prob. 1, show that

(a) $\displaystyle\int_0^\infty t^2 e^{-\alpha t^2} dt = \frac{\sqrt{\pi}}{2} \cdot \frac{1}{2} \alpha^{-\frac{3}{2}}$;

(b) $\displaystyle\int_0^\infty t^{2n} e^{-\alpha t^2} dt = \frac{\sqrt{\pi}}{2} \cdot \frac{1 \cdot 3 \cdot 5 \cdots (2n-1)}{2^n} \alpha^{-n-\frac{1}{2}}$.

3. From (189.5), prove that

$$\int_0^\infty t e^{-t} \sin xt \, dt = \frac{2x}{(1+x^2)^2}, \qquad x \geqq 0.$$

4. From $\displaystyle\int_0^\infty \frac{dx}{x^2 + \alpha} = \frac{\pi}{2} \alpha^{-\frac{1}{2}}$ $(\alpha > 0)$, prove that

$$\int_0^\infty \frac{dx}{(x^2 + \alpha)^{n+1}} = \frac{\pi}{2} \frac{1 \cdot 3 \cdot 5 \cdots (2n-1)}{2 \cdot 4 \cdot 6 \cdots (2n)} \alpha^{-n-\frac{1}{2}} .$$

5. If $I(\alpha) = \displaystyle\int_0^\infty e^{-x^2 - \alpha^2/x^2} dx$, show that $\dfrac{dI}{d\alpha} = -2I$, and hence $I = \frac{1}{2}\sqrt{\pi}\, e^{-2\alpha}$

191. Gamma Function. The *gamma function* is defined for positive values of x by the improper integral,

$$(1) \qquad\qquad \Gamma(x) = \int_0^\infty e^{-t}\, t^{x-1}\, dt, \qquad x > 0.$$

Let us write

$$I(x) = \int_0^1 e^{-t}\, t^{x-1}\, dt, \qquad J(x) = \int_1^\infty e^{-t}\, t^{x-1}\, dt.$$

The part I is proper when $x \geqq 1$, improper but absolutely convergent when $0 < x < 1$; for, as $t \to 0$,

$$t^{1-x} f(x, t) = e^{-t} \to 1 \qquad\qquad (144.2).$$

The part J also converges absolutely for all x; for, as $t \to \infty$,

$$t^2 f(x, t) = e^{-t}\, t^{x+1} \to 0 \qquad\qquad (142.2).$$

Thus $\Gamma(x)$ is well defined by (1) for $x > 0$.

$\Gamma(0+) = \infty$; for

$$I(x) > e^{-1} \int_0^1 t^{x-1}\, dt = \frac{1}{ex}.$$

Both I and J are uniformly convergent in any interval $0 < a \leqq x \leqq b$; for they are, respectively, less than the convergent integrals independent of x:

$$\int_0^1 e^{-t} t^{a-1}\, dt, \qquad \int_1^\infty e^{-t} t^{b-1}\, dt.$$

Thus $\Gamma(x)$ is continuous for $x > 0$.

We next deduce the functional equation for $\Gamma(x)$. An integration by parts gives

$$\int_\varepsilon^b e^{-t} t^{x-1}\, dt = \frac{t^x}{x} e^{-t} \Big|_\varepsilon^b + \frac{1}{x} \int_\varepsilon^b e^{-t} t^x\, dt.$$

As $b \to \infty$ and $\varepsilon \to 0$, the integrated part vanishes at both limits; and, since the integral on the right is $\Gamma(x+1)$,

$$(2) \qquad\qquad \Gamma(x+1) = x\Gamma(x), \qquad x > 0.$$

More generally, when n is a positive integer,

$$(3) \qquad \Gamma(x+n) = (x+n-1)(x+n-2)\cdots x\Gamma(x), \qquad x > 0;$$

By direct computation

$$\Gamma(1) = \int_0^\infty e^{-t}\, dt = 1.$$

Hence, on putting $x = 1$ in (3), we get

$$(4) \qquad\qquad \Gamma(n+1) = n!$$

Thus $\Gamma(x)$ is a continuous function that assumes factorial values for positive integral arguments.

We now extend the definition of $\Gamma(x)$ to negative nonintegral values of x by means of the functional equation. Thus we *define*

$$(5) \qquad \Gamma(x) = \frac{\Gamma(x+1)}{x}, \qquad -1 < x < 0,$$

$$\Gamma(x) = \frac{\Gamma(x+2)}{x(x+1)}, \qquad -2 < x < -1,$$

$$\cdots\cdots\cdots\cdots\cdots\cdots\cdots\cdots\cdots\cdots\cdots\cdots$$

$$(6) \qquad \Gamma(x) = \frac{\Gamma(x+n)}{x(x+1)\cdots(x+n-1)}, \qquad -n < x < -n+1.$$

Note that $\Gamma(x + n)$ in the interval $-n < x < -n + 1$ passes through the values of $\Gamma(x)$ in the interval $0 < x < 1$. $\Gamma(x)$ may be extended by the sole use of (5), being first used in $(-1, 0)$; then, since these values are known, being used again in $(-2, -1)$, and so on. Thus it is evident that the extended gamma function still satisfies the functional equation (2). This is another instance of the principle of *permanence of form* (§ 10).

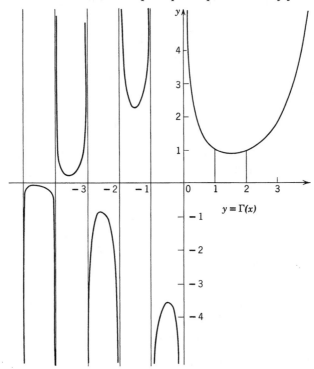

Fig. 191. Graph of the gamma function

Since $\Gamma(x) > 0$ in $(0, 1)$, (5) shows that $\Gamma(x) < 0$ in $(-1, 0)$, $\Gamma(x) > 0$ in $(-2, -1)$, changing sign each time we move a unit distance to the left (Fig. 191). $\Gamma(-n) = \pm \infty$, the sign depending on the direction in which $-n$ is approached.

The integral (1) can be put in various other forms by changing the variable of integration. Thus,

$$(7) \qquad\qquad \Gamma(x) = k^x \int_0^\infty e^{-ku} u^{x-1}\, du \qquad (t = ku, \quad k > 0);$$

$$(8) \qquad\qquad \Gamma(x) = 2 \int_0^\infty e^{-u^2} u^{2x-1}\, du \qquad (t = u^2).$$

In particular, when $x = 1/2$ in (8),

$$\text{(9)} \qquad \Gamma\left(\frac{1}{2}\right) = 2 \int_0^\infty e^{-u^2} \, du = \sqrt{\pi}. \qquad \text{(162.3)}.$$

From (148.1), the Laplace transform of x^p is

$$\mathscr{L}(x^p) = \int_0^\infty e^{-xt} t^p \, dt \qquad (p > -1, \quad x > 0).$$

When $-1 < p < 0$, the integral is of type II as well as type I; and, when $x > 0$, its convergence readily follows from

$$\lim_{t \to 0} t^{-p}(e^{-xt} \, t^p) = 1, \qquad \lim_{t \to \infty} t^2(e^{-xt} \, t^p) = 0.$$

The change of variable $xt = u$ now gives

$$\text{(10)} \qquad \mathscr{L}(x^p) = \frac{\Gamma(p+1)}{x^{p+1}} \qquad (p > -1, \quad x > 0).$$

192. Beta Function. The *beta function* is defined for positive values of x and y by the integral

$$\text{(1)} \qquad B(x, y) = \int_0^1 t^{x-1}(1-t)^{y-1} \, dt, \qquad x, y > 0.$$

This is a proper integral when $x, y \geq 1$, but is improper at the lower limit when $x < 1$, at the upper limit when $y < 1$. We therefore split it into two parts,

$$B(x, y) = \int_0^{1/2} + \int_{1/2}^1 t^{x-1}(1-t)^{y-1} \, dt.$$

The first converges when $0 < x < 1$, diverges when $x \leq 0$; for, as $t \to 0$,

$$t^{1-x} f(t) = (1-t)^{y-1} \to 1,$$

$$t f(t) = t^x (1-t)^{y-1} \to \begin{cases} 1 & (x = 0), \\ \infty & (x < 0). \end{cases}$$

If we make the change of variable $t = 1 - u$, the second integral reduces to the first with x and y interchanged. Hence we can draw the same conclusions as before with y in place of x. Consequently, $B(x, y)$ is well defined by (1) when $x, y > 0$. Moreover $B(x, y)$ is continuous, for, when improper, it converges uniformly when $x, y \geq a > 0$; in fact,

$$B(x, y) \leq M(t) = \int_0^1 t^{a-1}(1-t)^{a-1} \, dt, \qquad x, y \geq a.$$

The change of variable $t = 1 - u$ in (1) permutes x and y; hence,

$$\text{(2)} \qquad B(x, y) = B(y, x).$$

The integral (1) can be put in various forms by changing the variable of integration. Thus, with $t = \cos^2 u$,

(3) $$B(x, y) = 2 \int_0^{\pi/2} (\cos u)^{2x-1} (\sin u)^{2y-1} \, du;$$

and, in particular,

(4) $$B(\tfrac{1}{2}, \tfrac{1}{2}) = \pi.$$

If we put

$$t = \frac{u}{1+u}, \qquad 1 - t = \frac{1}{1+u}, \qquad dt = \frac{du}{(1+u)^2},$$

(5) $$B(x, y) = \int_0^\infty \frac{u^{x-1}}{(1+u)^{x+y}} \, du.$$

If we split this into two integrals between the limits 0, 1 and 1, ∞, the latter

$$\int_1^\infty \frac{u^{x-1}}{(1+u)^{x+y}} \, du = \int_0^1 \frac{v^{y-1}}{(1+v)^{x+y}} \, dv, \qquad v = \frac{1}{u}.$$

Thus, we have

(6) $$B(x, y) = \int_0^1 \frac{t^{x-1} + t^{y-1}}{(1+t)^{x+y}} \, dt,$$

a formula that puts the symmetry (2) in evidence.

193. Relation between Beta and Gamma Function. When either x or y is an integer, we can evaluate $B(x, y)$. Thus, when $y = 1$,

(1) $$B(x, 1) = \int_0^1 t^{x-1} \, dt = \frac{1}{x}.$$

If y is an integer $n > 1$, we integrate by parts to get

(2) $$B(x, n) = \frac{n-1}{x} \int_0^1 t^x (1-t)^{n-2} \, dt = \frac{n-1}{x} B(x+1, n-1).$$

If we use (2) successively until the second argument is reduced to 1, and then use (1), we find

(3) $$B(x, n) = \frac{(n-1)!}{x(x+1) \cdots (x+n-1)}.$$

Now multiply (3) above and below by $\Gamma(x)$ and use (191.3) and (191.4); then

(4) $$B(x, n) = \frac{\Gamma(x) \, \Gamma(n)}{\Gamma(x+n)}.$$

This relation is still true when n is replaced by any positive number y. We use the method of Ex. 162 in the proof. The double integral,

$$\int\int u^{2x-1}\, v^{2y-1}\, e^{-u^2-v^2}\, du\, dv = \int\int r^{2(x+y)-1}\, e^{-r^2}\, (\cos\theta)^{2x-1}\, (\sin\theta)^{2y-1}\, dr\, d\theta,$$

when we make the change of variable,

$$u = r\cos\theta, \qquad v = r\sin\theta.$$

The left-hand integral over the square $S(0 \leq x, y \leq a)$ in Fig. 162 equals

$$J(a) = \int_0^a u^{2x-1}\, e^{-u^2}\, du \int_0^a v^{2y-1}\, e^{-v^2}\, dv \text{ and } \to \frac{1}{4}\,\Gamma(x)\,\Gamma(y) \text{ as } a \to \infty$$

by virtue of (191.8). The right-hand integral over a circular quadrant of radius b is

$$\frac{1}{2}\, B(x, y) \int_0^b e^{-r^2}\, r^{2(x+y)-1}\, dr$$

by virtue of (192.3). Since the square S lies between the quadrants of radii a and $\sqrt{2}\, a$, and

$$u^{2x-1}\, v^{2y-1}\, e^{-u^2-v^2} \geqq 0,$$

we have

$$\frac{1}{2}\, B(x, y) \int_0^a e^{-r^2}\, r^{2(x+y)-1}\, dr < I(a) < \frac{1}{2}\, B(x, y) \int_0^{a\sqrt{2}} e^{-r^2}\, r^{2(x+y)-1}\, dr.$$

As $a \to \infty$, the extreme members approach

$$\tfrac{1}{4} B(x, y)\Gamma(x + y) \tag{191.8};$$

therefore,

$$(5) \qquad\qquad B(x, y) = \frac{\Gamma(x)\,\Gamma(y)}{\Gamma(x + y)}, \qquad x, y > 0.$$

Example. When $n = 0, 1, 2, \cdots$, we have from (192.3)

$$(6)\quad S_{2n} = \int_0^{\pi/2} \sin^{2n} u\, du = \frac{1}{2}\, B\left(\tfrac{1}{2}, n + \tfrac{1}{2}\right) = \frac{\Gamma(\tfrac{1}{2})\,\Gamma(n + \tfrac{1}{2})}{2\Gamma(n + 1)},$$

$$(7)\quad S_{2n+1} = \int_0^{\pi/2} \sin^{2n+1} u\, du = \frac{1}{2}\, B\left(\tfrac{1}{2}, n + 1\right) = \frac{\Gamma(\tfrac{1}{2})\,\Gamma(n + 1)}{2\Gamma(n + \tfrac{3}{2})}$$

and hence

$$\frac{S_{2n}}{S_{2n+1}} = \frac{\Gamma(n + \tfrac{1}{2})\,\Gamma(n + \tfrac{3}{2}),}{(n!)^2}$$

$$\frac{S_{2n-1}}{S_{2n+1}} = \frac{\Gamma(n)\,\Gamma(n + \tfrac{3}{2})}{\Gamma(n + \tfrac{1}{2})\,\Gamma(n + 1)} = \frac{n + \tfrac{1}{2}}{n}.$$

Now, since $0 \leq \sin u \leq 1$ in the interval $(0, \pi/2)$,

$$0 < S_{2n+1} < S_{2n} < S_{2n-1},$$

$$1 < \frac{S_{2n}}{S_{2n+1}} < \frac{S_{2n-1}}{S_{2n+1}} = 1 + \frac{1}{2n}.$$

As $n \to \infty$, the last ratio $\to 1$; hence, the central ratio

$$\left(\frac{\Gamma(n+\frac{1}{2})}{n!}\right)^2 \left(n + \frac{1}{2}\right) = \left(\frac{\frac{1}{2} \cdot \frac{3}{2} \cdots (n-\frac{1}{2})}{1 \cdot 2 \cdots n}\right)^2 (2n+1)\frac{\pi}{2} \to 1.$$

We thus obtain the infinite product of Wallis for

$$\frac{\pi}{2} = \lim_{n\to\infty} \left(\frac{2 \cdot 4 \cdot 6 \cdots 2n}{1 \cdot 3 \cdot 5 \cdots (2n-1)}\right)^2 \frac{1}{2n+1};$$

this is usually written

(8) $$\frac{\pi}{2} = \frac{2}{1}\frac{2}{3}\frac{4}{3}\frac{4}{5}\frac{6}{5}\frac{6}{7} \cdots = \prod_{n=1}^{\infty} \frac{2n}{2n-1} \cdot \frac{2n}{2n+1}.$$

From (8) we also have

(9) $$\sqrt{\pi} = \lim_{n\to\infty} \left(\frac{2 \cdot 4 \cdots 2n}{1 \cdot 3 \cdots 2n-1}\right) \frac{1}{\sqrt{n+\frac{1}{2}}} = \lim_{n\to\infty} \frac{(n!)^2\, 2^{2n}}{(2n)!\,\sqrt{n}}, \text{ for } \sqrt{n+\tfrac{1}{2}}/\sqrt{n} \to 1.$$

PROBLEMS

1. Compute: $\Gamma(\frac{3}{2})$, $\Gamma(\frac{5}{2})$, $\Gamma(\frac{7}{2})$, $\Gamma(\frac{9}{2})$.

2. Compute: $\Gamma(-\frac{1}{2})$, $\Gamma(-\frac{3}{2})$, $\Gamma(-\frac{5}{2})$, $\Gamma(-\frac{7}{2})$.

3. Evaluate:

(a) $\displaystyle\int_0^\infty e^{-t} t^{3/2}\, dt$;   (b) $\displaystyle\int_0^\infty e^{-4t} t^{3/2}\, dt$;   (c) $\displaystyle\int_0^\infty e^{-t^2} t^2\, dt$;

(d) $\displaystyle\int_0^\infty t^{-1/2} e^{-at}\, dt$;   (e) $\displaystyle\int_0^\infty t^{1/2} e^{-at}\, dt$;   (f) $\displaystyle\int_0^1 \left(\log\frac{1}{t}\right)^n dt$.

4. Show that $\Gamma\left(\dfrac{1}{2} + n\right) = \dfrac{(2n)!}{4^n\, n!}\sqrt{\pi}$.

5. Show that $\Gamma(\frac{1}{2} - n)\, \Gamma(\frac{1}{2} + n) = -\, \Gamma(\frac{1}{2} - n + 1)\, \Gamma(\frac{1}{2} + n - 1)$ and apply n times to prove

$$\Gamma(\tfrac{1}{2} - n)\, \Gamma(\tfrac{1}{2} + n) = (-1)^n \pi \qquad (n = 1, 2, 3, \cdots).$$

6. Prove that

$$\int_0^1 t^{x-1} \left(\log\frac{1}{t}\right)^{y-1} dt = \frac{\Gamma(y)}{x^y}, \qquad (x, y > 0).$$

Verify the following:

7. $\displaystyle\int_0^1 \frac{dt}{1 - t^{1/4}} = 4B\left(4, \frac{1}{2}\right) = \frac{128}{35}.$

8. $\displaystyle\int_{-\infty}^{\infty} \frac{dt}{(1 + t^2)^4} = B\left(\frac{1}{2}, \frac{7}{2}\right) = \frac{5}{16}\pi.$

9. $\displaystyle\int_0^{\infty} e^{-t^p}\, dt = \frac{1}{p}\, \Gamma\left(\frac{1}{p}\right) \qquad (p > 0).$

10. $\displaystyle\int_0^{\pi/2} \sin^x t\, dt = \frac{\sqrt{\pi}}{2}\, \frac{\Gamma\left(\dfrac{x + 1}{2}\right)}{\Gamma\left(\dfrac{x}{2} + 1\right)} \qquad (x > -1).$

11. $\displaystyle\int_0^1 t^x (1 - t^p)^y\, dt = \frac{1}{p}\, B\left(\frac{x + 1}{p}, y + 1\right) \qquad (x, y > -1,\ p > 0).$

12. $\displaystyle\int_0^1 \frac{dt}{\sqrt{1 - t^p}} = \frac{\sqrt{\pi}}{p}\, \frac{\Gamma(1/p)}{\Gamma(1/p + 1/2)} \qquad (p > 0).$

13. Prove that
$$\lim_{x \to 0+} x\Gamma(x) = 1; \qquad \lim_{x \to -n+0} (x + n)\, \Gamma(x) = (-1)^n/n!$$

14. Show that $\Gamma(x)$ has a continuous derivative when $x > 0$ which is given by
$$\Gamma'(x) = \int_0^{\infty} e^{-t}\, t^{x-1} \log t\, dt.$$

[If $f(x, t) = e^{-t}\, t^{x-1} \log t$, show that both $\displaystyle\int_0^1 f(x, t)\, dt$ and $\displaystyle\int_1^{\infty} f(x, t)\, dt$ are uniformly convergent in any interval $0 < \alpha \leq x \leq \beta$.]

15. Show that $\Gamma(x)$ has continuous derivatives of all orders when $x > 0$, and that
$$\Gamma^{(n)}(x) = \int_0^{\infty} e^{-t}\, t^{x-1} \log^n t\, dt, \qquad x > 0.$$
Prove that the curve $y = \Gamma(x)$ is concave upward when $x > 0$.

16. Prove that, when $x > 0$,

(a) $B(x, x) = 2^{1-2x} B(x, \tfrac{1}{2})$;

(b) $\sqrt{\pi}\, \Gamma(2x) = 2^{2x-1}\, \Gamma(x)\, \Gamma(x + \tfrac{1}{2}).$

$\left[B(x, x) = 2 \displaystyle\int_0^{1/2} (t - t^2)^{x-1}\, dt.\ \text{Put }\ t - t^2 = \frac{u}{4},\ \text{and use Prob. 11. As for } (b),\right.$

$\left.\text{use (193.4)}.\right]$

17. Show that $\displaystyle\int_0^{\pi/2} \sqrt{\tan t}\, dt = \frac{\pi}{\sqrt{2}}$ in two ways: (i) put $\tan t = \sin t/\cos t$, and use (192.3); (ii) put $u = \tan^2 t$, and use (192.5).

18. Prove that the n-dimensional sphere $x_1{}^2 + x_2{}^2 + \cdots + x_n{}^2 \leq a^2$ has the content $V(n)a^n$ where

$$V(n) = \pi^{\frac{1}{2}n}/\Gamma(\tfrac{1}{2}n + 1), \qquad n = 1, 2, 3, \cdots.$$

Verify for $n = 1, 2, 3$.

$$\left[\text{Prove } V(n) = B\!\left(\frac{1}{2}, \frac{n+1}{2}\right) V(n-1); \text{ hence,} \right.$$

$$V(n)\,\Gamma\!\left(\frac{n}{2}+1\right) - \sqrt{\pi}\,V(n-1)\,\Gamma\!\left(\frac{n-1}{2}+1\right) = 0.$$

Thus $f(n) = V(n)\,\Gamma(n/2 + 1)$ satisfies the difference equation

$$f(n) - \sqrt{\pi}\,f(n-1) = 0 \quad \text{with the condition} \quad f(1) = 2\Gamma(^3/_2) = \sqrt{\pi}. \Big]$$

194. Uniform Convergence. In this summary we use $\to\!\to$ to denote *uniform* convergence.

The *sequence* $f_n(x) \to\!\to f(x)$, $a \leq x \leq b$, if

$$|f(x) - f_n(x)| < \varepsilon \quad \text{when} \quad n > N \quad \text{(indep. of } x\text{)}.$$

Three basic theorems:

I. If $f_n(x) \to\!\to f(x)$ and the functions $f_n(x)$ are continuous in (a, b), then $f(x)$ is continuous in $a \leq x \leq b$.

II. If $f_n(x) \to\!\to f(x)$ and the functions $f_n(x)$ are integrable in (a, b), then $\displaystyle\int_a^b f_n(x)\,dx \to\!\to \int_a^b f(x)\,dx$ in $a \leq x \leq b$.

III. If $f_n(x) \to f(x)$ and $f_n{}'(x) \to\!\to \varphi(x)$ in (a, b), then $f'(x) = \varphi(x)$.

The *series* $\displaystyle\sum_{n=0}^{\infty} u_n(x) \to\!\to s(x)$ in $a \leq x \leq b$ if the sequence of partial sums $s_n(x) \to\!\to s(x)$ in (a, b). The theorems I, II, III applied to $s_n(x)$ give corresponding theorems for series.

If $|u_n(x)| \leq M_n$ in (a, b) and $\sum M_n$ converges, $\sum u_n(x) \to\!\to s(x)$ in (a, b) and $\sum u_n(x)$ is absolutely convergent. Hence this "M-test" only applies to absolutely convergent series.

A complex power series,

$$\sum_{n=0}^{\infty} a_n z^n \to\!\to f(z) \quad \text{when} \quad |z| \leq r < R = \frac{1}{\overline{\lim}\,|a_n|^{1/n}}.$$

Within the *circle of convergence* $|z| = R$, the power series represents a continuous function and may be integrated or differentiated term by term as often as desired, the resulting series always having the same R. When $\{a_n\}$ is real,

$$\sum_{n=0}^{\infty} a_n x^n \to\to f(x) \quad \text{when} \quad |x| \leq r < R;$$

and, if the series converges when $x = R$, $\lim_{x \to R} f(x) = f(R)$.

The *integral* $\int_a^\infty f(x, t)\, dt \to\to F(x)$, $\alpha \leq x \leq \beta$, if

$$\int_b^\infty f(x, t)\, dt < \varepsilon \quad \text{when} \quad b > B \quad \text{(indep. of } x)$$

Theorems I, II, III have obvious analogues, and there is a corresponding M-test applicable to absolutely convergent integrals.

If the *Laplace Transform* $\mathscr{L}\{f(x)\}$ converges for $x = r$,

$$\int_0^\infty e^{-xt} f(t)\, dt \to\to \mathscr{L}\{f(x)\}, \qquad x \geq r.$$

Moreover, $\mathscr{L}\{f(x)\} \to 0$ as $x \to \infty$; and $\mathscr{L}\{f(x)\}$ has derivatives of all orders which may be obtained by differentiating under the integral sign.

The *gamma function* $\Gamma(x)$, defined by

$$\Gamma(x) = \int_0^\infty e^{-t} t^{x-1}\, dt, \qquad x > 0,$$

satisfies the functional equation

$$\Gamma(x + 1) = x\Gamma(x).$$

This equation is used successively to define $\Gamma(x)$ within the intervals $(-1, 0)$, $(-2, -1)$, etc For positive integral n, $\Gamma(n + 1) = n!$, $\Gamma(1) = 1$; and $\Gamma(\frac{1}{2}) = \sqrt{\pi}$. $\Gamma(x) \to \infty$ when $x \to 0$, -1, -2, ... (Fig. 191); at points other than these, $\Gamma(x)$ is continuous and differentiable. The integral converges uniformly in any interval $0 < a \leq x \leq b$.

The *beta function*, defined by

$$B(x, y) = \int_0^1 t^{x-1}(1 - t)^{y-1}\, dt, \qquad x, y > 0,$$

may be expressed in terms of gamma functions:

$$B(x, y) = \frac{\Gamma(x)\Gamma(y)}{\Gamma(x + y)} = B(y, x).$$

The integral converges uniformly when $x, y \geq a > 0$.

CHAPTER 11

Functions of a Complex Variable

195. Rational Functions of a Complex Variable. The fundamental operations on complex numbers—addition, subtraction, multiplication, and division (except by zero)—were defined in §9. Thus a polynomial

$$P(z) = a_0 z^n + a_1 z^{n-1} + \cdots + a_n$$

with complex coefficients is a single-valued function of $z = x + iy$. The same is true of a rational function, the quotient of two polynomials:

$$f(z) = P(z)/Q(z), \qquad Q(z) \neq 0.$$

Since the limit operations of §16 apply without change when the variables are complex, we see that

$$\lim_{z \to \zeta} P(z) = a_0 \, \zeta^n + a_1 \, \zeta^{n-1} + \cdots + a_1 = P(\zeta),$$

and

$$\lim_{z \to \zeta} \frac{P(z)}{Q(z)} = \frac{P(\zeta)}{Q(\zeta)} \quad \text{if} \quad Q(\zeta) \neq 0.$$

In other words, complex polynomials are continuous at all points of the complex plane, whereas rational functions of a complex variable are continuous at all points where the denominator does not vanish. If $Q(z)$ is a polynomial of degree n, $Q(z)$ has exactly n linear factors, which, set equal to zero, yield the n roots $\zeta_1, \cdots, \zeta_n$ of $Q(z) = 0$. As $z \to \zeta_i$, $|P/Q| \to \infty$ if $P(\zeta_i) \neq 0$; hence, we write

$$\lim_{z \to \zeta_i} \frac{P(z)}{Q(z)} = \infty.$$

The n points ζ_i (some of which may be repeated) are called the *poles* of $f(z)$.

The *derivative* of a rational function is defined by

(1)
$$f'(z) = \lim_{\Delta z \to 0} \frac{f(z + \Delta z) - f(z)}{\Delta z}$$

where the complex increment approaches zero in an arbitrary manner. Thus we find that

$$P'(z) = a_0 n z^{n-1} + a_1(n-1)z^{n-2} + \cdots + a_{n-1};$$

and that a sum, product, and quotient of polynomials have derivatives which may be computed by the same rules that apply to real functions.

A rational function of z may be written in the form

$$(2) \qquad\qquad f(z) = P\bar{Q}/Q\bar{Q} = u + iv,$$

where $\bar{Q}$ is the conjugate of Q; for $Q\bar{Q}$ is real and we need only separate the polynomial $P\bar{Q}$ into its real and imaginary parts. Now

$$(3) \qquad\qquad f'(z) = \lim_{\Delta z \to 0} \frac{\Delta u + i\Delta v}{\Delta x + i\Delta y},$$

where $\Delta z = \Delta x + i\Delta y$. If in particular $\Delta y = 0$ and $\Delta x \to 0$,

$$(4) \qquad\qquad f'(z) = \frac{\partial u}{\partial x} + i\frac{\partial v}{\partial x};$$

and, if $\Delta x = 0$ and $\Delta y \to 0$,

$$(5) \qquad\qquad f'(z) = \frac{\partial v}{\partial y} - i\frac{\partial u}{\partial y}.$$

When $f(z)$ is a rational function of z, these expressions must be identical, and we obtain the *Cauchy-Riemann equations*:

$$(6) \qquad\qquad \frac{\partial u}{\partial x} = \frac{\partial v}{\partial y}, \qquad \frac{\partial u}{\partial y} = -\frac{\partial v}{\partial x}.$$

Now u, v, and all their partial derivatives are continuous except at the poles of $f(z)$; hence,

$$u_{xy} = u_{yx}, \qquad\qquad v_{xy} = v_{yx}$$

at all points where $f(z)$ is finite. In view of (6), these equations show that both u and v satisfy *Laplace's* equation:

$$(7) \qquad\qquad \frac{\partial^2 \varphi}{\partial x^2} + \frac{\partial^2 \varphi}{\partial y^2} = 0.$$

Example 1. $f(z) = z^2$ and $f'(z) = 2z$. Since

$$z^2 = (x + iy)^2 = x^2 - y^2 + i2xy,$$

$$u = x^2 - y^2, \qquad v = 2xy,$$

and the C–R equations are satisfied.

Example 2. $f(z) = 1/z \ (z \neq 0)$, and $f'(z) = -1/z^2$.
Since

$$\frac{1}{z} = \frac{1}{x + iy} = \frac{x - iy}{x^2 + y^2},$$

$$u = \frac{x}{x^2 + y^2}, \qquad v = \frac{-y}{x^2 + y^2},$$

and the C–R equations are satisfied.

196. Functions of a Complex Variable. Let the complex variable z range over a region R of the complex plane. Then, if we have a rule that associates a single definite complex number w with each value of z, we say that w is a *function* of z in R and write $w = f(z)$.

If we separate z and w into their real and imaginary parts,

$$z = x + iy, \qquad w = u + iv,$$

the relation $w = f(z)$ determines uniquely two real numbers u, v when x, y are given. Thus u and v are functions of x and y (§ 73), and we write

$$(1) \qquad w = f(z) = u(x, y) + iv(x, y).$$

Just as z may be regarded an ordered pair of real numbers (x, y), $f(z)$ may be regarded as an ordered pair of real functions $u(x, y)$, $v(x, y)$. Rational functions of z were defined in terms of the four basic operations performed on z. But, since z determines x and y, operations on these variables giving u and v uniquely also define functions of z. Thus the *absolute value* of z (§ 9),

$$(2) \qquad |z| = \sqrt{x^2 + y^2}, \qquad (u = \sqrt{x^2 + y^2}, \qquad v = 0),$$

and the *conjugate* of z (§ 8),

$$(3) \qquad \bar{z} = x - iy \qquad (u = x, \quad v = -y),$$

are functions of z over the entire complex plane. Since z determines $\bar{z}$, any function of z and $\bar{z}$ defines a function of z alone; for example,

$$(4) \qquad \frac{z}{\bar{z}} = \frac{z^2}{z\bar{z}} = \frac{x^2 - y^2 + i2xy}{x^2 + y^2}$$

defines a function of z when $z \neq 0$.

Limit and continuity are defined precisely as in the case of real variables (§§ 44, 45). Limits, however, are two-dimensional: *the approach is over any path in the complex plane.*

If $f(z)$ is defined in a neighborhood of the point ζ (except perhaps at ζ itself), we write

(5) $$\lim_{z \to \zeta} f(z) = \omega,$$

when for every $\varepsilon > 0$ we can find a number $\delta > 0$ such that

(6) $$\left| f(z) - \omega \right| < \varepsilon \quad \text{when} \quad 0 < \left| z - \zeta \right| < \delta.$$

THEOREM 1. *If $\zeta = a + ib$ and $\omega = A + iB$, the limit (5) holds when and only when*

(7) $$\lim_{\substack{x \to a \\ y \to b}} u(x, y) = A, \qquad \lim_{\substack{x \to a \\ y \to b}} v(x, y) = B.$$

Proof. Assume (5). When z lies in a circle of radius δ about ζ, $\left| x - a \right| < \delta$, $\left| y - b \right| < \delta$; then, from (6), $\left| u - A \right| < \varepsilon$, $\left| v - B \right| < \varepsilon$, and the limits (7) exist.

Assume (7). When (x, y) lies in a square of side δ centered at (a, b), $\left| x - a \right| < \tfrac{1}{2}\delta$, $\left| y - b \right| < \tfrac{1}{2}\delta$, and $\left| z - \zeta \right| < \delta$. Choose δ so small that $\left| u - A \right| < \tfrac{1}{2}\varepsilon$, $\left| v - B \right| < \tfrac{1}{2}\varepsilon$; then $\left| f(z) - \omega \right| < \varepsilon$, and the limit (5) exists.

If $f(z)$ is defined when $z = \zeta$, $f(z)$ is said to be continuous at ζ when

(8) $$\lim_{z \to \zeta} f(z) = f(\zeta).$$

THEOREM 2. *The function $f(z) = u + iv$ is continuous at $\zeta = a + ib$ when and only when the real functions, $u(x, y)$ and $v(x, y)$ are continuous at (a, b).*

Proof. From Theorem 1 the limit (8) implies the limits (7) with $A = u(a, b)$, $B = v(a, b)$; and conversely.

This theorem in combination with Theorems 75.2 and 75.3 now gives

THEOREM 3. *A function $f(z)$ which is continuous in a closed region R is uniformly continuous there; that is, for any $\varepsilon > 0$ we can choose a δ so that, if z, z' are any points of R for which*

(9) $$\left| z - z' \right| < \delta, \quad \text{then} \quad \left| f(z) - f(z') \right| < \varepsilon.$$

THEOREM 4. *A function $f(z)$ which is continuous in a closed region is bounded there.*

197. Analytic Functions. The derivative of a function $w = f(z)$ at a point z is defined by

(1) $$\frac{dw}{dz} = f'(z) = \lim_{\Delta z \to 0} \frac{f(z + \Delta z) - f(z)}{\Delta z}$$

when this limit exists, no matter how $\Delta z \to 0$. If we write

$$f(z) = u + iv, \qquad \Delta z = h + ik$$

(1) becomes

(2)
$$f'(z) = \lim_{\substack{h \to 0 \\ k \to 0}} \frac{\Delta u + i\Delta v}{h + ik}.$$

When this limit exists, we can show just as in § 195 that u and v must satisfy the Cauchy-Riemann equations,

(3)
$$u_x = v_y, \qquad u_y = -v_x.$$

Conversely, if $u(x, y)$ and $v(x, y)$ are *differentiable* functions (§ 77) at the point $z = x + iy$ and satisfy the C–R equations there, then $f(z) = u + iv$ has a derivative given by

(4)
$$f'(z) = u_x + iv_x = \frac{1}{i}(u_y + iv_y).$$

For from (77.3)

$$\Delta u = hu_x + ku_y + \eta_1 h + \eta_2 k,$$

$$\Delta v = hv_x + kv_y + \eta_3 h + \eta_4 k,$$

where the η's $\to 0$ as $h, k \to 0$; and, if we put $u_y = -v_x$, $v_y = u_x$,

$$\Delta u + i\Delta v = (u_x + iv_x)(h + ik) + \eta h + \eta' k,$$

where $\eta = \eta_1 + i\eta_3$, $\eta' = \eta_2 + i\eta_4$. Hence, as $h, k \to 0$,

$$\frac{\Delta u + i\Delta v}{h + ik} \to u_x + iv_x,$$

since

$$\frac{|h|}{|h + ik|} \leqq 1, \qquad \frac{|k|}{|h + ik|} \leqq 1.$$

The second expression for $f'(z)$ in (4) follows at once from the C–R equations.

A function $f(z) = u + iv$ which has a continuous derivative $f'(z)$ throughout a region R is said to be analytic in R.† The function is said to be *analytic at a point* if it is analytic in some neighborhood (§ 74) of that point. A function analytic in an *open* region R is analytic at each point of R.

We now have the important

THEOREM 1. *In order that the function*

$$f(z) = u(x, y) + iv(x, y)$$

---

† Goursat has proved that the mere existence of $f'(z)$ in R implies its continuity; hence $f(z)$ is analytic in R if it has a derivative at all points of R.

be analytic in a region R, it is necessary and sufficient that u_x, u_y, v_x, v_y be continuous in R and satisfy the Cauchy-Riemann equations.

Proof. When $f(z)$ is analytic in R, it has a continuous derivative (4) and the C–R equations are satisfied.

Conversely, the continuity of the four partial derivatives of u, v in R implies that they are differentiable in R (Theorem 78). Then the C–R equations show that $f'(z)$ exists, is given by (4), and consequently is continuous (Theorem 196.2).

It can be shown that, if $f(z)$ is analytic in R, it possesses continuous partial derivatives of all orders in R.† In particular, the second-order partial derivatives of u and v exist and are continuous in R; then we have from the C–R equations

$$u_{xx} + u_{yy} = v_{yx} - v_{xy} = 0,$$

$$v_{xx} + v_{yy} = -u_{yx} + u_{xy} = 0$$

from Theorem 82.1. We have thus proved

THEOREM 2. *If $f(z) = u + iv$ is analytic in R, both u and v must satisfy Laplace's equation,*

$$(5) \qquad \frac{\partial^2 u}{\partial x^2} + \frac{\partial^2 u}{\partial y^2} = 0, \qquad \frac{\partial^2 v}{\partial x^2} + \frac{\partial^2 v}{\partial y^2} = 0.$$

The solutions of Laplace's equation are called *harmonic functions*; and two such functions u, v which satisfy the C–R equations are called *harmonic conjugates.*‡ When a harmonic function $u(x, y)$ is given, a conjugate function $v(x, y)$ may be found by line integration; for from (3)

$$dv = v_x \, dx + v_y \, dy = u_x \, dy - u_y \, dx,$$

and

$$(6) \qquad v(x, y) = \int_{(x_0, y_0)}^{(x, y)} u_x \, dy - u_y \, dx \quad \text{(indep. of path)}.$$

The C–R equations (3) are equivalent to the vector equation,

$$(7) \qquad \nabla v = \mathbf{e}_3 \times \nabla u \quad \text{or} \quad \nabla u = \nabla v \times \mathbf{e}_3,$$

as we see on writing $\nabla u = u_x \mathbf{e}_1 + u_y \mathbf{e}_2$, $\nabla v = v_x \mathbf{e}_1 + v_y \mathbf{e}_2$. The form (7) is independent of the choice of coordinates. Thus, for polar coordinates r, φ, (7) becomes

$$u_r \mathbf{R} + u_\varphi \frac{\mathbf{P}}{r} = \left(v_r \mathbf{R} + v_\varphi \frac{\mathbf{P}}{r} \right) \times \mathbf{e}_3 = \frac{v_\varphi}{r} \mathbf{R} - v_r \mathbf{P},$$

---

† Cf. Knopp, Konrad, *Theory of Functions*, part I, Dover Publications, 1945, § 16. (Reissued by Dover in 1996 as *Theory of Functions, Parts I and II.* 0-486-69219-1)

‡ This use of "conjugate," applied to a pair of *real functions*, must not be confused with "conjugate" as applied to the pair of *complex numbers* $u \pm iv$ (§ 8).

since $\nabla r = \mathbf{R}$, $\nabla \varphi = \mathbf{P}/r$ (106.14). Hence the C–R equations in polar coordinates are

(8)
$$u_r = \frac{1}{r} v_\varphi, \qquad \frac{1}{r} u_\varphi = -v_r.$$

When they are satisfied, we may compute dw/dz along an outward ray of angle φ; then $\Delta z = (\cos \varphi + i \sin \varphi)\, \Delta r$ and

(9)
$$\frac{dw}{dz} = \lim_{\Delta r \to 0} \frac{\Delta w}{\Delta z} = \frac{1}{\cos \varphi + i \sin \varphi} \left(\frac{\partial u}{\partial r} + i \frac{\partial v}{\partial r} \right).$$

THEOREM 3. *If the derivative of an analytic function is everywhere zero in a region R, the function is constant in R.*

Proof. If $f'(z) = 0$, we have from (4)

$$u_x = v_x = 0, \qquad u_y = v_y = 0 \qquad \text{in } R.$$

Corollary. *If two analytic functions have the same derivative in R, they differ only by an additive constant.*

Example 1. If $u = ax^3 + bx^2y + cxy^2 + dy^3$ is a harmonic function, we must have $c = -3a$, $b = -3d$; thus,

$$a(x^3 - 3xy^2) + d(y^3 - 3x^2y)$$

is a harmonic function for any choice of a and d.

If we choose $u = x^3 - 3xy^2$, the harmonic conjugate given by (6) is

$$v = \int (3x^2 - 3y^2)\, dy + 6xy\, dx.$$

Integrating over the step path $(0, 0) - (x, 0) - (x, y)$, we find

$$v = 0 + \int_0^y (3x^2 - 3y^2)\, dy = 3x^2y - y^3.$$

The function

$$f(z) = u + iv = (x^3 - 3xy^2) + i(3x^2y - y^3) = (x + iy)^3$$

is analytic for all finite z.

Example 2. The function

$$w = \frac{\cos \varphi}{r} - i \frac{\sin \varphi}{r} = \frac{1}{z}$$

is analytic in any region not including the origin; for equations (8) are satisfied. From (9) we find

$$\frac{dw}{dz} = -\frac{1}{r^2} \frac{\cos \varphi - i \sin \varphi}{\cos \varphi + i \sin \varphi} = -\frac{1}{z^2}.$$

PROBLEMS

1. Find conjugates to the following functions:

(i) $x^2 - y^2$;   (ii) $e^x \cos y$;   (iii) $x/(x^2 + y^2)$,   (iv) $\cos x \sinh y$.

2. If $u(x, y)$ and $v(x, y)$ are conjugate functions, show that the integrals

$$U(x, y) = \int_{(x_0, y_0)}^{(x,y)} u \, dx - v \, dy, \qquad V(x, y) = \int_{(x_0, y_0)}^{(x,y)} v \, dx + u \, dy.$$

are independent of the path and that U and V are also conjugate functions. Apply this result to $u = x$, $v = y$. $[U = \frac{1}{2}(x^2 - y^2), V = xy.]$

3. If u and v are harmonic conjugates, show that

(i) $U = v, V = -u$;

(ii) $U = au - bv, V = bu + av$;

(iii) $U = u/(u^2 + v^2), V = -v/(u^2 + v^2)$,

are also harmonic conjugates.
[The functions $(a + ib)(u + iv)$ and $(u + iv)^{-1}$ are analytic.]

4. If u_1, v_1 and u_2, v_2 are harmonic conjugates, $U = u_1 u_2 - v_1 v_2$, $V = u_1 v_2 - v_1 u_2$ are also.
[Consider $(u_1 + iv_1)(u_2 + iv_2)$.]

198. Exponential Function. The exponential function for complex arguments is defined by the power series,

$$(1) \qquad e^z = 1 + z + \frac{z^2}{2!} + \cdots + \frac{z^n}{n!} + \cdots,$$

which converges throughout the complex plane; for

$$\tau = \lim_{n \to \infty} \frac{1/n!}{1/(n-1)!} = \lim_{n \to \infty} \frac{1}{n} = 0, \qquad R = \infty.$$

Thus e^z is continuous throughout the complex plane (Theorem 183.1); when $z = x$ (real), we obtain the familiar series for e^x. Moreover, since term-by-term differentiation does not alter the series (1), we have

$$(2) \qquad \frac{d}{dz} e^z = e^z \qquad \text{(Theorem 183.2)},$$

and all higher derivatives have the same value.
If ζ is a fixed but arbitrary value of z, the power series for

$$e^{\zeta + z} = e^\zeta \left(1 + z + \frac{z^2}{2!} + \cdots \right);$$

for the coefficient of the term in z^n is

$$\frac{1}{n!} f^{(n)}(0) = \frac{1}{n!} e^{\zeta} \qquad \text{(Theorem 183.3)}.$$

Hence, in view of (1)

(3) $e^{\zeta+z} = e^{\zeta} e^{z};$

thus the addition theorem for the exponential function is valid in the complex domain.

199. Sine and Cosine.   The sine and cosine for complex arguments are defined by the power series,

$$(1) \qquad\qquad \sin z = z - \frac{z^3}{3!} + \frac{z^5}{5!} - \cdots,$$

$$(2) \qquad\qquad \cos z = 1 - \frac{z^2}{2!} + \frac{z^4}{4!} - \cdots,$$

which converge throughout the complex plane. We still have the relations

(3) $\sin(-z) = -\sin z, \qquad \cos(-z) = \cos z;$

and, when $z = x$ (real), we get the familiar series for $\sin x$ and $\cos x$. Moreover, term-by-term differentiation shows that

$$(4) \qquad\qquad \frac{d}{dz}\sin z = \cos z, \qquad \frac{d}{dz}\cos z = -\sin z.$$

The series for e^{iz} converges absolutely and may be rearranged into two series, one of even and one of odd powers of iz (Theorem 36.1). Since the powers of i repeat in the cycle

$$i = i, \qquad i^2 = -1, \qquad i^3 = -i, \qquad i^4 = 1,$$

we find that

(5) $e^{iz} = \cos z + i \sin z.$

When we replace z by $-z$ and make use of (3), this becomes

(6) $e^{-iz} = \cos z - i \sin z.$

We can now solve (5) and (6) for $\sin z$ and $\cos z$:

$$(7)\,(8) \qquad \sin z = \frac{(e^{iz} - e^{-iz})}{2i}, \qquad \cos z = \frac{(e^{iz} + e^{-iz})}{2}.$$

These famous formulas of Euler express the circular functions in terms of the exponential function.

When we make use of (5), the addition theorem,

$$e^{i(\zeta+z)} = e^{i\zeta} e^{iz}$$

becomes

$$\cos(\zeta + z) + i\sin(\zeta + z) = (\cos\zeta + i\sin\zeta)(\cos z + i\sin z).$$

On equating the even and odd functions in both members, we obtain the addition theorems for the sine and cosine:

(9) $$\cos(\zeta + z) = \cos\zeta\cos z - \sin\zeta\sin z,$$

(10) $$\sin(\zeta + z) = \sin\zeta\cos z + \cos\zeta\sin z.$$

If we put $\zeta = -z$ in (9), we get the fundamental relation,

(11) $$\cos^2 z + \sin^2 z = 1.$$

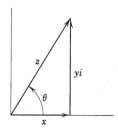

Fig. 199.   Complex vector $z = x + yi$

Finally we have

(12) $$e^z = e^{x+yi} = e^x(\cos y + i\sin y),$$

(13) $$\left| e^z \right| = e^x \neq 0.$$

Hence, $e^z \neq 0$ throughout the complex plane.

If n is an integer,

(14) $$e^{2n\pi i} = \cos 2n\pi + i\sin 2n\pi = 1.$$

Hence, from the addition theorem,

(15) $$e^{z+2n\pi i} = e^z;$$

thus e^z admits the imaginary period $2\pi i$. Moreover, all periods of e^z are included in $2n\pi i$; for, if

$$e^{\zeta+z} = e^\zeta, \qquad e^z = e^x(\cos y + i\sin y) = 1;$$

taking absolute values, we have $e^x = 1$, $x = 0$; hence,

$$\cos y + i\sin y = 1, \qquad y = 2n\pi, \qquad z = 2n\pi i.$$

If we plot the vector $z = x + yi$ in the complex plane (Fig. 199), we have $x = r\cos\theta$, $y = r\sin\theta$, and

(16) $$z = r(\cos\theta + i\sin\theta) = re^{i\theta}.$$

This is the *polar form* of z; $r = \sqrt{x^2 + y^2}$ is the absolute value or *modulus* of z, while θ is its *angle*.† Evidently $\theta + 2n\pi$ would also serve as the angle of z.

200. Hyperbolic Functions. The hyperbolic sine and cosine are defined by the formulas used when the variable is real:

$$(1) \qquad \sinh z = \frac{e^z - e^{-z}}{2}, \qquad \cosh z = \frac{e^z + e^{-z}}{2}.$$

Their power series may be obtained from the sine and cosine series by changing all minus signs to plus; they converge throughout the complex plane. The derivative of either function is the other:

$$(2) \qquad \frac{d}{dz} \sinh z = \cosh z, \qquad \frac{d}{dz} \cosh z = \sinh z.$$

In view of Euler's formulas, we have

$$(3) \qquad \sinh iz = i \sin z, \qquad \cosh iz = \cos z;$$
$$(4) \qquad \sin iz = i \sinh z, \qquad \cos iz = \cosh z.$$

Thus, every rational formula involving $\sin z$ and $\cos z$ gives a corresponding hyperbolic formula by changing z into iz. Hence the rule:

Replace $\sin z$, $\cos z$ *in a trigonometric formula by* $i \sinh z$, $\cosh z$ *to obtain the analogous hyperbolic result.*

Thus $\cos^2 z + \sin^2 z = 1$ gives

$$(5) \qquad \cosh^2 z - \sinh^2 z = 1;$$

and the addition theorems for the sine and cosine become

$$(6) \qquad \cosh (\zeta + z) = \cosh \zeta \cosh z + \sinh \zeta \sinh z,$$
$$(7) \qquad \sinh (\zeta + z) = \sinh \zeta \cosh z + \cosh \zeta \sinh z.$$

Making use of (3) and (4), we have

$$\sin (x + yi) = \sin x \cosh y + i \cos x \sinh y,$$
$$\cos (x + yi) = \cos x \cosh y - i \sin x \sinh y.$$

These equations give the absolute values of $\sin z$ and $\cos z$:

$$(8) \qquad | \sin z |^2 = \sin^2 x \cosh^2 y + \cos^2 x \sinh^2 y$$
$$= \sin^2 x + \sinh^2 y,$$
$$(9) \qquad | \cos z |^2 = \cos^2 x \cosh^2 y + \sin^2 x \sinh^2 y$$
$$= \cos^2 x + \sinh^2 y.$$

---

† θ is also called the *amplitude* or *argument* of z, and written amp z or arg z.

Since sinh y vanishes only at $y = 0$, the zeros of sin z and cos z are all real and precisely those of sin x and cos x.

We have seen that many equations between real functions carry over unchanged when the variable is complex. This, however, is not the case with *inequalities*. For example, while

$$|\sin x| \leq 1, \qquad |\cos x| \leq 1, \qquad (x \text{ real}),$$

this is not in general true when we replace x by $z = x + yi$. In fact, we see from (8) and (9) that, as $y \to \infty$, both $|\sin z|$ and $|\cos z| \to \infty$.

If we replace z by $iz = -y + ix$ in (8) and (9), we get

(10) $$|\sinh z|^2 = \sin^2 y + \sinh^2 x,$$

(11) $$|\cosh z|^2 = \cos^2 y + \sinh^2 x.$$

Since sinh x vanishes only at $x = 0$, we see that the zeros of sinh z and cosh z are all pure imaginary, namely, $n\pi i$ and $(n + \frac{1}{2})\pi i$, respectively.

201. Trigonometric Relations. The Euler relations (199.7–8) between the exponential and circular functions often suggest a method for obtaining a trigonometric identity. Thus, from $e^{inx} = (e^{ix})^n$ we have *DeMoivre's theorem*:

(1) $$\cos nx + i \sin nx = (\cos x + i \sin x)^n.$$

This enables us to express cos nx and sin nx in terms of powers of cos x and sin x by equating real and imaginary parts. For example,

$$\cos 3x = \cos^3 x - 3 \cos x \sin^2 x,$$
$$\sin 3x = 3 \cos^2 x \sin x - \sin^3 x.$$

From $e^{nx} = (e^x)^n$ we get the hyperbolic analogue of (1);

(1)' $$\cosh nx + \sinh nx = (\cosh x + \sinh x)^n.$$

Hence cosh nx and sinh nx equal, respectively, the even and odd parts (§ 68) of $(\cosh x + \sinh x)^n$. Thus when $n = 3$,

$$\cosh 3x = \cosh^3 x + 3 \cosh x \sinh^2 x,$$
$$\sinh 3x = 3 \cosh^2 x \sinh x + \sinh^3 x.$$

Again we may readily express a product such as $\cos^m x \sin^n x$ as a sum of sines and cosines of multiple angles by means of Euler's formulas. For example,

$$\sin^2 x \cos^3 x = \left(\frac{e^{xi} - e^{-xi}}{2i}\right)^2 \left(\frac{e^{xi} + e^{-xi}}{2}\right)^3$$
$$= -\tfrac{1}{32}(e^{2xi} - e^{-2xi})^2 (e^{xi} + e^{-xi})$$
$$= -\tfrac{1}{32}(e^{5xi} + e^{-5xi} + e^{3xi} + e^{-3xi} - 2e^{xi} - 2e^{-xi})$$
$$= -\tfrac{1}{16}(\cos 5x + \cos 3x - 2 \cos x).$$

We next obtain the sums $\displaystyle\sum_{k=1}^{n} \cos kx$ and $\displaystyle\sum_{k=1}^{n} \sin kx$ by taking the real

and imaginary parts of the geometric series,

$$\sum_{k=1}^{n} e^{kxi} = e^{xi}\, \frac{e^{nxi} - 1}{e^{xi} - 1}.$$

If we multiply above and below by $e^{-\frac{1}{2}xi}$, the right member becomes

$$\frac{e^{(n+\frac{1}{2})xi} - e^{\frac{1}{2}xi}}{e^{\frac{1}{2}xi} - e^{-\frac{1}{2}xi}} = \frac{\cos (n + \frac{1}{2})x + i \sin (n + \frac{1}{2})x - \cos \frac{1}{2}x - i \sin \frac{1}{2}x}{2i \sin \frac{1}{2}x},$$

Hence, when $x \neq 0$, we have

$$(2) \qquad \sum_{k=1}^{n} \cos kx = \frac{\sin (n + \frac{1}{2})x - \sin \frac{1}{2}x}{2 \sin \frac{1}{2}x},$$

$$(3) \qquad \sum_{k=1}^{n} \sin kx = \frac{\cos \frac{1}{2}x - \cos (n + \frac{1}{2})x}{2 \sin \frac{1}{2}x}.$$

On adding $1/2$ to both sides of (2), we obtain

$$(4) \qquad \frac{1}{2} + \cos x + \cos 2x + \cdots + \cos nx = \frac{\sin (n + \frac{1}{2})x}{2 \sin \frac{1}{2}x},$$

a summation useful in discussing Fourier series. If we replace $n + \frac{1}{2}$ and $\frac{1}{2}$ in (2) and (3) by $\frac{1}{2}(n + 1) + \frac{1}{2}n$, $\frac{1}{2}(n + 1) - \frac{1}{2}n$, and use the addition formulas for sine and cosine, we have

$$(5) \qquad \sum_{k=1}^{n} \frac{\cos}{\sin} kx = \frac{\sin \frac{1}{2}nx}{\sin \frac{1}{2}x} \left(\frac{\cos}{\sin} \tfrac{1}{2}(n + 1)x \right).$$

Lastly, the geometric series,

$$e^{xi} + e^{3xi} + \cdots + e^{(2n-1)xi} = e^{xi}\, \frac{e^{2nxi} - 1}{e^{2xi} - 1},$$

yields the formulas

$$(6) \qquad \cos x + \cos 3x + \cdots + \cos (2n - 1)x = \frac{\sin 2nx}{2 \sin x},$$

$$(7) \qquad \sin x + \sin 3x + \cdots + \sin (2n - 1)x = \frac{1 - \cos 2nx}{2 \sin x}.$$

202. Logarithm. The logarithm of the complex number z is defined as any one of the numbers w that satisfy the equation

$$(1) \qquad\qquad e^{w} = z.$$

If we put $w = u + iv$, $z = re^{i\varphi}$,

$$e^u e^{iv} = re^{i\varphi};$$

and, on taking absolute values,

(2) $$e^u = r, \qquad u = \log r.$$

Hence v must satisfy $e^{iv} = e^{i\varphi}$; this equation has infinitely many solutions owing to the period $2\pi i$ of the exponential function; hence,

(3) $$v = \varphi + 2k\pi, \quad k \text{ an integer.}$$

Thus

(4) $$\log z = \log r + i(\varphi + 2k\pi)$$

has an infinite number of values which differ by integral multiples of $2\pi i$. If α is a real constant, there is just one value of v in the range $\alpha < v \leq \alpha + 2\pi$. With this restriction on v, $\log z$ is a single-valued function of z; such values of $\log z$ form a *single-valued branch* of the multivalued logarithm. The *principal value* of $\log z$ is the one corresponding to the branch $-\pi < v \leq \pi$.

As $z \to 0$, the real part of $\log z$, namely, $\log r$, approaches $-\infty$; thus $\log 0$ does not exist. The logarithm of a negative number $-r$ has the principal value,

(5) $$\log(-r) = \log r + i\pi.$$

The functional equation for real logarithms,

(6) $$\log r_1 + \log r_2 = \log r_1 r_2,$$

must be modified when we deal with complex logarithms. Thus, if

$$z_1 z_2 = r_1 e^{i\theta_1} r_2 e^{i\theta_2} = r_1 r_2 e^{i(\theta_1 + \theta_2)},$$

(7) $$\log z_1 + \log z_2 = \log z_1 z_2$$

is true for logarithms belonging to the same branch only when θ_1, θ_2, $\theta_1 + \theta_2$ lie in the appropriate range. Thus, if $\log z_1$, $\log z_2$ denote principal values, (7) is valid only when $-\pi < \theta_1 + \theta_2 \leq \pi$. In general, (7) must be modified by adding a suitable multiple of $2\pi i$ to the right-hand member. For example, if $-a$, $-b$ are negative numbers,

$$\log(-a) + \log(-b) = \log(ab) + 2\pi i$$

when the logarithms are all principal values.

Since the values of u and v given in (2) and (3) satisfy the C–R equations (197.8), any branch of the logarithm has a unique derivative given by

(8) $$\frac{d}{dz} \log z = \frac{1}{re^{i\varphi}} = \frac{1}{z} \qquad (197.9).$$

PROBLEMS

1. If $f(z) = u(x, y) + iv(x, y)$ find the conjugate functions u and v when $f(z) = z^2$; iz^3; $1/z$; e^{iz}; $\log z$ (prin. branch); $\sin z$; $\cos z$; $\tan z$; $\sinh z$; $\cosh z$.

2. Show that the functions u, v found in Prob. 1 satisfy Laplace's equation (197.5).

3. Find the function $v(x, y)$ conjugate to $u = ax + by$; $x^3 - 3xy^2$; $e^x \sin y$.

4. If $f(x + iy) = u + iv$, show that $f(x - iy) = u - iv$ when $f(z) = e^z$, $\sin z$, $\cosh z$, $\log z$ (prin. branch).

5. Find the conjugate functions $u(r, \varphi)$, $v(r, \varphi)$ in polar form when

$$f(z) = z^2, 1/z, \log z, e^z, e^{iz}.$$

6. Find the principal values of

$$\log i; \quad \log(-2); \quad \log(1 + i); \quad \log(1 - \sqrt{3}\,i); \quad \log(-3 - 4i).$$

7. Show that $1/(x - iy)$ and e^{x-iy} are not analytic functions.

8. Which of the following functions is analytic when $r \neq 0$:

$$r + i\varphi; \quad e^{-i\varphi}/r; \quad re^{-i\varphi}; \quad \sqrt{r}\,e^{\frac{1}{2}i\varphi}?$$

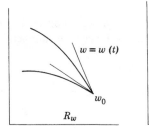 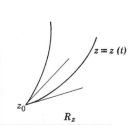

FIG. 203a.   Conformal transformation

9. Find the sum of the zeros of $z^{2n} + 1$ above the axis of reals when n is a positive integer.

10. What points of the complex plane satisfy $|e^{-2z}| < 2$? $|z-1| < x$?

11. Find the roots of least absolute value:

$$(a)\ \sin z = 2; \quad (b)\ \cos z = -2i.$$

203. Conformal Mapping.   Let

$$(1) \qquad\qquad w = f(z) = u(x, y) + i\,v(x, y)$$

be defined in a region R_z of the z-plane. Then every value of $z = x + iy$ in R_z corresponds to a single value $w = u + iv$ of a region R_w of the w-plane. We shall also suppose that each point w of R_w corresponds to just one value of z in R_z; then the mapping $R_z \leftrightarrow R_w$ is one-to-one. Then there is also a one-to-one correspondence between the curves of R_z,

$$x = x(t), \qquad y = y(t) \qquad (\text{or } z = z(t)),$$

and their maps in R_w,

$$u = u(t), \qquad v = v(t) \qquad \text{(or } w = w(t)\text{)}.$$

If each pair of curves in R_z cutting at an angle θ corresponds to a pair in R_w cutting at this same angle, the mapping is said to be *isogonal*. If in addition the *sense* of the angles is preserved, the mapping is said to be *conformal*. We now prove the fundamental

THEOREM 1. *Let the function* $w = f(z)$, *analytic in* R_z, *map* R_z *one-to-one on* R_w; *then, if* $f'(z) \neq 0$ *in* R_z, *the mapping is conformal and the magnification at any point is independent of the direction and in the ratio of* $|f'(z)| : 1$.

Proof. Let the smooth curve $z = z(t)$, issuing from the point z_0 in R_z, map into the curve $w = w(t)$ issuing from w_0 in R_w. The tangent vectors to these curves have the components

$$\left[\frac{dx}{dt}, \frac{dy}{dt} \right], \qquad \left[\frac{du}{dt}, \frac{dv}{dt} \right] \tag{101.2}$$

and are represented by the complex vectors,

$$\frac{dz}{dt} = \frac{dx}{dt} + i\frac{dy}{dt}, \qquad \frac{dw}{dt} = \frac{du}{dt} + i\frac{dv}{dt}.$$

By the chain rule,

$$\frac{dw}{dt} = \frac{dw}{dz}\frac{dz}{dt};$$

hence, if at the point z_0 we express dw/dz and dz/dt in the polar form (199.16),

$$\frac{dw}{dz} = ae^{i\alpha} \quad (a \neq 0), \qquad \frac{dz}{dt} = re^{i\theta},$$

we have, at the corresponding point w_0,

$$\frac{dw}{dt} = are^{i(\alpha+\theta)}.$$

Hence, the mapping multiplies all tangent vectors at z_0 by $a = |f'(z)|$ and turns them through the same angle α. Thus the angle between two tangent vectors at z_0 is preserved in magnitude and sense, and the magnification $|f'(z)|$ is independent of their direction.

A small curvilinear triangle with a vertex at z_0 is thus mapped upon a triangle of the same general shape (conformal) at w_0; for the angles at these points are equal and the corresponding sides are proportional. But, while the mapping is *locally* conformal, corresponding regions need not be similar in the large since the magnification varies in general from point to point.

The converse of Theorem 1 is also true.

THEOREM 2. *If the map formed by the differentiable functions,*

$$(2) \qquad\qquad u = u(x, y), \qquad v = v(x, y),$$

is conformal in a region R, the function $f(z) = u + iv$ is analytic in R.

Proof. Since the perpendicular lines $u = c_1$, $v = c_2$ in the w-plane correspond to orthogonal curves $u(x, y) = c_1$, $v(x, y) = c_2$ in the z-plane, ∇u and ∇v are perpendicular vectors. Moreover, the angle θ from the line $u = c_1$ to the line $au + bv = c_2$ is the angle between the normal vectors $\mathbf{e}_1$ and $a\mathbf{e}_1 + b\mathbf{e}_2$, and hence $\tan \theta = b/a$. In the z-plane the curves

$$u(x, y) = c_1, \qquad au(x, y) + bv(x, y) = c_2$$

have the normal vectors ∇u and $a\,\nabla u + b\,\nabla v$; and, since ∇u and ∇v are perpendicular, the corresponding angle is given by

$$\tan \theta = \frac{b}{a} \frac{|\nabla v|}{|\nabla u|}; \quad \text{hence} \quad |\nabla v| = |\nabla u|.$$

Thus ∇u and ∇v are perpendicular vectors of the same length; and, since the map is conformal, they are oriented like $\mathbf{e}_1$ and $\mathbf{e}_2$; namely, $\mathbf{e}_2 = \mathbf{e}_3 \times \mathbf{e}_1$; thus the vector C–R equation

$$(3) \qquad\qquad\qquad \nabla v = \mathbf{e}_3 \times \nabla u \qquad\qquad\qquad (197.7),$$

is satisfied and $f(z) = u + iv$ is analytic in R.

Corollary. If angles are preserved but reversed in sense, ∇u and ∇v are still perpendicular vectors of the same length, but

$$(4) \qquad\qquad \nabla v = \nabla u \times \mathbf{e}_3 \quad \text{or} \quad \nabla(-v) = \mathbf{e}_3 \times \nabla u.$$

Now $u - iv$, the *conjugate* of $f(z)$, is analytic in R.

The Jacobian of the transformation (2) is

$$(5) \qquad\qquad\qquad J = \frac{\partial(u, v)}{\partial(x, y)} = u_x v_y - u_y v_x.$$

According as $u + iv$ or $u - iv$ is an analytic function, we have from the C–R equations

$$J = \begin{cases} u_x^2 + u_y^2 = |f'(z)|^2 > 0, \\ -u_x^2 - u_y^2 < 0. \end{cases}$$

Thus in an isogonal transformation (2) the sense of angles is preserved or reversed according as the Jacobian (5) is positive or negative.

Example 1. Rotation–Stretch. If $ae^{i\alpha} \neq 0$ is a complex constant in polar form, then

$$(6) \qquad\qquad\qquad w = ae^{i\alpha}z \qquad (a, \alpha \text{ real})$$

maps the entire z-plane conformally upon the w-plane since $dw/dz \neq 0$. If we write $z = re^{i\varphi}$, $w = Re^{i\Phi}$, (6) is equivalent to

$$(6)' \qquad\qquad R = ar, \qquad \Phi = \alpha + \varphi.$$

Thus, if the w-plane is placed upon the z-plane so that the axes coincide, any point $P \neq 0$ is mapped on a point w obtained by revolving OP through an angle α into OP' and then stretching OP' in the ratio $a/1$ into OQ (Fig. 203b):

$$P(z) \to P'(e^{i\alpha}z) \to Q(ae^{i\alpha}z).$$

When $a = 1$, we have a pure *rotation* $w = e^{i\alpha}z$; when $\alpha = 0$, a pure stretch $w = az$.

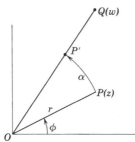

FIG. 203b.   Rotation—stretch

Example 2. *Translation.*   The function

$$(7) \qquad\qquad w = z + be^{i\beta}$$

maps the z-plane conformally on the w-plane by giving each point z the same vector displacement $be^{i\beta}$. Thus, in view of Ex. 1,

$$(8) \qquad\qquad w = ae^{i\alpha}z + be^{i\beta}$$

produces a conformal map by means of a rotation α, a stretch $a/1$, and a translation $be^{i\beta}$. If a and b denote complex constants, the general *rotation–stretch–translation* may be written more compactly as

$$(9) \qquad\qquad w = az + b \qquad (a, b \text{ complex}).$$

Example 3. *Reciprocal Transformation.*   The function

$$(10) \qquad\qquad w = 1/z$$

maps the z-plane conformally on the w-plane except in the neighborhood of $z = 0$ where dw/dz does not exist. Putting $z = re^{i\varphi}$, $w = Re^{i\Phi}$, we have $Re^{i\Phi} = e^{-i\varphi}/r$; hence,

$$R = \frac{1}{r}, \qquad \Phi = -\varphi.$$

Thus a point $z \neq 0$ is inverted in the unit circle to $z' = e^{i\varphi}/r$ then z' is reflected in the axis of reals to $w = e^{-i\varphi}/r$ (Fig. 203c). *The mapping $w = 1/z$ is a combination of inversion in the unit circle $|z| = 1$ followed by a reflection in the axis of reals.* Since inversion and reflection are isogonal transformations which reverse the sense of angles, their resultant is conformal.

The equation of any circle in the z-plane may be written

$$z\bar{z} - \bar{c}z - c\bar{z} + k = 0 \qquad (k \text{ real}),$$

where $\bar{z} = x - iy$ denotes the conjugate of z (Prob. 10.15): when $k = 0$, it passes through the origin. The transformation $w = 1/z$ maps the circle on another circle,

$$1 - \bar{c}w - cw + kw\bar{w} = 0 \quad \text{if} \quad k \neq 0,$$

or a straight line in case $k = 0$. Thus $w = 1/z$ maps circles into circles, except those through the origin that are mapped into straight lines. The unit circle $|z| = 1$ is reflected into itself.

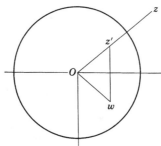

FIG. 203c.　Reciprocal transformation

Example 4. *Linear Fractional Transformation.* The nonsingular matrix $\begin{pmatrix} a & b \\ c & d \end{pmatrix}$ of complex constants corresponds to a transformation

(11) $$w = \frac{az + b}{cz + d}, \qquad D = \begin{vmatrix} a & b \\ c & d \end{vmatrix} \neq 0.$$

which is conformal except in the neighborhood of $z = -d/c$; for

$$\frac{dw}{dz} = \frac{D}{(cz + d)^2} \neq 0.$$

When $c = 0$, $d \neq 0$, (11) is a transformation of type (9). When $c \neq 0$,

$$w = \frac{a}{c} - \frac{D/c^2}{z + d/c} \, ;$$

hence (11) is the resultant of

$$z_1 = z + \frac{d}{c}, \qquad z_2 = \frac{1}{z_1}, \qquad z_3 = -\frac{D}{c^2}z_2, \qquad w = \frac{a}{c} + z_3,$$

namely a translation, an inversion–reflection, a rotation–stretch, and again a translation.

If we regard a straight line as a circle through the point ∞, each of these transformations maps circles on circles; hence the same is true of the general transformation (11).

The transformation (11) is uniquely determined by specifying the mapping of three distinct points into three distinct points, say $z_1 \leftrightarrow w_1$, $z_2 \leftrightarrow w_2$, $z_3 \leftrightarrow w_3$. In fact,

(12) $$\frac{w - w_1}{w - w_2} = k \frac{z - z_1}{z - z_2}$$

maps $z_1 \leftrightarrow w_1$, $z_2 \leftrightarrow w_2$, and putting $z = z_3$, $w = w_3$ determines k. When (12) is solved for w, we obtain the transformation in form (11) with nonzero determinant.

The resultant of (11) and

$$(13) \qquad w' = \frac{a'w + b'}{c'w + d'}, \qquad D' = \begin{vmatrix} a' & b' \\ c' & d' \end{vmatrix} \neq 0$$

is another linear fractional transformation of nonzero determinant. For, if we write (11) symbolically as

$$\frac{w}{1} = \begin{pmatrix} a & b \\ c & d \end{pmatrix} \frac{z}{1},$$

we find that

$$\frac{w'}{1} = \begin{pmatrix} a' & b' \\ c' & d' \end{pmatrix} \frac{w}{1} = \begin{pmatrix} a' & b' \\ c' & d' \end{pmatrix} \begin{pmatrix} a & b \\ c & d \end{pmatrix} \frac{z}{1}$$

has the determinant $D'D \neq 0$. With this notation the inverse of (11) is

$$(14) \qquad \frac{z}{1} = \begin{pmatrix} d & -b \\ -c & a \end{pmatrix} \frac{w}{1} ; \dagger$$

and $\begin{pmatrix} 1 & 0 \\ 0 & 1 \end{pmatrix}$ corresponds to the identity $w = z$. Evidently the linear fractional transformations of nonzero determinant form a *group* (§ 1).

Example 5. All linear fractional transformations of the form

$$(15) \qquad w = e^{i\alpha} \frac{z - z_0}{z - \bar{z}_0} \qquad (\alpha \text{ real}, \quad z_0 \neq \bar{z}_0)$$

map the axis of reals $y = 0$ on the circumference $|w| = 1$; for

$$|w| = \frac{|z - z_0|}{|z - \bar{z}_0|} = 1$$

when and only when $z = \bar{z}$, that is, when z is real. Since $z = z_0$ maps into $w = 0$, the *center* of the circle, the half-plane containing z_0 is mapped on (inside) the circle $|w| \leq 1$,

For example,

$$(16) \qquad w = \frac{1}{i} \frac{z - i}{z + i} \quad \text{or} \quad z = \frac{1}{i} \frac{w - i}{w + i}$$

maps the *upper* half-plane $y \geq 0$ on the circle $|w| \leq 1$. This remarkable transformation has an inverse of the same form; hence it also maps the upper half-plane $v \geq 0$ on the circle $|z| \leq 1$. Figure 203d shows how the planes correspond; the points ∞, -1, 0, 1, ∞ on the axis $y = 0$ and the points $-i$, 1, i, -1, $-i$ on the circumference $|z| = 1$ pass into points so marked in the w-plane. Moreover, the eight regions a, b, c, d, A, B, C, D of the z-plane map on the regions so lettered in the w-plane. *The transformation maps each upper half-plane on the unit circle in the other plane.*

---

† The matrix in (14) is $D \begin{pmatrix} a & b \\ c & d \end{pmatrix}^{-1}$. Note that the matrix of a transformation may be multiplied by a number k without altering the transformation, although its determinant D becomes $k^2 D$.

Example 6. The mapping $w = z^2$ is expressed by

$$R = r^2, \qquad \Phi = 2\varphi$$

in polar coordinates, or by

$$u = x^2 - y^2, \qquad v = 2xy$$

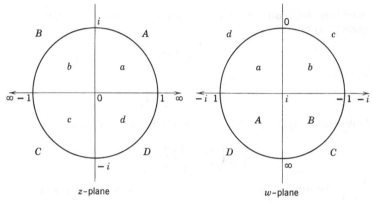

FIG. 203d. Mapping $w = \dfrac{1}{i}\dfrac{z-i}{z+i}$

in rectangular coordinates. The first quadrant of the z-plane is mapped conformally on the upper half of the w-plane except in the neighborhood of $z = 0$ (where $dw/dz = 0$).

In Fig. 203e the region *abc* bounded by

$$y = 0, \qquad x = 2, \qquad x^2 - y^2 = 1 \quad \text{(hyperbola)}$$

is mapped on *ABC* bounded by

$$v = 0, \qquad v^2 + 16u = 64 \quad \text{(parabola)}, \qquad u = 1.$$

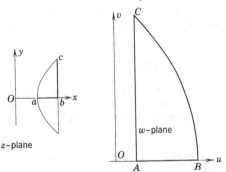

FIG. 203e. Mapping $w = z^2$

Note that the angles at *a*, *b*, *c* equal those at *A*, *B*, *C* in magnitude and sense. Although the transformation is everywhere conformal in the small, the regions *abc* and *ABC* do not have the same shape.

PROBLEMS

1. Find the angle through which the mapping $w = z^2$ revolves tangent vectors at the points $z_a = 1$, $z_b = 2$, $z_c = 2 + i\sqrt{3}$ in Fig. 203e.

Putting $w = u + iv$, $z = x + iy$, find u, v as functions of x, y. Show that the lines $x = a$, $y = b$ in the z-plane transform into orthogonal parabolas in the w-plane. What curves in the xy-plane transform into the lines $u = a$, $v = b$?

2. Find the magnification and rotation effected by the mapping $w = 1/z$ at the points $z = i$, $1 + i$, $1 - i$.

Putting $w = u + iv$, $z = x + iy$, find u, v as functions of x, y. Show that the lines $x = a$, $y = b$ in the z-plane transform into orthogonal circles in the w-plane; What curves in the xy-plane transform into the lines $u = a$, $v = b$?

3. Show that the mapping $w = \sin z$ transforms the coordinate lines $x = a$, $y = b$ in the z-plane into families of hyperbolas and ellipses respectively, having foci at $(\pm 1, 0)$. Prove that a half strip of the z-plane bounded by the lines $x = 0$, $x = 2\pi$, $y = 0$ is transformed into the entire w-plane,

4. Discuss the mapping $w = (z - 1)/(z + 1)$ writing $w = Re^{i\Phi}$, $z - 1 = r_1e^{i\varphi_1}$, $z + 1 = r_2e^{i\varphi_2}$. Prove that the coaxial circles $r_1/r_2 = a$, $\varphi_1 - \varphi_2 = b$ cut at right angles by finding their transforms.

5. Set up a linear fractional transformation which maps

    (a) $z = 1, i, -1$ into $w = -1, i, 1$;

    (b) $z = 0, -i, i$ into $w = -\frac{1}{2}, 1, -1$;

    (c) $z = 0, 2, -2$ into $w = -i, 1, -1$.

In part (b) show that the unit circle $|z| \leq 1$ maps on the unit circle $|w| \leq 1$.

6. If a, b, c, d are real and $D > 0$, show that (203.11) maps the upper half z-plane $y \geq 0$ on the upper half w-plane $v \geq 0$. Prove that these transformations form a group.

7. Show that the transformation

$$w = i\frac{z - 2i}{z + 2i}$$

maps the upper half z-plane $y \geq 0$ on the circle $|w| \leq 1$. Find the inverse transformation, and show that it maps the lower half w-plane $v \leq 0$ on the circle $|z| \leq 2$.

8. Show that $w = (a + z)/(a - z)$, where a is real, maps the left half of the z-plane $x < 0$ on the interior of the unit circle $|w| < 1$; and, as z describes the y-axis from $-\infty$ to $+\infty$, w describes the unit circle $|w| = 1$ counterclockwise. What curves in the z-plane map into the radial lines through $w = 0$? Take $a > 0$.

204. Definite Integrals. Let the function

$$w = f(z) = u(x, y) + iv(x, y)$$

be defined in a region R of the z-plane, and let C be a curve in R given by the parametric equations,

(1) $x = x(t)$, $y = y(t)$, $\alpha \leq t \leq \beta$.

Form a net

$$\alpha = t_0 < t_1 < t_2 < \cdots < t_n = \beta$$

of mesh δ over (α, β), and let x_j, y_j, z_j, u_j, v_j denote the values of $x, y, z,$ u, v when $t = t_j$. Then the equation

(2)
$$\int_C f(z)\, dz = \lim_{\delta \to 0} \sum_{j=1}^{n} f(z_j)(z_j - z_{j-1})$$

defines the complex line integral in the left member whenever the limit exists. Now

$$f(z_j) = u_j + iv_j, \qquad z_j - z_{j-1} = x_j - x_{j-1} + i(y_j - y_{j-1}),$$

and hence the sum in (2) equals

$$\sum_{j=1}^{n} \Big[u_j(x_j - x_{j-1}) - v_j(y_j - y_{j-1}) \Big]$$

$$+ i \sum_{j=1}^{n} \Big[v_j(x_j - x_{j-1}) + u_j(y_j - y_{j-1}) \Big].$$

If these real sums approach limits as $\delta \to 0$, we see from (150.2) that

(3)
$$\int_C f(z)\, dz = \int_C (u\, dx - v\, dy) + i \int_C (v\, dx + u\, dy).$$

Thus the complex line integral over C is a combination of two real line integrals over C. Since the continuity of $f(z)$ in R implies the continuity of $u(x, y)$ and $v(x, y)$ in R, the Theorem 150 gives the following important

THEOREM 1. *If $f(z)$ is continuous in a region R and C is a regular curve in R given by equations* (1),

(4)
$$\int_C f(z)\, dz = \int_\alpha^\beta (ux' - vy')\, dt + i \int_\alpha^\beta (vx' + uy')\, dt.$$

The curve C may be given by the single complex equation,

$$z = x(t) + i\, y(t) = z(t).$$

Along C we have

$$f\{z(t)\} = u(t) + i\, v(t),$$

and (4) may be written in the concise form,

(5)
$$\int_C f(z)\, dz = \int_\alpha^\beta f\{z(t)\}\frac{dz}{dt}\, dt.$$

Just as in the case of line integrals (§ 150) the definition (2) may be generalized by replacing $f(z_j)$ by $f(\bar{z}_j)$ where

$$\bar{z}_j = x(\bar{t}_j) + i\, y(\bar{t}_j), \qquad t_{j-1} \leqq \bar{t}_j \leqq t_j.$$

Here $\bar{z}$ is not the conjugate of z.

From (2) we obtain the following upper bound for the absolute value of a complex integral.

THEOREM 2. *If L is the length of the curve C and $|f(z)| \leqq M$ over C, then*

(6)
$$\left| \int f(z)\, dz \right| \leqq ML.$$

Proof. The integral (2) is the limit of

$$S_n = \sum_{j=1}^{n} f(z_j)\,(z_j - z_{j-1}).$$

Now

$$|S_n| \leqq \sum_{j=1}^{n} |f(z_j)|\,|z_j - z_{j-1}| \leqq M \sum_{j=1}^{n} |z_j - z_{j-1}|,$$

where the last sum gives the length of a broken line inscribed in C. Since the length L of the curve is the least upper bound of the lengths of all such broken lines (§ 128),

$$|S_n| \leqq ML.$$

This inequality holds for all nets t_i and hence holds on passing to the limit $\delta \to 0$.

Example 1. Compute the integral of $f(z) = 1/z$ over the unit circle C:

$$z = \cos t + i \sin t, \qquad 0 \leqq t \leqq 2\pi.$$

From (5) we have

$$\oint_C \frac{dz}{z} = \int_0^{2\pi} \frac{-\sin t + i \cos t}{\cos t + i \sin t}\, dt = i \int_0^{2\pi} dt = 2\pi i.$$

Since $L = 2\pi$ and $|z| = 1$ on the unit circle, (6) shows that the integral cannot exceed 2π in absolute value.

Example 2. Compute the integral

$$\int_C \bar{z}\, dz = \int_C (x - iy)\,(dx + i\, dy) = \int_C (x\, dx + y\, dy) + i \int_C x\, dy - y\, dx$$

for the five curves connecting $(0, 0)$ to $(1, 1)$ given in Ex. 150.1.

In all cases the first integral,

$$\int_C (x\, dx + y\, dy) = \frac{1}{2}\,(x^2 + y^2)\Big|_{(0,0)}^{(1,1)} = 1,$$

while

$$\int_C x\, dy - y\, dx = 0, 1, -1, \frac{1}{3}, \frac{\pi}{2} - 1,$$

as previously computed for the respective curves.

Example 3. Compute the integral

$$\int_C z^2 \, dz = \int_C (x^2 - y^2 + 2ixy) \, (dx + i \, dy)$$

$$= \int_C (x^2 - y^2) \, dx - 2xy \, dy + i \int_C 2xy \, dx + (x^2 - y^2) \, dy$$

over any curve from 0 to $a + ib$.

The vectors

$$[x^2 - y^2, \ -2xy], \qquad [2xy, \ x^2 - y^2]$$

are both irrotational (153.7) and hence gradient vectors. Thus both of the real line integrals are independent of the path. We may therefore compute the integral over the rectilinear path,

$$z = at + ibt = (a + ib)t, \qquad 0 \le t \le 1;$$

we thus obtain

$$\int_0^1 z^2 \, dt = (a + ib)^3 \int_0^1 t^2 \, dt = \frac{1}{3} \, (a + ib)^3.$$

Example 4. Compute the integral of $f(z) = (z - a)^n$, for any integer n, over the circle C with center at a and of radius b:

$$z = a + be^{it}, \qquad 0 \le t \le 2\pi.$$

Since $dz = ibe^{it} \, dt$, we have from (5)

$$\oint_C (z - a)^n \, dz = ib^{n+1} \int_0^{2\pi} e^{i(n+1)t} \, dt.$$

From (199.5) we have

$$e^{i(n+1)t} = \cos(n + 1)t + i \sin(n + 1)t;$$

hence,

(7) $$\oint_C (z - a)^n \, dz = \begin{cases} 2\pi i, & n = -1; \\ 0, & n \ne -1. \end{cases}$$

205. Cauchy's Integral Theorem. *If $f(z)$ is analytic in a simply connected region R, then*

(1) $$\oint f(z) \, dz = 0$$

over any closed regular curve in R; and

$$\int_{z_1}^{z_2} f(z) \, dz$$

is independent of the path in R joining z_1 and z_2.

Proof. If $f(z) = u + iv$, we have from (204.3)

(2) $$\oint f(z) \, dz = \oint u \, dx - v \, dy + i \oint v \, dx + u \, dy.$$

But, from the C–R equations

(3) $$u_y = -v_x, \qquad v_y = u_x,$$

and hence, from Theorem 160.2, both real integrals vanish. The corollary to this theorem shows that the integral from z_1 to z_2 is independent of the path.

Cauchy's theorem is fundamental in the theory of functions of a complex variable. It permits the following

Extensions. When $f(z)$ be analytic in a multiply connected region M, (1) may not hold for all regular closed curves C in M. But, if C can be embedded in a simply connected subregion of M, (1) obviously does hold.

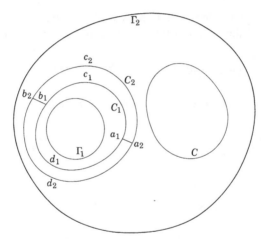

FIG. 205.   Multiply connected region

Thus let M be the annular region lying between the closed curves Γ_1 and Γ_2 (Fig. 205). Then (1) holds for the curve C, but not in general for C_1 or C_2 which surround the inner boundary Γ_1. However, if we join C_1 to C_2 by the segments a_1a_2, b_1b_2, equation (1) does hold for the closed regular curves $a_1a_2c_2b_2b_1c_1a_1$ and $a_1d_1b_1b_2d_2a_2a_1$; and, on adding the equations, we get

(4) $$\oint_{C_2} f(z)\, dz - \oint_{C_1} f(z)\, dz = 0$$

for the integrals over a_1a_2 and b_1b_2 cancel: hence

The circuit integrals taken over C_1 and C_2 in the same sense are equal.

When $f(z)$ is analytic within a closed curve Γ_2 except at a point P_1, $f(z)$ is analytic in the doubly connected region R within Γ_2 *and punctured* *at P_1.* Then all circuit integrals about curves in R that encircle P_1 have the same value. Thus (204.7) is true when C is any closed regular curve about the point $z = a$.

THEOREM 2. *If $f(z)$ is analytic in a simply connected region R, the integral*

$$(5) \qquad\qquad F_0(z) = \int_{z_0}^{z} f(z)\, dz$$

taken over any curve in R joining z_0 and z; and $F_0(z)$ is an analytic function whose derivative

$$(6) \qquad\qquad F_0'(z) = f(z).$$

Proof. When the lower limit z_0 is held fast, the integral (5), being independent of the path, is a function $F_0(z)$ of its upper limit. Now

$$F_0(z) = \int_{(x_0,y_0)}^{(x,y)} u\, dx - v\, dy + i \int_{(x_0,y_0)}^{(x,y)} v\, dx + u\, dy = U(x,y) + i\,V(x,y);$$

and, since both real integrals are independent of the path, Theorem 151.2 shows that

$$U_x = u, \qquad U_y = -v; \qquad V_x = v, \qquad V_y = u;$$

and hence

$$U_x = V_y, \qquad U_y = -V_x.$$

These C–R equations show that $F_0(z)$ is analytic in R; hence from (195.4)

$$F_0'(z) = U_x + iV_x = u + iv = f(z).$$

Corresponding to the fundamental theorem of the calculus (Theorem 120), we now have

THEOREM 3. *If $f(z)$ is analytic in a simply connected region R and $F(z)$ is any function having $f(z)$ as derivative in R, then*

$$(7) \qquad\qquad \int_{z_0}^{z_1} f(z)\, dz = F(z_1) - F(z_0).$$

Proof. Since both $F_0(z)$ and $F(z)$ have $f(z)$ as derivative in R, they differ by only an additive constant (Theorem 197.3, Cor.):

$$\int_{z_0}^{z} f(z)\, dz = F(z) + C.$$

Putting $z = z_0$, we have $C = -F(z_0)$; and, when $z = z_1$, we obtain (7).

Cauchy's theorem admits a converse known as

MORERA'S THEOREM. *If $f(z)$ is continuous in a simply connected region R and if*

$$(1) \qquad \oint f(z)\, dz = 0$$

over all regular closed paths in R, then $f(z)$ is analytic in R.

Proof. While the theorem is true under the conditions stated,† we shall only prove it under the additional hypothesis that u_x, u_y, v_x, v_y are continuous in R. From (2) we see that (1) implies that the real integrals,

$$\oint u\, dx - v\, dy = 0, \qquad \oint v\, dx + u\, dy = 0,$$

over all regular paths in R. Now from Theorem 160.4, the C–R equations (3) must hold in R, and hence $f(z)$ is analytic in R.

206. Cauchy's Integral Formula. *If the function $f(z)$ is analytic in a region R, its value at any point a of R is given by the circuit integral,*

$$(1) \qquad f(a) = \frac{1}{2\pi i} \oint_C \frac{f(z)}{z - a}\, dz,$$

taken over any regular closed curve in R having a in its interior.

Proof. We may write the right member of (1) in the form

$$\frac{1}{2\pi i} \oint_C \frac{f(a)}{z - a}\, dz + \frac{1}{2\pi i} \oint_C \frac{f(z) - f(a)}{z - a}\, dz = I_1 + I_2.$$

Both integrands are analytic functions of z in the doubly connected region consisting of C and its interior, but punctured at a; hence, if we replace C by a circle γ of radius r about a as center and lying within C, the values of both integrals are unaltered (§ 205). The equation of the circle γ,

$$(2) \qquad z = a + re^{i\theta} \quad \text{and} \quad dz = rie^{i\theta} d\theta;$$

hence, the first integral,

$$I_1 = \frac{f(a)}{2\pi i} \int_0^{2\pi} i\, d\theta = f(a).$$

If we choose r so small that

$$\left| f(z) - f(a) \right| < \varepsilon \quad \text{over } \gamma,$$

the absolute value of the second integral,

$$\left| I_2 \right| \leq \frac{1}{2\pi} \frac{\varepsilon}{r} 2\pi r = \varepsilon, \qquad (204.6)$$

and hence $I_2 = 0$. Thus $I_1 + I_2 = f(a)$, as stated in (1).

---

† Cf. Knopp, *op. cit.*, part I, Chap. 5, for proof.

Cauchy's formula (1) shows that, if a function is analytic on and within a regular closed curve C, the values of $f(z)$ along C completely determine its values at all interior points. In particular, if C is the circle γ, (1) reduces to

$$(3) \qquad f(a) = \frac{1}{2\pi} \int_0^{2\pi} f(a + re^{i\theta})\, d\theta.$$

In view of (119.7) we may state (3) in the form

If a function is analytic on and within a circle, its value at the center is the mean of its values on the circumference.

Example 1. On the circle $|z| = 1$,

$$\frac{1}{2\pi i} \oint \frac{\cos z}{z}\, dz = \cos 0 = 1.$$

Example 2. On the circle $|z| = 3$,

$$\oint \left(\frac{z}{z-1} + \frac{z^2}{z+2} \right) dz = 2\pi i(1 + 4) = 10\pi i.$$

Note that the circle includes $z = 1$ and $z = -2$.

Example 3. To evaluate

$$I = \oint \frac{dz}{z^4 - 1} \quad \text{over the circle} \quad |z| = 2,$$

we split $1/(z^4 - 1)$ into the partial fractions:

$$\frac{1}{z^4 - 1} = \frac{A}{z-1} + \frac{B}{z+1} + \frac{C}{z-i} + \frac{D}{z+i}.$$

On clearing of fractions and putting $z = 1, -1, i, -i$ in turn, we find

$$A = \frac{1}{4}, \qquad B = -\frac{1}{4}, \qquad C = -\frac{1}{4i}, \qquad D = \frac{1}{4i}.$$

Since the circle includes ± 1 and $\pm i$.

$$I = 2\pi i(A + B + C + D) = 0.$$

207. Complex Taylor Series. We have seen in § 183 that every complex power series,

$$(1) \qquad f(z) = \sum_{n=0}^{\infty} c_n(z - a)^n = \sum_{n=0}^{\infty} c_n \zeta^n \qquad (\zeta = z - a),$$

represents a continuous function $f(z)$ inside its circle of convergence $|z - a| = R$ and diverges outside. Within this circle (1) may be differentiated term by term to give the unique derivative.

$$(2) \qquad f'(z) = \sum_{n=1}^{\infty} nc_n(z - a)^{n-1};$$

hence $f(z)$ is an analytic function in the circle. The series (2) may be differentiated again, giving $f''(z)$; and by repeated differentiation we obtain series for the derivatives of all orders, all of which converge in the same circle $|z - a| = R$. Moreover, by Theorem 183.3, equation (1) implies that

$$(3) \qquad c_n = \frac{f^{(n)}(a)}{n!};$$

for $z = a$ when $\zeta = 0$. These facts, all of the first importance, are collected in

THEOREM 1. *A power series $\sum c_n(z - a)^n$ which converges in a circle of radius $R \neq 0$ about the point a represents an analytic function $f(z)$ within this circle; and the series is the Taylor expansion of $f(z)$ about a.*

We now prove the converse

THEOREM 2. *If $f(z)$ is an analytic function in the neighborhood of $z = a$, it may be expanded into a power series $\sum c_n(z - a)^n$ which converges within the circle of radius R about a that passes through the nearest "singular" point of $f(z)$—where $f'(z)$ fails to exist.*

Proof. By hypothesis $f(z)$ is analytic in a region having the point a in its interior. This region will certainly include the circle of radius R about the point a and passing through the *nearest* singular point of $f(z)$; for within this circle there are no points where $f(z)$ fails to be analytic. Let $h \neq 0$ be a complex number whose absolute value $|h| < R$; and choose the real number r between $|h|$ and R: $0 < |h| < r < R$. Then $f(z)$ is analytic on and within the circle C of radius r about a; namely, $|z - a| = r$. Since the point $a + h$ lies within the circle C, Cauchy's integral formula gives

$$(4) \qquad f(a + h) = \frac{1}{2\pi i} \oint_C \frac{f(z)}{z - a - h} \, dz. \qquad (206.1).$$

We now expand $1/(z - a - h)$ into an infinite geometric series,

$$(5) \qquad \frac{1}{z - a - h} = \frac{1}{z - a} + \frac{h}{(z - a)^2} + \frac{h^2}{(z - a)^3} + \cdots,$$

which converges since the absolute value of its ratio,

$$\left| \frac{h}{z - a} \right| = \frac{|h|}{r} < 1.$$

Moreover, the series (5) converges uniformly for all z on the circle C and remains uniformly convergent when its terms are multiplied by the continuous function $f(z)$.† Hence, if we substitute the series (5) in (4), we may integrate term by term to obtain

---

† For $f(z)$ is bounded on C: $|f(z)| < M$. See also Theorem 183.1.

(6) $\quad f(a+h)=\dfrac{1}{2\pi i}\left[\oint_C \dfrac{f(z)}{z-a}\,dz+h\oint_C \dfrac{f(z)}{(z-a)^2}\,dz+h^2\oint_C \dfrac{f(z)}{(z-a)^3}+\cdots\right]$

This power series in h must be the Taylor series for $f(z)$ about $z=a$; and its coefficients are

(7) $\qquad\qquad\qquad c_0=f(a)=\dfrac{1}{2\pi i}\oint_C \dfrac{f(z)}{z-a}\,dz,$

(8) $\qquad\qquad c_n=\dfrac{f^{(n)}(a)}{n!}=\dfrac{1}{2\pi i}\oint_C \dfrac{f(z)}{(z-a)^{n+1}}\,dz,\qquad n=1,2,\cdots.$

Equation (7) is Cauchy's integral formula; and equations (8) are Cauchy's formulas for the derivatives of an analytic function. Since $z=a+h$ is any complex number within the circle of radius R about a, the Taylor series,

(9) $\qquad f(z)=f(a)+\dfrac{f'(a)}{1!}(z-a)+\dfrac{f''(a)}{2!}(z-a)^2+\cdots,$

converges for all z within this circle.

Example 1. The geometric series,

$$\frac{1}{1+z^2}=1-z^2+z^4-z^6+\cdots,$$

converges to $1/(1+z^2)$ within a circle with center at $z=0$ and passing through the poles $\pm i$ of this function: that is, the circle $|z|=1$. When $z=x$, the corresponding real series converges in the interval $-1<x<1$; but *why* it diverges at $x=\pm1$ is not clear since the function $1/(1+x^2)$ shows no peculiarities at these points. The complex series clears up this point; for its circle of convergence cuts the real axes at $x=\pm1$.

Example 2. Since the functions e^z, $\sin z$, $\cos z$, $\sinh z$, $\cosh z$ are analytic throughout the complex plane, our theory guarantees the convergence of their Maclaurin series for all values of z.

208. Cauchy's Inequality. In formula (207.8) for derivatives, the circuit integral is taken about a circle centered at a; but, since $f(z)/(z-a)^{n+1}$ is analytic in a region *punctured at* a, the circle may be replaced by any closed curve in this region enclosing a (§ 205). Thus we have the

THEOREM 1. *If the function $f(z)$ is analytic in the neighborhood of a, its derivatives of all orders exist and are analytic in this region; and*

(1) $\qquad\qquad f^{(n)}(a)=\dfrac{n!}{2\pi i}\oint_C \dfrac{f(z)}{(z-a)^{n+1}}\,dz,\qquad n=1,2,\cdots,$

where C is any regular closed curve in this region enclosing a.

Note that (1) may be obtained by differentiating Cauchy's integral formula (206.1) n times with respect to a. Since (1) was deduced in another way, this differentiation under the integral sign is thereby justified.

When C is a circle of radius r about a and

$$|f(z)| \leq M \quad \text{on } C,$$

equation (1) yields *Cauchy's inequality*:

$$(2) \qquad |f^{(n)}(a)| \leq \frac{n!}{2\pi} \cdot \frac{M}{r^{n+1}} 2\pi r = \frac{n!M}{r^n} \qquad (204.6).$$

From (207.7) we also have $|f(a)| \leq M$ which is included in (2) if we put $n = 0, 0! = 1$.

From (2) we now deduce

LIOUVILLE'S THEOREM. *A function that is analytic and bounded for all complex values is a constant.*

Proof. Let $f(z)$ be analytic and $|f(z)| < M$ throughout the complex plane. Then, with $n = 1$ in (2), we have, for any point z and for any $\varepsilon > 0$,

$$|f'(z)| \leq \frac{M}{r} < \varepsilon$$

if we choose $r > M/\varepsilon$. Hence $f'(z) = 0$ and $f(z)$ is a constant (Theorem 197.3).

Liouville's theorem leads to a very simple proof of

THE FUNDAMENTAL THEOREM OF ALGEBRA. *If $f(z)$ is any nonconstant polynomial with real or complex coefficients, the equation $f(z) = 0$ has at least one root.*

Proof. If $f(z) \neq 0$ for all values of z, $1/f(z)$ is an analytic function throughout the complex plane which approaches zero as $z \to \infty$. Hence we can take a sufficiently large circle of radius R about the origin so that

$$1/|f(z)| < M \quad \text{when} \quad |z| > R.$$

But, since $1/f(z)$ is continuous in the closed circle $|z| \leq R$, it is also bounded there (Theorem 196.4). Thus $1/f(z)$ is analytic and bounded throughout the complex plane and is therefore a constant, contradicting the hypothesis.

209. Isolated Singularities. If $f(z)$ has no derivative when $z = a$ but $f'(z)$ exists in some punctured neighborhood of $z = a$, we call a an *isolated singular point* of $f(z)$. We shall show that all isolated singularities are of three different types.

1. *Removable Singularities.* The singular point a is said to be removable if $f(z)$ is bounded and analytic in some punctured neighborhood of a; then

$$(1) \qquad |f(z)| < M, \quad \text{when} \quad 0 < |z - a| < \delta.$$

The aptness of the term "removable" will be clear from

RIEMANN'S THEOREM. *If $f(z)$ is bounded and analytic in a punctured neighborhood of $z = a$, then*

$$\lim_{z \to a} f(z) = A;$$

and, if we define $f(a) = A$, then $f(z)$ is also analytic at a, so that the singularity at a is "removed."

Proof. Assume (1), and define the function

(2) $F(z) = (z - a)^2 f(z)$ $(z \neq a)$, $F(a) = 0$,

within the circle $|z - a| = \delta$. $F'(z)$ exists at all interior points; for

$$F'(z) = 2(z - a) f(z) + (z - a)^2 f'(z),   (z \neq a),$$

$$F'(a) = \lim_{z \to a} \frac{F(z) - F(a)}{z - a} = \lim_{z \to a} (z - a) f(z) = 0.$$

Hence, $F(z)$ is analytic at a, and its Taylor series,

(3) $F(z) = c_2(z - a)^2 + c_3(z - a)^3 + \cdots$ $(c_0 = c_1 = 0)$,

converges in the circle $|z - a| = \delta$. A comparison of (2) and (3) shows that

(4) $$f(z) = c_2 + c_3(z - a) + \cdots,   z \neq a,$$
and

$$\lim_{z \to a} f(z) = c_2.$$

Hence, if we define $f(a) = c_2$, the series (4) is the Taylor expansion of $f(z)$ and represents an analytic function within the circle $|z - a| = \delta$.

2. *Poles.* The function $f(z)$ is said to have a *pole* at a if

$$\lim_{z \to a} f(z) = \infty   \text{ over any path.}$$

Then, if G is any positive number, however large, we can always find a circle $|z - a| = \delta$ about a in which $|f(z)| > G$. Within this circle define the function

(5) $F(z) = 1/f(z)$ $(z \neq a)$, $F(a) = 0$;

Since $|F(z)| < 1/G$ and is therefore bounded in the circle, $F(z)$ can at most be a removable singularity at a; and to remove it we need only define

$$F(a) = \lim_{z \to a} \frac{1}{f(z)} = 0$$

by Riemann's theorem. But, since this *is* the definition adopted in (5), $F(z)$ is analytic at a and may be expanded about this point in a Taylor series,

$$F(z) = \sum_{n=1}^{\infty} a_n(z - a)^n, \qquad (a_0 = 0).$$

If $a_m \neq 0$ is the first nonzero coefficient in this series,

$$F(z) = (z - a)^m[a_m + a_{m+1}(z - a) + \cdots] = (z - a)^m G(z),$$

where $G(z)$ is the analytic function represented by the power series in brackets. From (5) we now have

$$(6) \qquad f(z) = \frac{\varphi(z)}{(z - a)^m}, \qquad \varphi(a) \neq 0,$$

where $\varphi(z) = 1/G(z)$ is analytic within the circle $|z - a| = \delta$. If we now expand $\varphi(z)$ in a Taylor series about a (its constant term is not zero) and divide each term by $(z - a)^m$ the result may be written

$$(7) \quad f(z) = \frac{c_{-m}}{(z - a)^m} + \frac{c_{-m+1}}{(z - a)^{m-1}} + \cdots + \frac{c_{-1}}{z - a}$$
$$+ c_0 + c_1(z - a) + c_2(z - a)^2 + \cdots,$$

where $c_{-m} \neq 0$.

When $f(z)$ may be put in the form (6), it is said to have a *pole of order* m at $z = a$; then $f(z)$ admits the series expansion (7) of which the fractional terms form the "principal part." Evidently m is the least positive integer for which

$$(8) \qquad (z - a)^m f(z) = \varphi(z)$$

is analytic at a and $\varphi(a) \neq 0$.

3. *Essential Singularities.* An isolated singularity which is neither removable or a pole is called *essential.* For such singularities we have

WEIERSTRASS' THEOREM. *In any neighborhood of an isolated essential singularity a of $f(z)$ the function comes arbitrarily close to any prescribed value.*

Proof. Let C be any prescribed complex number; then we shall prove that, in any punctured circle about a, $0 < |z - a| < \delta$, there is always a point ζ such that

$$(9) \qquad |f(\zeta) - C| < \varepsilon, \qquad \zeta \neq a.$$

For, if $|f(z) - C| \geq \varepsilon$ in the circle, then

$$\left| \frac{1}{f(z) - C} \right| \leq \frac{1}{\varepsilon}, \qquad 0 < |z - a| < \delta,$$

and the function $1/(f(z) - C)$ fulfills the condition (1) for a removable singularity. Hence, by Riemann's theorem

$$\lim_{z \to a} \frac{1}{f(z) - C} = A, \quad \text{a finite limit.}$$

If $A = 0$,

$$\lim_{z \to a} (f(z) - C) = \infty, \qquad \lim_{z \to a} f(z) = \infty,$$

and $f(z)$ would have a pole at a. If $A \neq 0$,

$$\lim_{z \to a} f(z) = C + \frac{1}{A},$$

and $f(z)$ would have a removable singularity at a. But an essential singularity specifically excludes these cases; and this contradiction proves the theorem.†

Finally let us consider the behavior of $f(z)$ at infinity. Since the transformation $w = 1/z$ carries each point of the z-plane outside of the circle $|z| = R$ into a point of the w-plane inside of the circle $|w| = 1/R$, we may regard $w = 0$ as the transform of the "point ∞." We therefore lay down the following

DEFINITION. *The function $f(z)$ is said to be analytic $z = \infty$ or to have a certain singularity there according as the function $f(1/w)$ is analytic or has that singularity at $w = 0$.*

Consider, for example, the rational function,

$$f(z) = \frac{P(z)}{Q(z)} = \frac{a_0 z^p + \cdots + a_p}{b_0 z^q + \cdots + b_q}, \qquad (a_0, b_0 \neq 0).$$

Then we readily see that, if

$p > q$, $f(z)$ has a pole of order $p - q$ at ∞;

$p = q$, $f(z)$ is analytic at ∞ and $f(\infty) = a_0/b_0$;

$p < q$, $f(z)$ is analytic at ∞ and has a zero of order $q - p$ there.

In particular, the polynomial $P(z)$ has a pole of order p at ∞ while $1/Q(z)$ has a zero of order q at ∞.

Example 1. The function

(10) $$f(z) = \frac{z}{\sin z} = 1 \Big/ \left(1 - \frac{z^2}{3!} + \frac{z^4}{5!} - \cdots \right), \qquad z \neq 0$$

has the limit 1 as $z \to 0$. Hence, if we define $f(0) = 1$, $f(z)$ is analytic in the neighborhood of $z = 0$.

Example 2. The functions

$$\operatorname{cosec} z = \frac{1}{\sin z}, \qquad \sec z = \frac{1}{\cos z}$$

† Picard has proved that in the neighborhood of an essential singularity an analytic function *assumes* every value except possibly one.

have simple poles at all points where

$$\sin z = 0 \quad (z = n\pi), \qquad \cos z = 0 \quad (z = (n - \tfrac{1}{2})\pi),$$

respectively. In both cases all poles lie on the real axis (§ 200).

The points $z = n\pi$ (n an integer) are poles of the first order of $1/\sin z$; for

$$\frac{z - n\pi}{\sin z} = \frac{(z - n\pi)\cos n\pi}{\sin z \cos n\pi} = (-1)^n \frac{z - n\pi}{\sin (z - n\pi)},$$

and hence

(11) $$\lim_{z \to n\pi} \frac{z - n\pi}{\sin z} = (-1)^n \qquad \text{(Ex. 1).}$$

Similarly the points $z = (n - \tfrac{1}{2})\pi$ are simple poles of $1/\cos z$; for

$$\frac{z - (n - \tfrac{1}{2})\pi}{\cos z} = \frac{(z + \pi/2) - n\pi}{\sin (z + \pi/2)},$$

and, if we put $w = z + \pi/2$,

(12) $$\lim_{z \to (n - \tfrac{1}{2})\pi} \frac{z - (n - \tfrac{1}{2})\pi}{\cos z} = \lim_{w \to n\pi} \frac{w - n\pi}{\sin w} = (-1)^n.$$

We shall see in § 213 that (11) and (12) state that the poles $n\pi$ of $1/\sin z$ and $(n - \tfrac{1}{2})\pi$ of $1/\cos z$ have *residues* of $(-1)^n$.

Example 3. The function $e^{1/z}$ has an essential singularity at $z = 0$; for the equations

(i) (ii) $$\lim_{x \to 0+} e^{1/x} = \infty, \qquad \lim_{x \to 0-} e^{1/x} = 0 \qquad \text{(Ex. 44.4)}$$

show that $z = 0$ is neither a removable singularity nor a pole.

We can readily verify Weierstrass' theorem for $e^{1/z}$. For, if $ae^{i\alpha}$ ($a > 0$) is an arbitrary complex number, the equation

$$e^{1/z} = ae^{i\alpha}$$

has an infinite number of roots,

$$z = \frac{1}{\log a + (\alpha + 2\pi n)i}, \qquad n = 0, \pm 1, \pm 2, \cdots,$$

in any neighborhood of the origin. In fact, the larger we take $|n|$, the closer the root approaches zero. Although $e^{1/z} = 0$ has no root, equation (ii) shows that in any neighborhood of $z = 0$ there are points x on the negative real axis at which $e^{1/z}$ comes arbitrarily close to 0.

This example also shows that e^z has an essential singularity at ∞. The same is true of e^{iz}, e^{-iz} and, consequently, of $\sin z$ and $\cos z$ (199.7–8).

Example 4. The real function,

$$f(x) = x \sin \frac{1}{x}, \qquad f(0) = 0,$$

is continuous (and therefore bounded) in the interval $(-1, 1)$ and has a derivative at every point except $x = 0$ (Ex. 50.4). Nevertheless it cannot be defined at $x = 0$ so that $f'(0)$ exists—in sharp contrast to Riemann's theorem for complex functions. In fact,

$$f(z) = z \sin \frac{1}{z}, \qquad f(0) = 0$$

has an essential singularity at $z = 0$; for

$$\lim_{x \to 0} f(x) = 0; \qquad \lim_{y \to 0} f(iy) = \lim_{y \to 0} y \sinh \frac{1}{y} = \infty \qquad (200.4).$$

Example 5. The equation

$$w^2 = z = re^{i\theta}$$

has two solutions (§ 9):

$$w_1 = \sqrt{r}e^{i(\theta/2)}, \qquad w_2 = \sqrt{r}e^{i(\pi + \theta/2)} = -w_1.$$

Hence, w_1 and w_2 are called the two *branches* of the multiple-valued w. Thus w (which is not a function of z according to the definition of § 43) gives use to *two* functions $w_1(z)$, $w_2(z)$, which are distinct except at $z = 0$. Since w_1 and w_2 coalesce at $z = 0$, the origin is called a *branch point* of w.

Neither w_1 nor w_2 is continuous in any neighborhood of the origin. At the point $z = r$ on the positive real axis $w_1 = \sqrt{r}$, $w_2 = -\sqrt{r}$; now, if z circles the origin counterclockwise,

$$w_1 \to \sqrt{r}e^{\pi i} = w_2, \qquad w_2 \to \sqrt{r}e^{2\pi i} = w_1;$$

that is, the branches interchange. To prevent this we may make a *branch cut* extending out from the origin to infinity. For example, the negative x-axis may be chosen as a branch cut; then $-\pi < \theta \leq \pi$, and w_1 and w_2 cannot interchange. Both w_1 and w_2 are now analytic functions in the open region formed by the cut z-plane.

210. Laurent Series. If $f(z)$ has an isolated singularity at $z = a$, $f(z)$ may be developed about a in a power series $\sum c_n(z - a)^n$ in which n takes on negative as well as positive integral values. When $z = a$ is a pole of $f(z)$, such a series is given in (209.7); in this case there is only a finite number m of negative powers, of which the highest $(z - a)^{-m}$ corresponds to the order m of the pole. We shall see that, when $f(z)$ has an essential singularity at a, the negative powers form an infinite series.

LAURENT'S THEOREM. *Let $f(z)$ be analytic in a ring $R_1 < |z - a| < R_2$ about a; then $f(z)$ may be expanded into a Laurent series,*

$$(1) \qquad\qquad f(z) = \sum_{n = -\infty}^{\infty} c_n(z - a)^n,$$

converging within the ring and whose coefficients,

$$(2) \qquad c_n = \frac{1}{2\pi i} \oint_C \frac{f(z)}{(z - a)^{n+1}} \, dz, \qquad n = 0, \pm 1, \pm 2, \cdots,$$

are obtained by circuit integration over any regular closed curve C encircling a and lying entirely in the ring.

Proof. Let $z = a + h$ be any complex number inside the ring, and choose two real numbers r_1, r_2 so that

$$R_1 < r_1 < |h| < r_2 < R_2.$$

Then $f(z)$ is analytic on and between the circles

$$C_1: \ |z - a| = r_1 \quad \text{and} \quad C_2: \ |z - a| = r_2.$$

Join these circles by a radial cross-cut $\alpha\beta$ which does not pass through the point $z = a + h$ (Fig. 210). Then the closed circuit $\alpha\beta C_2^+ \beta\alpha C_1^-$,

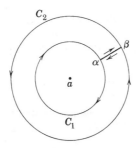

FIG. 210.   Ring of convergence for Laurent series

described in the positive sense encloses no singularity of $f(z)$, and from Cauchy's integral formula

$$(3) \qquad f(a + h) = \frac{1}{2\pi i} \oint_{C_2} \frac{f(z)}{z - a - h} \, dz - \frac{1}{2\pi i} \oint_{C_1} \frac{f(z)}{z - a - h} \, dz,$$

for the integrals over $\alpha\beta$ and $\beta\alpha$ cancel. In these integrals replace $1/(z - a - h)$ by the respective geometric series:

$$(4) \qquad \frac{1}{z - a - h} = \frac{1}{z - a} + \frac{h}{(z - a)^2} + \frac{h^2}{(z - a)^3} + \cdots,$$

$$(5) \qquad \frac{1}{z - a - h} = -\frac{1}{h} - \frac{z - a}{h^2} - \frac{(z - a)^2}{h^3} + \cdots.$$

These series converge absolutely and uniformly over C_2 and C_1, respectively; for the absolute values of their ratios,

$$\frac{|h|}{|z - a|} = \frac{|h|}{r_2} < 1, \qquad \frac{|z - a|}{|h|} = \frac{r_1}{|h|} < 1.$$

Moreover, the convergence remains uniform when the series are multiplied by the bounded function $f(z)$. Hence, when (4) is substituted in the first integral of (3), (5) in the second integral, we may integrate the resulting series term by term to obtain

$$(6) \qquad f(a + h) = \sum_{n=0}^{\infty} c_n h^n + \sum_{n=1}^{\infty} c_{-n} h^{-n},$$

where

(7)
$$c_n = \frac{1}{2\pi i} \oint_{C_2} \frac{f(z)}{(z-a)^{n+1}} \, dz,$$

(8)
$$c_{-n} = \frac{1}{2\pi i} \oint_{C_1} f(z)(z-a)^{n-1} \, dz.$$

Since these integrals are not altered when the circles C_1, C_2 are replaced by any other regular closed curve C lying within the ring $R_1 < |z-a| < R_2$, we see that *both* are included in (2); for, when n in (7) is replaced by $-n$, we obtain (8).

Suppose that $f(z)$ has just one singular point z_1 in the finite plane and that $|z_1 - a| = R_1$. Then $f(z)$ admits a Laurent expansion in the ring $R_1 < |z - a| < \infty$. If $R_1 = 0$, the "ring" becomes the punctured plane $0 < |z - a| < \infty$. If $R_1 \neq 0$, $f(z)$ also admits a Taylor expansion in the circle $|z - a| < R_1$.

If $f(z)$ has just two singular points z_1, z_2 in the finite plane and $|z_2 - a| = R_2 > R_1$, $f(z)$ admits Laurent expansions in the rings $R_1 < |z - a| < R_2$, $R_2 < |z - a| < \infty$. If $R_1 = 0$, the first ring becomes a punctured circle; but, when $R_1 \neq 0$, $f(z)$ also admits a Taylor expansion in the circle $|z - a| < R_1$.

Finally, if $f(z)$ has n singular points at different distances $R_1 < R_2 < \cdots < R_n$ from a, $f(z)$ will have Laurent expansions in the n rings

$$R_1 < |z - a| < R_2, \quad \cdots, \quad R_n < |z - a| < \infty;$$

and, if $R_1 \neq 0$, also a Taylor expansion in the circle $|z - a| < R_1$.

We note that, while $\log z$ has a singularity at $z = 0$, $\log z$ cannot be defined as a single-valued function in any punctured circle about the origin and therefore admits no Laurent expansion about $z = 0$. The origin is said to be a *branch point* for the multivalued function $\log z$.

THEOREM 2. *If the series*

(9)
$$\sum_{n=-\infty}^{\infty} b_n(z-a)^n = f(z)$$

converges in the ring $R_1 < |z - a| < R_2$, *it must be the Laurent series of* $f(z)$.

Proof. Since the Laurent series (1) and the series (9) both represent $f(z)$ in the ring,

(10)
$$\sum_{n=-\infty}^{\infty} b_n(z-a)^n = \sum_{n=-\infty}^{\infty} c_n(z-a)^n, \qquad R_1 < |z - a| < R_2.$$

Let C be a circle of radius r about a, and choose r_1 and r_2 so that

$$R_1 < r_1 < r < r_2 < R_2.$$

Both series in (10) converge uniformly in the ring $r_1 < |z - a| < r_2$; in the first, for example, we may take

$$M_n = |b_n| r_1{}^n \quad (n < 0), \qquad M_n = |b_n| r_2{}^n \quad (n \geq 0)$$

in the M-test. Hence, if we multiply (10) by $(z - a)^{-(k+1)}$ and integrate both sides term by term over the circle $|z - a| = r$, all integrals vanish except when $n = k$ (204.7); thus we find

$$2\pi i b_k = 2\pi i c_k \quad \text{or} \quad b_k = c_k.$$

Example 1. Since the Maclaurin series (198.1) for e^z converges for all z, the series obtained by replacing z by $1/z$ converges for all $z \neq 0$; thus,

$$(11) \qquad e^{1/z} = 1 + z^{-1} + \frac{z^{-2}}{2!} + \frac{z^{-3}}{3!} + \cdots$$

is a Laurent series converging in the "ring" $0 < |z| < \infty$. In the same way we obtain Laurent series for $\sin 1/z$ and $\cos 1/z$.

Example 2. The function $f(z) = e^{-1/z^2}$ has an essential singularity at $z = 0$; for

$$\lim_{x \to 0} f(x) = 0, \qquad \lim_{y \to 0} f(iy) = \infty.$$

This explains why the real function e^{-1/x^2} has no Taylor expansion about $x = 0$ although possessing continuous derivatives of all orders there (§ 66). However, we have the Laurent series,

$$(12) \qquad e^{-1/z^2} = 1 - z^{-2} + \frac{z^{-4}}{2!} - \frac{z^{-6}}{3!} + \cdots, \qquad 0 < |z| < \infty.$$

Example 3. Consider the function

$$f(z) = \frac{1}{(z - 1)(z - 2)} = \frac{1}{z - 2} - \frac{1}{z - 1}$$

with simple poles at $z = 1$ and $z = 2$. It admits a Taylor expansion about $z = 0$ in the circle $|z| < 1$:

$$f(z) = \frac{1}{1 - z} - \frac{1}{2} \frac{1}{1 - z/2} = \sum_{n=0}^{\infty} \left(1 - \frac{1}{2^{n+1}} \right) z^n.$$

$f(z)$ also admits a Laurent expansion about $z = 0$ in the ring $1 < |z| < 2$. To find it, we write $f(z)$ as the sum of two geometric series which converge in this ring:

$$f(z) = -\frac{1}{2} \frac{1}{1 - z/2} - \frac{1}{z} \frac{1}{1 - 1/z} = -\frac{1}{2} \sum_{n=0}^{\infty} \frac{z^n}{2^n} - \sum_{n=-1}^{-\infty} \frac{1}{z^n}.$$

Finally $f(z)$ admits a Laurent series in the ring $2 < |z| < \infty$:

$$f(z) = \frac{1}{z} \frac{1}{1 - 2/z} - \frac{1}{z} \frac{1}{1 - 1/z} = \sum_{n=1}^{\infty} \frac{2^n - 1}{z^{n+1}}.$$

If we expand $f(z)$ about $z = 1$ in the ring $0 < |z - 1| < 1$ formed by the punctured unit circle about $z = 1$,

$$f(z) = -\frac{1}{z-1} - \frac{1}{1-(z-1)} = -\frac{1}{z-1} - \sum_{n=0}^{\infty} (z-1)^n.$$

PROBLEMS

Expand the following functions in Laurent series valid in the neighborhood of $z = 0$; of $z = \infty$. In each case give the region of convergence:

1. $\dfrac{z}{z^2 + 1}.$

2. $\dfrac{1}{z(1 + z^2)}.$

3. $\dfrac{1}{z^2(1 - z)}.$

4. $\dfrac{1 + z^2}{1 - z}.$

5. $\dfrac{\sinh z}{z}.$

6. $\dfrac{e^{-z}}{z}.$

7. $\dfrac{e^z}{z(1 + z^2)}$ (4 terms).

8. $1/\sinh z$ (4 terms).

9. In Prob. 3 and 4 find the Laurent expansion about $z = 1$, and give the ring of convergence.

10. Expand $1/z$ in powers of $z - 1$.

211. Bernoulli Numbers.   The function

$$f(z) = \frac{z}{e^z - 1}$$

is not defined when $z = 0$; but from the exponential series (198.1) we have

$$f(z) = \frac{1}{1 + \dfrac{z}{2!} + \dfrac{z^2}{3!} + \cdots}, \qquad z \neq 0.$$

As $z \to 0$, the continuous function on the right $\to 1$; hence, if we define $f(0) = 1$, $f(z)$ is analytic in the neighborhood of $z = 0$ and may be expanded into a Maclaurin series which we write

(1)
$$\frac{z}{e^z - 1} = \sum_{n=0}^{\infty} \frac{B_n z^n}{n!}, \qquad B_0 = 1.$$

Since all the zeros of $e^z - 1$ are given by $z = 2\pi n i$ (199.14), the poles of $f(z)$ nearest to the origin are $\pm 2\pi i$; hence the series (1) converges in a circle of radius 2π about the origin (Theorem 207.2).

We may write (1) symbolically as

(2)
$$\frac{z}{e^z - 1} = e^{Bz},$$

provided that B^n is replaced by B_n in the series for e^{Bx}. With this symbolism the Cauchy product of the series,

$$e^{Az} = \sum_{n=0}^{\infty} \frac{A_n z^n}{n!}, \qquad e^{Bz} = \sum_{n=0}^{\infty} \frac{B_n z^n}{n!} \qquad (A_0 = B_0 = 1),$$

is represented symbolically by $e^{(A+B)z}$; for

$$e^{Az} e^{Bz} = 1 + (A_1 + B_1)z + \frac{A_2 + 2A_1 B_1 + B_2}{2!} z^2 + \cdots .†$$

Thus from (2) we have the symbolic equation,

$$(3) \qquad\qquad z = e^{(B+1)z} - e^{Bz}.$$

The constant term on the right is $1 - 1 = 0$; the coefficient of z is $(B_1 + 1) - B_1 = 1$; and the coefficients of all higher powers of z must be zero. Hence, the coefficient of $z^n/n!$ is

$$(4) \qquad\qquad (B+1)^n - B^n = 0, \qquad n = 2, 3, \cdots,$$

where, after canceling B^n, we write B_k instead of B^k. We now use (4) as a recurrence relation for computing the *Bernoulli numbers* B_k. Thus we have

$$2B_1 + 1 = 0, \qquad B_1 = -\tfrac{1}{2};$$
$$3B_2 + 3B_1 + 1 = 0, \qquad B_2 = \tfrac{1}{6};$$
$$4B_3 + 6B_2 + 4B_1 + 1 = 0, \qquad B_3 = 0;$$
$$5B_4 + 10B_3 + 10B_2 + 5B_1 + 1 = 0, \qquad B_4 = -\tfrac{1}{30};$$

and so on.

We shall now show that all Bernoulli numbers of odd index, except B_1, are zero. Then $-z/2$ will be the only odd power in the series (1); consequently, we must show that the function

$$\varphi(z) = \frac{z}{2} + \frac{z}{e^z - 1} = \frac{z}{2} \cdot \frac{e^z + 1}{e^z - 1} = \frac{z}{2} \coth \frac{z}{2}$$

is even; that is, $\varphi(-z) = \varphi(z)$. This property is readily verified; hence,

$$(5) \qquad\qquad B_{2n+1} = 0, \qquad n = 1, 2, \cdots,$$

---

† If $f(z) = e^{Az}$ (symbolically),

$$f(2z) = e^{2Az} = 1 + 2A_1 z + 4A_2 \frac{z^2}{2!} + \cdots,$$

whereas $\qquad f^2(z) = e^{(A+A)z} = 1 + 2A_1 z + (2A_2 + 2A_1^2)\frac{z^2}{2!} + \cdots.$

Thus, in general, $e^{(A+A)z} \neq e^{2Az}$; these expansions are the same only when $A_n = a$ (const), that is, when e^{Ax} is an actual exponential function e^{ax}.

and, consequently,

$$\frac{z}{2} \coth \frac{z}{2} = \sum_{n=0}^{\infty} \frac{B_{2n} z^{2n}}{(2n)!} = \cosh Bz$$

where $\cosh Bz$ is symbolic in the same sense as e^{Bz}. Hence, on replacing z by $2z$, we have

(6) $$z \coth z = \cosh 2Bz.$$

When z is replaced by iz, (6) becomes

(7) $$z \cot z = \cos 2Bz \qquad\qquad (200.3\text{–}4);$$

and, since the poles of $z \cot z$ are the zeros of $\sin z$ (except $z = 0$), the series (7) converges in a circle of radius π about the origin (§ 200).

In order to find the series for $\tan z$, we use the identity

$$2 \cot 2z = \frac{\cos^2 z - \sin^2 z}{\sin z \cos z} = \cot z - \tan z,$$

or

$$z \tan z = z \cot z - 2z \cot 2z.$$

Hence, from (7)

(8) $$z \tan z = \cos 2Bz - \cos 4Bz;$$

since the poles of $z \tan z$ are the zeros of $\cos z$, the series (8) converges in a circle of radius $\pi/2$ about the origin (§ 200).

The identity

$$\cot \frac{z}{2} - \cot z = \frac{1 + \cos z}{\sin z} - \frac{\cos z}{\sin z} = \frac{1}{\sin z}$$

or

$$\frac{z}{\sin z} = 2 \frac{z}{2} \cot \frac{z}{2} - z \cot z$$

gives, in view of (7),

(9) $$\frac{z}{\sin z} = 2 \cos Bz - \cos 2Bz.$$

The power series on the right converges in a circle of radius π about the origin and effects the division indicated in (209.10).

The Bernoulli numbers are also useful in dealing with finite sums. As an example we shall obtain Bernoulli's famous formula for obtaining the sum of pth powers,

(10) $$S_n^{(p)} = 1^p + 2^p + \cdots + n^p,$$

where p is a positive integer. Consider the symbolic binomial expansions,

$$(x+B+1)^{p+1} = x^{p+1} + (p+1)x^p(B+1)^1 + \binom{p+1}{2}x^{p-1}(B+1)^2 + \cdots,$$

$$(x+B)^{p+1} = x^{p+1} + (p+1)x^p B^1 + \binom{p+1}{2}x^{p-1}B^2 + \cdots.$$

On subtraction, we have

(11) $$(x+B+1)^{p+1} - (x+B)^{p+1} = (p+1)x^p,$$

in view of (4). Now put $x = 1, 2, \cdots, n$ in turn in (11), and add all the equations; owing to telescopic cancelation, only two terms survive on the left, and we obtain

(12) $$(n+B+1)^{p+1} - (B+1)^{p+1} = (p+1)S_n^{(p)}.$$

On expanding the left member, we have

$$n^{p+1} + (p+1)n^p(B+1)^1 + \binom{p+1}{2}n^{p-1}(B+1)^2 + \cdots + \binom{p+1}{p}n(B+1)^p,$$

or, since

$$B_1 + 1 = \frac{1}{2}, \quad (B+1)^2 = B_2, \quad (B+1)^3 = B_3, \quad \cdots,$$

(13) $$(p+1)S_n^{(p)} = n^{p+1} + \tfrac{1}{2}(p+1)n^p + \binom{p+1}{2}B_2 n^{p-1} + \cdots + \binom{p+1}{p}B_p n.$$

The special cases $p = 1, 2, 3$ give

$$2S_n^{(1)} = n^2 + n = n(n+1),$$

$$3S_n^{(2)} = n^3 + \tfrac{3}{2}n^2 + 3 \cdot \tfrac{1}{6}n = \tfrac{1}{2}n(n+1)(2n+1),$$

$$4S_n^{(3)} = n^4 + 2n^3 + 6 \cdot \tfrac{1}{6}n^2 + 0 = n^2(n+1)^2.$$

For reference we give the following values of the Bernoulli numbers:

(14) $$B_1 = -\tfrac{1}{2}, \quad B_2 = \tfrac{1}{6}, \quad B_4 = -\tfrac{1}{30}, \quad B_6 = \tfrac{1}{42}, \quad B_8 = -\tfrac{1}{30},$$

$$B_{10} = \tfrac{5}{66}, \quad B_{12} = -\tfrac{691}{2730}, \quad B_{14} = \tfrac{7}{6}, \quad B_{16} = -\tfrac{3617}{510}.$$

PROBLEMS

1. Prove the identity

$$\frac{z}{\sinh z} - \frac{z}{\cosh z} = \frac{4ze^z}{e^{4z}-1} = e^{(4B+1)z},$$

and deduce therefrom the symbolic expansions

$$\frac{z}{\sinh z} = \cosh(4B+1)z, \qquad \frac{z}{\cosh z} = -\sinh(4B+1)z,$$

$$\frac{z}{\sin z} = \cos(4B+1)z, \qquad \frac{z}{\cos z} = -\sin(4B+1)z.$$

2. For any polynomial $f(x) = c_0x^n + c_1x^{n-1} + \cdots + c_n$, show, by means of (11), that

$$f(x + B + 1) - f(x + B) = f'(x).$$

3. From (2) and (3), prove that

(15) $$e^{-Bz} = e^{(B+1)z} = z + e^{Bz},$$

and hence

$$\sinh Bz = -\tfrac{1}{2}z; \qquad B_1 = -\tfrac{1}{2}, \qquad B_3 = B_5 = \cdots = 0.$$

4. From (15) prove that

$$e^{-2Bz} = e^{(2B+1)z} e^z \quad \text{or} \quad e^{-(2B+1)z} = e^{(2B+1)z},$$

and hence

$$\sinh (2B + 1)z = 0; \qquad (2B + 1)^{2n+1} = 0, \qquad n = 0, 1, 2, \cdots.$$

Assuming that $B_{2n+1} = 0 \ (n = 1, 2, \cdots)$, compute B_2, B_4, B_6 from the last relation.

5. Prove that $z/\sinh z = \cosh (2B + 1)z$.
[Since $\sinh (2B + 1) = 0$ (Prob. 4),

$$\cosh (2B + 1)z = e^{(2B+1)z} = e^{2Bz} e^z = \frac{2ze^z}{e^{2z} - 1}.]$$

212. Reciprocal of a Function. Let the functions $f(z)$ and $1/f(z)$ have the Maclaurin series

(1) $$f(z) = \sum_{n=0}^{\infty} \frac{a_n}{n!} z^n = e^{az}, \qquad a_0 = 1,$$

(2) $$1/f(z) = \sum_{n=0}^{\infty} \frac{b_n}{n!} z^n = e^{bz}, \qquad b_0 = 1,$$

where e^{az}, e^{bz}, are symbolic expansions in the sense of § 211. Then, in their common circle of convergence, their product is given symbolically by

(3) $$e^{(a+b)z} = 1.$$

When the sequence $\{a_n\}$ is known, we may therefore compute the sequence $\{b_n\}$ from the symbolic recurrence relation,

(4) $$(a + b)^n = 0, \qquad n = 1, 2, 3, \cdots;$$

thus $b_1, b_2, b_3, \cdots$ are computed in turn from

$$a_1 + b_1 = 0, \qquad a_2 + 2a_1b_1 + b_2 = 0, \qquad \cdots,$$

(5) $$a_n + \binom{n}{1} a_{n-1}b_1 + \binom{n}{2} a_{n-2}b_2 + \cdots + \binom{n}{n} b_n = 0.$$

Example. Euler Numbers. Let

$$f(z) = \cosh z = \sum_{n=0}^{\infty} \frac{z^{2n}}{(2n)!} = e^{Az};$$

then $A_n = 0$ or 1 according as n is odd or even. Then $1/\cosh z = \operatorname{sech} z$ is also an even function whose symbolic expansion may be written

(6) $$\operatorname{sech} z = \sum_{n=0}^{\infty} \frac{E_{2n} z^{2n}}{(2n)!} = e^{Ez}.$$

Here $E_0 = 1$, $E_1 = E_3 = E_5 = \cdots = 0$; and from (5) we have the recurrence relation

(7) $$1 + \binom{n}{2} E_2 + \binom{n}{4} E_4 + \cdots + \binom{n}{n} E_n = 0, \qquad n = 2, 4, 6, \cdots.$$

This may be written more simply as

(8) $$(1 + E)^n = 0, \qquad n = 2, 4, 6, \cdots,$$

if we keep in mind that $E_n = 0$ when n is odd. We thus find

$$\begin{aligned}
1 + E_2 &= 0, & E_2 &= -1; \\
1 + 6E_2 + E_4 &= 0, & E_4 &= 5; \\
1 + 15E_2 + 15E_4 + E_6 &= 0, & E_6 &= -61; \\
1 + 28E_2 + 70E_4 + 28E_6 + E_8 &= 0, & E_8 &= 1385;
\end{aligned}$$

and so on.

The integers thus obtained are known as *Euler's numbers.* They are defined by $\operatorname{sech} z = e^{Ez}$, or, since the even part of e^{Ez} is $\cosh Ez$, by

(9) $$\operatorname{sech} z = \cosh Ez.$$

If we replace z by iz, we also have

(10) $$\sec z = \cos Ez.$$

Since the poles of $\operatorname{sech} z$ and $\sec z$ (zeros of $\cosh z$ and $\cos z$) nearest the origin are $\pm \pi i/2$ and $\pm \pi/2$, respectively, the series (9) and (10) converge in the circle $|z| < \pi/2$.

PROBLEMS

1. Obtain the recurrence relation (211.4) for the Bernoulli numbers by finding the reciprocal of

$$\frac{e^z - 1}{z} = \sum_{n=0}^{\infty} \frac{z^n}{(n + 1)!}$$

[Put $a_n = 1/(n + 1)$ in (5), and multiply by $(n + 1)$.]

2. Find the Maclaurin series for $z/\sin z$ as the reciprocal of $\sin z/z$. Check by (211.9).

3. From the identity

$$2 \cos az \cos bz = \cos (a + b)z + \cos (a - b)z$$

show that Euler's numbers satisfy the recurrence relation

$$(E + 1)^{2n} + (E - 1)^{2n} = 0, \qquad n = 1, 2, 3, \cdots.$$

[From (10), $\cos Ez \cos z = 1$.]

4. By comparing the expansion

$$z \operatorname{sech} z = -\sinh (4B + 1)z \qquad \text{(Prob. 211.1)}$$

with $\operatorname{sech} z = \cosh Ez$ in (9), show that

$$(2n + 1)E_{2n} = -(4B + 1)^{2n+1}, \qquad n = 0, 1, 2, \cdots .$$

213. Residues. If $f(z)$ is analytic in a region R, punctured at $z = a$ where $f(z)$ has an isolated singular point, we define the

(1)   *Residue* of $f(z)$ at $a = \dfrac{1}{2\pi i} \oint_C f(z)\, dz$,

where C is any closed regular curve in R enclosing a. We recall from § 205 that the value of this integral is the same for all such curves. If $f(z)$ has the Laurent expansion,

(2) $$f(z) = \sum_{n=-\infty}^{\infty} c_n (z - a)^n, \qquad 0 < |z - a| < R_2,$$

about a, choose C as a circle with center at a and radius $r < R_2$; since the series (2) converges uniformly along C we may compute the integral in (1) by integrating the series (2) term by term. But, from (204.7), the series of integrals reduces to the single term for $n = -1$; thus,

(3) $$\oint_C f(z)\, dz = \oint_C c_{-1}(z - a)^{-1}\, dz = 2\pi i c_{-1},$$

so that the residue (1) equals c_{-1}. We thus have the

THEOREM. *The residue of $f(z)$ at $z = a$ is the coefficient of $(z - a)^{-1}$ in the Laurent expansion of $f(z)$ in a punctured circle about a.*

When $f(z)$ has a pole of order m at $z = a$, it admits the Laurent expansion,

$$f(z) = \frac{c_{-m}}{(z - a)^m} + \cdots + \frac{c_{-1}}{z - a} + c_0 + c_1(z - a) + \cdots ,$$

about a (209.7); hence, the function

(4) $$g(z) = (z - a)^m f(z)$$

is analytic when $z = a$ and has the Taylor expansion,

$$g(z) = c_{-m} + \cdots + c_{-1}(z - a)^{m-1} + c_0(z - a)^m + \cdots .$$

Since c_{-1} is the coefficient of $(z - a)^{m-1}$ in this series, we have, from (207.3),

(5) $$\text{Res}\ (a) = \frac{g^{(m-1)}(a)}{(m - 1)!} = \frac{1}{(m - 1)!} \lim_{z \to a} D^{m-1}(z - a)^m f(z).$$

In particular, for a simple pole we have

(6) $$\text{Res } (a) = \lim_{z \to a} (z - a) f(z) \qquad (m = 1);$$

and for pole of the second order

(7) $$\text{Res } (a) = \lim_{z \to a} D(z - a)^2 f(z) \qquad (m = 2).$$

At a simple pole the residue $c_{-1} \neq 0$; but for poles of higher order c_{-1} may be zero. Thus $1/(z - a)^2$ has a pole of the second order at a whose residue is zero.

The quotient of two analytic functions $f(z) = P(z)/Q(z)$ has poles at all the zeros of $Q(z)$ that are not zeros of $P(z)$. If $Q(a) = 0$, let $P(z)$, $Q(z)$ have the Taylor series,

$$P(z) = a_0 + a_1(z - a) + \cdots, \qquad Q(z) = b_1(z - a) + b_2(z - a)^2 + \cdots.$$

If a_0 and $b_1 \neq 0$, $f(z)$ has a simple pole at a; and, from (6), its residue

(8) $$\text{Res } (a) = \frac{a_0}{b_1} = \frac{P(a)}{Q'(a)} \qquad (m = 1).$$

If $a_0 \neq 0$, $b_1 = 0$, $b_2 \neq 0$, $f(z)$ has a pole of the second order at a; then

$$Q(z) = (z - a)^2 R(z) \quad \text{where} \quad R(z) = b_2 + b_3(z - a) + \cdots,$$

and from (7)

$$\text{Res } (a) = D\left(\frac{P(z)}{R(z)}\right)_{z=a} = \frac{R(a) P'(a) - P(a) R'(a)}{R^2(a)};$$

that is,

(9) $$\text{Res } (a) = \frac{a_1 b_2 - a_0 b_3}{b_2^2} \qquad (m = 2).$$

Example 1. The function $e^{1/z}$ has an essential singularity at $z = 0$, and its Laurent series (210.11) shows that $\text{Res } (0) = c_{-1} = 1$.

Example 2. The function e^{-1/z^2} has an essential singularity at $z = 0$, and (210.12) shows that $\text{Res } (0) = c_{-1} = 0$.

Example 3. The function

$$f(z) = \frac{z}{\sin z}, \qquad f(0) = 1 \qquad\qquad \text{(Ex. 209.1)}$$

has a simple pole at $z = \pi$; and from (8)

$$\text{Res } (\pi) = \frac{z}{\cos z}\bigg|_{\pi} = -\pi.$$

Example 4. At the poles $n\pi$ of $1/\sin z$ (Ex. 209.2) equation (8) shows that

$$\text{Res } (n\pi) = \frac{1}{\cos z}\bigg|_{n\pi} = \frac{1}{\cos n\pi} = (-1)^n$$

in agreement with (209.11).

Similarly, at the poles $(n - \tfrac{1}{2})\pi$ of $1/\cos z$ we have

$$\operatorname{Res}(n - \tfrac{1}{2})\pi = \frac{1}{-\sin z}\bigg|_{(n-\frac{1}{2})\pi} = (-1)^n.$$

Example 5. The function

$$f(z) = \frac{z^2 + 1}{z(z - 1)^2}$$

has poles at 0 ($m = 1$) and 1 ($m = 2$). Hence, from (6) and (7):

$$\operatorname{Res}(0) = \frac{z^2 + 1}{(z - 1)^2}\bigg|_0 = 1; \ \operatorname{Res}(1) = D\left(\frac{z^2 + 1}{z}\right)\bigg|_1 = \left(1 - \frac{1}{z^2}\right)\bigg|_1 = 0.$$

If we put $z - 1 = w$,

$$f(z) = \frac{2 + 2w + w^2}{w^2 + w^3},$$

and $a_0 = a_1 = 2, b_2 = b_3 = 1$; now (9) shows that $\operatorname{Res}(1) = 0$.

Example 6. The function

$$f(z) = \frac{1}{z(e^z - 1)} = \frac{1}{z^2 + \frac{1}{2}z^3 + \cdots}.$$

has a pole of order 2 at $z = 0$ since $z^2 f(z) \to 1$ as $z \to 0$. Since $a_0 = 1, a_1 = 0, b_2 = 1,$ $b_3 = \frac{1}{2}$, (9) gives $\operatorname{Res}(0) = -\frac{1}{2}$.

PROBLEMS

1. Find the residue at each pole of the functions:

(a) $\dfrac{z + 1}{z^2 - z}$; (b) $\dfrac{z^2 + 1}{z^3 - z^2}$; (c) $\dfrac{e^z}{(z + 1)^2}$; (d) $\dfrac{e^z}{z^3}$.

2. Show that residues at the poles of $\tan z$ are all -1; at the poles of $\cot z$ are all 1.

3. Find the poles and residues of

(a) $\dfrac{1}{\sinh z}$; (b) $\tanh z$; (c) $\dfrac{1}{\cosh z}$; (d) $\coth z$.

4. Find $\operatorname{Res}(0)$ for $1/z \sin z$; $1/\sin^2 z$.

5. Find $\operatorname{Res}(0)$ for $(1 - \cos z)/z^5$.

6. Find the residues for $(z^2 - 2)/z(z^2 + 1)$ at all poles.

7. If $P(z), Q(z)$ are polynomials with real coefficients and $P(z)/Q(z)$ has simple poles $a \pm bi$, show that $\operatorname{Res}(a + bi)$ and $\operatorname{Res}(a - bi)$ are conjugate numbers. Is this true if $P(z) = e^z$; $\sin z$; $\sinh z$?

214. Residue Theorem. *Let the function $f(z)$ be analytic in a region R punctured where $f(z)$ is singular. Then, if the regular closed curve C in R encloses a finite number of singular points, $z_1, z_2, \cdots, z_k,$*

(1)
$$\oint_C f(z)\, dz = 2\pi i \sum_{n=1}^{k} \operatorname{Res}(z_n).$$

Proof. Since C lies *in* R, it must avoid the punctures. Let $C_1, \cdots, C_k$ be circles about $z_1, \cdots, z_k$ so small that none intersect and all are enclosed by C (Fig. 214). Draw cross-cuts from C to the circles, and integrate over

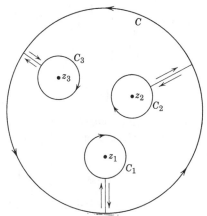

FIG. 214.   Path of integration in residue theorem

the closed path formed by C, the circles, and the cross-cuts; then, by Cauchy's theorem (§ 205),

$$\oint_C f(z)\, dz - \sum_{n=1}^{k} \oint_{C_n} f(z)\, dz = 0.$$

In view of the definition (213.1) of a residue, this equation is the same as (1).

Cauchy's integral formula,

(2) $$\oint_C \frac{f(z)}{z - a}\, dz = 2\pi i\, f(a) \qquad\qquad (206.1),$$

is a special case of this theorem; for, if $f(a) \neq 0$, $f(z)/(z - a)$ has a simple pole at a and Res $(a) = f(a)$ from (213.8); and, if $f(a) = 0$, $f(z)/(z - a)$ is analytic at a, and the integral is zero.

Example 1. If C is any closed curve about $z = 0$, but not through $z = 1$,

$$\oint_C \frac{z^2 + 1}{z(z - 1)^2}\, dz = 2\pi i;$$

for, from Ex. 213.5, Res $(0) = 1$, Res $(1) = 0$.

When $f(z)$ has but a finite number of singular points, $z_1, \cdots, z_k$, we can describe a circle $|z| = R$ about the origin which includes *all* of them in its interior. Then, if $z = Re^{i\theta}$,

(3) $$\oint_{|z|=R} f(z)\, dz = \int_0^{2\pi} f(Re^{i\theta})\, Rie^{i\theta}d\theta = 2\pi i \sum_{n=1}^{k} \text{Res}\, (z_n),$$

where we integrate with respect to the real variable θ. Now make the change of variable $z = 1/t$; if $t = re^{i\varphi}$, this means writing $R = 1/r$ and making the change of real variable $\theta = -\varphi$; hence,

$$\oint_{|z|=R} f(z)\, dz = \int_0^{-2\pi} f\left(\frac{1}{re^{i\varphi}}\right) \frac{1}{r}\, ie^{-i\varphi}(-d\varphi)$$

$$= \int_0^{2\pi} f\left(\frac{1}{re^{i\varphi}}\right) \frac{rie^{i\varphi}\, d\varphi}{r^2 e^{2i\varphi}}$$

$$= \oint_{|t|=r} f\left(\frac{1}{t}\right) \frac{dt}{t^2}\,.$$

Since $f(z)$ has no finite singularities *outside* of the circle $|z| = R$, $f(1/t)$ has none *inside* the circle $|t| = 1/R$ except possibly at $t = 0$. The residue theorem now gives

(4) $$\oint_{|z|=R} f(z)\, dz = 2\pi i\ \text{Res}\,(0)\ \text{of}\ \frac{1}{t^2} f\left(\frac{1}{t}\right)\,.$$

Since the circle $|z| = R$ surrounds *all* singular points of $f(z)$ in the finite plane, we have

THEOREM 2. *If $f(z)$ is analytic in the entire z-plane except for a finite number of singular points $z_1, z_2, \cdots, z_k$,*

(5) $$\sum_{n=1}^{k} \text{Res}\,(z_n)\ \text{of}\ f(z) = \text{Res}\,(0)\ \text{of}\ \frac{1}{t^2} f\left(\frac{1}{t}\right)\,.$$

If the integral (3) is taken in a *clockwise* sense, it may be regarded as a positive circuit of the region *exterior* to the circle $|z| = R$; for a person making the circuit has outer region to his left. In view of the definition (213.1), it is conventional to write

(6) $$-\oint_{|z|=R} f(z)\, dz = 2\pi i\ \text{Res}\,(\infty),$$

where we define

(7) $$\text{Res}\,(\infty)\ \text{of}\ f(z) = -\text{Res}\,(0)\ \text{of}\ \frac{1}{t^2} f\left(\frac{1}{t}\right)$$

in accordance with (5).

On adding (3) and (6), we have

(8) $$\sum_{n=1}^{k} \text{Res}\,(z_n) + \text{Res}\,(\infty) = 0.$$

We have thus proved

THEOREM 3. *If $f(z)$ is analytic in z-plane except for a finite number of singular points, the sum of their residues and Res (∞) is zero.*

The residue of $f(z)$ is zero at every *finite* point where $f(z)$ is analytic; but Res (∞) may not vanish even when $f(z)$ is analytic at ∞ (§ 209). Thus, when

$$f(z) = 1 + \frac{1}{z}, \qquad \frac{1}{t^2} f\left(\frac{1}{t}\right) = \frac{1+t}{t^2},$$

and, from (7), Res $(\infty) = -1$ (213.7).

Example 2. The function

$$f(z) = \frac{z^2 - 2}{z(z^2 + 1)}$$

has simple poles at 0, i, $-i$; and from (213.6)

$$\text{Res } (0) = -2, \qquad \text{Res } (i) = \tfrac{3}{2}, \qquad \text{Res } (-i) = \tfrac{3}{2}.$$

For the function

$$\frac{1}{t^2} f\left(\frac{1}{t}\right) = \frac{1 - 2t^2}{t + t^3}, \qquad \text{Res } (0) = \left.\frac{1 - 2t^2}{1 + t^2}\right|_0 = 1.$$

Thus we verify that the sum of the residues of $f(z)$, including Res $(\infty) = -1$, is zero.

PROBLEMS

1. Find $\oint \tan z \, dz$ over the circles: $|z| = 1$; $|z - 1| = 1$; $|z| = 2$.

2. Find $\displaystyle\oint \frac{dz}{\cosh z}$ over the circle $|z| = 2$.

3. Evaluate the following integrals about the circle $|z| = 1$:

$$\oint \frac{dz}{\sin z}; \quad \oint \frac{dz}{z \sin z}; \quad \oint \frac{dz}{z^2 \sin z}; \quad \oint \frac{e^{-z}}{z^2}; \quad \oint z^2 e^{1/z} \, dz.$$

4. Find the residues at the poles of

$$f(z) = \frac{1}{z^3(z + 2)} \quad \text{and at } \infty.$$

Compute $\oint f(z) \, dz$ over the circles $|z| = 1$: $|z| = 3$.

5. Find the residues at the poles of

$$f(z) = \frac{z^2 + 2}{(z - 1)(z^2 + 1)} \quad \text{and at } \infty.$$

Compute $\oint f(z) \, dz$ over the circles: $|z - 1| = 1$; $|z - i| = 1$; $|z| = 2$.

215. Evaluation of Definite Integrals. The residue theorem may be used to compute certain real definite integrals. We first consider integrals of the form

$$(1) \qquad\qquad I = \int_0^{2\pi} f(\cos \theta, \sin \theta) \, d\theta,$$

where $f(\sin \theta, \cos \theta)$ is a rational function of $\sin \theta$ and $\cos \theta$ which is continuous in the interval $(0, 2\pi)$.

The substitution

(2) $$z = e^{i\theta}, \qquad\qquad dz = iz\, d\theta,$$

$$\cos \theta = \frac{1}{2}\left(z + \frac{1}{z}\right), \qquad \sin \theta = \frac{1}{2i}\left(z - \frac{1}{z}\right) \qquad (199.7\text{–}8)$$

converts (1) into the complex integral $\oint \varphi(z)\, dz$ about the unit circle $|z| = 1$, where

(3) $$\varphi(z) = \frac{1}{iz} f\left(\frac{z^2 + 1}{2z}, \frac{z^2 - 1}{2iz}\right).$$

Now, if $\varphi(z)$ has the poles $z_1, z_2, \cdots, z_k$ inside this circle, we have from the residue theorem

(4) $$I = 2\pi i \sum_{n=1}^{k} \operatorname{Res}(z_n).$$

Example 1. The substitution (2) converts

$$I = \int_0^{2\pi} \frac{d\theta}{\cos \theta + a}, \qquad |a| > 1, \quad a \text{ real},$$

into the integral

$$I = \frac{2}{i} \oint \frac{dz}{z^2 + 2az + 1} \qquad \text{over} \quad |z| = 1.$$

The integrand has simple poles where $z^2 + 2az + 1 = 0$: namely,

$$z_1 = -a + \sqrt{a^2 - 1}, \qquad z_2 = -a - \sqrt{a^2 - 1}.$$

Only one of these lies within the unit circle: z_1 when $a > 1$, z_2 when $a < -1$. Now from (213.8)

$$\operatorname{Res}(z_1) = \frac{1}{i(z + a)}\bigg|_{z_1} = \frac{1}{i\sqrt{a^2 - 1}}, \qquad \operatorname{Res}(z_2) = \frac{1}{i(z + a)}\bigg|_{z_2} = \frac{-1}{i\sqrt{a^2 - 1}};$$

hence, from (4)

(5) $$I = 2\pi i \frac{\operatorname{sgn} a}{i\sqrt{a^2 - 1}} = \operatorname{sgn} a \frac{2\pi}{\sqrt{a^2 - 1}},$$

where $\operatorname{sgn} a$ denotes the sign of a.

Example 2. The substitution (2) converts

$$J = \int_0^{2\pi} \frac{d\theta}{(\cos \theta + a)^2}, \qquad |a| > 1, \quad a \text{ real},$$

into

$$J = \frac{4}{i} \oint \frac{z\, dz}{(z^2 + 2az + 1)^2} \qquad \text{over} \quad |z| = 1,$$

in which $\varphi(z)$ has the same poles as in Ex. 1. Now the poles are of the second order, and we find from (213.7) that

$$\text{Res}\,(z_1) = \text{Res}\,(z_2) = \frac{|a|}{i(a^2 - 1)^{3/2}}\,;$$

hence, from (4)

(6) $$J = \frac{2\pi\,|a|}{(a^2 - 1)^{3/2}}\,.$$

PROBLEMS

Verify the following integrals:

1. $\displaystyle\int_0^{2\pi} \frac{d\theta}{a + b\cos\theta} = \int_0^{2\pi} \frac{d\theta}{a + b\sin\theta} = \text{sgn}\,a\,\frac{2\pi}{\sqrt{a^2 - b^2}}$

when $a^2 > b^2$.

2. $\displaystyle\int_0^{2\pi} \frac{d\theta}{(a + b\cos\theta)^2} = \int_0^{2\pi} \frac{d\theta}{(a + b\sin\theta)^2} = \frac{2\pi\,|a|}{(a^2 - b^2)^{3/2}}$

when $a^2 > b^2$.

3. $\displaystyle\int_0^{2\pi} \frac{d\theta}{(a + b\cos^2\theta)^2} = \int_0^{2\pi} \frac{d\theta}{(a + b\sin^2\theta)^2} = \frac{2\pi\,|a + \frac{1}{2}b|}{(a^2 + ab)^{3/2}}$

when $a^2 + ab > 0$.

[Use $\cos^2 x = \frac{1}{2}(1 + \cos 2x)$, $\sin^2 x = \frac{1}{2}(1 - \cos 2x)$.]

4. $\displaystyle\int_0^{\pi} \frac{\cos 2\theta\,d\theta}{1 - 2\alpha\cos\theta + \alpha^2} = \frac{\pi\alpha^2}{1 - \alpha^2}$ $\qquad(\alpha^2 < 1)$.

216. Improper Real Integrals. We next consider *convergent* improper integrals,

(1) $$I = \int_{-\infty}^{\infty} \frac{P(x)}{Q(x)}\,dx, \qquad Q(x) \neq 0,$$

where $P(x)$ and $Q(x)$ are polynomials of degree p and q such that $q - p \geqq 2$ (cf. § 138). In the complex domain

$$\frac{P(z)}{Q(z)} = \frac{z^p}{z^q} \cdot \frac{a_0 + a_1/z + \ldots}{b_0 + b_1/z + \ldots} = z^{p-q}\,g(z),$$

and, as $z \to \infty$, $g(z) \to a_0/b_0$; hence, we can choose a positive number R so that

$$|g(z) - a_0/b_0| < \varepsilon \quad\text{when}\quad |z| > R.$$

Since $g(z)$ remains bounded as $z \to \infty$, we can write in the O-notation of § 64

(2) $$\frac{P(z)}{Q(z)} = O(z^{p-q}) \quad\text{as}\quad z \to \infty.$$

We shall now show that when $P(z)$ and $Q(z)$ have no zeros in common and $Q(z)$ has the zeros $z_1, z_2, \cdots, z_k$ *in the upper half-plane,*

(3)
$$\int_{-\infty}^{\infty} \frac{P(x)}{Q(x)}\, dx = 2\pi i \sum_{n=1}^{k} \text{Res}\, (z_n),$$

when the integral exists.

Integrate $P(z)/Q(z)$ along the closed path composed of the segment $-R \leq x \leq R$ of the real axis and the semicircle $|z| = R$ in the upper half-plane (Fig. 216a). Choose R so large that the semicircle S includes

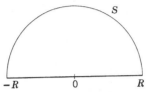

FIG. 216a.   Semicircular contour

all the zeros of $Q(z)$ in the upper half-plane; then from the residue theorem

(4)
$$\int_{-R}^{R} \frac{P(x)}{Q(x)}\, dx + \int_{S} \frac{P(z)}{Q(z)}\, dz = 2\pi i \sum_{n=1}^{k} \text{Res}\, (z_n).$$

Since $|P/Q| < M\,|z|^{p-q}$ from (2), we have

$$\left| \int_{S} \frac{P}{Q}\, dz \right| < M\, R^{p-q} \cdot \pi R = \frac{\pi M}{R^{q-p-1}} \leqq \frac{\pi M}{R} \; ;$$

hence, as $R \to \infty$ in (4), the second integral $\to 0$, and we obtain equation (3).

Even when $\int_{-R}^{R} f(x)\, dx$ has a limit as $R \to \infty$, the convergence of the improper integral is not ensured; Ex. 138.2 gives a case in point. However, if $f(x)$ is an *even* function,

$$\int_{-R}^{R} f(x)\, dx = 2 \int_{0}^{R} f(x)\, dx,$$

and, if the left-hand integral has a limit, the one on the right has also; in this case

(5)
$$\int_{0}^{\infty} \frac{P(x)}{Q(x)}\, dx = \pi i \sum_{n=1}^{k} \text{Res}\, (z_n).$$

Example 1. We shall use (5) to show that

$$I = \int_0^\infty \frac{dx}{x^4 + 1} = \frac{\pi\sqrt{2}}{4}.$$

The zeros of $z^4 + 1$ above the axis of reals are

$$z_1 = e^{\pi i/4} = \frac{1 + i}{\sqrt{2}}, \qquad z_2 = e^{3\pi i/4} = \frac{-1 + i}{\sqrt{2}}.$$

Now from (213.8)

$$\mathrm{Res}\,(z_1) + \mathrm{Res}\,(z_2) = \frac{1}{4}\left(\frac{1}{z_1{}^3} + \frac{1}{z_2{}^3}\right) = -\frac{1}{4}(z_1 + z_2) = -\frac{i}{2\sqrt{2}};$$

and, from (5), $I = \pi/2\sqrt{2}$.

In the derivative

$$\left(\frac{P}{Q}\right)' = \frac{QP' - PQ'}{Q^2},$$

the numerator and denominator are polynomials of degree $p + q - 1$ and $2q$, respectively; hence, from (2)

$$\left(\frac{P}{Q}\right)' = O(z^{p-q-1}).$$

We shall now show that, when $q - p \geq 1$ and $Q(z)$ has the same zeros as before,

(6) $$\int_{-\infty}^\infty \frac{P(x)}{Q(x)} e^{aix}\, dx = 2\pi i \sum_{n=1}^k \mathrm{Res}\,(z_n) \qquad (a > 0),$$

when the integral exists. On integrating $P(z)\, e^{aiz}/Q(z)$ over the same semicircular contour, we have

(7) $$\int_{-R}^R \frac{P(x)}{Q(x)} e^{aix}\, dx + \int_S \frac{P(z)}{Q(z)} e^{aiz}\, dz = 2\pi i \sum_{n=1}^k \mathrm{Res}\,(z_n).$$

Integrate the second integral by parts,

$$\int_S \frac{P}{Q} e^{aiz}\, dz = \frac{P}{Q} \frac{e^{aiz}}{ai}\Big|_R^{-R} - \frac{1}{ai} \int_S \left(\frac{P}{Q}\right)' e^{aiz}\, dz,$$

and let $R \to \infty$. Then, since

$$|e^{aiz}| = |e^{ai(x+iy)}| = e^{-ay} \leq 1,$$

the absolute value of the integrated part is less than

$$\frac{2M}{a} R^{p-q} = \frac{2M}{aR^{q-p}} \leq \frac{2M}{aR}$$

and therefore approaches zero. Moreover,

$$\left| \int_S \left(\frac{P}{Q}\right)' e^{aiz}\, dz \right| < MR^{p-q-1} \cdot \pi R = \frac{\pi M}{R^{q-p}} \to 0.$$

Hence, as $R \to \infty$ in equation (7), the integral over S approaches zero, and

$$(8) \qquad \int_{-\infty}^{\infty} \frac{P(x)}{Q(x)} \cos ax\, dx + i \int_{-\infty}^{\infty} \frac{P(x)}{Q(x)} \sin ax\, dx = 2\pi i \sum_{n=1}^{k} \text{Res}\, (z_n),$$

when the integrals exist. On equating the real and imaginary parts in (8), we obtain both real integrals.

Example 2. From (6) we find that

$$(9) \qquad \int_{-\infty}^{\infty} \frac{e^{ix}}{x^2 + a^2}\, dx = 2\pi i\, \text{Res}\, (ai) = 2\pi i \frac{e^{-a}}{2ai} = \frac{\pi e^{-a}}{a},$$

since $q - p = 2$. Moreover, the integral converges absolutely since $|e^{ix}| = 1$, and

$$\int_{-\infty}^{\infty} \frac{dx}{x^2 + a^2} = \frac{1}{a} \tan^{-1} \frac{x}{a} \Big|_{-\infty}^{\infty} = \frac{\pi}{a}.$$

On equating real parts in (9), we get

$$\int_{-\infty}^{\infty} \frac{\cos x}{x^2 + a^2}\, dx = 2 \int_0^{\infty} \frac{\cos x}{x^2 + a^2}\, dx = \frac{\pi e^{-a}}{a}.$$

Example 3. The integral

$$\int_{-\infty}^{\infty} \frac{x e^{ix}}{x^2 + a^2}\, dx = 2\pi i\, \text{Res}\, (ai) = 2\pi i\, \frac{e^{-a}}{2},$$

since $q - p = 1$. On equating imaginary parts, we get

$$\int_{-\infty}^{\infty} \frac{x \sin x}{x^2 + a^2}\, dx = 2 \int_0^{\infty} \frac{x \sin x}{x^2 + a^2}\, dx = \pi e^{-a}.$$

Theorem 145.2 shows that this integral converges conditionally.

A semicircular contour is not always suited to a problem. In the next example, we use a rectangular contour.

Example 4. We shall show that

$$(10) \qquad I = \int_{-\infty}^{\infty} \frac{e^{mx}}{1 + e^x}\, dx = \frac{\pi}{\sin m\pi}, \qquad 0 < m < 1,$$

by integrating $f(z) = e^{mz}/(1 + e^z)$ over the rectangle whose vertices are $-a$, a, $a + 2\pi i$, $-a + 2\pi i$ shown in Fig. 216b. This includes one pole πi of $f(z)$ whose

$$\text{Res}(\pi i) = \frac{e^{m\pi i}}{e^{\pi i}} = -e^{m\pi i}.$$

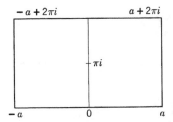

FIG. 216b.   Rectangular contour

The integral about the rectangle is therefore

(11) $$\int_{-a}^{a} + \int_{a}^{a+2\pi i} + \int_{a+2\pi i}^{-a+2\pi i} + \int_{-a+2\pi i}^{-a} \frac{e^{mz}}{1 + e^z} \, dz = -2\pi i e^{m\pi i}.$$

Call these integrals I_1, I_2, I_3, I_4; then

$$I_3 = \int_{a}^{-a} \frac{e^{m(x+2\pi i)}}{1 + e^{(x+2\pi i)}} \, dx = e^{2m\pi i} \int_{a}^{-a} \frac{e^{mx}}{1 + e^x} \, dx = -e^{2m\pi i} I_1,$$

$$\lim_{a \to \infty} (I_1 + I_3) = (1 - e^{2m\pi i}) \int_{-\infty}^{\infty} \frac{e^{mx}}{1 + e^x} \, dx.$$

Moreover,

$$I_2 = \int_{0}^{2\pi} \frac{e^{m(a+iy)}}{1 + e^{a+iy}} i \, dy, \qquad I_4 = \int_{2\pi}^{0} \frac{e^{m(-a+iy)}}{1 + e^{-a+iy}} i \, dy,$$

and, from (204.6),

$$|I_2| \leq \frac{2\pi e^{ma}}{e^a - 1} = \frac{2\pi e^{(m-1)a}}{1 - e^{-a}}, \qquad |I_4| \leq \frac{2\pi e^{-ma}}{1 - e^{-a}};$$

for when $|z_1| > |z_2|$,

(12) $$|z_1| - |z_2| < |z_1 + z_2|.$$

Since $0 < m < 1$, both I_2 and $I_4 \to 0$ as $a \to \infty$. Thus, on letting $a \to \infty$ in (11), we have

$$(1 - e^{2m\pi i})I = -2\pi i e^{m\pi i}, \qquad I = \frac{2\pi i}{e^{m\pi i} - e^{-m\pi i}}$$

which, in view of (199.8), gives (10).

PROBLEMS

Verify the following integrals:

1. $\displaystyle\int_{-\infty}^{\infty} \frac{dx}{x^2 + 2x + 2} = \pi.$

2. $\displaystyle\int_0^{\infty} \frac{x^2\,dx}{x^4 + 1} = \frac{\pi\sqrt{2}}{4}.$

3. $\displaystyle\int_0^{\infty} \frac{dx}{(x^2 + a^2)(x^2 + b^2)} = \frac{\pi}{2ab(a + b)}$ $(a, b > 0).$

4. $\displaystyle\int_0^{\infty} \frac{x^2\,dx}{x^6 + 1} = \frac{\pi}{6}.$

5. $\displaystyle\int_0^{\infty} \frac{x^2\,dx}{(x^2 + 4)^2(x^2 + 9)} = \frac{\pi}{200}.$

6. $\displaystyle\int_0^{\infty} \frac{x^4\,dx}{(x^2 + 1)^2(x^2 + 4)} = \frac{5\pi}{36}.$

7. $\displaystyle\int_{-\infty}^{\infty} \frac{\sin x\,dx}{(x + a)^2 + b^2} = -\frac{\pi \sin a}{be^b}.$

8. $\displaystyle\int_{-\infty}^{\infty} \frac{\cos x\,dx}{(x + a)^2 + b^2} = \frac{\pi \cos a}{be^b}.$

9. $\displaystyle\int_0^{\infty} \frac{\cos bx}{(x^2 + a^2)^2}\,dx = \frac{\pi}{4a^3}(1 + ab)e^{-ab}.$

10. $\displaystyle\int_0^{\infty} \frac{\cos bx}{x^4 + a^4}\,dx = \frac{\pi}{2\sqrt{2}\,a^3}\,e^{-ab/\sqrt{2}}\left(\cos\frac{ab}{\sqrt{2}} + \sin\frac{ab}{\sqrt{2}}\right)$

11. If x, a, b in Problems 9 and 10 have the dimensions L, L, L^{-1}, show that the answers check dimensionally. [Cf. Appendix 4.]

12. If x, a, b in Prob. 3 have the dimensions L, show that the answer checks dimensionally.

Owing to $\sin x$ and $\cos x$ in Problems 7 and 8, x, and therefore a and b can only be interpreted as pure numbers; hence these problems do not permit a dimensional check. Evaluate these integrals (by a change of variable) after replacing $\sin x$, $\cos x$ by $\sin cx$, $\cos cx$; then check the answers dimensionally after assigning x, a, b, c the dimensions L, L, L, L^{-1}.

13. $\displaystyle\int_0^{\infty} \frac{x \sin bx}{x^4 + a^4}\,dx = \frac{\pi}{2a^2}\,e^{-ab/\sqrt{2}}\sin\frac{ab}{\sqrt{2}}$ $(a, b > 0).$

14. $\displaystyle\int_0^{\infty} \frac{x^3 \sin bx}{x^4 + a^4}\,dx = \frac{\pi}{2}\,e^{-cb/\sqrt{2}}\cos\frac{ab}{\sqrt{2}}$ $(b > 0).$

15. $\displaystyle\int_0^\infty \frac{x^{2m}}{x^{2n}+1}\,dx = \frac{\pi}{2n \sin\left(\dfrac{2m+1}{2n}\,\pi\right)},$

where m, n are positive integers and $m < n$.

$\Big[$The poles above the axis of reals are $z_1 = e^{i\pi/2n}$, rz_1, r^2z_1, $\cdots$, $r^{n-1}z_1$, where $r = e^{i\pi/n}$. The residue at any pole is the value of

$$\frac{z^{2m}}{2nz^{2n-1}} = \frac{z^{2m+1}}{-2n} \quad \text{at that pole;}$$

hence, the sum of the residues is

$$-\frac{1}{2n}\,z_1^{\,2m+1} \sum_{k=0}^{n-1} r^{(2m+1)k} = -\frac{1}{n}\frac{z_1^{\,2m+1}}{1-r^{2m+1}}\cdot\Big]$$

16. Using (162.3), show that

$$\int_{-\infty}^\infty e^{-(x+ib)^2}\,dx = \int_{-\infty}^\infty e^{-x^2}\,dx = \sqrt{\pi}$$

by integrating e^{-z^2} about the rectangle whose vertices are $-a$, a, $a+bi$, $-a+bi$ and letting $a \to \infty$. Hence, deduce

$$\int_0^\infty e^{-x^2}\cos 2bx\,dx = \frac{\sqrt{\pi}}{2}\,e^{-b^2}.$$

17. Compute *Fresnel's integrals*

$$\int_0^\infty \cos x^2\,dx = \int_0^\infty \sin x^2\,dx = \frac{\sqrt{2\pi}}{4}$$

by integrating e^{-x^2} about the 45° sector of the circle $|z| = R$ included between the radii $y = 0$ and $y = x$, and making use of $\displaystyle\int_0^\infty e^{-x^2}\,dx = \frac{\sqrt{\pi}}{2}$.

$\Big[\displaystyle\oint e^{-z^2}\,dx = I_1 + I_2 + I_3 = 0$ where

$$I_1 = \int_0^R e^{-x^2}\,dx, \qquad I_3 = \int_R^0 e^{-ir^2}\,e^{i(\pi/4)}\,dr,$$

$$I_2 = \int_0^{\pi/4} e^{-R^2(\cos 2\theta + i \sin 2\theta)}(i\,Re^{i\theta}\,d\theta);$$

as $R \to \infty$,

$$|I_2| \le R\int_0^{\pi/4} e^{-R^2\cos 2\theta}d\theta = \frac{R}{2}\int_0^{\pi/2} e^{-R^2\sin\varphi}\,d\varphi \to 0.\Big]$$

217. Indented Contours.   The curve over which a circuit integral is taken should avoid the poles of $f(z)$, the integrand. If a curve suitable to the problem at hand *does* pass through a pole a of $f(z)$, we may avoid a by passing around it on a small circular arc s centered at a. The contour is then said to be *indented* at a.

Thus let s be the arc of the circle

(1) $$z = a + re^{i\theta}, \qquad \theta_1 \leq \theta \leq \theta_2,$$

of radius r about a (Fig. 217a). Then, if $f(z)$ has a simple pole at a, its Laurent expansion about a has the form

$$f(z) = \frac{\text{Res}\,(a)}{z - a} + g(z),$$

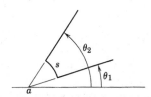

Fig. 217a.   Angle indented at vertex

where $g(z)$ is analytic in the neighborhood of a. The integral over the arc s is therefore

(2) $$\int_s f(z)\,dz = \text{Res}\,(a)\int_s \frac{dz}{z-a} + \int_s g(z)\,dz,$$

where

$$\int_s \frac{dz}{z-a} = \int_{\theta_1}^{\theta_2} \frac{rie^{i\theta}}{re^{i\theta}}\,d\theta = i\int_{\theta_1}^{\theta_2} d\theta = i(\theta_2 - \theta_1).$$

Since $g(z)$ is continuous at a, we can choose r so small that

$$\max |g(z)| \text{ on } s < |g(a)| + \varepsilon;$$

then, as $r \to 0$,

$$\left| \int_s g(z)\,dz \right| < [|g(a)| + \varepsilon]\,(\theta_2 - \theta_1)r \to 0.$$

Thus from (2) we obtain

(3) $$\lim_{r \to 0} \int_s f(z)\,dz = i(\theta_2 - \theta_1)\,\text{Res}\,(a),$$

a formula that enables us to deal with indentations at simple poles.

Example. To show that the integral

(4)
$$I = \int_0^\infty \frac{\sin z}{z}\, dz = \frac{\pi}{2},$$

we integrate $f(z) = e^{iz}/z$ over a semicircular contour indented at $z = 0$, a simple pole of $f(z)$ of residue 1 (Fig. 217b). Since the indented contour encloses no poles, the contour integral,

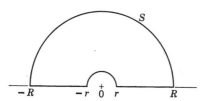

FIG. 217b.   Semicircle indented at center

(5)
$$\int_r^R + \int_S + \int_{-R}^{-r} + \int_s \frac{e^{iz}}{z}\, dz = 0,$$

where the integral over s is clockwise. Call these integrals I_1, I_2, I_3, I_4; then, with the substitution $z = -w$,

$$I_3 = \int_R^r \frac{e^{-iw}}{w}\, dw = -\int_r^R \frac{e^{-iz}}{z}\, dz,$$

$$I_1 + I_3 = \int_r^R \frac{e^{iz} - e^{-iz}}{z}\, dz = 2i \int_r^R \frac{\sin z}{z}\, dz.$$

Now, as $r \to 0$, $R \to \infty$ in (5),

$$I_1 + I_3 = 2i \int_0^\infty \frac{\sin z}{z}\, dz;$$

$$I_2 = \int_S \frac{e^{iz}}{z}\, dz = \frac{e^{iz}}{iz}\Big|_R^{-R} + \frac{1}{i}\int_S \frac{e^{iz}}{z^2}\, dz \to 0.$$

as in § 216; and from (3)

$$I_4 \to i(0 - \pi)\, \mathrm{Res}\,(0) = -\pi i.$$

Thus, the limiting process applied to (5) gives

$$2i \int_0^\infty \frac{\sin z}{z}\, dz - \pi i = 0,$$

and $I = \pi/2$ in agreement with (189.7).

PROBLEMS

1. Compute the integral in Ex. 1 by integrating over the rectangular contour of Fig. 216b indented at the origin.

2. By integrating $f(z) = z^{p-1}/(1 + z)$ around the contour of Fig. 217c composed of the circles $|z| = r$, $|z| = R$ and the segment $r \leqq x \leqq R$ traversed twice, show that

$$\int_0^\infty \frac{x^{p-1}}{1 + x} \, dx = \frac{\pi}{\sin p\pi} \qquad (0 < p < 1).$$

Put $x = e^t$ in the integral, and compare the result with Ex. 216.4.

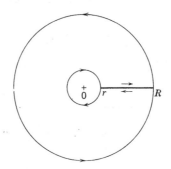

Fig. 217c.   Circle indented at center

218. Properties of Analytic Functions. A function $f(z)$ of a complex variable is said to be analytic in a region R if it has a derivative at every point of R. Then $f(z)$ is continuous in R and has continuous derivatives of all orders in R. In order that $f(z)$ be analytic in R it is necessary and sufficient that u and v have continuous first partial derivatives that satisfy the Cauchy-Riemann equations:

(1) $u_x = v_y, \qquad u_y = -v_x.$

Then u and v are harmonic functions.

When $f(z)$ is analytic in a region, the map $w = f(z)$ is *conformal*: the size and sense of angles are preserved, and the magnification at a point is independent of direction.

The integral of $f(z)$ over the curve C: $z = z(t), \alpha \leqq t \leqq \beta$, is computed as

(2) $$I = \int_C f(z) \, dz = \int_\alpha^\beta f\{z(t)\} \frac{dz}{dt} \, dt;$$

and, if $|f(z)| < M$ over a curve C of length L, $|I| \leqq ML$.

If $f(z)$ is analytic in a simply connected region R, then $\oint f(z) \, dz = 0$ over any closed regular curve in R. Moreover $\int_{z_0}^z f(z) \, dz$ is independent of the path connecting z_0 and z and defines an analytic function $F_0(z)$ whose derivative is $f(z)$. If $F(z)$ is *any* function whose derivative is $f(z)$,

(3) $$\int_{z_1}^{z_2} f(z) \, dz = F(z_2) - F(z_1).$$

If $f(z)$ is analytic on and within a regular closed curve C enclosing the point a,

$$(4) \qquad f(a) = \frac{1}{2\pi i} \oint_C \frac{f(z)}{z-a} \, dz.$$

If $\left| z - a \right| = R$ is a circle about a passing through the nearest singular point of $f(z)$, *Cauchy's integral formula* (4) gives

$$f(a+h) = \frac{1}{2\pi i} \oint_C \frac{f(z)}{z-a-h} \, dz = \frac{1}{2\pi i} \int_C \sum_{n=0}^{\infty} \frac{f(z) \, h^n}{(z-a)^{n+1}} \, dz, \quad (207.5)$$

where C is the circle $\left| z - a \right| = r < R$ and $\left| h \right| < r$. Since $\left| h \right| / \left| z - a \right| < 1$ on C, the geometric series converges absolutely and uniformly over C, and we may integrate it term by term to obtain the Taylor series,

$$(5) \qquad f(a+h) = \sum_{n=0}^{\infty} c_n h^n, \qquad c_n = \frac{1}{2\pi i} \oint_C \frac{f(z)}{(z-a)^{n+1}} \, dz,$$

which converges within the circle $\left| z - a \right| = R$. Moreover, on differentiating the series (5) n times and then putting $h = 0$, we get $c_n = f^{(n)}(a)/n!$; hence,

$$(6) \qquad f^{(n)}(a) = \frac{n!}{2\pi i} \oint_C \frac{f(z)}{(z-a)^{n+1}} \, dz;$$

and, if $\left| f(z) \right| < M$ when $\left| z \right| = r$,

$$(7) \qquad \left| f^{(n)}(a) \right| \leqq \frac{n! \, M}{r^n} \qquad \text{(Cauchy's inequality)}.$$

An isolated singular point a of $f(z)$ is *removable* if $f(z)$ is bounded in the neighborhood of a; a *pole* if $f(z) \to \infty$ as $z \to a$; an *essential singularity* in all other cases.

A function analytic in a ring $R_1 < \left| z - a \right| < R_2$ has a convergent *Laurent series* there:

$$(8) \qquad f(z) = \sum_{n=-\infty}^{\infty} c_n (z-a)^n, \qquad c_n = \frac{1}{2\pi i} \oint_C \frac{f(z)}{(z-a)^{n+1}} \, dz,$$

where C lies in the ring and encloses a. When a is an isolated singularity of $f(z)$, the point is a pole or an essential singularity according as the number of negative powers in (8) is finite or infinite; and c_{-1} is the *residue* of a. If the highest negative power is $(z-a)^{-m}$, a is a pole of order m; and

$$\text{Res}\,(a) = \frac{D^{m-1}(z-a)^m f(z)}{(m-1)!} \bigg|_{z=a}.$$

If the regular closed curve C encloses a finite number of singular points $z_1, \cdots, z_k$ of $f(z)$,

$$\int_C f(z)\, dz = 2\pi i \sum_{n=1}^{k} \text{Res}\,(z_n).$$

The function e^z is analytic at all finite points but has an essential singularity at ∞; hence its Taylor expansion about any point converges everywhere. The same is true of the circular functions,

$$\cos z = \frac{1}{2}\,(e^{iz} + e^{-iz}), \qquad \sin z = \frac{1}{2i}\,(e^{iz} - e^{-iz}),$$

and the hyperbolic functions,

$$\cosh z = \tfrac{1}{2}(e^z + e^{-z}), \qquad \sinh z = \tfrac{1}{2}(e^z - e^{-z}).$$

While e^z, $\cosh z$, $\sinh z$ have the period $2\pi i$, $\cos z$ and $\sin z$ have the real period 2π. The zeros of $\cos z$, $\sin z$ are all real, those of $\cosh z$, $\sinh z$ pure imaginary, while e^z has no finite zeros. The equations

$$\cos iz = \cosh z, \qquad \sin iz = i \sinh z,$$
$$\cosh iz = \cos z, \qquad \sinh iz = i \sin z,$$

show how to pass from the circular to the hyperbolic functions and vice versa.

CHAPTER 12

Fourier Series

219. Orthogonal Sets of Functions. The aggregate of all bounded functions of $f(x)$ integrable in a fundamental interval $a \leq x \leq b$ and whose *norm*,

$$(1) \qquad \text{Norm} f = \int_a^b [f(x)]^2 \, dx > 0$$

is said to constitute a certain *function-space* $\mathscr{F}$. Two functions $f(x)$, $g(x)$ of $\mathscr{F}$ are said to be orthogonal in the interval (a, b) when

$$(2) \qquad \int_a^b f(x) \, g(x) \, dx = 0.$$

The countably infinite ($\aleph_0$) set of functions $\varphi_1(x), \varphi_2(x), \cdots$ is said to be *orthogonal* when

$$(3) \qquad \int_a^b \varphi_i(x) \, \varphi_j(x) \, dx = \begin{cases} 0 & (i \neq j), \\ N_i^2 > 0 & (i = j). \end{cases}$$

Then the functions $\varphi_i(x)/N_i$ are also orthogonal and have norms of 1; such a set of functions is called *orthonormal*. Thus the functions $\varphi_n(x)$ form an orthonormal set when

$$(4) \qquad \int_a^b \varphi_i(x) \, \varphi_j(x) \, dx = \delta_{ij} \qquad \qquad (97.10).$$

We now consider the problem of representing a function $f(x)$ of $\mathscr{F}$ by means of an infinite series,

$$(5) \qquad f(x) = c_1\varphi_1(x) + c_2\varphi_2(x) + \cdots, \qquad a \leq x \leq b.$$

THEOREM 1. *If the series (5) converges uniformly to $f(x)$ in (a, b), the constants*

$$(6) \qquad c_n = \int_a^b f(x) \, \varphi_n(x) \, dx.$$

Proof. If we multiply (5) by the bounded function $\varphi_n(x)$, the series remains uniformly convergent and may be integrated term by term from a to b. Then by virtue of equations (4) we have

$$\int_a^b f(x)\,\varphi_n(x)\,dx = \sum_{i=1}^{\infty} c_i \int_a^b \varphi_i(x)\,\varphi_n(x)\,dx = c_n.$$

The numbers $\{c_n\}$ are called the *Fourier constants* of $f(x)$ relative to the orthonormal system $\{\varphi_n(x)\}$, irrespective of the behavior of the series (5). The series $\sum c_n \varphi_n(x)$ is called the *Fourier series corresponding to $f(x)$* in this system; and we write

(7)
$$f(x) \sim \sum_{i=1}^{\infty} c_i \varphi_i(x),$$

whether or not the series converges to $f(x)$.

Now let $a_1, a_2, \cdots, a_n$ be n arbitrary constants, and

$$t_n(x) = \sum_{i=1}^{n} a_i \varphi_i(x).$$

We propose to find the values of these constants that make $t_n(x)$ the best approximation to $f(x)$ in the sense of the method of least squares; that is, the values that give the integral

(8)
$$I = \int_a^b [f(x) - t_n(x)]^2\,dx \geqq 0$$

its minimum value. Since

$$[f(x) - t_n(x)]^2 = f^2 - 2\sum a_i f \varphi_i + \sum a_i^2 \varphi_i^2 + 2\sum_{i \neq j} a_i a_j \varphi_i \varphi_j,$$

we have from (4) and (6)

$$I = \int_a^b f^2\,dx - 2\sum_1^n a_i c_i + \sum_1^n a_i^2$$

$$= \int_a^b f^2\,dx + \sum_1^n (a_i - c_i)^2 - \sum_1^n c_i^2.$$

Thus I will assume its least value when $a_i = c_i$, the Fourier constants of $f(x)$; and, if we write

$$s_n(x) = \sum_{i=1}^{n} c_i \varphi_i(x)$$

for the nth partial sum of the Fourier series,

$$(9) \qquad I_{min} = \int_a^b [f(x) - s_n(x)]^2 \, dx = \int_a^b f^2(x) \, dx - \sum_{i=1}^n c_i^2;$$

and hence

$$\int_a^b [f(x) - t_n(x)]^2 \, dx \geq \int_a^b [f(x) - s_n(x)]^2 \, dx \geq 0.$$

We may therefore state

THEOREM 2. *For any given n, the sum $t_n(x)$ is the best approximation to $f(x)$ in the sense of least squares when the constants a_i are the Fourier constants of $f(x)$.*

It is noteworthy that the terms in the best approximation $s_n(x)$ for a given n are all retained in the best approximation for all larger values.

Since $I_{min} \geq 0$ and n can be taken arbitrarily large, (9) also proves *Bessel's inequality*:

$$(10) \qquad \sum_{i=1}^\infty c_i^2 \leq \int_a^b f^2(x) \, dx$$

and at the same time shows that the terms $\{c_i^2\}$ form a null sequence (§ 21).

THEOREM 3. *The series of squares of the Fourier constants of any function $f(x)$ converge to a sum not exceeding the norm of $f(x)$, and the constants themselves form a null sequence.*

Example 1. Fourier series in the strict sense are expansions in terms of the orthogonal set

$$(11) \qquad\qquad 1, \quad \cos nx, \quad \sin nx \qquad (n = 1, 2, \cdots)$$

over the interval $(-\pi, \pi)$. Since the norms are, respectively, $2\pi, \pi, \pi$, the corresponding orthonormal set is

$$(12) \qquad\qquad \frac{1}{\sqrt{2\pi}}, \quad \frac{\cos nx}{\sqrt{\pi}}, \quad \frac{\sin nx}{\sqrt{\pi}} \qquad (n = 1, 2, \cdots).$$

The orthogonality relations follow easily from the equations

$$\int_{-\pi}^\pi e^{imx} e^{inx} \, dx = 0, \qquad \int_{-\pi}^\pi e^{imx} e^{-inx} \, dx = \begin{cases} 0 & m \neq n, \\ 2\pi, & m = n \end{cases}$$

on taking real and imaginary parts. Due to the period 2π of the sine and cosine the set (11) is also orthogonal with respect to any interval $(a, a + 2\pi)$.

Example 2. The Legendre polynomials,

$$(13) \qquad P_0(x) = 1, \qquad P_n(x) = \frac{1}{2^n n!} D^n(x^2 - 1)^n, \qquad n = 1, 2 \cdots,$$

where $D = d/dx$, are polynomials of degree 0, 1, 2, $\cdots$ which are orthogonal over the interval $(-1, 1)$. If $m > n$, we find by n integrations by parts,

$$\int_{-1}^{1} D^m(x^2 - 1)^m \cdot D^n(x^2 - 1)^n \, dx = (-1)^n \int_{-1}^{1} D^{m-n}(x^2 - 1)^m \cdot D^{2n}(x^2 - 1)^n \, dx$$

$$= (-1)^n(2n)! \int_{-1}^{1} D^{m-n}(x^2 - 1)^m \, dx = 0.$$

Moreover, by n integrations by parts,

$$\int_{-1}^{1} P_n{}^2 \, dx = \frac{(-1)^n (2n)!}{(2^n n!)^2} \int_{-1}^{1} (x^2 - 1)^n dx;$$

and from (193.7)

$$\int_{-1}^{1} (1 - x^2)^n \, dx = \int_{0}^{\pi} \sin^{2n+1} u \, du = \frac{n!}{(n + \frac{1}{2})(n - \frac{1}{2}) \cdots \frac{1}{2}}$$

$$= \frac{2}{2n + 1} \cdot \frac{(2^n n!)^2}{(2n)!},$$

so that

$$\int_{-1}^{1} P_n{}^2 \, dx = \frac{2}{2n + 1}.$$

220. Closed and Complete Orthonormal Sets. Consider a class of functions $\mathscr{F}$ having the Fourier constants c_i relative to the orthonormal set $\varphi_i(x)$. The *function-space* $\mathscr{F}$ consists of all functions of nonzero norm having some other property, such as being *continuous*, or *piecewise continuous*, or *integrable*, or *absolutely integrable*—to name a few classes of constantly wider scope. Then, if

(1) $$\lim_{n \to \infty} \int_{a}^{b} \left[f(x) - \sum_{i=1}^{n} c_i \varphi_i(x) \right]^2 dx = 0$$

for all functions of $\mathscr{F}$, the set $\{\varphi_i\}$ is said to be *closed* with respect to $\mathscr{F}$. In view of (219.9) the limit (1) equals

$$\int_{a}^{b} f^2(x) \, dx - \sum_{i=1}^{\infty} c_i{}^2;$$

hence, we have

PARSEVAL'S THEOREM. *Let the set of orthonormal functions* $\{\varphi_i(x)\}$ *in the interval* (a, b) *be closed with respect to a class of functions* $\mathscr{F}$. *Then, for any functions* $f(x), \bar{f}(x)$ *of* $\mathscr{F}$, *we have*

(2) $$\sum_{i=1}^{\infty} c_i{}^2 = \int_{a}^{b} f^2(x) \, dx,$$

(3) $$\sum_{i=1}^{\infty} c_i \bar{c}_i = \int_{a}^{b} f(x) \bar{f}(x) \, dx.$$

Proof. Equation (2) follows at once from (1). To prove (3) we apply (2) to the function $f(x) + \bar{f}(x)$ whoseOurier constants are $c_i + \bar{c}_i$; thus,

$$\sum_{i=1}^{\infty} (c_i{}^2 + 2c_i\bar{c}_i + \bar{c}_i{}^2) = \int_a^b [f^2(x) + 2f(x)\bar{f}(x) + \bar{f}^2(x)]\, dx.$$

On subtracting (2) and the corresponding equation for $\bar{f}(x)$ from this equation, we obtain (3).

Equations (2) and (3) are analogous to

$$f_1{}^2 + f_2{}^2 + f_3{}^2 = \mathbf{f} \cdot \mathbf{f}, \qquad f_1 g_1 + f_2 g_2 + f_3 g_3 = \mathbf{f} \cdot \mathbf{g},$$

where $\mathbf{f}$ and $\mathbf{g}$ are vectors in 3-space. Just as the numbers $f_i = \mathbf{f} \cdot \mathbf{e}_i$ are the components of $\mathbf{f}$ relative to the orthogonal set of unit vectors $\mathbf{e}_1$, $\mathbf{e}_2$, $\mathbf{e}_3$, the Fourier constants,

$$c_i = \int_a^b f(x)\, \varphi_i(x)\, dx,$$

may be regarded as the components ($\aleph_0$ in number) of the function $f(x)$ relative to the orthonormal set $\{\varphi_i(x)\}$.

An orthogonal set $\{\varphi_i(x)\}$ is said to be *complete* if no function of $\mathscr{F}$ is orthogonal to all the functions of the set. We now have the

THEOREM 1. *Every closed orthogonal set is complete.*

Proof. If $f(x)$ were orthogonal to all the functions φ_i, its Fourier constants c_i would all be zero, and hence norm $f = 0$ from (2); hence $f(x)$ is not a member of the function space $\mathscr{F}$.

A function $f(x)$ uniquely determines its Fourier series, but the converse is not true; for, if the value of $f(x)$ is altered at a finite number of points in (a, b), the new function is still integrable and will have the same Fourier constants as $f(x)$. However, in a *closed* system we have the

THEOREM 2. *If $g(x)$ and $h(x)$ are continuous at a point ξ of (a, b) and $g(\xi) \neq h(\xi)$, their Fourier series relative to a closed set $\{\varphi_i\}$ will differ.*

Proof. Let $f(x) = g(x) - h(x)$ have the Fourier constants c_i. If $g(x)$ and $h(x)$ have the same Fourier series, all $c_i = 0$ and norm $f = 0$ from (2). But this is impossible, since $f(x)$ is continuous at ξ, $f(\xi) \neq 0$, and $f(x)$ will keep the same sign in a suitably restricted neighborhood of ξ, so that $\int_a^b f^2\, dx > 0$.

Thus in a closed system the correspondence between continuous functions and their Fourier series is one-to-one.

221. Fourier Series. Using the orthonormal set (219.12)

$$\frac{1}{\sqrt{2\pi}}, \qquad \frac{\cos nx}{\sqrt{\pi}}, \qquad \frac{\sin nx}{\sqrt{\pi}}$$

the Fourier series corresponding to $f(x)$ is given by

$$(1) \qquad f(x) \sim c_0 \frac{1}{\sqrt{2\pi}} + \sum_{n=1}^{\infty} \left(c_n \frac{\cos nx}{\sqrt{\pi}} + c_n' \frac{\sin nx}{\sqrt{\pi}} \right),$$

where the Fourier constants (219.6) are

$$c_0 = \frac{1}{\sqrt{2\pi}} \int_{-\pi}^{\pi} f(x) \, dx, \qquad \begin{matrix} c_n \\ c_n' \end{matrix} = \frac{1}{\sqrt{\pi}} \int_{-\pi}^{\pi} f(x) \begin{matrix} \cos nx, \\ \sin nx \end{matrix} \, dx.$$

The series is more conveniently written

$$(2) \qquad f(x) \sim \frac{1}{2} a_0 + \sum_{n=1}^{\infty} (a_n \cos nx + b_n \sin nx),$$

where

$$(3) \qquad \begin{matrix} a_n \\ b_n \end{matrix} = \frac{1}{\pi} \int_{-\pi}^{\pi} f(x) \begin{matrix} \cos nx \\ \sin nx \end{matrix} \, dx, \qquad n = 0, 1, 2, \cdots.$$

Since

$$c_0 = \sqrt{\frac{\pi}{2}} a_0, \qquad c_n = \sqrt{\pi} \, a_n, \qquad c_n' = \sqrt{\pi} \, b_n,$$

Bessel's inequality (219.11) becomes

$$(4) \qquad \frac{a_0^2}{2} + \sum_{n=1}^{\infty} (a_n^2 + b_n^2) \leq \frac{1}{\pi} \int_{-\pi}^{\pi} f^2(x) \, dx.$$

Since the sine-cosine orthonormal set can be shown to be closed for all absolutely integrable functions, (4) may be written as an *equation—Parseval's equation.*[†]

In any case the series in (4) converges, and hence $\{a_n\}$ and $\{b_n\}$ are null sequences. We thus have

RIEMANN'S THEOREM. *For all absolutely integrable functions $f(x)$*

$$(5) \qquad \lim_{n \to \infty} \int_{-\pi}^{\pi} f(x) \begin{matrix} \cos nx \\ \sin nx \end{matrix} \, dx = 0.$$

Every term of the Fourier series (2) has the period 2π; hence, when the series converges to $f(x)$, $f(x)$ must be a periodic function of period 2π:

$$(6) \qquad f(x + 2\pi) = f(x).$$

[†] Cf. Rogosinski, *Fourier Series*, Chelsea, 1950, p. 53.   Note that, if $f(x)$ is integrable, so is $|f(x)|$, but not conversely (§ 118).

Therefore we shall assume once for all that, when $f(x)$ is given in any interval $c \leq x < c + 2\pi$ of length 2π, its values at all other points are given by (6). Consequently the constants a_n, b_n in (3) may be computed by integrating between an arbitrary lower limit c and $c + 2\pi$ (123.4).

When $f(x)$ is an *even* function, $f(x) \cos nx$ is even, $f(x) \sin nx$ odd; hence (Ex. 123.2)

$$(7) \qquad a_n = \frac{2}{\pi} \int_0^\pi f(x) \cos nx \, dx, \qquad b_n = 0,$$

and the corresponding series is a pure cosine series.

When $f(x)$ is an *odd* function, $f(x) \cos nx$ is odd, $f(x) \sin nx$ is even, and

$$(8) \qquad a_n = 0, \qquad b_n = \frac{2}{\pi} \int_0^\pi f(x) \sin nx \, dx,$$

and the corresponding series is a pure sine series.

When $f(x)$ is only defined in the interval $(0, \pi)$ we may extend its definition to $(-\pi, 0)$ so that the function is even or odd. *Thus an integrable function over $(0, \pi)$ may be associated with either a pure cosine or a pure sine series.*

Example 1. The odd *signum function* (Prob. 44.2)

$$sgn\ x = \begin{cases} -1 & (-\pi < x < 0) \\ 0 & (x = 0) \\ 1 & (0 < x < \pi) \end{cases}$$

corresponds to a sine series. Hence, from (8),

$$\frac{\pi}{2} b_n = \int_0^\pi \sin nx \, dx = \frac{1 - \cos n\pi}{n} = \begin{cases} \dfrac{2}{n}, & n \text{ odd,} \\ 0, & n \text{ even;} \end{cases}$$

and, in $(-\pi < x < \pi)$,

$$(9) \qquad sgn\ x = \sim \frac{4}{\pi} \left(\sin x + \frac{1}{3} \sin 3x + \frac{1}{5} \sin 5x + \cdots \right).$$

This is the sine series for 1 in $0 < x < \pi$; the cosine series for 1 is the single term 1.

Example 2. The odd function $f(x) = x$ $(-\pi < x < \pi)$ has a sine series for which

$$\frac{\pi}{2} b_n = \int_0^\pi x \sin nx \, dx = -\left. \frac{x \cos nx}{n} \right|_0^\pi + \frac{1}{n} \int_0^\pi \cos nx \, dx$$

$$= -\frac{\pi}{n} \cos (n\pi) = -\frac{\pi}{n} (-1)^n,$$

$$b_n = \frac{2}{n} (-1)^{n+1};$$

hence, in the interval $-\pi < x < \pi$,

$$(10) \qquad \frac{x}{2} \sim \sin x - \frac{1}{2} \sin 2x + \frac{1}{3} \sin 3x - \cdots.$$

Example 3. The even function $f(x) = |x|$ has a cosine series. Hence, from (7),

$$\frac{\pi}{2} a_0 = \int_0^\pi x \, dx = \frac{\pi^2}{2}, \qquad \tfrac{1}{2}a_0 = \frac{\pi}{2};$$

$$\frac{\pi}{2} a_n = \int_0^\pi x \cos nx \, dx = \frac{x \sin nx}{n} \Big|_0^\pi - \frac{1}{n} \int_0^\pi \sin nx \, dx$$

$$= \frac{1}{n^2}(\cos n\pi - 1) = \begin{cases} -\dfrac{2}{n^2} & n \text{ odd}, \\[2mm] 0 & n \text{ even}; \end{cases}$$

hence, in the interval $(-\pi < x < \pi)$,

(11) $$|x| \sim \frac{\pi}{2} - \frac{4}{\pi}\left(\cos x + \frac{\cos 3x}{3^2} + \frac{\cos 5x}{5^2} + \cdots \right).$$

In the interval $(0 \leqq x < \pi)$, (10) and (11) are the sine and cosine series for $\tfrac{1}{2}x$ and x.

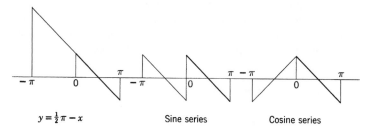

$y = \tfrac{1}{2}\pi - x$ Sine series   Cosine series

FIG. 221.   Odd and even functions corresponding to a function defined
over $(0, \pi)$

Example 4. The function $f(x) = \tfrac{1}{2}\pi - x$ has a "mixed" series since it is neither even nor odd.   In the interval $-\pi < x < \pi$,

(12) $$\frac{1}{2}\pi - x \sim \frac{1}{2}\pi - 2\left(\sin x - \frac{\sin 2x}{2} + \frac{\sin 3x}{3} - \cdots \right),$$

in view of (10).

The sine series for $\tfrac{1}{2}\pi - x$ is the difference of the sine series for $\tfrac{1}{2}\pi$ and x.   From (9) and (10).

$$\tfrac{1}{2}\pi \sim 2(\sin x + \tfrac{1}{3} \sin 3x + \tfrac{1}{5} \sin 5x + \cdots),$$

$$x \sim 2(\sin x - \tfrac{1}{2} \sin 2x + \tfrac{1}{3} \sin 3x - \tfrac{1}{4} \sin 4x + \cdots),$$

and, hence, in the interval $0 < x < \pi$,

(13) $$\tfrac{1}{2}\pi - x \sim \sin 2x + \tfrac{1}{2} \sin 4x + \tfrac{1}{3} \sin 6x + \cdots.$$

The cosine series for $\tfrac{1}{2}\pi - x$ is the difference of the cosine series for $\tfrac{1}{2}\pi$ and x.   The former is $\tfrac{1}{2}\pi$, the latter is given by (11); hence, in the interval $0 < x < \pi$,

(14) $$\frac{1}{2}\pi - x \sim \frac{4}{\pi}\left(\cos x + \frac{\cos 3x}{3^2} + \frac{\cos 5x}{5^2} + \cdots \right).$$

The graphs of the three functions represented by (12), (13), (14) in the interval $-\pi < x < \pi$ are shown in Fig. 221.   Of course, all agree in $(0, \pi)$.

PROBLEMS

Obtain the following Fourier series in the interval $(-\pi, \pi)$.

1. $h(x) = \begin{cases} 0 & (-\pi < x < 0), \\ \frac{1}{2} & (x = 0), \\ 1 & (0 < x < \pi). \end{cases} \qquad \sim \frac{1}{2} + \frac{2}{\pi} \sum_{n=1}^{\infty} \frac{\sin(2n-1)x}{2n-1},$

2. $x^2 \sim \frac{\pi^2}{3} + 4 \sum_{n=1}^{\infty} (-1)^n \frac{\cos nx}{n^2}.$

3. $e^x \sim \frac{2\sinh \pi}{\pi} \left[\frac{1}{2} + \sum_{n=1}^{\infty} \frac{(-1)^n}{1+n^2} (\cos nx - n \sin nx) \right].$

4. Obtain the Fourier series for $f(x) = \begin{cases} 0 & (-\pi < x \leq 0) \\ x & (0 \leq x < \pi) \end{cases}$

from (10) and (11).

5. Show that, in the interval $(0, \pi)$,

$$\sin x \sim \frac{2}{\pi} - \frac{4}{\pi} \sum_{n=1}^{\infty} \frac{\cos 2nx}{4n^2 - 1}.$$

6. Find the sine series for x^2 in $(0, \pi)$.

7. When $f(\pi - x) = f(x)$, show that $a_{2n+1} = 0$, $b_{2n} = 0$ in the Fourier series (2).

8. When $f(x + \pi) = f(x)$, show that $a_{2m+1} = b_{2m+1} = 0$ in the Fourier series (2); and that

$$\begin{aligned} a_{2m} \\ b_{2m} \end{aligned} = \frac{2}{\pi} \int_0^\pi f(x) \begin{aligned} \cos 2mx \\ \sin 2mx \end{aligned} dx.$$

9. When $f\left(x + \frac{\pi}{2}\right) = f(x)$, show that $a_n = b_n = 0$ in the Fourier series (2) except when $n = 4m$; and that

$$\begin{aligned} a_{4m} \\ b_{4m} \end{aligned} = \frac{4}{\pi} \int_0^{\frac{\pi}{2}} f(x) \begin{aligned} \cos 4mx \\ \sin 4mx \end{aligned} dx.$$

10. If $f(x) = x \ (0 \leq x < \pi)$ and $f(x + \pi) = f(x)$, graph the function and show that

$$f(x) = \frac{1}{2}\pi - \sum_{n=1}^{\infty} \frac{\sin 2nx}{n}.$$

11. If $f(x) = x \ (0 \leq x < \frac{1}{2}\pi)$ and $f(x + \frac{1}{2}\pi) = f(x)$, graph the function and show that

$$f(x) = \frac{1}{4}\pi - \sum_{n=1}^{\infty} \frac{\sin 4nx}{2n}.$$

222. Convergence Theorem. We shall now formulate conditions under which the Fourier series of $f(x)$ converges to $f(x)$. Write the sum of the first $2n+1$ terms of the series

$$s_n(x) = \frac{1}{2} a_0 + \sum_{k=1}^{n} (a_k \cos kx + b_k \sin kx).$$

Making use of the formulas (221.3), we may write this as an integral:

$$s_n(x) = \frac{1}{\pi} \int_{-\pi}^{\pi} f(t) \left[\frac{1}{2} + \sum_{k=1}^{n} (\cos kt \cos kx + \sin kt \sin kx) \right] dt$$

$$= \frac{1}{\pi} \int_{-\pi}^{\pi} f(t) \left[\frac{1}{2} + \sum_{k=1}^{n} \cos k(t - x) \right] dt$$

$$= \frac{1}{\pi} \int_{-\pi-x}^{\pi-x} f(u + x) \left[\frac{1}{2} + \sum_{k=1}^{n} \cos ku \right] du,$$

on putting $t = u + x$. The summation formula (201.4),

(1) $$\frac{\sin (n + \frac{1}{2})u}{2 \sin \frac{1}{2}u} = \frac{1}{2} + \cos u + \cos 2u + \cdots + \cos nu,$$

now gives the compact form,

(2) $$s_n(x) = \frac{1}{\pi} \int_{-\pi}^{\pi} f(u + x) \frac{\sin (n + \frac{1}{2})u}{2 \sin \frac{1}{2}u} du.$$

Here the limits have been altered in accordance with (123.4); for the integrand is a function of u with the period 2π since $f(x + u)$ and also $\sin (n + \frac{1}{2})u/2 \sin \frac{1}{2}u$ have this period, as we can see from (1).

On integrating (1) between $-\pi$ and π, we find that

(3) $$1 = \frac{1}{\pi} \int_{-\pi}^{\pi} \frac{\sin (n + \frac{1}{2})u}{2 \sin \frac{1}{2}u} du;$$

and, on multiplying (3) by $f(x)$ and subtracting from (2),

(4) $$s_n(x) - f(x) = \frac{1}{\pi} \int_{-\pi}^{\pi} [f(u + x) - f(x)] \frac{\sin (n + \frac{1}{2})u}{2 \sin \frac{1}{2}u} du.$$

This formula enables us to prove a convergence theorem for certain functions that are continuous in a period interval (of length 2π) except for a finite number of finite jumps. We shall call such functions *piecewise continuous*.

THEOREM. *If $f(x)$ is piecewise continuous in the interval $-\pi \leq x \leq \pi$, its Fourier series converges to $f(x)$ at every point where $f(x)$ is continuous and has right- and left-hand derivatives.*

Proof. If we put

$$F(u) = \frac{f(x+u) - f(x)}{u} \cdot \frac{\tfrac{1}{2}u}{\sin \tfrac{1}{2}u},$$

$$\sin (n + \tfrac{1}{2})u = \sin nu \cos \tfrac{1}{2}u + \cos nu \sin \tfrac{1}{2}u,$$

the integral in (4) equals

$$\int_{-\pi}^{\pi} F(u) \cos \tfrac{1}{2} u \cdot \sin nu \, du + \int_{-\pi}^{\pi} F(u) \sin \tfrac{1}{2} u \cdot \cos nu \, du.$$

The discontinuities of $F(u)$ are those of $f(x+u)$ with possibly another at $u = 0$. The latter is at most a finite jump for

$$\lim_{u \to 0+} F(u) = f'_+(x) \cdot 1, \qquad \lim_{u \to 0-} F(u) = f'_-(x) \cdot 1.$$

Thus $F(u) \cos \tfrac{1}{2}u$ and $F(u) \sin \tfrac{1}{2}u$ are integrable functions, and, as $n \to \infty$, both integrals approach zero (Theorem 221). Since the same is true of the integral in (4),

$$\lim_{n \to \infty} s_n(x) = f(x);$$

that is, the Fourier series converges to $f(x)$.

When $f(x)$ is continuous at $\pm\pi$ its periodicity demands that $f(-\pi) = f(\pi)$. If

$$\lim_{x \downarrow -\pi} f(x) = p, \qquad \lim_{x \uparrow \pi} f(x) = q \neq p,$$

as x approaches $\pm\pi$ from inside the interval $(-\pi, \pi)$, $f(x)$ has the finite jump,

$$f(\pi+) - f(\pi-) = p - q \quad \text{at } \pm\pi.$$

223. Convergence at Discontinuities. At all points c where $f(x)$ has a finite jump,

(1) $$\delta = f(c+) - f(c-) \neq 0,$$

we shall redefine $f(x)$ when necessary so that

(2) $$f(c) = \frac{f(c-) + f(c+)}{2}.$$

These changes in $f(x)$ at a finite number of points will not alter the values of the integrals (221.3) giving the coefficients of the Fourier series for $f(x)$.

From Ex. 221.1 we see that the function

(3) $$g(x) = \begin{cases} -1 & (-\pi < x < 0) \\ 0 & (x = 0, \pm \pi), \\ 1 & (0 < x < \pi) \end{cases} \quad g(x + 2\pi) = g(x),$$

has the convergent Fourier series,

(4) $g(x) = \dfrac{4}{\pi} (\sin x + \dfrac{1}{3} \sin 3x + \dfrac{1}{5} \sin 5x + \cdots);$

for Theorem 222 guarantees the convergence when $x \neq 0$, while the convergence to $g(0) = 0$ when $x = 0$ is obvious since every term is zero.

To consider the behavior of the Fourier series for $f(x)$ at a point of discontinuity c, form the function

(5) $F(x) = f(x) - \dfrac{\delta}{2} g(x - c).$

Then $F(x)$ is continuous at c; for

$$F(c) = f(c) - \dfrac{\delta}{2}(0) = f(c),$$

$$F(c-) = f(c-) - \dfrac{\delta}{2}(-1) = f(c),$$

$$F(c+) = f(c+) - \dfrac{\delta}{2}(+1) = f(c),$$

where $f(c)$ is defined by (2). Moreover, when $h > 0$,

$$\dfrac{F(c + h) - F(c)}{h} = \dfrac{f(c + h) - \delta/2 - f(c)}{h} = \dfrac{f(c + h) - f(c+)}{h},$$

$$\dfrac{F(c - h) - F(c)}{-h} = \dfrac{f(c - h) + \delta/2 - f(c)}{-h} = \dfrac{f(c - h) - f(c-)}{-h}.$$

As $h \to 0$, we see that

(6) $F'_+(c) = \lim\limits_{h \to 0} \dfrac{f(c+h) - f(c+)}{h},$ $F'_-(c) = \lim\limits_{h \to 0} \dfrac{f(c-h) - f(c-)}{-h},$

provided the limits on the right exist. These limits represent limiting slopes of the graph of $f(x)$ to the right and to the left of $x = c$. These limiting slopes must not be confused with the right- and left-hand derivatives of $f(x)$ at c. For example, the function $g(x)$ defined in (3) has the limiting slopes 0 to the right and left of 0, but the right- and left-hand derivatives do not exist for

$$\lim\limits_{h \to 0} \dfrac{1 - 0}{h} = \infty, \qquad \lim\limits_{h \to 0} \dfrac{-1 - 0}{-h} = \infty.$$

We are now able to prove the

THEOREM. *If $f(x)$ is piecewise continuous in the interval $-\pi \leqq x \leqq \pi$, its Fourier series converges to*

$$f(x) = \tfrac{1}{2}\{f(x+) + f(x-)\}$$

at every point where $f(x)$ has right- and left-hand slopes.

Proof. Let $f(x)$ make the finite jump δ at c. Then the function $F(x)$ defined in (5) is continuous at c and has the right- and left-hand derivatives given by (6). Therefore, by Theorem 221 the Fourier series of $F(x)$ converges to $F(c)$ when $x = c$. But from (5) the Fourier series for $f(x)$ is the sum of the series for $F(x)$ and that for $\frac{1}{2}\delta\, g(x - c)$. Since the latter converges to zero when $x = c$, the Fourier series for $f(x)$ converges to

$$F(c) = f(c) = \frac{f(c+) + f(c-)}{2}.$$

Note that this theorem includes Theorem 222 as a special case. For, if $f(x)$ is continuous when $x = c$,

$$f(c) = f(c+) = f(c-),$$

and the right- and left-hand slopes become right- and left-hand *derivatives*.

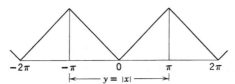

FIG. 223a. Sawtooth function

Example 1. From Ex. 221.1 we have

$$(7) \qquad \sin x + \frac{\sin 3x}{3} + \frac{\sin 5x}{5} + \cdots = \begin{cases} -\pi/4 & (-\pi < x < 0), \\ 0 & (x = 0, \pm\pi), \\ \pi/4 & (0 < x < \pi). \end{cases}$$

At $x = 0$ and $x = \pm\pi$ the series converges to $\frac{1}{2}(\pi/4 - \pi/4) = 0$.

When $x = \frac{1}{2}\pi$, we get Gregory's series (181.4);

$$(8) \qquad 1 - \frac{1}{3} + \frac{1}{5} - \frac{1}{7} + \cdots = \frac{\pi}{4}.$$

Example 2. In Ex. 221.3 the series (11) converges to $|x|$ throughout $-\pi \le x \le \pi$, for the periodic function $|x|$ in the interval $(-\pi, \pi)$ is continuous at 0 and at $\pm\pi$ and has right- and left-hand derivatives (Fig. 223a). Hence, we have

$$(9) \qquad \frac{\pi^2}{8} - \frac{\pi}{4}|x| = \cos x + \frac{\cos 3x}{3^2} + \frac{\cos 5x}{5^2} + \cdots \qquad (-\pi \le x \le \pi).$$

When $x = 0$, this gives

$$(10) \qquad \frac{\pi^2}{8} = 1 + \frac{1}{3^2} + \frac{1}{5^2} + \frac{1}{7^2} + \cdots.$$

Example 3. If $f(x) = \frac{1}{2}(\pi - x)$ in $0 < x < 2\pi$ and has the period 2π, $f(x)$ is an odd function (Fig. 223b). Hence, $a_n = 0$ and, on integrating by parts,

$$b_n = \frac{2}{\pi} \int_0^\pi \frac{\pi - x}{2} \sin nx \, dx = \frac{1}{n}.$$

Thus we have

(11) $\dfrac{1}{2}(\pi - x) = \sin x + \dfrac{\sin 2x}{2} + \dfrac{\sin 3x}{3} + \cdots$ $(0 < x < 2\pi).$

At $x = 0$ and 2π the function is discontinuous, and the series converges to $\frac{1}{2}(\frac{1}{2}\pi - \frac{1}{2}\pi)$ $= 0.$

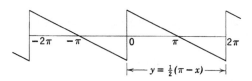

FIG. 223b, Zigzag function

Example 4. For the function

$$h(x) = \begin{cases} 0 & (-\pi < x < 0), \\ \frac{1}{2} & (x = 0, \pm\pi), \\ 1 & (0 < x < \pi), \end{cases}$$

we find $a_0 = \dfrac{1}{\pi} \int_0^\pi dx = 1,$ $a_n = \dfrac{1}{\pi} \int_0^\pi \cos nx \, dx = 0,$

$$b_n = \frac{1}{\pi} \int_0^\pi \sin nx \, dx = \frac{1 - \cos n\pi}{n\pi} = \begin{cases} \dfrac{2}{n\pi} & n \text{ odd}, \\ 0 & n \text{ even}; \end{cases}$$

(12) $h(x) = \dfrac{1}{2} + \dfrac{2}{\pi}\left(\sin x + \dfrac{\sin 3x}{3} + \dfrac{\sin 5x}{5} + \cdots\right).$

This result also follows at once from (7); for, if $f(x)$ denotes the function on the right of (7), $\dfrac{1}{2} + \dfrac{2}{\pi} f(x) = h(x).$

Example 5. For the even function x^2 we have

$$a_0 = \frac{2}{\pi} \int_0^\pi x^2 \, dx = \frac{2\pi^2}{3} \qquad a_n = \frac{2}{n} \int_0^\pi x^2 \cos nx \, dx = (-1)^n \frac{4}{n^2};$$

(13) $x^2 = \dfrac{\pi^2}{3} - 4\left(\dfrac{\cos x}{1^2} - \dfrac{\cos 2x}{2^2} + \dfrac{\cos 3x}{3^2} - \cdots\right),$ $(-\pi \leqq x \leqq \pi).$

Note that the periodic function x^2 is continuous at $\pm\pi$ and has left- and right-hand derivatives. When $x = 0$, this gives

(14) $\dfrac{\pi^2}{12} = 1 - \dfrac{1}{2^2} + \dfrac{1}{3^2} - \dfrac{1}{4^2} + \cdots;$

and, when $x = \pi$,

(15) $\dfrac{\pi^2}{6} = 1 + \dfrac{1}{2^2} + \dfrac{1}{3^2} + \dfrac{1}{4^2} + \cdots.$

On adding (14) and (15) we get (10).

Example 6. Parseval's equation (221.4) can be used to deduce equations such as (10), (14), (15). When applied to pure cosine or sine series in the interval $(-\pi, \pi)$, it has one of the forms

(16) $$\frac{1}{2}a_0{}^2 + \sum_{n=1}^{\infty} a_n{}^2 = \frac{2}{\pi}\int_0^{\pi} f^2(x)\,dx, \qquad \sum_{n=1}^{\infty} b_n{}^2 = \frac{2}{\pi}\int_0^{\pi} f^2(x)\,dx.$$

Thus, from (9), we have

(17) $$\sum_1^{\infty} \frac{1}{(2n-1)^4} = \frac{2}{\pi}\frac{\pi^2}{64}\int_0^{\pi} (\pi - 2x)^2\,dx = \frac{\pi^4}{96};$$

and, from (13),

$$\frac{2\pi^4}{9} + 16\sum_1^{\infty} \frac{1}{n^4} = \frac{2}{\pi}\int_0^{\pi} x^4\,dx = \frac{2}{5}\pi^4,$$

(18) $$\sum_1^{\infty} \frac{1}{n^4} = \frac{\pi^4}{90}.$$

Equations (15) and (18) are special cases of the general formula

$$\zeta(2n) = \frac{2^{2n-1}\,|B_{2n}|}{(2n)!}\pi^{2n}, \qquad n = 1, 2, 3, \cdots,$$

where $\zeta(x)$ is Riemann's zeta function (26.9) and B_{2n} a Bernoulli number (211.14); thus

$$\zeta(2) = \pi^2/6, \qquad \zeta(4) = \pi^4/90, \qquad \zeta(6) = \pi^6/945.$$

No analogous closed result for $\zeta(2n + 1)$ is known at present.

PROBLEMS

1. With $x = \pi/6$ in series (7), deduce that

$$1 + \frac{1}{5} - \frac{1}{7} - \frac{1}{11} + \frac{1}{13} + \frac{1}{17} - \cdots = \frac{\pi}{3}.$$

2. From the series for $h(x)$ in Prob. 221.1 obtain the series for

$$k(x) = \begin{cases} 1 & (-\pi < x < 0) \\ \tfrac{1}{2} & (x = 0, \pm\pi) \\ 0 & (0 < x < \pi) \end{cases} = \frac{1}{2} - \frac{2}{\pi}\sum_{n=1}^{\infty} \frac{\sin(2n-1)x}{2n-1}$$

3. Find without integration the Fourier series for

$$f(x) = \begin{cases} a & (-\pi < x < 0), \\ \tfrac{1}{2}(a+b) & (x = 0, \pm\pi), \\ b & (0 < x < \pi). \end{cases}$$

4. Use Parseval's equation to deduce (10) from series (7); and (15) from series (11).

5. From Prob. 221.3 show that

$$\cosh x = \frac{\sinh \pi}{\pi}\left\{1 + 2\sum_{n=1}^{\infty} \frac{(-1)^n}{1+n^2}\cos nx\right\} \qquad (-\pi \leqq x \leqq \pi);$$

$$\sinh x = 2\frac{\sinh \pi}{\pi}\sum_{n=1}^{\infty} \frac{n(-1)^{n+1}}{1+n^2}\sin nx, \qquad (-\pi < x < \pi).$$

Hence, prove that

$$\frac{\pi}{\sinh \pi} = 2 \sum_{n=2}^{\infty} \frac{(-1)^n}{1 + n^2}, \qquad \frac{\pi}{\tanh \pi} = 1 + 2 \sum_{n=1}^{\infty} \frac{1}{1 + n^2}.$$

6. Obtain the sine series for the function

$$f(x) = \begin{cases} x & (0 \leq x \leq \tfrac{1}{2}\pi) \\ \pi - x & (\tfrac{1}{2}\pi \leq x \leq \pi) \end{cases} = \frac{4}{\pi} \sum_{n=1}^{\infty} (-1)^{n+1} \frac{\sin (2n - 1)x}{(2n - 1)^2}.$$

Deduce (17) from this series.

7. From the Taylor series,

$$f(z) = a_0 + a_1 z + a_2 z^2 + \cdots, \qquad |z| < R,$$

with real coefficients a_n we can derive two Fourier series by putting

$$z = r(\cos \theta + i \sin \theta), \qquad f(z) = u(r, \theta) + i v(r, \theta),$$

which converge when $r < R$. Find these series when $f(z) = (1 - z)^{-1}$.

7. From $\log (1 + z) = \displaystyle\sum_{n=1}^{\infty} (-1)^{n+1} \frac{z^n}{n}$ deduce the series

$$\frac{1}{2} \log (1 + 2r \cos \theta + r^2) = \sum_{n=1}^{\infty} (-1)^{n+1} r^n \frac{\cos n\theta}{n}, \qquad r < 1;$$

$$\tan^{-1} \frac{r \sin \theta}{1 + r \cos \theta} = \sum_{n=1}^{\infty} (-1)^{n+1} r^n \frac{\sin n\theta}{n}, \qquad r < 1.$$

8. From the series of Prob. 7 use Abel's theorem to deduce the series

$$\log \cos \frac{\theta}{2} = -\log 2 + \cos \theta - \frac{1}{2} \cos 2\theta + \frac{1}{3} \cos 2\theta - \cdots,$$

$$\frac{\theta}{2} = \sin \theta - \frac{1}{2} \sin 2\theta + \frac{1}{3} \sin 3\theta - \cdots,$$

valid when $-\pi < \theta < \pi$. Check these series when $\theta = 0$ and $\tfrac{1}{2}\pi$.

224. Resolution of cot πx into Partial Fractions. The Fourier constants for the even, continuous, periodic function $\cos \gamma x (-\pi \leq x \leq \pi)$, where γ is not an integer, are $b_n = 0$, and

$$a_0 = \frac{1}{\pi} \int_0^{\pi} \cos \gamma x \, dx = \frac{2}{\gamma \pi} \sin \gamma \pi,$$

$$a_n = \frac{2}{\pi} \int_0^{\pi} \cos \gamma x \cos nx \, dx$$

$$= \frac{1}{\pi} \int_0^{\pi} [\cos (\gamma + n)x + \cos (\gamma - n)x] \, dx$$

$$= \frac{1}{\pi} \left[\frac{\sin (\gamma + n)\pi}{\gamma + n} + \frac{\sin (\gamma - n)\pi}{\gamma - n} \right]$$

$$= (-1)^n \frac{2\gamma \sin \gamma \pi}{\gamma^2 - n^2};$$

therefore,

$$(1) \qquad \cos \gamma x = \frac{2\gamma \sin \gamma \pi}{\pi} \left(\frac{1}{2\gamma^2} - \frac{\cos x}{\gamma^2 - 1^2} + \frac{\cos 2x}{\gamma^2 - 2^2} - \cdots \right).$$

This series converges at $\pm \pi$ due to the continuity of $\cos \gamma x$; hence, if we put $x = \pi$ in (1), divide by $\sin \gamma \pi$, and write x instead of γ, we get

$$(2) \qquad \cot \pi x = \frac{2x}{\pi} \left(\frac{1}{2x^2} + \frac{1}{x^2 - 1^2} + \frac{1}{x^2 - 2^2} + \cdots \right),$$

a formula that puts the poles of $\cot \pi x$ in evidence.

If we write this series in the form

$$(3) \qquad \pi \cot \pi x - \frac{\pi}{\pi x} = \frac{-2x}{1^2 - x^2} + \frac{-2x}{2^2 - x^2} + \cdots,$$

both sides vanish as $x \to 0$. Moreover, if $0 \leq x \leq a < 1$, the series (3) converges uniformly in $(0, a)$, for we can take $M_n = 2/(n^2 - x^2)$ in the M-test (§ 179). We can therefore integrate (3) term by term from 0 to x to obtain

$$\begin{aligned}
\log \frac{\sin \pi x}{\pi x} &= \lim_{n \to \infty} \sum_{j=1}^{n} \log \left(1 - \frac{x^2}{j^2} \right) \\
&= \lim_{n \to \infty} \log \prod_{j=1}^{n} \left(1 - \frac{x^2}{j^2} \right) \\
&= \log \prod_{j=1}^{\infty} \left(1 - \frac{x^2}{j^2} \right).
\end{aligned}$$

We have thus obtained the famous infinite product,

$$(4) \qquad \sin \pi x = \pi x \prod_{n=1}^{\infty} \left(1 - \frac{x}{n} \right) \left(1 + \frac{x}{n} \right),$$

which puts all the zeros of $\sin \pi x$ $(0, \pm 1, \pm 2, \cdots)$ in evidence and is in effect a factorization of $\sin \pi x$ into $\aleph_0$ linear factors.

On putting $x = {}^1/_2$ in (4), we obtain Wallis' product for $\pi/2$:

$$(5) \qquad \frac{\pi}{2} = \prod_{n=1}^{\infty} \frac{2n}{2n-1} \cdot \frac{2n}{2n+1} = \frac{2}{1} \cdot \frac{2}{3} \cdot \frac{4}{3} \cdot \frac{4}{5} \cdot \frac{6}{5} \cdot \frac{6}{7} \cdots,$$

in agreement with (193.8).

225. Approximation Theorems. We shall next show that, if $y = g(x)$ represents a broken line such that $g(-\pi) = g(\pi)$, the Fourier series for $g(x)$ is uniformly convergent in $(-\pi, \pi)$. Let the k segments of the broken line connect $k + 1$ vertices whose abscissas are

$$x_0 = -\pi, \quad x_1, x_2, \cdots, x_{k-1}, \quad x_k = \pi.$$

Since $g(x)$ is continuous and piecewise smooth, its Fourier series converges everywhere in $(-\pi, \pi)$. The Fourier constants may be computed by piecewise integration. Thus, if λ_j is the slope of $g(x)$ between x_{j-1} and x_j, we have

$$\int_{x_{j-1}}^{x_j} g(x) \cos nx\, dx = \frac{g(x) \sin nx}{n}\Big|_{x_{j-1}}^{x_j} - \frac{1}{n}\int_{x_{j-1}}^{x_j} \lambda_j \sin nx\, dx$$

$$= \frac{g(x) \sin nx}{n} + \frac{\lambda_j}{n^2}\cos nx \Big|_{x_{j-1}}^{x_j}.$$

When the integrals are added for $j = 1$ to $j = k$, the sine terms give zero, for the end terms vanish and the rest cancel in pairs; and the cosine terms give

$$\pi a_n = \frac{1}{n^2}\sum_{j=1}^{k} \lambda_j(\cos nx_j - \cos nx_{j-1}).$$

Hence, if λ denotes the largest of the slopes λ_j, $\pi\,|\,a_n\,| < 2k\lambda/n^2$. In the same way we can show that $\pi\,|\,b_n\,| < 2k\lambda/n^2$; in this case the cosine terms add up to zero, for the end terms

$$-\frac{1}{n}[g(\pi)\cos n\pi - g(-\pi)\cos(-n\pi)] = 0,$$

while the rest cancel in pairs. Thus the Fourier constants of the broken-line function $g(x)$ are numerically less than C/n^2, where C is a constant independent of n, and

$$|\,a_n \cos nx + b_n \sin nx\,| < \frac{2C}{n^2}.$$

Hence, the series converges uniformly to $g(x)$ by the Weierstrass M-test (§ 179).

We are now in position to prove

THEOREM 1 (Weierstrass). *Any continuous function $f(x)$ of period 2π can be uniformly approximated by a finite trigonometric sum $t_n(x)$ to any assigned degree of accuracy: that is,*

(1) $|\,f(x) - t_n(x)\,| < \varepsilon, \qquad -\pi \leq x \leq \pi.$

Proof. Since $f(x)$ is continuous in $(-\pi, \pi)$ it is uniformly continuous there (Theorem 46.1). Hence, for any assigned $\varepsilon > 0$ we can choose δ so that

$$|\,f(x) - f(x')\,| < \varepsilon/4 \quad \text{when} \quad |\,x - x'\,| < \delta.$$

Now inscribe a broken line $y = g(x)$ in the graph of $y = f(x)$ whose segments have horizontal projections $< \delta$; then, if x lies in the interval (x_j, x_{j+1}),

$$|g(x) - g(x_j)| \leq |g(x_{j+1}) - g(x_j)| < \varepsilon/4,$$

(2) $\qquad |f(x) - g(x)| \leq |f(x) - f(x_j) + g(x_j) - g(x)| < \tfrac{1}{2}\varepsilon.$

throughout $(-\pi, \pi)$. Now the Fourier series for $g(x)$ converges uniformly to $g(x)$ in $(-\pi, \pi)$; hence, we can choose n so large that the nth partial sum $t_n(x)$ satisfies

(3) $\qquad |g(x) - t_n(x)| < \tfrac{1}{2}\varepsilon, \qquad -\pi \leq x \leq \pi.$

Then $t_n(x)$ is a finite trigonometric sum that satisfies (1); for (1) follows from (2) and (3).

THEOREM 2. (Weierstrass). *Any continuous function $f(x)$ in the interval $a \leq x \leq b$ can be uniformly approximated by a polynomial $P(x)$ to any assigned degree of accuracy: that is,*

(4) $\qquad |f(x) - P(x)| < \varepsilon, \qquad a \leq x \leq b.$

Proof. If necessary we first make a linear change of variable $x' = cx + d$ which maps the interval (a, b) on one interior to $(-\pi, \pi)$. Thus we can assume that $-\pi < a < b < \pi$ and extend the function $f(x)$ so that it is continuous in $(-\pi, \pi)$ and also $f(-\pi) = f(\pi)$.† We may now approximate this extended function by a trigonometric sum $T(x)$ such that

(5) $\qquad |f(x) - T(x)| < \tfrac{1}{2}\varepsilon, \qquad -\pi \leq x \leq \pi.$

Now $T(x)$ can be expanded into a Taylor series about $x = 0$ which converges everywhere (since the sine and cosine series have this property) and hence converges uniformly in $(-\pi, \pi)$. Thus, when n is sufficiently large, the nth partial sum of this series is a polynomial $P(x)$ which satisfies

(6) $\qquad |T(x) - P(x)| < \tfrac{1}{2}\varepsilon, \qquad -\pi \leq x \leq \pi.$

The inequality (4) now follows from (5) and (6).

226. Parseval's Theorem. Let $f(x)$ be a continuous function of period 2π having the Fourier series,

(1) $\qquad f(x) \sim \dfrac{1}{2} a_0 + \displaystyle\sum_{k=1}^{\infty} (a_k \cos kx + b_k \sin kx).$

---

† For example, we may extend the graph of $f(x)$ in (a, b) by straight lines to the points $(-\pi, 0)$, $(\pi, 0)$.

If we write

$$s_n(x) = \frac{1}{2} a_0 + \sum_{k=1}^{n} (a_k \cos kx + b_k \sin kx)$$

for the partial sums of this series, while $t_n(x)$ denotes a trigonometric sum of like index in which the constants are other than the Fourier constants of $f(x)$, we have from (219.10)

(2) $$0 \le \int_{-\pi}^{\pi} [f(x) - s_n(x)]^2 \, dx \le \int_{-\pi}^{\pi} [f(x) - t_n(x)]^2 \, dx.$$

Now Theorem 225.1 shows that we can make the last integral $< \varepsilon$ by choosing a sum $t_n(x)$ of sufficiently high index n so that

$$|f(x) - t_n(x)| < \sqrt{\varepsilon/2\pi}.$$

Therefore,

(3) $$\lim_{n \to \infty} \int_{-\pi}^{\pi} [f(x) - s_n(x)]^2 \, dx = 0,$$

and, consequently, the orthonormal set,

$$\frac{1}{\sqrt{2\pi}}, \quad \frac{\cos nx}{\sqrt{\pi}}, \quad \frac{\sin nx}{\sqrt{\pi}} \qquad (n = 1, 2, \cdots),$$

is closed and complete with respect to continuous functions of period 2π (§ 220). We may therefore apply Theorem 220 to Fourier series:

PARSEVAL'S THEOREM FOR FOURIER SERIES. *For all continuous functions* $f(x), \bar{f}(x)$ *of period* 2π,

(4) $$\frac{1}{2} a_0^2 + \sum_{n=1}^{\infty} (a_n^2 + b_n^2) = \frac{1}{\pi} \int_{-\pi}^{\pi} f^2(x) \, dx,$$

(5) $$\frac{1}{2} a_0 \bar{a}_0 + \sum_{n=1}^{\infty} (a_n \bar{a}_n + b_n \bar{b}_n) = \frac{1}{\pi} \int_{-\pi}^{\pi} f(x) \bar{f}(x) \, dx.$$

Proof. With the notation for Fourier constants (c_0, c_n, c_n') used in § 221, the series (220.3) becomes

(6) $$c_0 \bar{c}_0 + \sum_{n=1}^{\infty} (c_n \bar{c}_n + c_n' \bar{c}_n') = \int_{-\pi}^{\pi} f(x) \bar{f}(x) \, dx.$$

But, since

$$c_0 = \sqrt{\pi/2} \, a_0, \qquad c_n = \sqrt{\pi} \, a_n, \qquad c_n' = \sqrt{\pi} \, b_n \qquad (\S \, 221),$$

the series (6) divided by π becomes (5); and, when $\bar{f} = f$, (5) becomes (4).

PROBLEMS

1. If $f(x)$ is a continuous even function in $-\pi \leqq x \leqq \pi$ and

$$\int_0^\pi f(x) \cos nx \, dx = 0, \qquad n = 0, 1, 2, \ldots,$$

show that $f(x) = 0$ in $(-\pi, \pi)$. [Use (226.4).]

2. If $f(x)$ is continuous in $a \leqq x \leqq b$ and

$$\int_a^b f(x) \, x^n \, dx = 0, \qquad n = 0, 1, 2, \ldots,$$

show that $f(x) = 0$ in (a, b).

[If $P(x)$ is any polynomial, $\int_a^b f(x) P(x) \, dx = 0$; use Theorem 225.2 to show that

$$\int_a^b f^2(x) \, dx = \int_a^b f(x) \, (f(x) - P(x)) \, dx = 0.]$$

227. Integration of Fourier Series. Let $f(x)$ be a piecewise continuous function in $(-\pi, \pi)$ which has the Fourier series,

(1) $$f(x) \sim \frac{1}{2} a_0 + \sum_{n=1}^\infty (a_n \cos nx + b_n \sin nx).$$

We do not assume that this series represents $f(x)$ or even converges. Nevertheless, we shall show that the series may be integrated term by term to give the expected result.

THEOREM. *If c and x are two points of $(-\pi, \pi)$, the Fourier series for a piecewise continuous function $f(x)$ may be integrated term by term to give*

$$\int_c^x f(x) \, dx = \frac{1}{2} a_0(x - c) + \sum_{n=1}^\infty \int_c^x (a_n \cos nx + b_n \sin nx) \, dx.$$

Proof. The function

$$F(x) = \int_c^x \left[f(x) - \frac{1}{2} a_0 \right] dx$$

is continuous in $(-\pi, \pi)$ and has the derivative $f(x)$ at all points where $f(x)$ is continuous; and, at a point c where $f(x)$ makes a finite jump,

$$F'_+(c) = f(c+), \qquad F'_-(c) = f(c-).$$

Moreover,

(2) $$F(\pi) - F(-\pi) = \int_{-\pi}^\pi \left[f(x) - \frac{1}{2} a_0 \right] dx = \pi a_0 - \pi a_0 = 0.$$

Theorem 222 shows that $F(x)$ may be expanded into a convergent Fourier series,

$$F(x) = \frac{1}{2} A_0 + \sum_{n=1}^{\infty} (A_n \cos nx + B_n \sin nx).$$

Remembering that $F'(x) = f(x)$ except at the finite jumps of $f(x)$, we now compute A_n and B_n by integration by parts. Thus,

(3) $$A_n = \frac{1}{\pi} \int_{-\pi}^{\pi} F(x) \cos nx \, dx = \frac{-1}{n\pi} \int_{-\pi}^{\pi} f(x) \sin nx \, dx = -\frac{b_n}{n},$$

(4) $$B_n = \frac{1}{\pi} \int_{-\pi}^{\pi} F(x) \sin nx \, dx = \frac{1}{n\pi} \int_{-\pi}^{\pi} f(x) \cos nx \, dx = \frac{a_n}{n};$$

where the integrated part in (3) vanishes at both limits but in (4) vanishes by virtue of (2).† We thus have

(5) $$F(x) = \frac{1}{2} A_0 + \frac{1}{n} \sum_{n=1}^{\infty} (-b_n \cos nx + a_n \sin nx),$$

and, on putting $x = c$,

(6) $$0 = \frac{1}{2} A_0 + \frac{1}{n} \sum_{n=1}^{\infty} (-b_n \cos nc + a_n \sin nc).$$

On subtracting (6) from (5), we obtain

$$\int_c^x f(x) \, dx - \frac{1}{2} a_0(x - c) = \sum_{n=1}^{\infty} a_n \frac{\sin nx - \sin nc}{n} - \sum_{n=1}^{\infty} b_n \frac{\cos nx - \cos nc}{n},$$

which is precisely the result of integrating (1) (with $\sim$ replaced by $=$) term by term.

Example 1. From (221.10) we have

(7) $$\frac{x}{2} = \sum_{n=1}^{\infty} (-1)^{n+1} \frac{\sin nx}{n} \qquad (-\pi < x < \pi),$$

while at $-\pi, \pi$ the series converges to 0. The series therefore converges nonuniformly in any interval containing one of these points (Theorem 180). Nevertheless it may be integrated term by term, say from 0 to x, even when $x = \pm\pi$: thus,

$$\frac{x^2}{4} = \sum_{n=1}^{\infty} (-1)^{n+1} \frac{1 - \cos nx}{n^2} = \frac{\pi^2}{12} - \sum_{n=1}^{\infty} (-1)^{n+1} \frac{\cos nx}{n^2},$$

---

† We may regard (3) and (4) as the result of piecewise integration between the jumps of $f(x)$ followed by addition (§122).

in view of (223.14). The absolute convergence of this series permits the rearrangement of terms (Theorem 36.1). Hence, we have

$$(8) \qquad \sum_{n=1}^{\infty} (-1)^{n+1} \frac{\cos nx}{n^2} = \frac{\pi^2}{12} - \frac{x^2}{4} \qquad (-\pi \leq x \leq \pi),$$

a series valid at $\pm\pi$, where the left member becomes $-\pi^2/6$ (223.15).

If we integrate again from 0 to x,

$$(9) \qquad \sum_{n=1}^{\infty} (-1)^{n+1} \frac{\sin nx}{n^3} = \frac{\pi}{12} x (\pi^2 - x^2) \qquad (-\pi \leq x \leq \pi).$$

When $x = \frac{1}{2}\pi$, (9) gives

$$(10) \qquad 1 - \frac{1}{3^3} + \frac{1}{5^3} - \frac{1}{7^3} + \cdots = \frac{\pi^3}{32}.$$

Example 2. From (223.7) we have

$$(11) \qquad \sum_{n=1}^{\infty} \frac{\sin (2n - 1)x}{2n - 1} = \frac{\pi}{4} \qquad (0 < x < \pi);$$

since the left member is zero at $-\pi, 0, \pi$, this series converges nonuniformly in any interval containing $-\pi, 0,$ or π. Integrate (11) from x to $\frac{1}{2}\pi$; then,

$$(12) \qquad \sum_{n=1}^{\infty} \frac{\cos (2n - 1)x}{(2n - 1)^2} = \frac{\pi}{8} (\pi - 2x) \qquad (0 \leq x \leq \pi),$$

since the cosines vanish at the upper limit. The series converges to $\pi^2/8$ at 0, to $-\pi^2/8$ at π, as we may verify from (223.10).

Now integrate (12) from 0 to x:

$$(13) \qquad \sum_{n=1}^{\infty} \frac{\sin (2n - 1)x}{(2n - 1)^3} = \frac{\pi}{8} (\pi x - x^2) \qquad (0 \leq x \leq \pi).$$

Finally integrate (13) from x to $\frac{1}{2}\pi$:

$$(14) \qquad \sum_{n=1}^{\infty} \frac{\cos (2n - 1)x}{(2n - 1)^4} = \frac{\pi}{96} (\pi^3 - 6\pi x^2 + 4x^3) \qquad (0 \leq x \leq \pi).$$

The reader may verify these results by putting $x = 0, \frac{1}{2}\pi, \pi$ (223.17).

PROBLEMS

1. Integrate the series for x^2 in Prob. 221.2 to obtain

$$\sum_{n=1}^{\infty} (-1)^n \frac{\sin nx}{n^3} = \frac{1}{12} x(x^2 - \pi); \qquad \sum_{n=1}^{\infty} \frac{1}{n^6} = \frac{\pi^6}{945}.$$

2. Integrate the series in Prob. 223.6 to obtain

$$\sum_{n=1}^{\infty} (-1)^{n+1} \frac{\cos (2n-1)x}{(2n-1)^3} = \begin{cases} \dfrac{\pi}{32} (\pi - 2x)(\pi + 2x) & (0 \le x \le \tfrac{1}{2}\pi), \\[2ex] \dfrac{\pi}{32} (2x - \pi)(2x - 3\pi) & (\tfrac{1}{2}\pi \le x \le \pi). \end{cases}$$

3. From the series (223.11)

$$\sum_{n=1}^{\infty} \frac{\sin nx}{n} = \frac{\pi - x}{2} \qquad (0 < x < 2\pi),$$

integrate term by term to obtain successively

(i) $\displaystyle\sum_{n=1}^{\infty} \frac{\cos nx}{n^2} = \frac{\pi^2}{6} - \frac{\pi}{2} x + \frac{x^2}{4}$,

(ii) $\displaystyle\sum_{n=1}^{\infty} \frac{\sin nx}{n^3} = \frac{\pi^2}{6} x - \frac{\pi}{4} x^2 + \frac{x^3}{12}$,

(iii) $\displaystyle\sum_{n=1}^{\infty} \frac{\cos nx}{n^4} = \frac{n^4}{90} - \frac{\pi^2}{12} x^2 + \frac{\pi}{12} x^3 - \frac{x^4}{48}$,

all valid in $0 \le x \le 2\pi$. From these results deduce

$$\sum_{n=1}^{\infty} \frac{(-1)^{n+1}}{n^2} = \frac{\pi^2}{12}, \qquad \sum_{n=1}^{\infty} \frac{(-1)^{n+1}}{(2n-1)^3} = \frac{\pi^3}{32}, \qquad \sum_{n=1}^{\infty} \frac{(-1)^{n+1}}{n^4} = \frac{7}{720} \pi^4.$$

4. Integrate the series for $\cosh x$ in Prob. 223.5 to obtain the series for $\sinh x$ given in that problem.

228. Uniform Convergence of Fourier Series. When $f(x)$ has discontinuities in a period interval, its Fourier series obviously cannot converge uniformly in the interval (Theorem 180). However, we have the following

THEOREM. *The Fourier series of any function of period 2π which is continuous and piecewise smooth in $(-\pi, \pi)$ converges absolutely and uniformly in this interval.*

Proof. Since $f(x)$ is piecewise smooth (§ 129), $f'(x)$ is piecewise continuous and has the Fourier coefficients,

(1) $$a'_0 = 0, \qquad a'_n = nb_n, \qquad b'_n = -na_n;$$

for example, piecewise integration between the discontinuities of $f'(x)$ followed by addition shows that

$$a'_n = \frac{1}{\pi} \int_{-\pi}^{\pi} f'(x) \cos nx \, dx = \frac{1}{\pi} f(x) \cos nx \Big|_{-\pi}^{\pi} + \frac{n}{\pi} \int_{-\pi}^{\pi} f(x) \sin nx \, dx,$$

whence $a'_n = nb_n$, since the integrated part vanishes. Hence Bessel's inequality (221.4) applied to $f'(x)$ gives

$$(2) \qquad \sum_{k=1}^{\infty} k^2(a_k^2 + b_k^2) \leqq \frac{1}{\pi} \int_{-\pi}^{\pi} \{f'(x)\}^2 \, dx.$$

We can now show that the Fourier series,

$$f(x) = \frac{1}{2} a_0 + \sum_{k=1}^{\infty} (a_k \cos kx + b_k \sin kx),$$

converges absolutely and uniformly in $(-\pi, \pi)$. Since

$$| a_k \cos kx + b_k \sin kx | \leqq \sqrt{a_k^2 + b_k^2},\dagger$$

we need only show that the series of constants $\sum \sqrt{a_k^2 + b_k^2}$ converges. Now, from Schwarz's inequality$\ddagger$

$$\sum_{k=1}^{n} \sqrt{a_k^2 + b_k^2} = \sum_{k=1}^{n} \frac{1}{k} \cdot k\sqrt{a_k^2 + b_k^2} \leqq \left(\sum_{k=1}^{n} \frac{1}{k^2}\right)^{1/2} \left(\sum_{k=1}^{n} k^2(a_k^2 + b_k^2)\right)^{1/2};$$

and, as $n \to \infty$, both series on the right converge, and hence the series on the left also.

$\dagger$ The maximum value of $a \cos \theta + b \sin \theta$ is $\sqrt{a^2 + b^2}$.

$\ddagger$ If the a_k, b_k are real,

$$\sum_{k=1}^{n} (a_k + \lambda b_k)^2 = \sum_{k=1}^{n} a_k^2 + 2\lambda \sum_{k=1}^{n} a_k b_k + \lambda^2 \sum_{k=1}^{n} b_k^2 \geqq 0$$

for all real values of λ. Hence the quadratic equation in λ, $\Sigma(a_k + \lambda b_k)^2 = 0$, cannot have two different real roots, and its discriminant is negative or zero. This gives *Schwarz's inequality*,

$$(3) \qquad \left(\sum_{k=1}^{n} a_k b_k\right)^2 \leqq \left(\sum_{k=1}^{n} a_k^2\right) \left(\sum_{k=1}^{n} b_k^2\right),$$

where the equal sign corresponds to the case of equal roots ($\lambda = -a_k/b_k$, $k = 1, 2, \cdots, n$).

. If $f(x)$ and $g(x)$ are integrable in $a \leqq x \leqq b$, the same considerations applied to $\int_a^b [f(x) + \lambda g(x)]^2 \, dx$ will establish *Schwarz's inequality for integrals*:

$$(4) \qquad \left(\int_a^b f(x) g(x) \, dx\right)^2 \leqq \int_a^b f^2(x) \, dx \int_a^b g^2(x) \, dx.$$

229. Gibbs' Phenomenon. From (223.7) we have

$$
(1) \qquad \sin x + \frac{\sin 3x}{3} + \frac{\sin 5x}{5} + \cdots =
\begin{cases}
-\pi/4 & (-\pi < x < 0), \\
0 & (x = 0, \pm\pi), \\
\pi/4 & (0 < x < \pi).
\end{cases}
$$

We shall show that the approximation curve,

$$
(2) \qquad y = s_{2n-1}(x) = \sin x + \frac{\sin 3x}{3} + \cdots + \frac{\sin (2n-1)x}{2n-1},
$$

in the neighborhood of zero, approaches the lines $y = -\tfrac{1}{4}\pi\,(x < 0)$, $y = \tfrac{1}{4}\pi \; (x > 0)$, together with the segment of the x-axis $-k\dfrac{\pi}{4} \leqq y \leqq k\dfrac{\pi}{4}$ (Fig. 229) where

$$
(3) \qquad k = \frac{2}{\pi}\int_0^\pi \frac{\sin x}{x}\, dx = 1.179 \qquad\qquad \text{(Ex. 130.3).}
$$

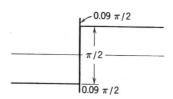

FIG. 229. Limiting curve in Gibbs' phenomenon

That fact that the last segment extends beyond the points $y = \pm\pi/4$ is known as *Gibbs' phenomenon*; for the limiting behavior of the approximation curves was first accurately described by J. Willard Gibbs, with reference to the series (223.11), in the English magazine *Nature*, vol. 59, 1899, p. 606.

We first prove that the approximation curve $y = s_{2n-1}(x)$ has a series of maxima of which the greatest occurs at $x = \pi/2n$. From (2) we have

$$
(4) \qquad \frac{dy}{dx} = \cos x + \cos 3x + \cdots + \cos (2n-1)x = \frac{\sin 2nx}{2 \sin x},
$$

on making use of (201.6); hence the curve (2) has extremes where $\sin 2nx = 0$. The maxima occur where $2nx = \pi, 3\pi, 5\pi, \cdots$; the minima where $2nx = 2\pi, 4\pi, 6\pi, \cdots$. The *greatest maximum* is the first, at $x = \pi/2n$; for, from (4),

$$
(5) \qquad s_{2n-1}(x) = \int_0^x \frac{\sin 2nx}{\sin x}\, dx = \int_0^{\pi/2n} + \int_{\pi/2n}^{2\pi/2n} + \int_{2\pi/2}^{3\pi/2n} + \cdots + \int^x,
$$

where the integrals form an alternating series of positive and negative terms. The first of these is clearly the largest since $\sin x$, in the denominator, ranges over its smallest values in $(0, \pi/2n)$; and this integral is also the height of the greatest wave:

$$(6) \qquad s_{2n-1}\left(\frac{\pi}{2n}\right) = \int_0^{\pi/2n} \frac{\sin 2nx}{\sin x}\, dx.$$

We must now find the limiting height of the greatest wave as $n \to \infty$, To this end we interpret $s_{2n-1}(\pi/2n)$ as a Riemann sum. Writing $h = \pi/n$. $s_{2n-1}(\tfrac{1}{2}h)$ is given by (2):

$$(7) \qquad s_{2n-1}\left(\frac{\pi}{2n}\right) = \frac{1}{2}h\left(\frac{\sin \tfrac{1}{2}h}{\tfrac{1}{2}h} + \frac{\sin \tfrac{3}{2}h}{\tfrac{3}{2}h} + \cdots + \frac{\sin \dfrac{2n-1}{2}h}{\dfrac{2n-1}{2}h}\right).$$

This is a Riemann sum for the integrand $\tfrac{1}{2}(\sin x)/x$, over the interval $0 \leq x \leq \pi$ divided into n parts h; and as $n \to \infty (h \to 0)$,

$$s_{2n-1}\left(\frac{\pi}{2n}\right) \to \frac{1}{2}\int_0^{\pi} \frac{\sin x}{x}\, dx = k\frac{\pi}{4},$$

where k is given in (3). Thus the first and highest wave of the approximation curves to the right of the origin tends to the segment $0 \leq y \leq k\pi/4$ of the y-axis. Similarly the lowest wave of the approximation curves to the left of the origin tends to a segment $-k\pi/4 \leq y \leq 0$. Thus the approximation curves approach the segment $-k\pi/4 \leq y \leq k\pi/4$, the "jump" $\pi/2$ being stretched in the ratio $k/1$.

The Fourier series of every piecewise continuous function $f(x)$ exhibits the same behavior at points of discontinuity. If 0 is such a point, the jump is $\delta = f(0+) - f(0-)$. Now, if $\varphi(x)$ denotes the function in (1), we form a new function,

$$F(x) = f(x) - \varphi(x)\frac{\delta}{\tfrac{1}{2}\pi},$$

which is continuous at 0 since

$$F(0+) = F(0-) = \tfrac{1}{2}[f(0+) + f(0-)].$$

But, since the Fourier series for $F(x)$ converges uniformly near 0 (§ 228), its approximation curves approach the curve $y = F(x)$ *uniformly*. Hence the approximation curves of $y = f(x)$ must have precisely the same behavior as those of $y = \varphi(x)\delta/\tfrac{1}{2}\pi$, whose jump δ on the y-axis is stretched to $k\delta$. Hence, the approximation curves for $y = f(x)$ in the neighborhood of 0 tend to the curve $y = f(x)$ plus a segment of the y-axis of length $k\delta$ centered at the point $\frac{1}{2}[f(0+) + f(0-)]$†. Since $k = 1.179$, the waves extend the jump by about 18%. Thus the high wave strikes the y-axis at a point about 0.09δ above $f(0+)$, the low wave about 0.09δ below $f(0-)$.

Although the Fourier series approaches a *discontinuous function*, the approximation curves approach a *continuous curve* made of the portions described above. This paradoxical result led Michelson to maintain that Fourier series such as (1) really represent continuous *functions* (*Nature*, vol. 58, 1898, p. 544). This gave rise to a lively controversy in the pages of *Nature*, vols. 58 to 60. Thus even a great scientist may be misled by a lack of sharp concepts and precise definitions!

230. Properties of Fourier Series. If $f(x)$ is integrable in $(-\pi, \pi)$, of nonzero norm, and of period 2π, its Fourier series is

$$(1) \qquad f(x) \sim \frac{1}{2} a_0 + \sum_{n=1}^{\infty} (a_n \cos nx + b_n \sin bx),$$

where the Fourier constants,

$$(2) \qquad \begin{matrix} a_n \\ b_n \end{matrix} = \frac{1}{\pi} \int_{-\pi}^{\pi} f(t) \begin{matrix} \cos nt \\ \sin nt \end{matrix} dt, \qquad n = 0, 1, 2, \cdots.$$

When $f(x)$ is absolutely integrable, $\{a_n\}$ and $\{b_n\}$ form null sequences. The series contains only cosines when $f(x)$ is an even function, only sines when $f(x)$ is odd. If $f(x)$ is only defined in $(0, \pi)$, it can be extended to $(-\pi, \pi)$ so that $f(-x) = f(x)$ or $f(-x) = -f(x)$; thus, in $(0, \pi)$ the Fourier series of $f(x)$ may be a cosine series or a sine series.

The integrability of $f(x)$, or even its continuity, does not ensure the convergence of $f(x)$. But, if $f(x)$ is piecewise continuous in $(-\pi, \pi)$ (which permits a finite number of finite jumps) its Fourier series converges to $\frac{1}{2}\{f(x+) + f(x-)\}$ at every point where $f(x)$ has right- and left-hand slopes. If $f(x)$ is continuous at such a point, $f(x+) = f(x-)$, and the series converges to $f(x)$.

In the neighborhood of a point x_0 where $f(x)$ makes a finite jump δ, the approximation curves $y = s_n(x)$ approach a continuous curve consisting of the adjacent graph of $y = f(x)$ plus a vertical segment of the line $x = x_0$ extended above and below the graph by about 0.09δ (*Gibbs' phenomenon*).

---

†See Knopp, *Theory and Application of Infinite Series,* London, 1928, p. 379, for an accurate graph of an approximation curve to the function of Fig. 223b. (*Theory and Application of Infinite Series* now available from Dover, 0-486-66163-2)

The sine-cosine orthonormal set,

$$\frac{1}{2\pi}, \quad \frac{\cos nx}{\pi}, \quad \frac{\sin nx}{\pi} \qquad (n = 1, 2, 3, \cdots),$$

is closed and complete with respect to continuous functions of period 2π; and for any such function $f(x)$ having Fourier constants a_n, b_n we have Parseval's equation:

$$\frac{1}{2} a_0{}^2 + \sum_{n=1}^{\infty} (a_n{}^2 + b_n{}^2) = \frac{1}{\pi} \int_{-\pi}^{\pi} f^2(x)\, dx.$$

The Fourier series of a piecewise continuous function may be integrated term by term just as if (1) were an *equation*; and this is true whether or not the series represents $f(x)$, or even if it diverges.

If $f(x)$ is continuous and piecewise smooth in $(-\pi, \pi)$, and $f(-\pi) = f(\pi)$, its Fourier series converges absolutely and uniformly in $(-\pi, \pi)$.

APPENDIX 1

Cluster Points

The following proof of Theorem 13.2 presents the least and greatest cluster points of a set as binary fractions (§ 6).

THEOREM. *Every bounded, infinite set of points has a least and a greatest cluster point, which may, in particular, coincide.*

Proof. Let the point set S lie in the segment gG of the number axis. Choose the points g and G to represent the numbers 0 and 1, respectively. Let P_1 bisect gG; then at least one of the halves gP_1, P_1G will contain an infinite number of set points. If the left half gP_1 has this property, let P_2 bisect it; otherwise, let P_2 bisect the right half P_1G. One of the quarters to the left and right of P_2 will contain an infinite number of set points; if the left quarter has this property, let P_3 bisect it; otherwise, let P_3 bisect the right quarter. Obviously we can proceed in this manner indefinitely. After n steps the length of the subintervals is $1/2^n$; and, as $n \to \infty$, $1/2^n \to 0$. We are thus led to the *least* cluster point expressed as a dual fraction $\alpha = 0 \cdot a_1 a_2 a_3 \cdots$, where $a_n = 0$ or 1 according as the subinterval to the *left* of P_n contains an infinite number of points of the set or not.

If we replace the word "left" by "right" in the above process, we are led to the *greatest* cluster point of the set $\beta = 0 \cdot b_1 b_2 b_3 \cdots$, provided $b_n = 1$ or 0 according as the subinterval to the *right* of P_n contains an infinite number of points of the set or not.

Evidently $a_n = b_n$ as long as only *one* subinterval contains an infinite number of set points; if this is *always* the case, $\alpha = \beta$, and the least and greatest cluster points coincide. But, if at the nth stage *both* subintervals contain an infinite number of set points, $a_n = 0$, $b_n = 1$, and $\alpha < \beta$.

For example the point set

$$1,\ 1/2,\ 3/4,\ 5/8,\ 11/16,\ \cdots \qquad \text{(Ex. 15.6)},$$

in which each number after the second is the mean of the two preceding, is contained in the interval $0 \leq x \leq 1$. By continued bisection we now find $a_1 = 1$, $a_2 = 0$, $a_3 = 1$, $a_4 = 0$, $\cdots$, so that the least cluster point is given by the binary fraction

$$\alpha = 0{\cdot}101010 \cdots = \frac{\frac{1}{2}}{1 - \frac{1}{4}} = \tfrac{2}{3}.$$

At each stage only one of the halves contains an infinite number of set points; hence $a_n = b_n$, and the greatest cluster point β of the set is also 2/3.

We next state a property that is shared by *every* cluster point of a set:

THEOREM. *If ξ is any cluster point of a set, we can always select from the set a sequence of points x_n such that $x_n \to \xi$.*

Proof. Let ε_1, ε_2, $\cdots$ be any decreasing null sequence. Since any neighborhood of ξ contains an infinite number of set points, choose x_n as any set point for which

$$\xi - \varepsilon_n < x_n < \xi + \varepsilon_n; \quad \text{then} \quad \left| x_n - \xi \right| < \varepsilon_n.$$

Now choose ε at pleasure. Since $\{\varepsilon_n\}$ is a *null* sequence we can pick N so that $\varepsilon_N < \varepsilon$; and, since $\{\varepsilon_n\}$ is also *decreasing*,

$$\left| x_n - \xi \right| < \varepsilon \quad \text{when} \quad n \geq N.$$

The least and greatest cluster points of a sequence $\{x_n\}$, namely,

$$\liminf x_n = \underline{\xi}, \qquad \limsup x_n = \bar{\xi},$$

are the cluster points characterized by the property

(1) $$\underline{\xi} - \varepsilon < x_n < \bar{\xi} + \varepsilon \quad \text{when} \quad n \geq N,$$

where N is a suitably larger integer.

PROBLEMS

1. Prove and then deduce corresponding results for *lim inf*:

(a) $\limsup (-x_n) = -\liminf x_n$;

(b) $\limsup 1/x_n = 1/\liminf x_n$, $\quad x_n > 0$, $\quad \underline{\xi} \neq 0$;

(c) $\limsup (x_n + y_n) \leq \limsup x_n + \limsup y_n$.

2. If $\{a_n\}$ is an arbitrary positive sequence, prove

(a) $\limsup \sqrt[n]{a_n} \leq \limsup \dfrac{a_{n+1}}{a_n}$; and hence

(b) $\liminf \dfrac{a_{n+1}}{a_n} \leq \liminf \sqrt[n]{a_n}$.

3. If $\{a_n\}$ is an arbitrary positive sequence, show that

$$\limsup \left(\frac{a_1 + a_{n+1}}{a_n} \right)^n \geq e \qquad \text{(Putnam 1949)}.$$

$$\left[\limsup \left\{ \frac{n(a_1 + a_{n+1})}{(n+1)a_n} \right\} < 1 \text{ contradicts } a_n > 0. \right]$$

APPENDIX 2

Difference Equations

A sequence $\{x_n\}$ is completely determined by x_1, x_2, and a *difference equation* (recurrence relation) connecting three successive terms, say,

$$(1) \qquad x_{n+2} + 2ax_{n+1} + bx_n = 0, \qquad n = 1, 2, 3, \cdots.$$

When a and $b \neq 0$ are constants, the following procedure will give the general term x_n. By substituting $x_n = k^n$ in (1), we find that k^n will be a particular solution of (1) when and only when k is a root of the *characteristic equation*

$$(2) \qquad k^2 + 2ak + b = 0, \qquad b \neq 0.$$

Thus, if k_1 and k_2 are distinct roots (nonzero since $b \neq 0$) of (2), $C_1 k_1^n$, $C_2 k_2^n$ are independent solutions of (1). Moreover all solutions of (1) are given by

$$(3) \qquad x_n = C_1 k_1^n + C_2 k_2^n$$

where C_1, C_2 are arbitrary constants. When x_1 and x_2 are given, the equations

$$x_1 = C_1 k_1 + C_2 k_2, \qquad x_2 = C_1 k_1^2 + C_2 k_2^2$$

determine C_1 and C_2 uniquely; for their determinant $k_1 k_2 (k_2 - k_1) \neq 0$.

When $b = a^2$, the quadratic (2) has equal roots $k_1 = k_2 = -a$, and k_1^n and nk_1^n are independent particular solutions. The former is known; as to nk_1^n, substitution in (1) gives

$$n(k_1^2 + 2ak_1 + a^2) + 2k_1(k_1 + a) = 0,$$

for both parentheses are zero. The general solution is now

$$(4) \qquad x_n = (C_1 + C_2 n)k_1^n.$$

When $b > a^2$, the quadratic (2) has conjugate complex roots

$$k_1 = r(\cos \alpha + i \sin \alpha), \qquad k_2 = r(\cos \alpha - i \sin \alpha) \qquad (9.2)$$

when a and b are real. The solution (3) is still valid and gives, on using DeMoivre's theorem (9.9),

$$(5) \qquad x_n = r^n(C_1 \cos n\alpha + C_2 \sin n\alpha).$$

The constants C_1 and C_2, again determined by x_1 and x_2, are real when x_1, x_2 are real.

The nonlinear difference equation

$$(6) \qquad x_{n+1} = \frac{ax_n + b}{cx_n + d}, \qquad c \neq 0, \qquad D = \begin{vmatrix} a & b \\ c & d \end{vmatrix} \neq 0$$

may be reduced to the form

$$(7) \qquad y_{n+1} = 2\alpha - \frac{\beta}{y_n}, \qquad 2\alpha = \frac{a+d}{c}, \qquad \beta = \frac{D}{c^2}$$

by the substitution $x_n = y_n - d/c$. On putting $y_n = z_{n+1}/z_n$, (7) becomes the linear recurrence relation

$$(8) \qquad z_{n+2} - 2\alpha z_{n+1} + \beta z_n = 0,$$

which may be solved as above.

Example 1. The difference equation

$$(i) \qquad x_{n+1} = 2 - \frac{2}{x_n}$$

is reduced to the linear form

$$(ii) \qquad y_{n+2} - 2y_{n+1} + 2y_n = 0$$

by the substitution $x_n = y_{n+1}/y_n$. The characteristic equation of (ii),

$$k^2 - 2k + 2 = 0$$

has the conjugate complex roots

$$1 + i = \sqrt{2}\left(\cos\frac{\pi}{4} \pm i\sin\frac{\pi}{4}\right).$$

Since the general solution of (ii) is

$$y_n = 2^{n/2}\left(C_1 \cos\frac{n\pi}{4} + C_2 \sin\frac{n\pi}{4}\right),$$

the general solution of (i) is

$$(iii) \qquad x_n = \sqrt{2}\ \frac{C_1 \cos(n+1)\frac{\pi}{4} + C_2 \sin(n+1)\frac{\pi}{4}}{C_1 \cos n\frac{\pi}{4} + C_2 \sin n\frac{\pi}{4}}.$$

If we determine γ in the interval $-\frac{1}{2}\pi < \gamma \leq \frac{1}{2}\pi$, so that

$$\frac{\cos \gamma}{C_1} = \frac{\sin \gamma}{C_2} \quad \text{or} \quad \gamma = \tan^{-1} \frac{C_2}{C_1},$$

(iii) takes the form

(iv) $$x_n = \sqrt{2} \frac{\cos \left[\left(n + 1 \right) \frac{\pi}{4} - \gamma \right]}{\cos \left(n \frac{\pi}{4} - \gamma \right)} = 1 - \tan \left(n \frac{\pi}{4} - \gamma \right).$$

in which γ plays the role of arbitrary constant. Since $\tan \theta$ has the period π, x_n has the period 4: $x_{n+4} = x_n$.

Example 2. The difference equation

(v) $$x_{n+1} = 4 - \frac{4}{x_n}$$

becomes

(vi) $$y_{n+2} - 4y_{n+1} + 4y_n = 0$$

on putting $x_n = y_{n+1}/y_n$. Now

$$k^2 - 4k + 4 = (k - 2)^2 = 0$$

has the equal roots 2, 2; and the general solution of (vi) is

$$y_n = (C_1 + C_2 n)2^n.$$

Hence (v) has the general solution

$$x_n = 2 \frac{C_1 + C_2(n + 1)}{C_1 + C_2 n} = 2 + \frac{2C_2}{C_1 + C_2 n}.$$

Evidently $x_n \to 2$, irrespective of the value of x_1.

PROBLEMS

Find the general solution of the following difference equations, giving the period if x_n is periodic or $\lim x_n$ if the limit exists:

1. $x_{n+1} = 1 - \dfrac{1}{x_n}$;   **2.** $x_{n+1} = 1 + \dfrac{2}{x_n}$, $x_1 = 1$;

3. $x_{n+1} = 3 - \dfrac{3}{x_n}$;   **4.** $x_{n+2} + x_n = 0$.

5. Show that the general solution of

(9) $$x_{n+2} + 2a\, x_{n+1} + b\, x_n = f(n)$$

is given by adding any particular solution of (9) to the general solution of (1).

6. Find x_n to satisfy

$$x_{n+2} - x_{n+1} - 2x_n = f(n), \qquad x_1 = 1, x_2 = 2,$$

when (a) $f(n) = 4$;   (b) $f(n) = 4n$;   (c) $f(n) = 2 - 4n^2$.

[Try a, $an + b$, $an^2 + bn + c$ respectively as particular solutions.]

APPENDIX 3

The Difference Calculus

Let $f(n)$ be a function defined for the integral values $n = 0, 1, 2, \cdots$. The *difference* $\Delta f(n)$ is defined as

$$(1) \qquad \Delta f(n) = f(n+1) - f(n).$$

For example the difference of the *factorial square* $n^{(2)} = n(n-1)$ is

$$\Delta n(n-1) = (n+1)n - n(n-1) = 2n.$$

The reader can readily verify the following

TABLE OF DIFFERENCES

$f(n)$	$\Delta f(n)$
C (const)	0
r^n	$(r-1)r^n$
$n^{(k)}$	$kn^{(k-1)}$
$n^{(-k)}$	$-kn^{(-k-1)}$
$\sin n\alpha$	$2 \cos (n + \tfrac{1}{2})\alpha \sin \tfrac{1}{2}\alpha$
$\cos n\alpha$	$-2 \sin (n + \tfrac{1}{2})\alpha \sin \tfrac{1}{2}\alpha$

The *factorial powers* $n^{(k)}$ and $n^{(-k)}$, where k is a positive integer, are defined by the equations

$$(2) \qquad n^{(k)} = n(n-1)(n-2) \cdots (n-k+1),$$

$$(3) \qquad n^{(-k)} = 1/(n+1)(n+2) \cdots (n+k).$$

In particular $n^{(1)} = n$, $n^{(-1)} = 1/(n+1)$; and we define $x^{(0)} = 1$.†

If k is *any* integer we have the general formula

$$(4) \qquad n^{(k)} = kn^{(k-1)}$$

analogous to the derivative of a power.

† See Brand, L., Binomial Expansions in Factorial Powers, *Am. Math. Monthly*, vol. 67, no. 10.

The proofs of the differences given are all straightforward. For example

$$\Delta \sin n\alpha = \sin (n + 1)\alpha - \sin n\alpha$$

$$= \sin (n + \tfrac{1}{2} + \tfrac{1}{2})\alpha - \sin (n + \tfrac{1}{2} - \tfrac{1}{2})\alpha$$

which gives the tabular result on using the addition formula for the sine.

In the differential calculus, we call $F(x)$ an antiderivative of $f(x)$ if $DF(x) = f(x)$; and we sometimes write $F(x) = D^{-1}f(x)$. Similarly in the difference calculus we call $F(n)$ an *antidifference* of $f(n)$ if $\Delta F(n) = f(n)$; and we write $F(n) = \Delta^{-1} f(n)$. From the difference table we can readily form a

TABLE OF ANTIDIFFERENCES

$f(n)$	$\Delta^{-1} f(n)$
0	C (const)
r^n	$\dfrac{r^n}{r-1} \qquad (r \neq 1)$
$n^{(k)}$	$\dfrac{n^{(k+1)}}{k+1}$
$n^{(-k)}$	$\dfrac{n^{(1-k)}}{1-k} \qquad (k \neq 1)$
$\cos n\alpha$	$\dfrac{\sin (n - \tfrac{1}{2})\alpha}{2 \sin \tfrac{1}{2}\alpha}$
$\sin n\alpha$	$-\dfrac{\cos (n - \tfrac{1}{2})\alpha}{2 \sin \tfrac{1}{2}\alpha}$

Just as *antiderivatives* are used to compute *integrals* by using the fundamental theorem of the integral calculus (§ 120), we may use *antidifferences* to compute *sums* by using the

FUNDAMENTAL THEOREM.   *If $F(n)$ is any function having $f(n)$ as difference, then*

$$(4) \qquad \sum_{n=p}^{q} f(n) = \Delta^{-1}f(n) \Big|_{p}^{q+1} = F(q + 1) - F(p).$$

Proof. By hypothesis the sum in (4) equals

$$\sum_{n=p}^{q} \Delta F(n) = F(p + 1) - F(p) + F(p + 2) - F(p + 1) + \cdots$$
$$+ F(q) - F(q - 1) + F(q + 1) - F(q)$$

in which all terms except the second and second last cancel.

We now use the table of antidifferences to compute some important sums.

Example 1. The *geometric series* (*N* terms)

$$\sum_{n=0}^{N-1} ar^n = \left.\frac{ar^n}{r-1}\right|_0^N = a\frac{r^N-1}{r-1}$$

in agreement with (22.2).

Example 2. The *arithmetic series* (*N* terms)

$$\sum_{n=0}^{N-1}(a+bn) = \left. an + \tfrac{1}{2}bn^{(2)}\right|_0^N$$

$$= aN + \tfrac{1}{2}bN(N-1).$$

Example 3. The *series of squares* (*N* terms)

$$\sum_{n=1}^{N} n^2 = \sum_{n=1}^{N}(n^{(1)} + n^{(2)}) = \left.\frac{n^{(2)}}{2} + \frac{n^{(3)}}{3}\right|_1^{N+1}$$

$$= \tfrac{1}{2}(N+1)N + \tfrac{1}{3}(N+1)N(N-1)$$

$$= \tfrac{1}{6}N(N+1)(2N+1).$$

Example 4. The *series of cosines* (*N* terms)

$$\sum_{n=1}^{N} \cos n\alpha = \left.\frac{\sin(n-\tfrac{1}{2})\alpha}{2\sin\tfrac{1}{2}\alpha}\right|_1^{N+1}$$

$$= \frac{\sin(N+\tfrac{1}{2})\alpha - \sin\tfrac{1}{2}\alpha}{2\sin\tfrac{1}{2}\alpha}.$$

in agreement with (201.2).

Example 5. Any polynomial $P(n)$ of degree k may be expressed in terms of the factorial powers $n^{(0)}, n^{(1)}, \ldots, n^{(k)}$ by means of the formula

(5) $$P(n) = P(0) + P'(0)n + \frac{P''(0)}{2!} n^{(2)} + \ldots + \frac{P^{(k)}(0)}{k!} n^{(k)}$$

where $P^{(k)}(0)$ denotes the value of the kth *difference* of $P(n)$ when $n = 0$. This is an obvious analogue of the polynomial of (63.5). To prove (5), write

$$P(n) = a_0 + a_1 n + a_2 n^{(2)} + \ldots + a_k n^{(k)},$$

difference the equation k times, and in each result put $n = 0$. For example, we find

$$n^3 = n^{(1)} + 3n^{(2)} + n^{(3)},$$

$$n^4 = n^{(1)} + 7n^{(2)} + 6n^{(3)} + n^{(4)}.$$

When the "Maclaurin" expansion (5) is known, we can readily compute polynomial sums $\Sigma P(n)$ with the aid of the formula

$$\Delta^{-1} n^{(k)} = \frac{n^{(k+1)}}{k+1}.$$

For example

$$\sum_{n=1}^{N} n^3 = \frac{n^{(2)}}{2} + n^{(3)} + \frac{n^{(4)}}{4} \Bigg|_{1}^{N+1}$$

$$= \tfrac{1}{4} n^2 (n-1)^2 \Bigg|_{1}^{N+1} = \tfrac{1}{4}(N+1)^2 N^2.$$

PROBLEMS

1. Prove formula (201.3).

2. Compute $\displaystyle\sum_{n=1}^{N} \frac{1}{n(n+1)}$, and let $N \to \infty$.

3. Compute $\displaystyle\sum_{n=1}^{N} \frac{1}{n(n+1)(n+2)}$, and let $N \to \infty$.

4. Show that

$$\Delta^{-1}(nr^n) = \frac{r^n}{r-1}\left(n - \frac{r}{r-1}\right), \qquad r \neq 1.$$

5. Prove that $\displaystyle\sum_{n=1}^{N} (n+1)2^n = N\, 2^{N+1}$.

6. Prove that $\displaystyle\sum_{n=1}^{N} \frac{n+1}{2^n} = 3 - \frac{N+3}{2^N}$ and let $N \to \infty$.

7. Prove that $\displaystyle\sum_{n=1}^{\infty} nr^n = \frac{r}{(1-r)^2}$, $r \neq 1$. [Cf. Prob. 4.]

8. If a is constant and k an integer, prove that
   (a) $\Delta(n+a)^{(k)} = k(n+a)^{(k-1)}$
   (b) $\Delta^{-1}(n+a)^{(k)} = (n+a)^{(k+1)}/(k+1)$, $k \neq -1$.

9. Show that $(n^2 - 1)^{-1} = (n-1)^{(-2)} + (n-2)^{(-2)}$.

10. Show that $\displaystyle\sum_{n=2}^{\infty} \frac{1}{n^2 - 1} = \frac{3}{4} - \frac{201}{20200}$. [Cf. Prob. 9.]

APPENDIX 4

Dimensional Checks

The ideas of dimensional analysis may often be profitably employed in problems of pure mathematics. Thus, if the coordinates x, y and the constants a, b are regarded as having the dimensions of length L, the equations of the curves

$$\frac{x^2}{a^2} + \frac{y^2}{b^2} = 1, \qquad y^2 = 4ax, \qquad x^3 + y^3 - 3axy = 0$$

are all written in "dimensional form"; that is, each term has the same dimensions in L, (0, 2, 3, respectively). The derivative $dy/dx = y'$ is then a pure number, and y'' has the dimensions L^{-1}. From the equation of the ellipse above, we find

$$y' = -b^2x/a^2y, \qquad y'' = -b^4/a^2y^3, \qquad \text{(Ex. 83.1)}$$

equations which check dimensionally. Thus certain errors in algebra, such as the omission of a letter or an improper exponent, are readily detected.

In vector formulas, the position vector $\mathbf{r}$ has the dimension L. If t denotes the time (dimension T), the velocity $\dot{\mathbf{r}} = d\mathbf{r}/dt$ and acceleration $\ddot{\mathbf{r}} = d^2\mathbf{r}/dt^2$ have the dimensions LT^{-1} and LT^{-2}. Thus, in the formulas for curvature and torsion of § 104, both κ and τ have the dimensions L^{-1}.

It is sometimes advisable to assign different dimensions to x, y. Thus, if $x \sim L$, $y \sim L^k$, all terms of the differential equation

$$(x - y^2)\, dx + (y^3 - 2xy)\, dy = 0$$

have the dimensions L^2, provided $k = \frac{1}{2}$. A method of solving such isobaric (*isos* equal, *barys* heavy) equations consists in introducing a dimensionless variable v by the substitution $y = vx^k$; in fact, this effects a separation of variables.

When transcendental functions, such as the sine, logarithm, or exponential, appear in an equation written in dimensional form, their arguments must be regarded as dimensionless; otherwise their Maclaurin expansions would contain terms of different dimensions. Consider, for example, the formulas of Prob. 62.12 which give the speed v and the distance x for a body falling in air; the arguments of the functions tanh, log cosh and exp are gt/V and gx/V^2, both of which are dimensionless.

It is frequently advantageous to put a formula containing transcendental functions in dimensional form. Suppose that we wish to compute the integral

$$I = \int_0^\infty \frac{\cos x}{(x^2 + 1)^2}\, dx.$$

Consider instead the simpler integral

$$\int_0^\infty \frac{\cos bx}{x^2 + a^2}\, dx$$

If $a \sim L$, $b \sim L^{-1}$, this has the dimension L^{-1}. From (216.10), we have

$$\int_0^\infty \frac{\cos t}{t^2 + a^2 b^2}\, dt = \frac{\pi}{2ab} e^{-ab} \qquad .$$

where t is a pure number. Putting $t = bx$, where $x \sim L$, we obtain

(1) $$\int_0^\infty \frac{\cos bx}{x^2 + a^2}\, dx = \frac{\pi}{2a} e^{-ab}.$$

If we differentiate (1) with respect to a (§ 190), we find that

(2) $$\int_0^\infty \frac{\cos bx}{(x^2 + a^2)^2}\, dx = \frac{\pi}{4a^3} (ab + 1)e^{-ab}.$$

As a check we note that both sides have the dimension L^{-3}. This gives $I = \pi/2e$ when $a = b = 1$.

As an extra dividend we may obtain two other integrals from (1) by successive differentiation with respect to b and a, namely:

(3) $$\int_0^\infty \frac{x \sin bx}{x^2 + a^2}\, dx = \frac{\pi}{2} e^{-ab},$$

(4) $$\int_0^\infty \frac{x \sin bx}{(x^2 + a^2)^2}\, dx = \frac{\pi b}{4a} e^{-ab}.$$

Note again the dimensional checks, and that (3) agrees with (216.11) when $b = 1$. By this procedure we have avoided computing the residues at poles of the second order.

Comprehensive Test

1. If n is a positive odd integer, find all the roots of the equation

$$\left(1 + \frac{ix}{n}\right)^n = \left(1 - \frac{ix}{n}\right)^n$$

2. Show that the general term of the Fibonacci sequence $1, 1, 2, 3, 5, 8, \cdots$, is given by

$$f_n = \frac{\cosh n\alpha}{\cosh \alpha} \quad (n \text{ odd}), \qquad f_n = \frac{\sinh n\alpha}{\cosh \alpha} \quad (n \text{ even}),$$

where $\sinh \alpha = \frac{1}{2}$, and that $\lim f_{n+1}/f_n = e^\alpha$.

3. If $f(x) = x \sin 1/x(x \neq 0)$, $f(0) = 0$, does Rolle's theorem guarantee a root of $f'(x)$ in the interval $0 \leq x \leq 1/\pi$? Show that $f'(x)$ has an infinite number of roots $x_1 > x_2 > x_3 > \cdots$ in the given interval which may be put in one-to-one correspondence with the roots of $\tan y = y$ in the interval $\pi \leq y < \infty$. Calculate x_1 to three decimal places.

4. Deduce the value of $\displaystyle\sum_{n=1}^{\infty} (2n-1)^{-2}$ from $\displaystyle\sum_{n=1}^{\infty} n^{-2} = \pi^2/6$.

5. If $\quad a_n = 1 + \dfrac{1}{3} + \dfrac{1}{5} + \cdots + \dfrac{1}{2n-1}$, prove that the alternating series

$a_1 - \dfrac{a_2}{2} + \dfrac{a_3}{3} - \dfrac{a_4}{4} + \cdots$ converges to $\pi^2/16$ by considering the power series for

$$f(x) = \int_0^x \frac{\tan^{-1} x}{1 + x^2} \, dx, \qquad |x| < 1.$$

Cite all theorems relevant to your argument.

6. Prove that

$$\int_0^{\pi/2} \frac{dx}{1 + \tan^r x} = C \quad (\text{const})$$

for any real value of r. Calculate C, and test your result for $r = 0, 2, -2, \infty, -\infty$.

551

7. Show that the Maclaurin series for $(1 + x + x^2)^{-1}$ and $(1 - x + x^2)^{-1}$ have, respectively, the coefficients 1, -1, 0 and 1, 1, 0, -1, -1, 0 periodically repeated. From these series deduce that

$$\frac{\pi}{3\sqrt{3}} = 1 - \frac{1}{2} + \frac{1}{4} - \frac{1}{5} + \frac{1}{7} - \frac{1}{8} + \cdots,$$

$$\frac{2\pi}{3\sqrt{3}} = 1 + \frac{1}{2} - \frac{1}{4} - \frac{1}{5} + \frac{1}{7} + \frac{1}{8} - \cdots,$$

and show that these results are consistent.

8. If $f(z) = \log(1 + e^z) - az$ is an even function, show that $a = \frac{1}{2}$ and $f(z) = \log(2 \cosh z/2)$. Obtain the expansions

$$f'(z) = \frac{1}{2} \tanh \frac{z}{2} = \sum_{n=1}^{\infty} \frac{(4^n - 1) B_{2n}}{(2n)!} z^{2n-1},$$

$$\log \cosh \frac{z}{2} = \sum_{n=1}^{\infty} \frac{(4^n - 1) B_{2n}}{(2n)!} \frac{z^{2n}}{2n},$$

and give their circle of convergence.

9. If $z = re^{i\theta} \ (-\pi < \theta \leq \pi)$, $\sqrt{z} = \sqrt{r} \, e^{i\theta/2}$ is defined as the *principal branch* of the two-valued square root. Compute $\int dz/\sqrt{z}$ over the upper and lower halves of the unit circle $|z| = 1$. Define the principal branch $z^{1/3}$ of the cube root in similar fashion, and compute $\int z^{1/3} \, dz$ over the same paths.

10. Compute the circuit integral $\oint (z^3 + a^3)^{-1} \, dz$ over the rectangle whose vertices are $(0, \pm a)$, $(a, \pm a)$, and check the result dimensionally. Hence deduce that

$$\oint (z^3 + a^3)^{-2} \, dz = -\frac{4 \pi i}{9 \, a^5}$$

over the same rectangle by differentiating with respect to a under the integral sign. Check by a direct computation of the last integral.

11. From (225.2) show that

$$\pi z \cot \pi z = 1 - \sum_{k=1}^{\infty} \sum_{n=1}^{\infty} \frac{z^{2n}}{k^{2n}} = 1 - \sum_{n=1}^{\infty} \zeta(2n) z^{2n}$$

using Weierstrass' theorem on double series (Knopp, *Infinite Series*, p. 430). Since $\pi z \cot \pi z = \cos 2B\pi z$ (211.7) deduce that

$$\zeta(2n) = (-1)^{n-1} \frac{2^{2n-1} B_{2n} \pi^{2n}}{(2n)!}.$$

12. In the symbolism of §211, show from (211.11) that

$$\Delta^{-1} x^p = \frac{(x + B)^{p+1}}{p + 1};$$

and hence $\Delta^{-1} f(x) = F(x + B)$ if $f(x)$ is any polynomial having $F(x)$ as primitive. Is $\Delta^{-1} e^{ax} = \frac{1}{a} e^{a(x+B)}$? $\Delta^{-1} \cos ax = \frac{1}{a} \sin a(x + B)$?

ANSWERS TO PROBLEMS

§ 10 (CHAPTER 1)

1. $x = a^{-1}ca^{-1}b^{-1}$.

2. Groups: (a), (b), (c) under addition, (d), (e).

9. $\pm 1, \pm e^{\pi i/3}, \pm e^{2\pi i/3}$. $[e^{i\varphi} = \cos \varphi + i \sin \varphi]$

10. $i(-1 \pm \sqrt{5})/2$.

11. $\sqrt{2} \, e^{\pi i/12}, \sqrt{2} \, e^{3\pi i/4}, \sqrt{2} \, e^{17\pi i/12}$.

17. $16(1 + i\sqrt{3})$.

§ 13 (CHAPTER 2)

1. (Starred numbers belong to the set.)

(a) $m^* = 1/2$, $M^* = 2$, $\alpha = \beta = 1$.

(b) $m^* = -2$, $M^* = 3$, $\alpha^* = -2$, $\beta^* = 3$.

(c) $m^* = 0$, $M^* = 5/3$, $\alpha = 1/2$, $\beta = 3/2$.

(d) $m^* = -2$, $M^* = 2$, $\alpha = -1$, $\beta = 1$.

(e) $m^* = 1$, $M^* = 2$, $\alpha = \beta = 5/3$.

4. x_n: $m = a$, $M = b$, $\alpha = a$, $\beta = b$.

y_n: $m^* = a - \frac{1}{2}$, $M^* = b + 1$, $\alpha = a$, $\beta = b$.

$x_n + y_n$: $m^* = a + b - 1$, $M^* = a + b + 2$, $\alpha = \beta = a + b$.

§ 15

2. 0.

3. $\log_2 x_n = \frac{2}{3} + \frac{1}{3}(-\frac{1}{2})^n \to \frac{2}{3}$; $x_n = \sqrt[3]{4}$.

8. $(\sqrt{5} - 1)/2$.

9. (a) π; (b) 0, 1.

11. (a) $x_n \to 2$; (b) 6, 5/2, 9/5, 4/3, 3/4, −1.

§ 29

3. Div; div; conv; div; div; conv.

4. Conv; conv; conv $c > 1$, div $c < 1$; conv.

5. Conv; conv; conv; conv $c > 1$, div $c < 1$; div.

10. Conv.

§ 31

1. Conv $\alpha > 1$, div $\alpha \leqq 1$.

2. Conv.

3. Div $p = 1$; no test $p = 2$; conv $p = 3$.

4. Div.

§ 34

1. Div.

2. Conv $a = b$; div $a \neq b$.

3. Div.

4. Div.

5. Conv $p > 1$;   conv cond $0 < p \leqq 1$.

6. Div.

7. Div.

8. Div.

9. Conv.

§ 41

1. 1.

2. $1 - 2x + x^2$.

3. $1 + x^2 + x^4 + x^6 + \cdots$.

5. (a) $0 < x \leqq 2$.

    (b) $x > \frac{1}{2}$.

    (c) $x < -1$ or $x > 0$.

    (d) $x > 0$.

    (e) $-4 < x < 4$.

    (f) $-\frac{1}{4} \leqq x < \frac{1}{4}$

14. Div when $z = -1$.   Real series: $\displaystyle\sum_1^\infty (-1)^{n-1} \frac{\sin n\theta}{n}, \sum_1^\infty (-1)^{n-1} \frac{\cos n\theta}{n}$.

§ 44 (CHAPTER 3)

3. 0; 0; 0; a.

§ 47

4. (a) 0, -1.

    (b) none.

    (c) $n\pi + \frac{1}{4}\pi$ (n an integer);

    (d) $n\pi - \pi/3$.

    (e) none.

    (f) $n + 1/2$.

§ 53

1. 1; 1; $\frac{3}{4}$; 0; 1.

2. $f'(0)$ does not exist; $f_+'(0) = \infty$; $f_-'(0) = 0$; $f'(0+) = 0$; $f'(0-) = 0$.

3. 1; -1; $1/x$ ($x \neq 0$); $1/x$ ($x \neq 0$); $-1/\sqrt{1 - x^2}$; $-\frac{1}{2}$; $1/(1 + x^2)$.

4. $f'(0) = 1$.   Yes.

5. $f'(x) = (ad - bc)/(cx + d)^2$, $x \neq -d/c$.

6. $\varphi'(y) = (ad - bc)/(cy - a)^2$.   When $ad - bc > 0$, $y \neq a/c$.

10. $dy/dx = 1/\sqrt{x^2 + 1}$; $x = \sinh y$, $dx/dy = \cosh y = \sqrt{x^2 + 1}$.

13. No.   Yes. $f(\infty+) = f(\infty-) = \frac{1}{2}$; $f'(\infty+) = f'(\infty-) = 0$.

§ 54

4. $(-1)^n n!/(x + a)^{n+1}$; $(-1)^n n!/(x - 1)^{n+1}$;

    $\frac{1}{2}(-1)^n n! \, [(x - 1)^{-n-1} - (x + 1)^{-n-1}]$;

    $\frac{1}{2}(-1)^n n! \, [3(x - 1)^{-n-1} - (x + 1)^{-n-1}]$.

§ 58

3. $x_3 = 0.16744$

4. $x_3 = 3.1038$

5. 4.4933 rad.

§ 62

1. $1/2$; $1/2$; $\log{(a/b)}$; -2.

2. π; 3; $\pi\sqrt{3}/6$.

3. $-\frac{1}{3}$; $\frac{1}{3}$; 0.

4. 1; 1.

5. e^{ab}.

6. 0.

7. $a^2/2$.

9. 1.

10. 1 [Cf. *Am. Math. Monthly*, vol. 61, no. 3, p. 189.]

§ 70

2. $(1 - x^2)^{-1} = 1 + x^2 + x^4 + \cdots$; $x(1 - x^2)^{-1} = x + x^3 + x^5 + \cdots$.

5. $e^{-x^2} = 1 - x^2 + \dfrac{x^4}{2!} + \dfrac{x^6}{3!}e^{-\theta x^2}$; $\log{(1 - x)} = -x - \dfrac{x^2}{2} - \dfrac{x^3}{3} - \dfrac{x^4}{4}(1 - \theta x)^{-4}$.

§ 71

1. $(1/e, -1/e)$ min; $(1, 0)$ max.

2. (a) $(\frac{2}{3}, 0)$ pt. inf.; $(\frac{7}{6}, \frac{3}{2})$ max; $(\frac{3}{2}, 0)$ min.

   (b) $(-1, 54)$ max; $(\frac{9}{5}, -\frac{6}{3125})$ min; $(2, 0)$ pt. inf.

   (c) $(2, 0)$ pt. inf.

   (d) $(-1, 4)$ max; $(0, 2)$ pt. inf.; $(1, 0)$ min.

3. (a) $(-3, -\frac{1}{2})$ min; $(\frac{1}{3}, 3)$ max.

   (b) None.

   (c) $(-a, -\frac{1}{2}a)$ min; $(a, \frac{1}{2}a)$ max.

   (d) $(\frac{3}{4}, \frac{5}{4})$ max.

4. Max $y = \tan^{-1}\frac{3}{4}$ at $x = 2$.

10. $D > 0$: extremes at $x = (b \pm D)/a$; if $a > 0$, $+$ gives min, $-$ max;

   $D = 0$: hor. pt. inf. at $x = -b/a$;

   $D < 0$: no extremes or hor. pt. inf.

§ 73 (CHAPTER 4)

1. (a) The entire plane exterior to the square with vertices at $(\pm2, 0)$, $(0, \pm2)$, including its boundary.

   (b) The entire plane within the angle to the right and left of both lines $x + y - 1 = 0$, $x - y + 1 = 0$, the lines excluded; no region.

   (c) The interior of quadrants I and III of the unit circle $x^2 + y^2 = 1$, the exterior of quadrants II and IV; the axes $x = 0$, $y = 0$ are excluded, but the four quadrantal arcs between the axes included; no region.

   (d) The closed region inside the circle $x^2 + y^2 = 1$.

   (e) The closed annular region between the circles $x^2 + y^2 = 1$, $x^2 + y^2 = 4$.

   (f) The open half-plane above and to the right of the line $x + y = 0$.

§ 80

3. (a) $df = \dfrac{x\,dx + y\,dy}{x^2 + y^2}$;

   (b) $df = \dfrac{2y\,dx - 2x\,dy}{(x + y)^2}$;

   (c) $df = e^{-x}(\cos y\,dy - \sin y\,dx)$;

   (d) $df = \dfrac{x\,dx + y\,dy}{x^2 + y^2}$;

   (e) $df = yx^{y-1}\,dx + x^y \log x\,dy$;

   (f) $df = a^{xy} \log a\,(x\,dy + y\,dx)$.

5. $du/dt = -1$; $u = \tan^{-1}\cot t = \frac{1}{2}\pi - t$.

6. See Ex. 87.2.

§ 81

1. (a) $\frac{2}{3}$; (c) 0; (d) 1; (e) 1; (g) 0; (h) 1; (j) 0

§ 84

2. (a) $y' = -x/y$; $y'' = -a^2/y^3$.

(b) $y' = 2a/y$; $y'' = -4a^2/y^3$.

(c) $y' = -x^{n-1}/y^{n-1}$; $y'' = -(n-1)a^n x^{n-2}/y^{2n-1}$.

(d) $y' = 2ax/3y^2$; $y'' = -2a/9y^2$.

(e) $y' = -a/2y$; $y'' = -a^2/4y^3$.

(f) $y' = 1$, $y'' = 0$.

8. $f_{xy} = f_{yx} = \dfrac{-x^2 + y^2 - 4xy}{(x^2 + y^2)^2}$; $f_{xx} + f_{yy} = 0$.

§ 87

1. $u' = \begin{vmatrix} -u - 1 & u \\ 1 - v & 2v + x \end{vmatrix} / D$, $v' = \begin{vmatrix} x + v - 1 & -u - 1 \\ -1 & 1 - v \end{vmatrix} / D$,

$D = \begin{vmatrix} x + v - 1 & u \\ -1 & 2v + x \end{vmatrix}$.

2. $\left(\dfrac{\partial u}{\partial x}\right)_y = -\dfrac{f_x}{f_u}$; $\left(\dfrac{\partial u}{\partial x}\right)_v = \dfrac{g_x f_v - f_x g_v}{f_u g_v}$.

3. $\dfrac{\partial z}{\partial x} = \dfrac{yz - x^2}{z^2 - xy}$; $\dfrac{\partial z}{\partial y} = \dfrac{xz - y^2}{z^2 - xy}$.

4. If $D = u^2 + v^2$, $u_x = v/D$, $u_y = -u/D$, $v_x = u/D$, $v_y = v/D$.

5. If $D = 4uv + 1$, $u_x = (4vx + 1)/D$, $u_y = 2(y - v)/D$,

$v_x = 2(u - x)/D$, $v_y = (4uy + 1)/D$.

6. If $D = 2(u + v)$, $u_x = (3x^2 + 4v)/D$, $u_y = (4yv - 3)/D$,

$v_x = (4u - 3x^2)/D$, $v_y = (4yu + 3)/D$.

7. If $D = 9u^2v^2 - xy$, $u_x = -(3v^3 + x)/D$, $u_y = (3v^2 + ux)/D$,

$v_x = (3u^2 + vy)/D$, $v_y = -(3u^3 + y)/D$.

8. Matrix $\dfrac{\partial(u, v, w)}{\partial(x, y, z)} = \begin{pmatrix} 1 & 1 & 1 \\ v + w & w + u & u + v \\ vw & wu & uv \end{pmatrix}^{-1}$

$J = \dfrac{\partial(u, v, w)}{\partial(x, y, z)} = u^2(v - w) + v^2(w - u) + w^2(u - v)$

and $u_x = u^2(v - w)/J$, etc.

9. Matrix $\dfrac{\partial(u, v, w)}{\partial(x, y, z)} = \begin{pmatrix} 1 & 1 & 1 \\ 2u & 2v & 2w \\ 3u^2 & 3v^2 & 3w^2 \end{pmatrix}^{-1}$,

$J = \dfrac{\partial(u, v, w)}{\partial(x, y, z)} = 6(u - v)(v - w)(w - u)$ and $u_x = vw/(u - v)(u - w)$, etc.

10. (i) $2x$; (ii) $2(x + tyz)$; (iii) $2(x - t^{-1}x^{-2}yz)$.

12. $(a^2 + b^2)z_{uu} + 2(a^2 - b^2)z_{uv} + (a^2 + b^2)z_{vv}$.

13. $z_{rr} + r^{-2}z_{\theta\theta} + r^{-1}z_r$.

14.
$$u' = - \begin{vmatrix} f_x & f_v & f_w \\ g_x & -1 & g_w \\ h_x & h_v & -1 \end{vmatrix} : \begin{vmatrix} -1 & f_v & f_w \\ g_u & -1 & g_w \\ h_u & h_v & -1 \end{vmatrix}.$$

16. $w_x = (u - v)/(u + v)$, $w_y = (u^2 + v^2)/(u + v)$.

§ 89

1. (a) $v = \tan u$; (b) $v = 2 \cosh u$; (c) $w = u^2 - 2v$.

§ 92

1. Min $u = -9$ at $(3, 0)$.

2. Min $z = -2$ at $(1, 0)$; saddle pt. at $(-1, 0)$.

3. Min $f = -8$ at $(\sqrt{2}, -\sqrt{2})$ and $(-\sqrt{2}, \sqrt{2})$.

4. Max $f = 0$ at $(0, 0)$; min $f = -9/8$ at $(\sqrt{3}/2, -\sqrt{3}/2)$ and $(-\sqrt{3}/2, \sqrt{3}/2)$.

5. Min $f = -1$ at $(1, 1)$; saddle pt. at $(0, 0)$.

6. Min $f = a^6/432$ at $(\tfrac{1}{2}a, \tfrac{1}{3}a)$.

7. Min $f = -3$ at $(\tfrac{3}{2}\pi, \tfrac{3}{2}\pi)$;

Max $f = \tfrac{3}{2}$ at $(\tfrac{1}{6}\pi, \tfrac{1}{6}\pi)$ and $(\tfrac{5}{6}\pi, \tfrac{5}{6}\pi)$;

Saddle points at $(\tfrac{1}{2}\pi, \tfrac{1}{2}\pi)$, $(\tfrac{1}{2}\pi, -\tfrac{1}{2}\pi)$, $(\tfrac{3}{2}\pi, -\tfrac{3}{2}\pi)$.

8. Min $f = -6$ at $(0, -3)$.

9. $\sqrt{2}(2 - \tfrac{1}{2}\sqrt{5})$.

10. $\tfrac{8}{13}$.

11. Plane $x/x_1 + y/y_1 + z/z_1 = 3$, min vol. $= \tfrac{9}{2}x_1y_1z_1$.

12. $x/a = y/b = z/c = k/(a + b + c)$.

§ 94

1. (a) $x = a/\sqrt{2}, y = b/\sqrt{2}$; max area $= 2ab$.

(b) $x = a^2/\sqrt{a^2 + b^2}, y = b^2/\sqrt{a^2 + b^2}$; max perimeter $= 4\sqrt{a^2 + b^2}$.

3. $y = mx$ where $bm^2 + (a - c)m - b = 0$.

5. $\lambda = 1$, $[2, 3, 6]$; $\lambda = \sqrt{2}$, $[6, 2, -3]$;

6. $x/a = y/b = z/c = 2K/(a^2 + b^2 + c^2)$.

7. $x/a = y/b = z/c = 1/\sqrt{3}$, max $V = 8abc/3\sqrt{3}$.

§ 99 (CHAPTER 5)

2. $-\tfrac{4}{3}$; $-\tfrac{2}{1}$.

3. $Q(-\tfrac{5}{9}, -\tfrac{5}{3})$; $R(-\tfrac{9}{10}, -\tfrac{6}{5})$.

4. $\mathbf{p} = \tfrac{1}{4}(3\mathbf{a} + \mathbf{c}) = \tfrac{1}{2}(\mathbf{b} + \mathbf{d})$.

6. 10; $[10, -4, 8]$; $[16, -2, 4]$; $[-17, 9, -7]$;

22; $[-36, -58, 16]$; $[10, -24, -52]$.

7. $4x + 2y - 3z - 3 = 0.$

9. $2x + 4y - 3z + 18 = 0.$

10. $-\frac{60}{13}.$

13. $\frac{14}{13}.$

14. 3.

15. $P(\frac{2}{3}, 1, \frac{4}{3}).$

16. $\mathbf{r} = \dfrac{\alpha\mathbf{b} + \mathbf{a} \times \mathbf{c}}{\mathbf{a} \cdot \mathbf{b}}.$

§ 104

10. $\mathbf{v} = [2, 2, 1]$, $\mathbf{a} = [2, 4, 0]$;

$\mathbf{T} = \frac{1}{3}[2, 2, 1]$, $\mathbf{N} = \frac{1}{3}[-1, 2, -2]$, $\mathbf{B} = \frac{1}{3}[-2, 1, 2]$;

$\kappa = \tau = \frac{2}{9}$; $a_t = 4$, $a_n = 2$.

§ 106

1. $\mathbf{n} = [-f'(u) \cos v, -f'(u) \sin v, 1]/\sqrt{f'^2(u) + 1}.$

4. $x = a \sin u \cos v$, $y = b \sin u \sin v$, $z = c \cos u$.

7. (a) $[-2, 2]$; (b) 0; (c) $135°$, $2\sqrt{2}.$

8. (a) $[4, 4, 2]$; (b) $10/\sqrt{3}$; (c) 6; (d) $2x + 2y + z = 7$; $x - 1 = y - 1 = 2z - 6.$

§ 109

6. div $\mathbf{f} = 3$, rot $\mathbf{f} = [1, 1, 1]$; div $\mathbf{g} = 2(x + y + z)$, rot $\mathbf{g} = \mathbf{0}.$

§ 111

1. $\mathbf{a}^1 = [1, -1, 0]$, $\mathbf{a}^2 = [0, 1, -1]$, $\mathbf{a}^3 = [0, 0, 1]$;

$\mathbf{u} = -\mathbf{a}_1 - \mathbf{a}_2 + 3\mathbf{a}_3 = \mathbf{a}^1 + 3\mathbf{a}^2 + 6\mathbf{a}^3$; $\mathbf{u} \cdot \mathbf{u} = 14.$

§ 117 (CHAPTER 6)

1. 1.

§ 122

1. (a) $\frac{1}{2}x(\cos \log x + \sin \log x).$

(b) $x - (1 + e^{-x}) \log (1 + e^x)$

(c) $\frac{1}{3}x^3 \tan^{-1} x - \frac{1}{6}x^2 + \frac{1}{6} \log (1 + x^2)$;

(d) $-e^{-x}/(x + 1).$

2. (a) $E_n = x^n e^x - nE_{n-1}.$

(b) $L_n = x(\log x)^n - nL_{n-1}.$

3. $S_n = -x^n \cos x + nC_{n-1}$, $C_n = x^n \sin x - nS_{n-1}.$

§ 123

1. See a table of integrals.

2. (a) $-\frac{1}{2}\sqrt{4 - x^4}.$

(b) $\frac{1}{10}(2x^3 - 3)(x^3 + 1)^{2/3}.$

(c) $)\frac{2}{3}[\sqrt{x^3 - 4} - 2 \tan^{-1} \frac{1}{2}\sqrt{x^3 - 4}].$

3. (a) $\frac{1}{16}(\log 3)^2$; (b) $2(e^2 + 1).$

5. $-(2x^2 + 1)/4(x^2 + 1)^2.$

6. $\dfrac{1}{n} \log \dfrac{x^n}{1 + x^n}.$

7. $1/(n + 1)(n + 2).$

§ 124

4. Put $x = 1/t^2(1 - t)$, $y = 1/t(1 - t).$

§ 129

1. $12\frac{1}{27}$.

2. $\frac{1}{2}a\pi^2$.

3. $6a$.

4. $\frac{3}{2}\pi a$.

5. $8a$.

6. $t^2 + \log t - 1$.

7. 5.

8. $\sqrt{2}\sinh t$.

9. (a) $a\log\dfrac{c}{b} + \dfrac{1}{8a}(c^2 - b^2)$.

   (b) $59a/24$.

§ 131

5. 1.0896; 1.0894.

§ 133

1. (a) $-\sin x$.

   (b) $(\frac{1}{2}\pi - \log 2)x$.

   (c) $(3x - 1/x^2)e^{x^3} + (1/x^2 - 2)e^{x^2}$.

   (d) $8x^3 - 3x^2 - 2x$.

   (e) $1/x^2$.

   (f) $(2x + 1)/(x^2 + x + 1)$.

2. $\log(\alpha + 1)$.

3. (a) $(1 + t)\sin x(1 + t) - \sin x$.

   (b) $\dfrac{1}{x}\sqrt{x^2 - t^2} - 1$.

7. $\varphi_a = -\dfrac{2a + b}{a^2(a + b)^2}\dfrac{\pi}{2}$, $\quad \varphi_b = \dfrac{-1}{a(a + b)^2}\dfrac{\pi}{2}$.

8. $2x - 1$.

§ 138 (CHAPTER 7)

1. Div.

2. -1.

3. Div.

4. $\pi/4$.

5. $\pi/4$.

6. $5\pi/18$.

7. $b/(a^2 + b^2)$.

8. $a/(a^2 + b^2)$.

9. Div.

10. $\frac{1}{2}\log 3 - \tan^{-1}2 + \frac{1}{2}\pi$.

11. $\displaystyle\sum_{k=1}^{n}\frac{1}{k}$.

12. $1/2a$.

13. $-(n + 1)^{-2}$.

14. $2\pi/\sqrt{3}$.

16. $V = 2\pi$, $x^* = \frac{1}{4}$, $y^* = 0$.

§ 144

1. Conv.

2. Conv.

3. Conv.

4. Conv.

5. Div.

6. Div.

7. $1/e$, $\pi/2e^2$.

8. ∞, π.

9. 1, $\pi/3$.

13. Put $x = \tan t$; π.

§ 148

2. $2ax^{-3} + bx^{-2} + cx^{-1}$; $2ax^{-3} - (a + b)x^{-2} + abx^{-1}$.

14. $\mathscr{L}(y) = kx/(x^2 + b^2)$, $\quad y = k\cos bx$.

15. $\mathscr{L}(y) = k(x + 2a)/(x^2 + 2ax + b^2)$,

$$y = ke^{-ax}\left[\cos\sqrt{b^2 - a^2}\,x + \frac{a}{\sqrt{b^2 - a^2}}\sin\sqrt{b^2 - a^2}\,x\right].$$

§ 151 (CHAPTER 8)

1. $\frac{7}{3}$; $\frac{17}{6}$; $\frac{11}{3}$; $\frac{5}{2}$; $(22 + 3\pi)/12$.

2. $-2\pi r^2(x_0 + y_0)$; 0; -2.

3. 0; π; 2π.

4. $\frac{1}{2}\pi^4$, 0.

5. $\frac{1}{2}x^2 + xy^2$.

6. $4\frac{1}{2}$; $4\frac{1}{2}$.

7. $3.75 - 2\log 4$.

8. (a) to (c): $-\frac{1}{4}$.

9. $77\frac{1}{3}$, $94\frac{2}{3}$, $88\frac{2}{3}$.

10. (a) $I = \mathbf{f} \cdot (\mathbf{r_2} - \mathbf{r_1})$, $F = ax + by + cz$.

    (b) $I = \frac{1}{2}(r_2{}^2 - r_1{}^2)$, $F = \frac{1}{2}(x^2 + y^2 + z^2)$.

    (c) $I = \dfrac{1}{r_2} - \dfrac{1}{r_1}$, $F = (x^2 + y^2 + z^2)^{-1/2}$.

§ 153

1. $F = \dfrac{1}{2}\left(1 - y^2 - \dfrac{y^2 + 1}{x^2}\right)$; $F(2, 1) - F(1, 0) = -1/4$.

2. $F = \dfrac{1}{2}x^2 + \dfrac{x}{y} + 2\log y$; $\frac{3}{2} + 2\log 2$.

3. $\varphi = y + 2z - xyz^2$.

4. $W = \varphi(1, 2, 3) - \varphi(3, 5, -1) = 2$.

5. $\varphi = -x^2 - 2xy + yz^2$.

7. $v_2{}^2 = v_1{}^2 - 2\gamma M\left(\dfrac{1}{r_1} - \dfrac{1}{r_2}\right)$.

8. $\varphi = \frac{1}{2}mkr^2$.

9. $h = v_0{}^2/2g$.

10. (a) $\tan^{-1} y/x$.   (b) $e^x \sin y$.   (c) $-y/(x^2 + y^2)$.   (d) $-\sin x \cosh y$.

§ 154

3. (a) $8a^2/15$.   (b) $2a^3/3b$.   (c) $a^2(2 - \frac{1}{2}\pi)$.

§ 159 (CHAPTER 9)

1. $A = \frac{8}{3}$; $I_x = \frac{344}{105}$, $I_y = \frac{8}{15}$.

2. $A = \frac{2}{3}ab$; $x^* = \frac{3}{5}a$, $y^* = \frac{3}{8}b$; $I_x = \frac{2}{15}ab^3$, $I_y = \frac{2}{7}a^3b$.

3. $I_x = \frac{1}{12}ab^3$.

4. $I_x = \frac{1}{4}\pi ab^3$, $I_y = \frac{1}{4}\pi a^3b$, $I_z = \frac{1}{4}\pi ab(a^2 + b^2)$; Mean $r^2 = \frac{1}{4}(a^2 + b^2)$.

5. $A = 9$; $x^* = \frac{8}{5}$, $y^* = 1$; $I_x = 25\frac{1}{5}$, $I_y = 30\frac{3}{14}$.

6. $a^2/6$.

8. Mean $(x + y) = 2 = x^* + y^*$.

§ 160.

1. 4π.   **2.** $\frac{8}{3}$.   **3.** $-\frac{1}{12}$.

4. $4ab(a^2 + b^2)$.   **5.** $0, 0$.   **6.** $80\frac{2}{3}$.

§ 162.

1. Value of $\partial(x, y)/\partial(u, v)$: (a) 1; (b) ab; (c) 1; (d) $\alpha\delta - \beta\gamma$; (e) $-(u^2 + v^2)^{-2}$.

2. $\pi(1 - e^{-a^2})$.   **5.** $\frac{1}{4}(e - e^{-1})$.

6. (a) $\pi(\frac{1}{4} - 1/3\sqrt{3})$; (b) $\frac{1}{2}\pi$.

§ 163.

1. (a) $2\pi a \sin \beta$; (b) $\dfrac{\pi}{4}\sqrt{2}a$; (c) $a\sqrt{2}\,E(1/\sqrt{2})$.

§ 164.

1. ma^2. 2. $8a^2$. 3. $2\pi a^2\sqrt{2}$.

4. $2\pi ac$. 5. $28a^2/3$.

§ 166.

3. 2π. 5. $\int \mathbf{n}\cdot \mathbf{e}_3\, dS = \oint x\, dy;\quad -\oint y\, dx$.

§ 170

1. $\tfrac{3}{2}\pi a^3$. 2. $\tfrac{1}{3}a^3$. 6. $\tfrac{2}{3}(3\pi - 4)a^3$.

§ 171

2. $8\pi/3$. 4. $g(r) = r^m/(m+3)$.

§ 172

2. $[\tfrac{1}{2}\mathbf{a}\times\mathbf{b} + \tfrac{1}{3}(\mathbf{b}-\mathbf{a})\times\mathbf{r}]\times\mathbf{r}$.

3. (a) $\mathbf{g}_0 = xz[1, -1, 0]$, $\mathbf{g} = \tfrac{1}{3}[xz + yz, -2xz, xy - x^2]$.

  (b) $\mathbf{g}_0 = [xz - \tfrac{1}{2}z^2,\ xy + yz - \tfrac{1}{2}x^2 - \tfrac{1}{2}z^2,\ 0]$, $\mathbf{g} = \tfrac{1}{3}\mathbf{f}\times\mathbf{r}$.

  (c) $\mathbf{g}_0 = -yz[y, x, 0]$, $\mathbf{g} = \tfrac{1}{4}\mathbf{f}\times\mathbf{r}$.

  (d) $\mathbf{g}_0 = [xz,\ x^2 - \tfrac{1}{2}z^2,\ 0]$, $\mathbf{g} = \tfrac{1}{3}\mathbf{f}\times\mathbf{r}$.

4. (a) $\tfrac{1}{6}\mathbf{f}\times\mathbf{r}$. (b) $\mathbf{g} = \tfrac{1}{6}[3z + 2yz, -4xz, 2xy - 3x]$.

§ 178 (CHAPTER 10)

In problems 1–6, the limits of $f_n(x)$, $\displaystyle\int_0^x f_n(x)\, dx$, $f'_n(x)$ are given in turn.

1. $f(0) = 1, f(x) = 0\ (x \neq 0);\ 0{:}\ 0$.

2. $f(0) = 1, f(x) = x\ (x \neq 0);\ \tfrac{1}{2}x^2;\ 1\ (x \neq 0)$.

3. $f(x) = 0;\ 0;\ 0\ (x \neq 0)$.

4. $f(x) = -1, 0, 1$ when $x < 0, x = 0, x > 0$; $|x|$; $0\ (x \neq 0)$.

5. $f(x) = 0;\ \tfrac{1}{4}\pi\ (x \neq 0);\ 0\ (x \neq 0)$.

6. $f(x) = 1;\ x;\ 0$.

9. Max pt. $[(1 + 2n)^{-1}, 2n^2(1 + 2n)^{-3/2}]$.

§ 179

1. $1/x(x + 1)$, unif. 6. $(1 + x)/(1 + x + x^2)$, not unif. 11. unif.

2. 1, not unif. 7. $(1 + x)/(1 + x + x^2)$, unif. 12. $1/2n^{3/2}$.

3. 1, unif. 8. 1, not unif. 16. $\zeta(\infty) = 0$.

4. $0\ (x = 0)$ or 1, not unif. 9. 1, unif.

5. 1, unif. 10. unif.

§ 182

1. $f(0) = 0, f(x) = 1\ (0 < x \leq 1)$; no; yes. 3. $g(n) = \displaystyle\sum_{k=1}^{n}\frac{1}{k}$.

2. $f(0+) = 0, f(\infty) = \pi^2/6$.

§ 183

1. $\sin^{-1} x = x + \dfrac{1}{2}\dfrac{x^3}{3} + \dfrac{1 \cdot 3}{2 \cdot 4}\dfrac{x^5}{5} + \cdots \; (-1 \leqq x \leqq 1)$;

$(1 - x^2)^{-3/2} = 1 + \dfrac{3}{2} x^2 + \dfrac{3 \cdot 5}{2 \cdot 4} x^4 + \cdots \; (-1 < x < 1)$.

2. Change signs of terms 2, 4, 6, $\cdots$ in Prob. 1.

3. $\tanh^{-1} x = x + \tfrac{1}{3}x^3 + \tfrac{1}{5}x^5 + \cdots (-1 \leqq x < 1)$;

$(1 - x^2)^{-2} = 1 + 2x^2 + 3x^4 + \cdots (-1 < x < 1)$.

4. (a) $(1 - x)^{-1}$;   (c) $2(1 - x)^{-3}$;   (e) $x(1 - x)^{-2}$;

(b) $(1 - x)^{-2}$;   (d) $(1 + x)^{-1}$;   (f) $x(x + 1)(1 - x)^{-3}$.

§ 185

2. (a) 0;   (b) 0;   (c) 0;   (d) 1;   (e) $-\infty$;   (f) ∞.

3. (a) All x;   (b) All $x \neq 2m\pi$;   (c) All $x \neq 2m\pi \; (m \neq 0)$.

6. $\sqrt{2}/2$.

7. (a) 1/2;   (b) 1/3;   (c) 2/3;   (d) 1/4.

§ 188

9. $F(0) = 0$, $F(x) = 1 \; (x > 0)$.

10. $F(0) = 0$, $F(x) = \tfrac{1}{2}\pi \; (x > 0)$, $F(x) = -\tfrac{1}{2}\pi \; (x < 0)$

§ 193

1. $\tfrac{1}{2}\sqrt{\pi}, \tfrac{3}{4}\sqrt{\pi}, \tfrac{15}{8}\sqrt{\pi}, \tfrac{105}{16}\sqrt{\pi}$.

2. $-2\sqrt{\pi}, \tfrac{4}{3}\sqrt{\pi}, -\tfrac{8}{15}\sqrt{\pi}, \tfrac{16}{105}\sqrt{\pi}$.

3. (a) $\tfrac{3}{4}\sqrt{\pi}$.   (b) $\tfrac{3}{128}\sqrt{\pi}$.   (c) $\tfrac{1}{4}\sqrt{\pi}$.   (d) $\dfrac{1}{2a}\sqrt{\dfrac{\pi}{a}}$;   (e) $\sqrt{\dfrac{\pi}{a}}$;   (f) $n!$.

§ 202 (CHAPTER 11)

1. $u = x^2 - y^2$, $v = 2xy$;

$u = y^3 - 3x^2 y$, $v = x^3 - 3xy^2$;

$u = x/(x^2 + y^2)$, $v = -y/(x^2 + y^2)$;

$u = e^{-y} \cos x$, $e^{-y} \sin x$;

$u = \tfrac{1}{2} \log (x^2 + y^2)$, $v = \tan^{-1} \dfrac{y}{x} \; (x > 0)$;

$u = \sin x \cosh y$, $v = \cos x \sinh y$;

$u = \cos x \cosh y$, $v = -\sin x \sinh y$;

$u = \dfrac{\tan x \, \text{sech}^2 y}{1 + \tan^2 x \tanh^2 y}, \; v = \dfrac{\tanh y \sec^2 x}{1 + \tan^2 x \tanh^2 y}$;

$u = \sinh x \cos y$, $v = \cosh x \sin y$;

$u = \cosh x \cos y$, $v = \sinh x \sin y$.

3. $ay - bx$; $3x^2 y - y^3$; $-e^x \cos y$.

5. $u = r^2 \cos 2\varphi, v = r^2 \sin 2\varphi;$

$u = \dfrac{1}{r} \cos \varphi, v = -\dfrac{1}{r} \sin \varphi;$

$u = \log r, v = \varphi\,(-\pi < \varphi \leqq \pi);$

$u = e^r \cos \varphi \cos (r \sin \varphi), v = e^r \cos \varphi \sin (r \sin \varphi).$

$u = e^{-r \sin \varphi} \cos (r \cos \varphi), v = e^{-r \sin \varphi} \sin (r \cos \varphi).$

6. $\frac{1}{2}\pi i;$ $\log 2 + \pi i;$ $\frac{1}{2} \log 2 + \frac{1}{4}\pi i;$ $\log 2 - \frac{1}{3}\pi i.$ $\log 5 + (\tan^{-1} \frac{4}{3} - \pi)i.$

8. $e^{-i\varphi}/r = 1/z,$ $\sqrt{r}e^{\frac{1}{2}i\varphi} = \sqrt{z}.$ **9.** $i / \sin \dfrac{\pi}{2n}.$

10. Points to right of line $x = -\frac{1}{2} \log 2;$
points within parabola $y^2 = 2x - 1.$

11. (a) $\frac{1}{2}\pi \pm i \cosh^{-1} 2;$ (b) $\pm(\frac{1}{2}\pi + i \sinh^{-1} 2).$

§ 203

1. $0, 0, \tan^{-1} \sqrt{3}/2;$ $u = x^2 - y^2, v = 2xy.$

2. $1, 0;$ $\frac{1}{2}, \frac{1}{2}\pi;$ $\frac{1}{2}, \frac{3}{2}\pi;$ $u = x(x^2 + y^2)^{-1}, v = -y(x^2 + y^2)^{-1}.$

5. (a) $w = -\dfrac{1}{z};$ (b) $w = -\dfrac{2z + i}{z + 2i};$ (c) $z = i\,\dfrac{z - 2i}{z + 2i}.$

7. $z = \dfrac{2}{i}\,\dfrac{w + i}{w - i}.$

§ 210

1. $\displaystyle\sum_{n=0}^{\infty} (-1)^n z^{2n+1}, |z| < 1;$ $\displaystyle\sum_{n=0}^{\infty} (-1)^n z^{-2n-1}, \quad |z| > 1.$

2. $\displaystyle\sum_{n=0}^{\infty} (-1)^n z^{2n-1}, 0 < |z| < 1;$ $\displaystyle\sum_{n=0}^{\infty} (-1)^n z^{-(2n+3)}, \quad |z| > 1.$

3. $\displaystyle\sum_{n=0}^{\infty} z^{n-2}, 0 < |z| < 1;$ $-\displaystyle\sum_{n=0}^{\infty} z^{-(n+3)}, \quad |z| > 1.$

4. $(1 + z^2)\displaystyle\sum_{n=0}^{\infty} z^n, |z| < 1;$ $-(z^{-1} + z)n = \displaystyle\sum_{0}^{\infty} z^{-n}, \quad |z| > 1.$

5. $\displaystyle\sum_{n=0}^{\infty} \dfrac{z^{2n}}{(2n + 1)!},$ all $z.$

6. $\displaystyle\sum_{n=0}^{\infty} (-1)^n \dfrac{z^{n-1}}{n!}, \quad |z| > 0.$

7. $z^{-1} + 1 - \frac{1}{2}z - \frac{5}{6}z^2 - \cdots, 0 < |z| < 1;$ the Laurent series in the ring $|z| > 1$
has the coefficients

$c_{-2} = \sin 1, c_{-1} = 1 - \cos 1, c_0 = 1 - \sin 1, c_1 = \cos 1 - \frac{1}{2}, c_2 = \sin 1 - \frac{5}{6}.$

8. $z^{-1} - \dfrac{1}{6} z + \dfrac{7}{360} z^3 - \dfrac{31}{15,120} z^5 + \cdots, \quad 0 < |z| < \pi. \quad (211.8).$

9. $(t = z - 1), \; -t^{-1} + 2 - 3t + 4t^2 - 5t^3 + \cdots, 0 < |t| < 1;$
$-2t^{-1} - 2 - t, \quad |t| > 0.$

10. $(t = z - 1), \; 1 - t + t^2 - t^3 + \cdots, \quad |t| < 1.$

§ 212

2. $\dfrac{z}{\sin z} = 1 + \dfrac{1}{3 \cdot 2!} x^2 + \dfrac{7}{15 \cdot 4!} x^4 + \dfrac{31}{21 \cdot 6!} x^6 + \cdots.$

§ 213

1. (a) Res $(0) = -1$, Res $(1) = 2$.   (c) Res $(-1) = e^{-1}$.

   (b) Res $(0) = -1$, Res $(1) = 2$.   (d) Res $(0) = \frac{1}{2}$.

3. (a) Res $(n\pi i) = (-1)^n$.   (c) Res $(n - \frac{1}{2})\pi i = (-1)^n i$.

   (b) Res $(n\pi i) = 1$.   (d) Res $(n - \frac{1}{2})\pi i = 1$.

4. 0; 0.   **5.** $-\frac{1}{24}$.

6. Res $(0) = -2$, Res $(i) = \frac{3}{2}$, Res $(-i) = \frac{3}{2}$.

§ 214

1. 0; $-2\pi i$; $-4\pi i$.   **2.** 0.

3. $2\pi i$; 0; $\pi i/3$; $-2\pi i$; $\pi i/3$.

4. Res $(0) = \frac{1}{8}$, Res $(-2) = -\frac{1}{8}$, Res $(\infty) = 0$; $\pi i/4$, 0.

5. Res $(1) = \frac{3}{2}$, Res $(i) = \frac{1}{4}(i - 1)$, Res $(-i) = -\frac{1}{4}(i + 1)$,

    Res $(\infty) = -1$; $3\pi i$, $-\dfrac{\pi}{2} (1 + i)$, $2\pi i$.

§ 221 (CHAPTER 12)

4. $f(x) = \frac{1}{2}(x + |x|).$

6. $x^2 = 2\pi \displaystyle\sum_{n=1}^{\infty} (-1)^{n+1} \dfrac{\sin nx}{n} - \dfrac{8}{\pi} \sum_{n=1}^{\infty} \dfrac{\sin (2n - 1)x}{(2n - 1)^3}.$

§ 223

3. $\dfrac{a + b}{2} + \dfrac{2}{\pi} (b - a) \displaystyle\sum_{n=1}^{\infty} \dfrac{\sin (2n - 1)x}{2n - 1}.$

7. $\dfrac{1 - r \cos \theta}{1 - 2r \cos \theta + r^2} = \displaystyle\sum_{n=0}^{\infty} r^n \cos n\theta, \quad r < 1;$

$\dfrac{r \sin \theta}{1 - 2r \cos \theta + r^2} = \displaystyle\sum_{n=1}^{\infty} r^n \sin n\theta, \quad r < 1.$

APPENDIX 2

1. $x_n = \frac{1}{2} - \frac{1}{2}\sqrt{3}\tan\left(\dfrac{n\pi}{3} - \gamma\right)$; period 3.

2. $x_n = \{2^{n+1} - (-1)^{n+1}\}: \{2^n - (-1)^n\};\ x_n \to 2.$

3. $x_n = \frac{3}{2} - \frac{1}{2}\sqrt{3}\tan\left(\dfrac{n\pi}{6} - \gamma\right)$; period 6.

4. $x_n = C_1 \cos \frac{1}{2}n\pi + C_2 \sin \frac{1}{2}n\pi$; period 4.

6. (a) $x_n = \frac{7}{6} 2^n - \frac{2}{3}(-1)^n - 2$;

   (b) $x_n = \frac{11}{6} 2^n - \frac{1}{3}(-1)^n - 2n - 1$;

   (c) $x_n = 2n^2 + 2n + 3 - \frac{19}{6} 2^n - \frac{1}{3}(-1)^n.$

APPENDIX 3

2. $1 - \dfrac{1}{N+1} \to 1.$

3. $\dfrac{1}{4} - \dfrac{1}{2(N+1)(N+2)} \to \dfrac{1}{4}.$

COMPREHENSIVE TEST

1. $n \tan k\pi/n,\ k = 0, 1, \cdots, n-1.$

3. $x_1 = 0.2225.$

4. $\pi^2/8.$

6. $C = \pi/4.$

8. Circle of convergence: $|z| = \pi.$

9. $-2 \pm 2i;\ -3(3 \pm i\sqrt{3})/8.$

10. $-2\pi i/3a^2.$

Index